D1694945

Feline Dermatology

Chiara Noli · Silvia Colombo
Editors

Feline Dermatology

 Springer

Editors
Chiara Noli
Servizi Dermatologici Veterinari
Peveragno
Italy

Silvia Colombo
Servizi Dermatologici Veterinari
Legnano
Italy

ISBN 978-3-030-29835-7   ISBN 978-3-030-29836-4   (eBook)
https://doi.org/10.1007/978-3-030-29836-4

This Springer imprint is published by the registered company Springer Nature Switzerland AG
The registered company address is: Gewerbestrasse 11, 6330 Cham, Switzerland

Foreword for *Feline Dermatology*

In 1980, Danny Scott (James Law Professor Emeritus, Section of Dermatology at Cornell University, New York, USA) published a monograph in the *Journal of the American Animal Hospital Association*, entitled *Feline dermatology 1900-1978: A Monograph*. This was the first comprehensive survey of skin diseases in the domestic cat; there had previously been other small descriptive articles and booklets. This was the first attempt to review all that was known at the time in veterinary science. Since 1980, cats and their skin conditions have always been part of standard textbooks of veterinary dermatology and veterinary science; Danny Scott published several more monographs. In 1999, Merial published a book entitled *A Practical Guide to Feline Dermatology*, devoted to cats and compiled by a multinational large author group.

While cats make popular pets, they have always been challenging to examine, to study, to investigate, and to treat. For example, they don't readily take to diet trials or accept long courses of oral medications; furthermore, trying to identify clinically significant allergens remains something of a dark art. They have always been remarkably independent creatures, and we never really "own them" as a pet. We remain enthralled with them in part because of their aloof persona as well as their engaging personalities. In some respects, the level of understanding of their skin diseases has always lagged behind other domestic animals, especially the dog (with the old adage that a cat is not a small dog).

Some 20 years after the publication of the book *A Practical Guide to Feline Dermatology*, we have this new book *Feline Dermatology*. The editors have emulated the *guide* and assembled a large international community of authors who share their experience and expertise in studying and caring for cats and their skin diseases. The book comprises three sections. The first section introduces the structure and function of the skin – the fundamental building blocks that help students and veterinarians to understand the pathogenesis of skin diseases. It is nice to see a chapter on coat color genetics – a topic often left for other publications and not included with clinical dermatology textbooks.

The next section provides a series of chapters that discuss the various clinical presentations of skin diseases in cats with, for example, an approach to the skin diseases associated with alopecia and so on. Given that cats can present with different cutaneous presentations to the same underlying aetiology, this section is followed by the third section that covers a large array of skin diseases organized by aetiology.

The coverage is comprehensive and so there are chapters that will be useful for students, veterinarians in general practice, residents in training programs, including dermatology, for veterinarians working in referral centers, and maybe even some cat owners. There are many illustrations and clinical images that befit the discipline of veterinary dermatology, which is so reliant on visual representation to appreciate and understand clinical lesions and the cutaneous reaction patterns that may be presented.

The editors are to be congratulated on assembling such an array of chapters and topics, demonstrating that our knowledge and understanding of feline dermatology has come a long way since the first monographs. This is the first major book on feline dermatology in many years and should prove to be a useful reference text for veterinarians for many years to come. Veterinary clinicians may gain a lot of knowledge from the internet but printed books remain popular with publishers and veterinarians. This is one book you ought to have on your shelf.

Aiden P. Foster
Bristol Veterinary School
University of Bristol
Langford, UK

Preface

The world of veterinary dermatology is growing rapidly year after year, as is true for our knowledge of all animal diseases. The cat is currently receiving great attention in veterinary medicine: many feline-specific textbooks have been published in recent years; we have now feline-specific scientific journals, and "feline specialists" are more and more numerous.

Being veterinary dermatologists with a particular interest in cats, we felt the need for a feline dermatology textbook. Our aim was to dedicate the appropriate attention to the cat's skin and its diseases, which are often peculiar and totally different from the counterparts described in dogs. A long time has passed by since two previous feline dermatology books *A Practical Guide to Feline Dermatology* by Eric Guaguère and Pascal Prélaud and *Skin Diseases of the Cat* by Sue Paterson, which were both published in 1999. After 20 years, it was time for a new feline dermatology textbook and here it is!

This book will hopefully serve both as an essential practical guide for the busy practitioner, to quickly and surely tackle cats with dermatological conditions, and a current and complete reference tool for the feline veterinarian and the veterinary dermatologist.

The most important feline skin diseases such as dermatophytosis and allergic diseases are described in dedicated chapters. We decided to select different authors for the majority of the chapters, in order to provide readers with the best possible review for each subject, written by experts in their specific fields. Each chapter is greatly enriched with many beautiful colour pictures, which are indispensable to properly describe a skin disease.

We are very grateful to Springer Nature and all their team for supporting our project with enthusiasm. Last but not least, we want to say a huge thanks to all the authors who contributed to this book.

Dedicated to:
Emma, Ada, Luca and all the cats of our lives.

Peveragno, Italy   Chiara Noli
Legnano, Italy   Silvia Colombo

Contents

Part III   Feline Skin Diseases by Etiology

Contributors

David J. Argyle The Royal (Dick) School of Veterinary Studies, University of Edinburgh, Easter Bush, Midlothian, UK

Frane Banovic University of Georgia, College of Veterinary Medicine, Department of Small Animal Medicine and Surgery, Athens, GA, USA

Špela Bavčar The Royal (Dick) School of Veterinary Studies, University of Edinburgh, Easter Bush, Midlothian, UK

Sonya V. Bettenay Tierdermatologie Deisenhofen, Deisenhofen, Germany

Petra Bizikova North Carolina State University, College of Veterinary Medicine, Raleigh, NC, USA

Silvia Colombo Servizi Dermatologici Veterinari, Legnano, Italy

Maria Cristina Crosta Clinica Veterinaria Gran Sasso, Milan, Italy

Alison Diesel College of Veterinary Medicine and Biomedical Sciences, Texas A&M University, College Station, TX, USA

Alessandra Fondati Veterinaria Trastevere - Veterinaria Cetego, Roma, RM, Italy
Clinica Veterinaria Colombo, Camaiore, LU, Italy

Vet Dominique Heripret CHV Fregis, Arcueil, France
CHV Pommery, Reims, France

Hans S. Kooistra Department of Clinical Sciences of Companion Animals, Faculty of Veterinary Medicine, Utrecht University, Utrecht, The Netherlands

Gary Landsberg CanCog Technologies, Fergus, ON, Canada

Julie D. Lemetayer Veterinary Medical Teaching Hospital, University of California, Davis, CA, USA

Federico Leone Clinica Veterinaria Adriatica, Senigallia (Ancona), Italy

Keith E. Linder College of Veterinary Medicine, North Carolina State University, Raleigh, NC, USA

Ken Mason Specialist Veterinary Dermatologist, Animal Allergy & Dermatology Service, Slacks Creek, QLD, Australia

Karen A. Moriello School of Veterinary Medicine, University of Wisconsin-Madison, Madison, WI, USA

Ralf S. Mueller Centre for Clinical Veterinary Medicine, München, Germany

John S. Munday Massey University, Palmerston North, New Zealand

Chiara Noli Servizi Dermatologici Veterinari, Peveragno, Italy

Tim Nuttall Royal (Dick) School of Veterinary Studies, University of Edinburgh, Roslin, UK

Carolyn O'Brien Melbourne Cat Vets, Fitzroy, Victoria, Australia

Catherine Outerbridge University of California, Davis, Davis, CA, USA

Maria Grazia Pennisi Dipartimento di Scienze Veterinarie, Università di Messina, Messina, Italy

Michelle L. Piccione School of Veterinary Medicine, University of Wisconsin-Madison, Madison, WI, USA

Philippa Ann Ravens Small Animal Specialist Hospital, North Ryde, NSW, Australia

Hock Siew Han The Animal Clinic, Singapore

C. Siracusa Department of Clinical Sciences and Advanced Medicine, School of Veterinary Medicine, University of Pennsylvania, Philadelphia, PA, USA

Andrew H. Sparkes Simply Feline Veterinary Consultancy, Shaftesbury, UK

Jane E. Sykes Veterinary Medical Teaching Hospital, University of California, Davis, CA, USA

Linda Jean Vogelnest University of Sydney, Sydney, NSW, Australia
Small Animal Specialist Hospital, North Ryde, NSW, Australia

Sylvie Wilhelm Vet Dermatology GmbH, Richterswil, Switzerland

Part I
Introductory Chapters

Structure and Function of the Skin

Keith E. Linder

Abstract

Knowledge of skin anatomy and function is fundamental for understanding the clinical manifestations and impacts of skin diseases. While true for any organ, this is especially true for the skin, because clinicians can see, touch, and otherwise interrogate the anatomy of this organ directly. Importantly, skin diseases result from deleterious agents or processes that disrupt specific anatomic components of the skin, and induce physiological responses that distort it, to create skin lesions. Recognition of skin lesion significance, and thus diseases, is based upon identifying alterations in normal skin anatomy, including the particular anatomical components that are targeted. Furthermore, the impacts of skin diseases and treatment choices are understood through knowledge of normal skin functions and the consequences of its dysfunction. This chapter reviews basic aspects of feline skin structure and function, with citations from the literature where available, and draws heavily on the comparative information available for humans and dogs.

The Skin Organ

The skin is organized into multiple, discrete, thin layers that are stacked to create a sheetlike organ that covers the entire body [1]. Starting externally, the epidermis is supported by the dermis and then by the panniculus, which connects via fascia to the underlying musculature or periosteum, for example, in the extremities (Fig. 1). Nerves and sensory nerve endings invest all three layers variably, whereas blood

K. E. Linder (✉)
College of Veterinary Medicine, North Carolina State University, Raleigh, NC, USA
e-mail: kelinder@ncsu.edu

© Springer Nature Switzerland AG 2020   3
C. Noli, S. Colombo (eds.), *Feline Dermatology*,
https://doi.org/10.1007/978-3-030-29836-4_1

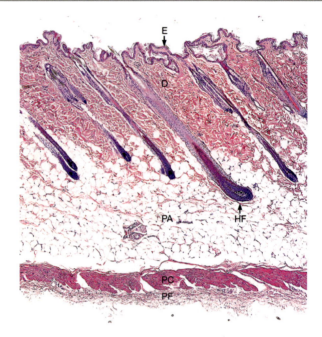

Fig. 1 Dorsal mid back, cat. The skin is organized into sheet-like tissue layers. The very thin epidermis (E) is on the surface and is supported below by the collagenous dermis (D). The panniculus is deepest and is composed of three parts, in areas of the body where all three are present. The panniculus adiposus (PA) is composed of lobules of adipose, and its most superficial part, the superficial adipose tissue, is shown here. The collagenous panniculus fibrosis (PF, superficial fascia) supports the panniculus carnosus (PC), which is composed of striated skeletal muscle. Adnexa are added into these layers, of which hair follicles (HF) are most visible at this magnification. 4X magnification. Hematoxylin and eosin

vessels are found only in the dermis and panniculus. The skin adnexa (appendages) are "little organs" added into these three layers multifocally during development and include, for example, hair follicles, skin glands, and claws. All three skin layers are highly modified to create discrete anatomical structures like the planum nasale and footpads.

The skin thickness, made of the dermis and epidermis together, varies by body region and is only generally 0.4–2.0 millimeters thick in the cat, being thicker on the dorsal body and proximal limbs and thinner on the ventral body, distal limbs, and ears [2]. These layers are the thickest on the footpads and planum nasale [2]. The panniculus varies greatly in thickness, being absent to >2 centimeters, depending on the degree of adiposity of the patient and the anatomical region of the body; it is generally thickest on the ventrum, especially in obese patients, thinner on the dorsum, and progressively thins to become mostly absent on the extremities.

Epidermis

The epidermis is remarkably thin (Fig. 2) and measures only 10–25 micrometers in truncal areas, but is thicker on footpads (Fig. 3) and planum nasale [2, 3]. In most body areas, the viable epidermis contains only three to five keratinocyte layers. The superficial nonviable epidermis, the stratum corneum, contains more numerous cell layers composed of very thin cells, called corneocytes, which are less than 1 micrometer thick (Fig. 2). Haired areas tend to have thinner epidermis than non-haired areas.

The epidermis is a stratified cornifying epithelium composed of keratinocytes (85%) arranged in four layers based on morphology: the stratum basale, stratum spinosum, stratum granulosum, and stratum corneum (Fig. 2) [1]. Keratinocytes continually proliferate in the basal epidermal layer, then migrate, and differentiate to form the upper epidermal layers and finally shed (desquamate) from the skin surface. The epidermis also contains resident Langerhans cells, migrating T-lymphocytes, and uncommon neuroendocrine Merkel cells (<1%) [1]. Melanocytes are present in pigmented epidermis and are absent in areas of white spotting. In the cat, mast cells are rare in the epidermis but can move into the epidermis in greater numbers during inflammatory diseases such as allergic skin diseases. Nerves extend into the epidermis but blood vessels do not.

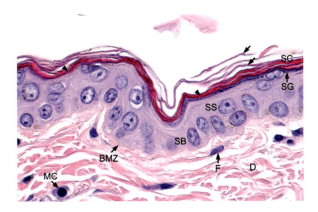

Fig. 2 Face, cat. The epidermis is composed of four morphological layers: stratum basale (SB), stratum spinosum (SS), stratum granulosum (SG), and stratum corneum (SC). The deep stratum corneum, called the stratum compactum (arrowheads), is very thin and formed by compact orthokeratosis. The superficial stratum corneum, called the stratum dysjunctum (arrows), is expanded mostly by histology artifact into a basket weave pattern of orthokeratosis. The basement membrane zone (BMZ; location of the ultramicroscopic basement membrane) connects the epidermis to the dermis (D). Fibrocytes (F) and mast cells (MC) are in the dermis. 100X magnification. Hematoxylin and eosin

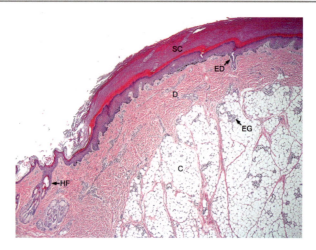

Fig. 3 Carpal footpad, cat. Footpads (including digital pads) have a thick epidermis, with a robust stratum corneum (SC), and thick dermis (D). Hair follicles (HF) and sebaceous glands are absent in footpads but are present in haired skin at the footpad margin (left of the image). Footpad cushions (C) are discrete modifications of the panniculus and contain small lobules of adipose with robust fibrous septa. Eccrine glands (EG) are embedded in the cushion, and eccrine ducts (ED) exit directly through the dermis and epidermis to empty onto the footpad surface. 4X magnification. Hematoxylin and eosin

The deepest epidermal layer, the stratum basale (stratum germinativum), contains epidermal stem cells with mitotic activity and continually supplies new keratinocytes to all epidermal layers (Fig. 2) [1, 4]. Basal layer keratinocytes are smaller and more cuboidal, with less cytoplasm, and attach the epidermis to its basement membrane and thus to the dermis. Moving up, the stratum spinosum is named for the spinous projections observed on keratinocyte membranes with paraffin section histology – an artifact of tissue processing that accentuates desmosomal attachments between cells. Spinous layer cells are larger due to more abundant cytoplasm, are polyhedral, and have more visible cytoplasmic keratin intermediate filaments. Next, the stratum granulosum is named for cytoplasmic, basophilic keratohyalin granules that are visible with hematoxylin and eosin (H&E) staining and store mostly proteins, like profilaggrin, needed for cornification [4]. Lamellar bodies, not visible with paraffin histology, form in this layer as well and deliver lipids, enzymes, and other key components to the extracellular surface during cornification [4]. The stratum corneum, the external most layer of the epidermis, forms by terminal differentiation called cornification, which creates nonviable corneocytes from the viable granular layer keratinocytes below [4]. During this process, keratinocytes loose most of their cytoplasmic water and organelles and flatten to become very thin (less than 1 micrometer) discoid cells with linear faceted (5–6) margins. The nucleus is also lost, and thus the cornification is orthokeratotic. On paraffin histology, the deep corneocytes are densely compacted into a discrete layer called the stratum compactum, and the superficial corneocytes are separated on their faces, because of processing artifact, in an open basket weave pattern in a layer called the stratum dysjunctum (Fig. 2) [5]. Corneocytes are continually shed from the body in a process called desquamation.

In the stratum corneum, corneocytes are stacked into many layers, approximately 10 to 15 on the trunk and 50+ on the footpads and planum nasale, and are sealed by intercellular lipids [3]. Fewer corneocyte layers are present in haired, low friction areas and more are in high friction areas such as palmar and plantar footpad surfaces. On the trunk, corneocytes are stacked into uniform vertical columns, overlapping only slightly at their margins, whereas on the footpads, corneocyte stacking is nonuniform and cells overlap extensively and variably, creating greater cell-to-cell surface contact, which is thought to increase adhesion. Intercellular lipids, delivered by lamellar bodies, are highly organized into a stack of lipids called the lipid envelop, which seals the entire extracellular space and creates the most important barrier preventing external water loss from the skin.[4] These lipids are composed of ceramides, cholesterol, and fatty acids. Certain lipids, like linoleic acid, are essential and are very important for lipid envelop formation and function. Corneocytes are continually shed from the skin surface by desquamation. Desquamation occurs because the normal physiochemical environment (pH, hydration, etc.) of outer stratum corneum promotes activation of numerous intercellular enzymes to cleave corneodesmosomes and degrade intercellular lipids, which allows corneocytes to separate away [4].

Clinically, a buildup of the stratum corneum on the skin surface, either due to increased production of corneocytes or altered desquamation, is called scaling. Partial loss of the epidermis leads to an erosion, which causes water loss from the skin surface. Eroded epidermis appears smooth and slightly moist due to the missing stratum corneum, which is responsible for normal epidermal surface architecture and barrier function. Eroded epidermis lacks hemorrhage as the epidermis does not contain blood vessels. In contrast, complete loss of the epidermis and the basement membrane is an ulcer, which appears moist to wet and granular (because of collagen exposure and recruited leukocytes and fibrin), and it often contains hemorrhage because of exposed dermal blood vessels.

Epidermal Basement Membrane

The epidermal basement membrane (basal lamina) is composed of numerous filamentous proteins and proteoglycans that bind together to form an ultrathin, mesh-like sheet that supports the basal cells and blankets the dermis [6]. Basal cells are structurally connected by hemidesmosomes to the basement membrane, which in turn is connected to the dermis by anchoring fibrils composed of collagen VII. Basement membrane zone is used to refer to this structure on light microscopic histology because it is too thin to be directly visualized (Fig. 2).

Epidermal strength results from physical interconnections between cytoskeletal proteins, cellular adhesion complexes (desmosomes and hemidesmosomes), and the epidermal basement membrane [6]. The cytoskeleton of each keratinocyte is linked by desmosomes, and in basal keratinocytes, it is linked to the basement membrane by hemidesmosomes. The keratinocyte cytoskeleton contains large amounts of keratin intermediate filaments, which are bundled together like ropes to form tonofilaments with high tensile strength. The buildup of specialized keratin filaments in each epidermal layer is called keratinization, a key part of cellular differentiation in the epidermis. Desmosomes of the stratum granulosum are modified by addition of corneodesmosin, and by other changes, to become corneodesmosomes in the stratum corneum [4]. Many diseases of epidermal fragility, i.e., mechanobullous diseases and pustular diseases, cause skin lesions by disrupting desmosomes, hemidesmosomes, or the basement membrane.

Dermis

The dermis (corium) is a thick, discrete, organized layer of extracellular matrix (collagens, etc.) that provides structure, toughness, and flexibility to the skin, and it supports the epidermis and adnexa as well as blood vessels, lymphatic vessels, and nerves found within it (Fig. 2) [1]. The dermis is divided into a thin superficial papillary layer with more loosely arranged matrix and finer collagen bundles and a thicker deep reticular layer that is more densely packed with coarser collagen bundles. The dermis is composed primarily of collagen, mostly types I and III, for strength, elastin for elasticity, and proteoglycans, like hyaluronic acid, for hydration and turgor pressure. In cats, the dermis has a scalloped deep margin (Fig. 1) with projections that connect to the lobular septa of the panniculus below. Dermal vessels are arranged in three sheet-like plexuses of arteries and veins that are located just below the epidermis, in the mid dermis, and in the deep dermis at the junction with the panniculus [7]. The dermis contains microscopic bundles of smooth muscle attached to hair follicles, called erector pili muscles, and free bundles in the dermis of teats (nipples) and scrotum [1, 2]. Also in the scrotum, the dartos tunic of the testis extends to the panniculus where it contributes smooth muscle and collagenous stroma. Small bundles of skeletal muscle extend to the dermis in facial and perineal areas only. Adipocytes are not normal constitutes of cat dermis and are part of the panniculus.

Mesenchymal cells maintain the dermal matrix and include fibrocytes (Fig. 2), which are spread individually throughout the dermis, as well as pericytes and Schwann cells that are localized around blood vessels and nerves, respectively. Low numbers of immune cells such as mast cells, dermal dendritic cells, lymphocytes, and basophils can be found in a healthy dermis where they are usually individualized and localized more to superficial perivascular and less to interstitial areas. Mast cells are common in the dermis of cats with 4 to 20 mast cells visible per 400x microscopic field on histology (Fig. 2) [8]. Neither neutrophils nor eosinophils are found in normal dermis or epidermis.

Panniculus

The panniculus (hypodermis, subcutis) is composed of discrete sheetlike layers of adipose, muscle, and fascia (Fig. 1) [1, 2, 9]. Immediately below the dermis, the panniculus adiposus (called the superficial adipose tissue) contains adipose arranged into lobules by thin fibrous septa (Fig. 1) [9]. Deeper, the panniculus fibrosus (superficial fascia) is a thin, variably discrete, sheet of fibrous tissue that connects to the lobular septa of the panniculus adiposus. Coursing within the fascia is a thin layer of striated muscle called the panniculus carnosus (cutaneous trunci) [1, 2]. The panniculus carnosus is more developed dorsally on the trunk (Fig. 1), neck, and proximal limbs and tapers away on the ventral abdomen (Fig. 4) and limbs to become absent on the extremities. Depending on the body region, like the extremities, the panniculus fibrosus merges with the deep fascia that surrounds the muscle of the skeleton or periosteum [8]. However, in some areas, like the ventral trunk, another layer of lobular adipose (called the deep adipose tissue) is present below the

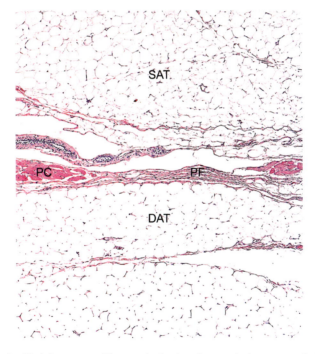

Fig. 4 Ventral mid abdomen, cat. The panniculus has three main layers: panniculus adiposus, panniculus carnosus, and panniculus fibrosus. The panniculus adiposus is composed of lobules of adipose just below the dermis, called the superficial adipose tissue (SAT), and in some body regions, a second layer which is deeper, called the deep adipose tissue (DAT), is also present. The panniculus fibrosus (PF) is a sheet of fibrous tissue (superficial fascia) that connects to the thin fibrous septa of adipose and supports the panniculus carnosus (PC), which is diminished ventrally. 40X magnification. Hematoxylin and eosin

panniculus fibrosus that is an additional deeper portion of the panniculus adiposus (Fig. 4) [9]. The panniculus adiposus is the thickest on the trunk, especially on the ventrum of the cat, where it can be measured in centimeters in obese patients, and it is mostly absent in the extremities. The panniculus is specialized to form cushions in footpads (Fig. 3), which are composed of lobules of adipose with thickened fibrous septa.[1,2] Arteries, veins, nerves, and lymphatic vessels are present in the panniculus and pass through to the dermis above.

Skin Adnexa (Skin Appendages)

Hair Follicles

Hair follicles produce hairs that cover nearly the entire body of the cat except for small areas like the mucocutaneous junctions, external genitalia, teats, planum nasale, and footpads [1, 2]. The density of hairs on the cat is higher, 25,000 per square centimeter, compared to the dog, 9000 per square centimeter, but the density varies by breed and anatomical location. Most hair follicles in the cat are a compound type in which several hair follicles share a single follicle opening (follicular ostium), while fewer are of a simple type with one hair follicle per ostium. Primary hair follicles are larger and produce larger hair shafts (guard hairs, outer coat hairs), while secondary hair follicles are smaller and produce smaller hair shafts (undercoat hairs). In cats, most hair follicles are grouped such that a single, simple, large primary hair follicle (central primary hair) is surrounded by two to five compound follicles, each with a primary hair(s) (lateral primary hair(s)) and 3–12 secondary hairs, with numbers partly depending on age [1, 2, 10, 11]. The tail lacks this arrangement and hair follicles are larger [2]. Primary hairs of cats are much thinner, 40–80 micrometer in diameter, compared to dogs, 80–140 micrometer, whereas the secondary hairs of cats are 10–20 micrometer and those of the dog are 20–70 micrometer. Most hair follicles lay in the skin at an angle to the epidermal surface such that their hair shafts all point caudally on the head and trunk and distally on the limbs (Fig. 1). In the dermis, the ectal side (outside) of the hair follicle is closer to the epidermis (acute angle), and the ental side (inside) is further from the epidermis (obtuse angle). Specialized sinus hair follicles (vibrissae or whiskers) are very large simple hair follicles surrounded by a blood sinus and have complex innervation and sensory tactile function (slow-adapting mechanoreceptor) (Fig. 5) [1, 2, 8]. Sinus follicles produce vibrissae that are large tactile hairs that vary greatly in length. Sinus hairs are found on the face (muzzle, eyebrows, lips), and palmar carpus, and variably on the neck, forelimbs, and paws of cats and are arranged individually, in small clusters, or in short rows in some areas on the face. A second tactile hair type, the tylotrich hair, arise from follicles that are slightly larger and more richly innervated than those of primary hair follicles and contact an adjacent sensory tylotrich pad (touch dome) when compressed [1, 2]. Tylotrich hairs are scattered individually, in low density, throughout most of the haired skin.

Fig. 5 Sinus hair follicle (vibrissae follicle), face, cat. The sinus follicle is a very large simple hair follicle that produces a large hair shaft (HS), or whisker, and is named for a large blood-filled sinus (S) that surrounds the follicle. The sebaceous gland (SG) and dermal papilla (DP) are annotated. 4X magnification. Hematoxylin and eosin

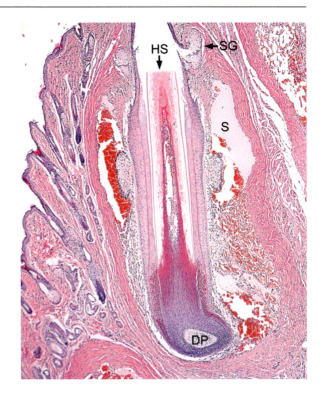

Hair follicles form during development as specialized epithelial down-growths of the epidermis (ectodermal origin) interacting with clusters of specialized mesenchymal cells (mesodermal origin) called the dermal papilla [10]. A fully formed hair follicle is a linear, layered, tubular epithelial structure that opens superficially at the follicular ostium and forms a solid bulb at its deep base with an invagination that encircles the dermal papilla (during anagen only) [1]. The erector pili muscle is a smooth muscle, and it originates in the dermis from the epidermal basement membrane and inserts on the ental side of the hair follicle [1]. This muscle elevates the hair shaft on the skin surface, for example, in behavior responses and with cold temperatures to trap more insulating air in the hair coat. The hair follicle epithelium is encased in a basement membrane (the glassy membrane) that is surrounded by a thin layer of collagen and specialized dermal fibrocytes, called the dermal root sheath or fibrous sheath [1]. The perifollicular dermis is richly supplied by small blood vessels that branch from all three dermal plexi but most prominently the mid dermal plexus [7]. Hair follicles in the anagen phase can extend into the panniculus adiposus (Fig. 1).

Hair follicles continually cycle to produce, hold, and shed hairs [10, 12]. The hair growing phase (anagen, Fig. 6) transitions through a short involution phase (catagen) and then ends in a resting phase (telogen, Fig. 6) in which a hair is retained or kenogen in which a hair is not retained (also called hairless telogen) [12]. A resting hair shaft is actively shed (exogen) usually when the cycle begins again. Hair

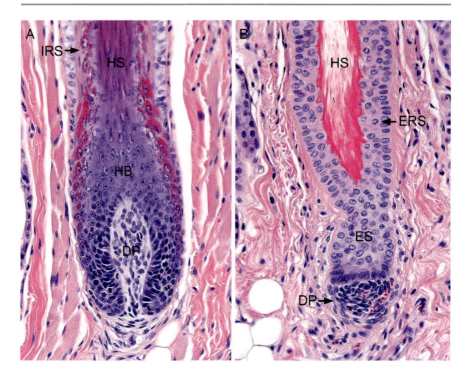

Fig. 6 Rostral chin, cat. Large primary hair follicles in the growing anagen phase and the resting telogen phase of the hair follicle cycle. A) In fully developed anagen, the hair bulb (HB) encases the dermal papilla (DP) and actively produces the hair shaft (HS) and inner root sheath (IRS). B) In fully developed telogen, the hair bulb and internal root sheath are absent and the external root sheath (ERS) regresses to surround the hair shaft, while the dermal papilla (DP), remains connected by only an epithelial strand (ES). The hair shaft stops growing, and its pointed end (club hair) is sealed by brightly eosinophilic trichilemmal cornification. 20X magnification. Hematoxylin and eosin

shedding in the cat is mosaic (nonsynchronous) [10]. The duration of phases varies by age, breed, season, etc. [13] For example, the length of the hair shaft depends on the length of the anagen phase – longer hair is due to a longer anagen phase.

The hair follicle has three zones (segments) called the infundibulum, isthmus, and inferior portions [12]. The infundibulum is the superficial, permanent, non-cycling segment that morphologically resembles the epidermis and attaches to it [1]. The isthmus and the deeper inferior portions change morphologically with the hair follicle cycle and have five main components, some only being present during the anagen phase (Fig. 6) [1, 12]. First, the inner root sheath surrounds the central follicle lumen and has its own three layers, an inner cuticle, Huxley's layer, and outer Henle's layer. The raised exposed edges of overlapping cuticle cells point internally (toward the hair bulb) and interlock with opposite facing hair shaft cuticle cells. The inner root sheath is only present during anagen (Fig. 6) when its keratinocytes continually migrate up in concert with the growing hair shaft, cornify, and shed to the infundibular lumen. Second, the companion layer is a single layer of cells that separates the inner rooth sheath from the external root sheath. Third, the external root sheath is several keratinocytes thick, it encases the inner root sheath, and it is contiguous with the infundibulum. Forth, the hair bulb, forms during anagen and is composed of hair matrix cells arranged in concentric layers that generate each of the layers of the inner root sheath, the companion layer, and the hair shaft (Fig. 6). Finally, the fourth part, the derma papilla (follicular papilla), is encased by an invagination of the hair bulb (Fig. 6). The dermal papilla is composed of mesenchymal spindle cells, blood vessels, and nerves, and its molecular communication with the hair matrix cells partly controls follicle cycling, hair shaft formation, and hair shaft pigmentation.

The hair shaft is formed by cornification of hair bulb cells (hair matrix cells), which makes it rigid, and contains three concentric layers, the outer cuticle, the cortex, and the inner medulla [1]. The hair shaft cuticle is a single layer of overlapping flattened cells in which exposed cell edges point outward (away from the hair bulb). The cortex is compacted and is nonpigmented or variably pigmented. The medullary cells have an open structure that in some follicles highlights an empty nuclear profile – primary hairs have a medulla but secondary hairs do not. The medulla may be pigmented or non pigmented. The outer end of the hair shaft is pointed in a long thin taper, while the inner end (hair root) is either connected to the soft viable hair bulb during anagen or is sealed-off in telogen by trichilemmal cornification to form a short, rigid, pointed taper with a rough surface (club hair) (Fig. 6).

Clinically, hair shafts are epilated and examined microscopically (trichogram) to identify the stage of the hair follicle cycle, primary or secondary status, and any hair shaft abnormalities. Anagen phase hair bulbs indicate active hair growth and, on trichogram, are recognized to be soft, flexible, rounded, and often axially deviated and pigmented when hair is pigmented. Telogen phase hairs (club hair) indicate resting hair follicles and have short tapered ends that are rough externally, rigid, and not axially deviated and are nonpigmented in hair that is pigmented or nonpigmented.

Skin Glands

In the cat, sebaceous glands are small, simple or compound, lobulated alveolar glands that connect to the lower infundibular lumen of hair follicles (pilosebaceous unit) by a very short duct lined by stratifed and cornifying epithelium [1, 2, 14]. At the edge of lobules (peripheral zone), a single thin layer of cuboidal reserve cells divides and differentiates to form larger polygonal lipid-vacuolated cells called sebocytes centrally (maturational zone), which shed to the lumen (holocrine secretion) to form sebum. In the cat, cytoplasmic vacuoles of sebocytes are very small and very uniform in size. Larger, often multilobulated, sebaceous glands are present on the face, especially the chin (Fig. 7; submental organ), ear base, dorsum, anal-rectal junction, palmar carpus (carpal gland), and interdigital paw skin. Meibomian glands (tarsal glands) are large sebaceous glands of the eyelid margin, especially the upper eyelid (Fig. 8) [2]. Sebaceous glands are not found in the planum nasale or the footpads.

Apocrine glands (epitrichial sweat glands) and eccrine glands (atrichial sweat glands) in cats are simple coiled tubular glands that secrete via a duct to the deep

Fig. 7 Rostral chin, cat. Sebaceous glands (SG) in the chin area (submental organ) are very large and multilobulated, and the dermis (D) is expanded to support the larger sebaceous glands in addition to the apocrine glands (AG), hair follicles (HF), and epidermis (E). 4X magnification. Hematoxylin and eosin

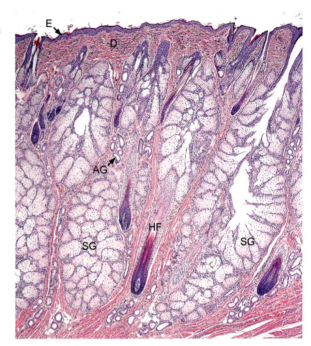

Fig. 8 Upper eyelid, cat. The upper eyelid is lined by haired skin (HS) externally and by mucosa (M) of the palpebral conjunctiva internally. Meibomian glands (MG) are enlarged sebaceous glands that align in a single long row that tracks the mucocutaneous junction. 4X magnification. Hematoxylin and eosin

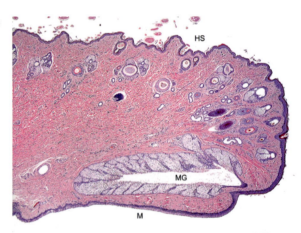

infundibulum of primary hair follicles (epitrichial) (Fig. 7) and to the footpad surface (atrichial) (Fig. 3) [1, 2, 14]. Glands are lined by cuboidal to low columnar cells that secrete by the release of apical blebs of the cytoplasm to the gland lumen and then to a thin duct lined by a bilayer of short cuboidal cells. A few myoepithelial cells surround the gland. Ceruminous glands are modified apocrine glands in the external ear canal (see below). Eccrine glands are not found on the planum nasale.

Fig. 9 Proximal dorsal tail, cat. The dorsal tail gland (DTG) of the cat, also called the supracaudal gland, is non-discrete and is composed of hepatoid glands on hair follicles along most of the dorsal tail. Erector pili muscles (M) are largest on the proximal dorsal tail and originate from the basement membrane of the epidermis (E) and insert on hair follicles in the dermis (D). 4X magnification. Hematoxylin and eosin

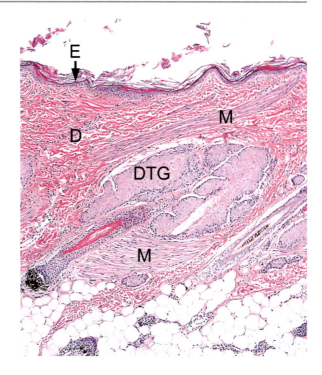

In the cat, the dorsal tail gland (Fig. 9, supracaudal gland) is formed by hepatoid glands located on hair follicles of the dorsal tail, especially proximally [15]. Feline hepatoid glands are a mixed lipid and protein secretion type and thus appear very pale, eosinophilic, and moderately vacuolated in hematoxylin and eosin-stained histology sections compared to brightly eosinophilic and non-vacuolated hepatoid glands (circumanal glands) of the dog, which produce primarily protein [15]. This is also the situation for hepatoid glands present on the anal sacs of cats.

Anal sacs (perianal sinuses) are paired in the cat and are located in the subdermal tissue of the perineum bilaterally (Fig. 10) [1, 2]. The anal sac and its short, duct-like, narrow opening to the anal-rectal skin junction are thin walled and lined by stratified squamous epithelium with an orthokeratotic pattern of cornification, which are supported by a thin layer of dermal matrix. Apocrine glands and large hepatoid glands of the anal sac (Fig. 10) are grouped in this matrix along the anal sac margin and empty to it [2, 14].

The mammary gland is a compound tubulo-alveolar gland arranged by septa into lobules and lobes where each gland empties via a branched ductular system to a teat (nipple) [1]. Glandular secretions pass through intralobular, to interlobular ducts, to lactiferous ducts, and then to a teat sinus (teat cistern), all of which are lined by either a simple layer or a bilayer of cuboidal cells. In the cat, the teat sinus empties externally through four to seven papillary ducts that are lined by stratified squamous epithelium. The teat dermis contains free bundles of smooth muscle but scant other adnexa. In the cat, four mammary glands are organized in linear mammary chains on the right and left side of the ventral abdomen.

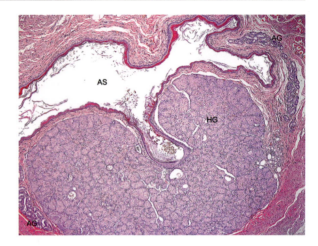

Fig. 10 Anal sac and associated glands, cat. The anal sac (AS) is thin walled and lined by stratified squamous epithelium. Multifocal apocrine glands (AG) and hepatoid glands (HG) empty to the anal sac. 4X magnification. Hematoxylin and eosin

Claws

The cat claw is a very specialized and complex structure composed of a cornified claw sheath (claw horn, claw) that is formed by stratified epithelium and supported by specialized dermis (corium) [16]. The claw in the cat is a highly tapered (sharply pointed) curved, and ventrally flattened cone that has a rounded wall dorsally, flattened blades on its slides, and a narrow cutting ridge ventrally. Proximally, a band of claw matrix cells continually divide and differentiate to supply keratinocytes for growth of the claw. More distally, the claw bed (claw plate) cells provide sliding adhesion that allows epithelial cells to move distally and to cornify into the rigid claw. Cats sharpen their claw tips by repeat shedding of a cornified horn cap that is promoted by scratching [16] – shed horn caps are sometimes mistaken for a sloughed claw. The claw encases the claw dermis and the closely apposed unguicular process of the third phalanx. The hard cornification of the rigid claw is bordered by soft cornification where it is contiguous with the skin fold (claw fold). The claw fold is large on the dorsal and lateral margins of the claw and minimal below. Below the claw centrally, a small sole first merges with the narrow skin fold, which merges with palmar or plantar digital pad [16]. The claw fold is modified and elaborated (claw sac, etc.) in the cat to allow claw retraction.

External Ear (Chapter, Otitis)

The skin of the pinna and external ear canal (external acoustic meatus) are lined by stratified squamous epithelium with an orthokeratotic pattern of cornification, which is supported by a thin dermis [2, 17]. Skin adnexa are present on all surfaces but are smaller and less densely placed in the inner ear pinna (concave pinna) and especially in the external ear canal in comparison to the outer pinna (convex pinna) (Figs. 11 and 12). Hair follicles and sebaceous glands are in all of these locations

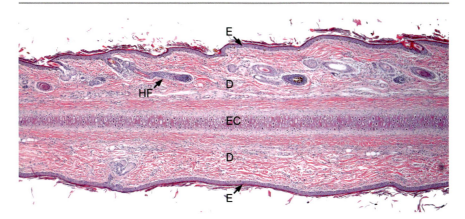

Fig. 11 External pinna, cat. The external pinnae contain a sheet of elastic cartilage (EC) centrally that is lined by dermis (D) and epidermis (E) on the convex (top of image) and concave (bottom of image) sides. Adnexa, including hair follicles (HF), sebaceous glands, and apocrine glands, are more numerous and larger on the convex surface compared to the concave surface. 4X magnification. Hematoxylin and eosin

Fig. 12 Horizontal ear canal (cross section), cat. The ear canal is lined by a thin epidermis (E) and dermis. The inset demonstrates the small sparse adnexa of the ear canal in the dermis (D), including a hair follicle (HF), a sebaceous gland (SG), and a ceruminous gland (CG). 4X magnification and 20x magnification (inset). Hematoxylin and eosin

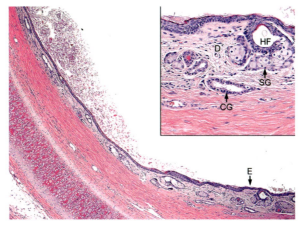

but are in a low density in the external ear canal. Apocrine glands are present on the convex and concave pinna. Modified apocrine glands, called ceruminous glands, are in the portion of the external ear canal that is supported by the annular cartilage (Fig. 12) and are more numerous in the deep one third of the canal [17–20]. These glands connect to the sparse follicles or directly to the epidermis [17]. Ceruminous gland secretions mix with sebum, epidermal surface lipids, and desquamated corneocytes to form a waxy protective material called cerumen. Centrifugal epithelial migration of keratinocytes off of the external tympanic membrane and on to the external ear canal helps to clear cerumen from the surface of the tympanic membrane, which is located at the deep extent of the ear canal [19].

Skin Pigmentation

The skin derives its color from pigmentation (melanin pigment); blood in vessels (red heme pigment); the endogenous reflective properties of the epidermis, dermis, and adnexa; as well as the quality of light being reflected [8]. Melanin pigment is composed of two types, brown to black eumelanin and red to yellow pheomelanin, that are expressed variably to produce a color range. From neural crest, melanocytes migrate during development into the epidermis, follicles, and claws and then produce melanin pigment in membrane-bound cytoplasmic organelles called melanosomes. Through dendritic processes, a melanocyte transfers pigmented melanosomes to a certain number of local keratinocytes, and together, they are called an epidermal melanin unit (or follicular melanin unit in the hair bulb). The amount and type of melanin produced and the degree of dispersal of melanosomes to keratinocytes affect the pigment intensity, ranging from a pale dilute to a very dark color. Melanocytes in the anagen hair bulb deliver melanosomes to growing hair shafts, continuously or episodically, and the latter produces color banding (agouti, etc.), which is partly controlled by the dermal papilla. Many coat color variations (Chapter, Coat Color Genetics) in cats are due to inheritance of alleles that alter melanocyte distribution or presence/absence, melanosome dispersal, and/or the amount and type of melanin produced. Clinically, the loss of pigment (leukoderma, leukotrichia) is due to disruption of the epidermal melanin unit, and/or follicular melanin unit, and thus can result from diseases that injure melanocytes and/or keratinocytes.

General Skin Functions

The skin has many important functions that, when compromised by disease, have significant consequences for the patient [8].

Physical Barrier Function

The skin protects the body from physiochemical injury. It prevents entry of foreign materials, parasites, and infectious agents while, at the same time, preventing loss of water and fluid components (electrolytes, macromolecules, etc.) from the body. To do this, the epidermis and dermis provide the toughness of the skin, and hair reduces frictional injury. The panniculus provides a cushion to injury especially in footpads. Skin pigment and hair block damaging solar radiation. The stratum corneum, especially the lipid envelop, seals the epidermis to water loss, whereas deeper epidermal layers also contribute, for example, via tight junctions in the stratum granulosum. Corneocytes are shed continuously from the skin surface, eliminating attached microorganisms. Claws even serve as offensive physical defense against attack of other animals and as tools needed by cats for climbing and handling of prey.

Immune Defense

More than just a passive physical barrier, the skin immune system actively identifies, blocks, and eliminates pathogens through actions of the innate (keratinocytes, mast cells, basophils, natural killer cells, dendritic cells, sebum, etc.) and acquired immune skin systems (T-cells, B-cells, dendritic cells, etc.). Additional cells, neutrophils, eosinophils, and macrophages are recruited to the skin through its vasculature and contribute to skin defense and immune function. Sebum and stratum corneum constituents contribute to skin surface pH, and fatty acid composition favors skin colonization by beneficial bacterial commensals and limits pathogens. Interestingly, the skin helps to maintain peripheral immune tolerance, by assisting the thymus in educating the acquired immune system on self and non-self antigens.

Thermoregulation

The skin is a key organ of thermoregulation that both works to prevent heat loss and to promote it, as needed, to optimize core body temperature. The hair coat and adipose are the main thermal insulating barriers, and the former can be modified by erector pili muscles moving the hair and controlling its density. Skin blood flow is actively promoted, restricted, and/or shunted to alter core heat transfer to the skin, especially in the distal limbs and ears. Pigment in hair and epidermis absorbs light energy, leading to heating of the skin. Sweating promotes cooling through evaporation.

Metabolic Functions

The skin has numerous metabolic functions; many maintain skin homeostasis, while others also serve systemic functions. For example, vitamin D is activated in the epidermis via exposure to sunlight. And, after further activation in the liver and kidney, vitamin D impacts epidermal proliferation and differentiation in the skin as well as contributes to calcium homeostasis of blood and bone, among many other functions, systemically. Expression of p450 enzymes in the epidermis means that xenobiotic compounds can be processed there. The panniculus adiposus contributes much of the bodies' capacity to store energy in the form of lipids. Similarly, dermal collagen is a protein reservoir. The epidermis, hair follicles, and skin glands produce useful substances but also eliminate endogenous and exogenous metabolic constituents, such as some toxins (lead in hair).

Communication

The skin glands produce scents that are important for olfactory communication in carnivores. Erector pili muscles, especially along the back and tail, elevate the hair, changing the physical hair coat appearance, to visually communicate behavior status and warning signals to other animals, and to disperse pheromones. Skin and hair coat pigmentation, although modified dramatically in many domestic cats by human selection pressures, provide camouflage important for carnivores while hunting.

Sensory Perception

The skin is a major organ of sensation, and its sensory nerve endings distinguish temperature (hot and cold), pain, pruritus, burning, touch, etc.

References

1. Monteiro-Riviere N. Integument. In: Eurell JA, Frappier BL, editors. Dellmann's textbook of veterinary histology. 6th ed. Iowa: Blackwell Publishing Professional; 2006. p. 320–49.
2. Strickland JH, Calhoun ML. The integumentary system of the cat. Am J Vet Res. 1963;24:1018–29.
3. Monteiro-Riviere NA, Bristol DG, Manning TO, et al. Interspecies and interregional analysis of the comparative histologic thickness and laser Doppler blood flow measurements at five cutaneous sites in nine species. J Invest Dermatol. 1990;95:582–6.
4. Matsui T, Amagai M. Dissecting the formation, structure and barrier function of the stratum corneum. Int Immunol. 2015;27:269–80.
5. Bowser PA, White RJ. Isolation, barrier properties and lipid analysis of stratum compactum, a discrete region of the stratum corneum. Br J Dermatol. 1985;112:1–14.
6. Hammers CM, Stanley JR. Mechanisms of disease: pemphigus and bullous pemphigoid. Annu Rev Pathol. 2016;11:175–97.
7. Meyer W, Godynicki S, Tsukise A. Lectin histochemistry of the endothelium of blood vessels in the mammalian integument, with remarks on the endothelial glycocalyx and blood vessel system nomenclature. Ann Anat. 2008;190:264–76.
8. Miller WH, Griffin CE, Campbell K. Muller & Kirk's small animal dermatology. 7th ed. St. Louis: Elsevier; 2013. p. 1–56.
9. Stecco C. Subcutaneous tissue and superficial fascia. In: Functional atlas of the human fascia. Philadelphia: Elsevier; 2015. p. 21–30.
10. Meyer W. Hair follicles in domesticated mammals with comparison to laboratory animals and humans. In: Mecklenburg L, Linek M, Tobin D, editors. Hair loss disorders in domestic animals. Iowa: Wiley-Blackwell; 2009. p. 43–61.
11. Zanna G, Auriemma E, Arrighi S, et al. Dermoscopic evaluation of skin in health cats. Vet Dermatol. 2015;26:14–7.
12. Welle MM, Wiener DJ. The hair follicle: a comparative review of canine hair follicle anatomy and physiology. Toxicol Pathol. 2016;44:564–74.
13. Ryder Ryder ML. Seasonal changes in the coat of the cat. Res Vet Sci. 1976;21:280–3.
14. Jenkinson DM. Sweat and sebaceous glands and their function in domestic animals. In: von Tscharner C, Halliwell REW, editors. Advances in veterinary dermatology, vol. 1. Philadelphia: Bailliere Tindall; 1990. p. 229.
15. Shabadash SA, Zelikina TI. Detection of hepatoid glands and distinctive features of the hepatoid acinus. Biol Bull. 2002;29:559–67.
16. Homberger DG, Ham K, Ogunbakin T, et al. The structure of the cornified claw sheath in the domesticated cat (Felis catus): implications for the claw-shedding mechanism and the evolution of cornified digital end organs. J Anat. 2009;214:620–43.

17. Strickland JH, Calhoun ML. The microscopic anatomy of the external ear of Felis domesticus. Am J Vet Res. 1960;21:845–50.
18. Fernando SDA. Microscopic anatomy and histochemistry of glands in the external auditory meatus of the cat (Felis domesticus). Am J Vet Res. 1965;26:1157–61.
19. Njaa BL, Cole LK, Tabacca N. Practical otic anatomy and physiology of the dog and cat. Vet Clin North Am Small Anim Pract. 2012;42:1109–26.
20. Tobias K. Anatomy of the canine and feline ear. In: Gotthelf L, editor. Small animal ear diseases, an illustrated guide. 2nd ed. St. Louis: Elsevier-Saunders; 2005. p. 1–21.

Coat Color Genetics

Maria Cristina Crosta

Abstract

The various breeds of cats differ considerably one from the other, not only in terms of their different morphological features but also in terms of the colour, length, structure and texture of their coat. The feline coat has various functions, such as aesthetic and mimetic, heat regulation, perception of the body position through vibrissae and tylotrich pads, social and sexual communication, and acts as a barrier against mechanical, physical and chemical insults. In the first part of this chapter, the morphology and cycle of the hair are briefly introduced, including melanin synthesis. In the second part, the genetics of hair length, structure, texture, colour and colour patterns are detailed, providing a good description to understand the specific functional aspects of the feline coat.

The various breeds of cats differ considerably one from the other, not only in terms of their different morphological features but also in terms of the colour, length, structure and texture of their coat.

The Coat

Cat show organisers classify cat breeds according to three main categories:

- *Longhair cats*, the representatives being the Persian (in all colour shades and varieties), the British Longhair, the Selkirk Rex and the Highland Fold

Images by Lia Stein

M. C. Crosta (✉)
Clinica Veterinaria Gran Sasso, Milan, Italy

- *Medium-length hair cats*, e.g. Norwegian Forest, Maine Coon, Balinese, Birman
- *Shorthair cats*, e.g. European, Chartreux, Russian Blue, British Shorthair

Within these categories, the coat is then classified according to pattern, colour and colour distribution.

Function

The coat has various functions:

- Aesthetic and mimetic;
- Heat regulation, in function of the length, thickness and density of the coat, as well as of its colour and shine (a light-coloured coat reflects light better and allows to keep body temperature constant);
- Perception of body position (vibrissae, tylotrich pads);
- Social and sexual communication, thanks to both the visual effect and as support to pheromones;
- Barrier against mechanical, physical and chemical insults.

Whatever its length, a cat's coat is made to protect the animal and help it adapt to its environment, like in the case of cats that live in very cold climates. The coat of these cats (Maine Coon, Norwegian Forest cats) consists of long primary hairs (basic coat) and a thick undercoat.

The hair of a Norwegian Forest cat is water-repellent. This trait makes its coat especially suited for the adverse weather conditions of its country of origin. In shows, sometimes judges assess this feature by dripping some water over its coat. Another example is the Turkish Van, a cat that has a very thick coat in winter and that sheds in a spectacular manner in summer. In fact, in this season it loses almost all of its fur, to the point of looking like a shorthair cat. This breed has adapted to the climate of Central Anatolia, its region of origin, where there is a great difference in temperature between winter (-20 °C) and summer (+40 °C).

Morphology

Macroscopically speaking, regardless of length, cat hair can be classified as:

1. Primary hair (guard hair)
2. Secondary hair (down hair).

Like all carnivores, cats have compound hair follicles. This means that the coat is comprised of many small units. Each unit consists of two to five larger hairs (primary hairs) surrounded by clusters of smaller hairs (secondary hairs). To each primary hair are associated from five to twenty secondary hairs.

Each primary hair has its own sebaceous gland, a sweat gland and an arrector pili muscle. A primary hair emerges through the surface of the skin through its own independent infundibulum. Secondary hairs are associated only to one sebaceous gland and emerge from a common infundibulum.

It is estimated that, in cats, for each square cm of skin there are anywhere from 800 to 1600 of these units.

From a functional point of view, hair can also be classified as:

1. Protection hair: these hairs are straight and thicker.
2. Intermediate hair: these hairs are thinner than protection hairs and more variable in cross diameter. They grow in the opposite direction of the underhair and play an isolating and protective role together with the underhair.
3. Underhair: these hairs are short and thin and have a crimped appearance, some-times curly. In winter, they trap warm air and create a veritable insulating barrier against the cold, while in summer they limit the absorption of heat coming from the outside.

The proportions vary according to breed:

- *All three hair types can be present,* but they can be highly modified (e.g. Devon Rex).
- *One can be missing* (e.g. protection hair in the Cornish).
- *One can be more abundant than the others* (e.g. underhair in Persians).
- *One can be very scarce* (e.g. underhair in the Korat).

There are two types of tactile hairs:

- *Vibrissae*: they grow on the muzzle, around the eyes, in the throat region and on the palmar face of the carpi. These hairs are thick and contain specialised nerve structures.
- *Tylotrich hairs*: they are distributed all over the body and consist of a larger than normal hair follicle containing a single short hair and surrounded by a capsule of neurovascular tissue at the level of the sebaceous gland. These hairs are believed to be slow-adapting mechanoreceptors.

The hair shaft is made up of three concentric structures, the medulla, the cortex and the cuticle.

The medulla is the inner structure of the hair and consists of longitudinal rows of cells that are solid close to the root and that progressively fill with air and glycogen as they rise towards the tip.

The cortex forms the middle structure of the hair and is formed by hard and fuse-shaped cells with the longer axis parallel to the hair's axis. These cells contain the pigment that gives the hair its colour.

The cuticle is the outermost structure of the hair and is made up of scales (in humans they are imbricate, like the tiles of a roof, while in cats they are triangular with a spinous edge and with the hooked free edge facing the direction of the hair tip).

Hair Growth Cycle

The hair follicle is the structure from which the hair grows. It consists of an upper section called 'infundibulum', an intermediate section called 'isthmus', and a deep section that is the 'bulb'. The infundibulum and the isthmus are the permanent portions of the hair, while the bulb is present only in the active growth phase.

The hair bulb consists of matrix cells (that generate the hair itself and the internal sheath that contains the root) and of pigment-producing cells called melanocytes.

The hair follicle has an active, cyclic growth phase called anagen and a rest phase called telogen. They are separated by a transition phase called catagen.

The duration of the *anagen phase* is hereditary and determines the final length of the hair. In this phase, the dermal papilla is very well developed and the cells of the bulb's matrix actively multiply to form the hair. The bulb's melanocytes actively produce pigment (melanin) and distribute it to the hair's cells that progressively migrate towards the surface of the skin.

During the transition phase called *catagen*, pigment production stops completely and the production of cells by the matrix progressively slows down to a stop. The last cells produced therefore are entirely pigment-less and this explains why in this phase of the cycle the section of the hair closest to the skin is also the lightest.

During *telogen*, the follicle, in a resting phase, has shrunk to one third of its length and the dermal papilla has transformed into a small mass of undifferentiated cells. Hair shedding does not occur simultaneously throughout the coat but according to a 'mosaic' shedding pattern. This is because neighbouring hair follicles are all at different stages of growth. The hair grows to a pre-set length that can vary according to the body region and is genetically determined.

The active growth phase and therefore hair growth speed is highest in summer and lowest in winter. Indeed, it is thought that in summer 50% of the follicles are in telogen, while in winter this percentage rises to 90%.

Melanin Synthesis

Melanin is the pigment responsible for colouring skin and hair. This is not its only function, however. By distributing through the cytoplasm, it protects the cells of

Fig. 1 The synthesis of
eumelanin and
pheomelanin

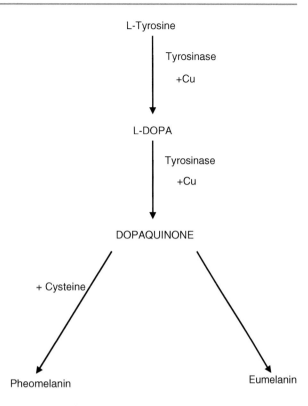

the epidermis and of the deeper skin layers from ionizing radiation and from ultraviolet (UV) light. Melanin also eliminates the toxic free radicals generated by the skin's cells following exposure to sunlight and in the course of inflammatory processes.

Melanin synthesis is genetically determined (Fig. 1). Its production may be stimulated by several factors, such as exposure to ultraviolet radiation from the sun, and may be affected by hormone unbalance. There are various types of melanin, but the basic types are eumelanin and pheomelanin. Eumelanin granules are contained in melanosomes and are responsible for the brown-blackish-black colour. Pheomelanin granules are also contained in melanosomes and impart a yellow-brownish-red colour. There are many intermediate variations between these two types. Pheomelanin features a higher sulphur content compared to eumelanin. Although different, eumelanin and pheomelanin have a common metabolic life. An enzyme called tyrosinase in the presence of oligoelements, such as copper, transforms tyrosine first into DOPA and then into dopaquinone and from there, through a sequence of oxidations, into the various kinds of melanin pigment. The importance of the role played by this enzyme in pigment synthesis is proven by the fact that the mutation of the structural gene of tyrosinase is responsible for many forms of albinism in humans and in animals. The synthesis of eumelanin requires high concentrations of tyrosinase, while that of pheomelanin needs lower concentrations but requires cysteine. Tyrosinase is heat-sensitive, which means that its concentration decreases in proportion to the increase in temperature. In body areas where the temperature is lower (legs, for example) thanks to the increase in enzyme activity there is a higher deposit of eumelanin granules, and this gives the coat a darker coloration. This explains why, when the coat of a black cat is observed closely in the trunk area, a lighter coloration is seen at the base, while in the muzzle areas and on the legs (that are colder areas), the colour is darker.

The Genes Controlling Hair Length, Structure and Texture

Although from a morphological viewpoint we distinguish three kinds of cat coats – long, semi-long and short – from a genetic viewpoint, in terms of hair length, the coats are two – shorthair and longhair (Figs. 2, 3 and 4). The genes described in this chapter are summarized in Table 1. A definition of genetic terms is provided in Box 1.

Fig. 2 Long-haired cat

Fig. 3 Semilong-haired cat

Fig. 4 Short-haired cat

Table 1 Main genes controlling coat colour in cats

Genes controlling distribution of coat colour	
A Agouti	a non-agouti
Original wild colour, hair with alternated light and dark bands	The colour bands on the single hairs disappear, and the result appears solid colour
Genes controlling colour	
B Black	b brown (or chocolate)
	bl light brown (or cinnamon)
o non-orange	O Orange – sex-linked because it is located on the X chromosome
w normal colour	W White dominant epistasis
Genes controlling colour intensity	
C full Colour – even intensity all over body	cb burmese
	cs siamese
	ca albino blue eyes
	c albino pink eyes
Genes controlling colour density	
D Dense – normal hair colour density	d dilution or maltesing
Genes controlling hair colour development	
i complete development of pigment in hair	I Inhibition of development of pigment in hair
wb (or ch) no tipping	Wb (or Ch) tipping
Genes controlling white spot distribution	
s normal colour distribution in coat – no white spots	S piebald white Spotting – more or less extensive white spots
G normal colour distribution in coat – no gloves	g 'gloves' of the birman
Genes controlling tiger tabby patterns	
T mackerel Tabby (wild type)	Ta abyssinian tabby
	tb blotched tabby (or classic tabby)
Genes controlling and modifying hair length, structure and texture	
L shorthair	l longhair
R normal coat	r Cornish Rex
Re normal coat	re Devon Rex
Ro normal coat	ro Oregon Rex
rd normal coat	Rd Dutch Rex
rs normal coat	Rs Selkirk Rex
Hr normal coat	hr Sphynx, Bambino, Elf, Dwelf
hrbd normal coat	Hrbd Don Sphynx, Peterbald, Levkoi
wh normal coat	Wh Wirehair

Box 1
- **Homozygote/Heterozygote**

A homozygote is an individual having two identical alleles for one same trait (one coming from the mother and the other from the father). In fact, for each single trait each individual receives a pair of 'corresponding' genes called **alleles,** one inherited from each parent. If the individual inherits two identical alleles, that individual is **homozygous** for that specific trait (e.g. BB or bb). Conversely, if the individual inherits two different alleles from its parents, the individual will be a **heterozygote** (Bb).

- **Dominant and Recessive**

When an allele manages to express itself (phenotypic expression) both in a homozygous individual (BB) and in a heterozygous individual (Bb or Bbl), it is said to be dominant. If an allele expresses itself only in the homozygote, it is called recessive and is indicated with the lower case letter (bb).

The dominant trait does not allow the recessive trait (b) to express itself. Two cats, one BB and the other Bb, shall both have a black coat, but BB is homozygous black while Bb is a black carrier of the chocolate colour that, being recessive to B, cannot be expressed. This is why in a genotype, 2 letters are used to indicate a trait (e.g. BB, Bb). If instead only one letter is used followed by a dash (e.g. B-), it means that we don't know whether the individual is homozygous (BB) or heterozygous (Bb) for that same trait. In genetics, the dominant allele is indicated with an upper case letter, and usually uses to be the first letter of the gene to which it refers (B for Black, D for Dense).

- **Polygenes**

These are a group of genes (also called 'modifiers') the single action of which is often not quantifiable, but when working together have a cumulative effect and can change the action of the main gene. They affect quantitative traits (size, hair length, etc.), often quite significantly.

- **Epistasis**

Some genes have the capacity to prevent other genes from expressing themselves. Pheomelanin, for example, masks eumelanin; the non-agouti gene covers the tabby patterns; the W gene (dominant white) masks the expression of all of the other genes responsible for coloration and colour distribution (Fig. 31).

Sometimes, the epistatic effect of the W gene is not totally effective. One often sees white kittens with a spot of colour (black, blue cream, etc.) on their head. This spot of colour, that totally disappears by the age of ten months or so, is nothing but the hidden colour the kitten will pass on to its progeny as an adult. All cats, independent of colour, are genetically tabby, i.e. they possess striping in their genotype (Fig. 32). In 'self' (solid colour) cats, striping is present but not visible because the **a** gene (non-agouti) does not allow striping to express itself (epistatic effect). The **O** gene (orange) transforms colour pigments into pheomelanin and through epistasis deactivates the loci coding for the production of eumelanin (Fig. 33). Moreover, the **a** gene (non-agouti) makes striping disappear (epistasis) only in eumelanic coats and has no effect on pheomelanic coats. This is why, in red cats, striping is always present. It is difficult to obtain an evenly red-coloured coat with intense red coloration (breeders have a hard task in doing this) because quite often residual stripes show up on the muzzle, tail and legs. Sometimes, in the attempt to diminish the undesired striping, varieties of red coats with excessively bleached coloration are selected. Unlike shorthair cats, in which the stripes are more evident, in the Persian these 'defects' are corrected by the hair length hair. Just like the other solid colour coats, the red coat must be even, i.e. each single hair must have the same intensity from the root to the tip, and its coloration must be as red as possible. Next to solid colour coats, also known as self, breeders select red tabbies in which the stripes are actually highlighted, in a curious game of contrast between the red background of the coat and the intense and prominent red of the pattern.

- **Density and Dilution**

Colour density is given by the dominant gene D responsible for dense pigmentation. Pigment granules are deposited one by one and evenly along the cortex and medulla of the hair. The entire surface of the granule reflects light, giving the hair a darker coloration. Coat coloration dilution is given by the recessive d gene (or Maltese gene) that causes a different distribution in space of the pigment granules without changing their shape. This different layout causes less light refraction and therefore the colour appears lighter.

- **Incomplete Dominance**

Incomplete dominance occurs when, in a pair of alleles, one allele does not totally dominate the other and the resulting individual shows intermediate traits (e.g. Tonkinese).

- **Agouti and Non-agouti**

This is an Indian term used to indicate a South American rodent. In genetics, it is used to describe the wild coloration of some mammals. The Agouti gene codes for the multiple banding of each single hair with yellow-greyish bands and a darker hair tip (ticking). Agouti allows the tabby alleles to express themselves. Agouti is the background colour against which one sees the stripes of a tabby coat (Fig. 34). The non-agouti gene codes for the masking of the bands of yellow-greyish colour on each single hair, making it still banded, but dark and very dark. Therefore, the coat appears to the eye of a single colour. It has an epistatic effect on tabby alleles.

- **Eumelanin, Pheomelanin and Tyrosinase**

The eumelanin granules determine brown, blackish or black coloration (B, bb, blbl). The pheomelanin granules determine red, yellow or orange coloration (O-). Eumelanin and pheomelanin are formed starting from the amino acid tyrosine. This process occurs thanks to the action of the enzyme tyrosinase (heat-sensitive) that oxidates tyrosine into various intermediate compounds (DOPA, dopaquinone). The C gene (colour intensity) codes for the enzyme's correct structure, which means that its inactivation by high temperature is much slower than its production and, therefore, melanin is regularly produced and can provide full coloration of the entire coat. The Albino alleles (cb and cs) progressively cause a structural change in tyrosinase, making it especially heat-sensitive. In the warmer areas of the body, due to a lower influence of tyrosinase, there will be less pigment deposit (the body is warmer and therefore the coat colour will be paler), while in the cooler areas of the body (the extremities), there will be a greater deposit of pigment and therefore darker coloration. In the Burmese (cb), the enzyme's structural modification causes the coat to change from black to dark brown and the eyes to become yellow or amber. In the Siamese coat, the cs gene makes the tyrosinase even more sensitive to heat, and therefore the difference between the colour of the body and that of the points (extremities) is much more evident and the eyes become blue (Fig. 35). The ca and c genes, instead, cause the destruction or lack of production of the enzyme, and therefore the coat is totally white and the eyes are blue and pink, respectively.

Length Genes

- short hair: L (dominant)
- long hair: l (recessive)

The original coat is short and is governed by a dominant gene labelled **L**. Long hair is given by its recessive mutant allele and is labelled **l**. Gene l is responsible not only for the long hair in Persian cats but also for the semi-long coats of the Maine Coon cat, Norwegian Forest cat, Siberian cat and Burmese cat. The various coat lengths are due to the presence of polygenes or modifier genes. These are minor genes the single effect of which is too small to be observed. However, when they are acting together with other genes, they produce observable effects because together they are variably capable of changing the action of the main gene. Hair length is not the only trait to be taken into consideration. Structure and texture, too, are important elements of evaluation when examining the various kinds of coats. The coat can be more or less thick and abundant, and the three types of hair (protection hair, intermediate hair and underhair) can be normal and all present at the same time. Sometimes, like in the Devon Rex, there are significant changes, or one of the three types of hair can be missing, such as the protection hair in the Cornish Rex. The types of hair can be present in different proportions, like in the Korat that has almost no underhair or in the Persian that abounds in underhair, compared to the amounts of protection and intermediate hair.

Structure and Texture

- r (recessive) Cornish Rex/German Rex
- re (recessive) Devon Rex
- ro (recessive) Oregon Rex
- Rd (dominant) Dutch Rex
- Rs (dominant) Selkirk Rex
- Wh (dominant) American Wirehair
- hr (recessive) Sphynx/Bambino/Elf/Dwelf
- Hrbd (dominant) Don Sphynx/Peterbald/Levkoy

The most significant changes regarding hair structure and texture involve the **r**, **h**, **Wh** and **Hrbd** genes.

r genes

Cornish Rex
A cat famous for its coat is the Cornish Rex. Its peculiar coat is due to the presence of the **r** gene. This gene codes for the lack of protection hairs and for deep changes in intermediate hair and underhair. This cat's fur is very soft, dense, coarse to the touch, wavy and so curly that the coat looks like that of a sheep's fleece. Even its facial and supraorbital vibrissae are curled.

Devon Rex
This cat's coat is due to a recessive gene called **re** (Fig. 5). All three hair types are present but deeply modified. The coat is less wavy and curly than that of the Cornish and in general its coat has a more sparse appearance. The facial and supraorbital vibrissae may be broken or even absent. The **r** and **re** genes are recessive mutant

Fig. 5 Devon Rex cat

genes located on different loci of the chromosome, and by crossing a Cornish Rex with a Devon Rex one can obtain non-rex coated kittens.

There are other modifications regarding the **r** genes, especially those of the Oregon Rex, a cat with a coat deriving from the presence of the **ro** gene, a recessive mutant allele that causes the protection hair to disappear. The coat of the German Rex is caused by the presence of the **r** gene, identical to that of the Cornish. In the Dutch Rex and in the Selkirk Rex, instead, the genes are two dominant mutant alleles: **Rd** and **Rs,** respectively. The Dutch Rex is currently not being bred. The Selkirk rex has a thick, plush and curly coat which can be short or long.

h genes

Sphynx/Bambino/Elf/Dwelf

The Sphynx is a cat that has no protection or intermediate hairs due to the presence of the recessive **hr** gene. The sparse underhair, that can even be totally absent, is located on the muzzle, on the outside base of the ears, on the feet, scrotum and tail. The skin is very soft to the touch, resembling suede, and wrinkles are present on the face, between the pinnae and on the shoulders. The cat is medium size, muscular, with wide chest and rounded abdomen. The cross-breeding of the Sphynx with other breeds has resulted in the Bambino, Elf and Dwelf breeds. The Bambino cat is the result of the cross between the Sphynx and the Munchkin ('Sausage cat'), a short-legged cat. The Bambino is a smaller version of the Sphynx has no hair, a long chest and a rounded abdomen. The hind legs are longer than the front legs, the muzzle is triangular with wide and tall pinnae. The cross between the Sphynx and the American Curl has generated the Elf breed, a naked, tall and muscular cat with prominent cheek bones, like the Sphynx cat. This cat also presents curled pinnae, like the American Curl breed. The breeding of the Elf with the Munchkin/Bambino has given rise to the Dwelf breed (the name is a blend of 'dwarf' and 'elf', the mythical creature with pointed ears). This cat is small, naked, short-legged and has curled pinnae like the Elf cat.

Hrbd genes

Donskoy/Peterbald/Levkoy

This group includes the Donskoy (Don Sphynx), the Peterbald and the Levkoy. The gene responsible for these breeds is **Hrbd**, a dominant gene located on a locus different from that of the **hr** gene. The Donskoy is a middle-sized cat with cuneiform head, wide ears with rounded tips positioned high on the head and medium to long legs. The Donskoy cat is preferable naked; however it occasionally may have hairs called 'flock' when less than 2 millimetres long and 'brush' when more than 2 millimetres long. The hair is sparse and hard all over the body, with naked areas on the head, upper neck or on the back. These cats, with residual hair, cannot participate in cat shows but are successfully employed for reproduction. The Peterbald derives from the cross of the Don Sphynx with the Siamese/Oriental Shorthair. It has all the morphologic features of the Siamese/Oriental Shorthair (long, elegant and slender body with long legs) but carries all the cutaneous features of the Donskoy, with wrinkles on the face, between the ears and on the shoulder. As for the Donskoy, the naked cats are preferred, but 'haired' cats can be used for reproduction. The Levkoy comes from the cross between the Donskoy and the Scottish Fold. The Levkoy may be naked or have residual hair and its pinnae are folded forward like the ones of the Scottish Fold. Again, the naked ones are preferred, but kittens and younger cats can sometimes present some residual hair. Crossing between these breeds and the Sphynx is not allowed.

Wh Genes

Wirehair

This group includes a cat breed that has a very particular coat, the American Wirehair, which owes its coat to the dominant mutant gene **Wh**. All three hair types are present but they appear modified and curly, which makes the Wirehair's coat hard and coarse to the touch.

The Genes Controlling Coat Coloration and Patterns

Tabby

The transmission of cat coat colours follows precise genetic rules according to Mendel's laws. If when speaking of hair length the shorthair is the original coat, when speaking of colour patterns all cat coats derive from the tabby. The tabby coat is the most commonly found in nature because it is highly mimetic. Indeed, tabby is the original wild coat, the primordial coloration from which all other coat colours have descended by mutation. The name 'tabby' derives from 'Attabi', the characteristic Baghdad district famous for making the precious striped silk cloth called 'taffetà'. This name, later shortened to 'tabby', was then used to describe the striped coat of cats. In tabby cats, the stripes seem drawn against the coat's background that is commonly called Agouti.

Agouti and Non-agouti

Agouti is an Indian word used to indicate a rodent that lives in the rain forests of Central and South America, subsequently used in genetics to describe the wild coloration of the hare and rabbit. Agouti (**A**) in the genotype determines a 'striped' or 'banded' coloration of the coat via a pigment synthesis system called **on-off**. Darker colour bands produced during the **on** phase alternate with lighter colour bands produced during the **off** phase. In this way, each single hair is not of a single colour but is marked by alternated dark and light colour bands and has a dark-coloured tip (Fig. 6). Its recessive mutant allele, non-agouti, represented by the symbol **a**, suppresses the light-color banding, which is substituted by a dark colored band different from the first one. The hair appears to the eye as self-coloured (solid) because the bands cannnot be distinguished. The difference between a tabby coat and a "self" (solid) coat is well represented by the coat of the leopard and the black panther. As many will know, the leopard and the black panther are the same animal. However, in the leopard the black patches are very visible on the yellow coat, while in the black panther the patches are black on a black background and therefore cannot be appreciated. The non-agouti gene manages to make the light colour bands disappear only with eumelanic colours, but has no effect on pheomelanic colours (in practice, a red non-agouti cat has a striped coat). Agouti allows the striping coat to appear, which means that what one sees in a striped coat is a complex coloration deriving from two colour components and controlled by two different gene groups: **Agouti + Tabby.**

Fig. 6 Agouti hair coat

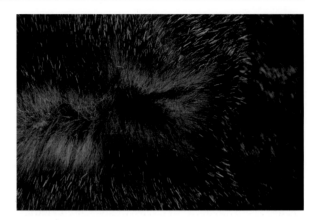

Gene **A** not only affects coat coloration but also that of the skin and nose. In eumelanic cats, the nose leather is not one solid colour, like in solid coloured cats, but rather brick red/pink/old rose coloured, rimmed with the coat's basic colour.

Patterns

Agouti is the basic colour, i.e. the background on which the pattern seems to be drawn. There are four main patterns associated with the three tabby genes:

- Ticked tabby or Abyssinian
- Spotted tabby
- Mackerel tabby
- Blotched tabby

The tabby genes are **T, Ta** and **tb**, and they are autosomic (three different alleles on the same locus). They are capable of generating striped coat patterns that are called **markings**:

- **Ta** is responsible for the ticked tabby coat (or Abyssinian coat).
- **T** is responsible for the mackerel tabby and for the spotted tabby coat.
- **tbtb** is responsible for the Blotched tabby or classic tabby coat.
- **Ta** is considered dominant over **T**, **T** is dominant over **tb,** and **tb is** recessive to both. These coats in heterozygous specimens (such as **Ttb** or **TaT**) are not as well defined and precise as homozygous coats (**TaTa** or **TT**).

T is responsible for both the mackerel and the spotted tabby patterns. There are many theories to explain this. Some claim that the spotted coat is derived from a polygenic action; others state it derives from the presence of other genes capable of breaking up and rounding the stripes of mackerel coats.

Ticked Tabby or Abyssinian (Ta Gene)
In this coat, the Agouti is distributed all over the coat and for this reason all of the coat appears evenly 'ticked'. Each hair has regularly alternating bands of different

colour (Fig. 7). The root is apricot colour while the tip is of the so-called basic colour, that can be black, chocolate, blue, cinnamon or fawn. The more bands on the hair, the more appreciated the coat. This colour is most frequent in the cats of the savannah and in wild cats living in arid and desert areas, while in selected breeds it is typical of the Abyssinian, of the Singapura and of the Ceylon cat. In some breeds, the presence of stripes on legs, neck, muzzle and tail are considered defects, as in the Abyssinian, while in others they are indispensable, as in the Singapura.

Mackerel Tabby (T Gene)

Mackerel is the term used to indicate the tabby coat with uninterrupted vertical lines (Fig. 8). 'Mackerel' is a fish that has thin parallel stripes that descend from its back to its midline. The cat's coat has a straight and uninterrupted black line along the spine,

Fig. 7 Abyssinian cat

Fig. 8 Mackerel tabby cat

running from the back of the head to the base of the tail. On its sides, shoulders and thighs, there are distinct narrow, continuous and parallel stripes. Its legs, tail and neck are well banded. These cats have an 'M' on their forehead and two or three lines that follow the profile of the cheeks. They have many small spots on their underside, from the throat to the belly. The 'M' on the forehead has inspired quite a few legends. One tells of the baby Jesus trembling in the manger despite being wrapped in blankets. Mary called in all the animals to warm Him but Jesus still trembled. Then a tabby cat appeared and snuggled up to Him in the manger, covered Him with his body and warmed Him. As a sign of her gratitude, Mary drew an M on its forehead. Another legend states that a snake crawled into the sleeve of the Prophet Mohammed's robe, and a tabby cat killed it immediately. From that moment onwards, all tabby cats were born with the M on their forehead to remind everyone that these cats deserve respect.

Spotted Tabby (T Gene)

The cat with a spotted coat has many small round or oval spots on its coat, separated one from the other and evenly distributed (Fig. 9). A thin, straight and uninterrupted

Fig. 9 Spotted tabby cat

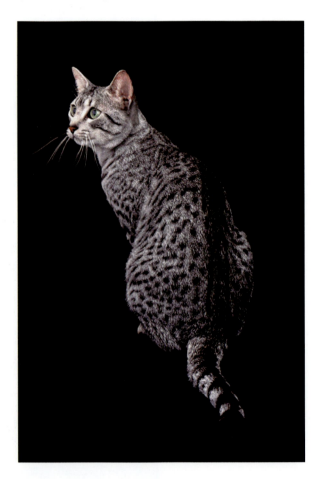

black line may be present from the back of the head to the base of the tail. An M is designed on the forehead and two or three lines follow the profile of the cheeks. The neck shows two uninterrupted stripes, while the legs and tail are banded. They have many small dots on their abdomen, from the neck to the belly. This coat is also called 'maculate' and is typical of the Egyptian Mau and of the Ocicat. The 'resetting' on the coat of the Bengal can be considered a modified spotted trait.

Blotched Tabby (tb Gene)

Also known as classic tabby. This is the showiest and most spectacular of coats because the Agouti background is marked by a butterfly-shaped pattern, the upper and lower wings of which are clearly designed on the flanks and shoulders of the cat (Fig. 10). Along the spine, from the back of the head to the base of the tail, there are three large stripes, a central one flanked by another two, distinctly separated and parallel to the first. The forehead bears an M and two or three lines follow the profile of the cheeks. The neck is decorated by two uninterrupted stripes while the legs and tail are banded. They have many small spots on their abdomen, from the neck to the belly. The coat of the Marbled Bengal is considered a classic tabby modified.

Differences Between Tabby and Self-Coloured

- **Nose:** in self-coloured cats, the colour of the nose is solid. In tabby cats, the nose colour is brick red, pink, old rose, and rimmed with the coat's basic colour.
- **Chin:** in tabby cats, the colour of the chin is lighter than in self cats.

Fig. 10 Blotched tabby cat

- **Eyes:** the eyes of tabby cats are rimmed with the coat's basic colour (and sometimes the lips are too) and a halo of lighter colour immediately around the eye.
- **Ears:** in self-coloured cats, the whole ear is evenly coloured, while in the tabby, and especially in tabby Points, a 'thumbmark' (or *pouce*) is present, i.e. a lighter area of colour on the external base of the ear.

Self or Solid Colours

The original colour is the tabby coat due to the association of the Agouti gene with the tabby alleles. All that was needed was the mutation of the Agouti gene (**A**) into non-agouti (**a**) to make the grey-yellowish bands on every single hair to disappear and generate the solid colour coat. Non-tabby cats, usually called self or solid, are cats that have coat of a single colour with hair that is evenly coloured from the root to the tip. The colour of each single hair is defined by the colour genes. Gene **B** is the colour-coding gene that allows melanosomes to produce eumelanin granules that give the hair a dark coloration. Gene **B** produces the hair's black pigmentation.

Gene **B** has two recessive mutant alleles called **b** (brown or chocolate) and **bl** (light brown or cinnamon). In these mutations, the pigment granules become deformed until they take on oval (**b**) or even more elongated (**bl**) shapes. The granules deformed in this way reflect the light, giving the hair a lighter colour: **b** produces the colour chocolate while **bl** the colour cinnamon. Black, **B**, is the dominant shape, while **b** and **bl** are both recessive to gene **B**, and **bl** is recessive to **b** (**B** > **b** > **bl**).

Dilution

These colours exist in the dilute form. In fact, thanks to the action of the dilution gene also known as the Maltese gene (**d**), the pigment granules inside the cortex aggregate and take on a different distribution. By doing this, they reflect the light and the coat appears of a lighter colour, allowing black to become blue, chocolate to become lilac and cinnamon to become fawn. Solid coat colours thus are six:

Non-diluted	Diluted
Black	Blue
Chocolate	Lilac
Cinnamon	Fawn

Black
The hair should be black from the root up and not contain any traces of brown or have any white hairs or grey underhair. Often the hair tips when exposed to the sun,

or the collar that is easily soiled by food and water, tend to become reddish or brown coloured. Nose leather and paw pads are black (Fig. 11).

Blue

The coat that ranges from a very light grey colour to slate grey is called blue. The most desirable is the lighter colour, as even as possible, from the tip to the root, without black tips or white hairs. Nose leather and paw pads should be blue. It is a dilution of black (Fig. 11).

Chocolate

The coat is milk chocolate coloured with a warm and even hue from the tip to the root of the hairs, without stripes or hairs of other colours. The nose leather is milk chocolate coloured, while the paw pads range from milk chocolate to cinnamon rose. Chocolate is a mutation of black (Fig. 12).

Lilac

Also called lavender or frost, lilac is a dilution of chocolate. The coat appears evenly pinkish-light grey without any kind of striping (Fig. 13). Nose leather and paw pads are coloured pinkish lavender.

Cinnamon

The brown coat is very light: this is a mutation of black (Fig. 14). In the Abyssinian and in the Somali, this colour is called sorrel.

Fawn

It is a dilution of cinnamon (Fig. 15).

Red and Tortoiseshell

Gene **O** transforms colour pigments into pheomelanin and deactivates the eumelanin production loci. Pheomelanins are granules that produce a red/orange colour. The orange gene is located on the X chromosome and for this reason is defined 'sex-linked'. Pheomelanic colours do not express all of the changes in hues seen

Fig. 11 Black and blue kittens

Fig. 12 Chocolate cat

Fig. 13 Lilac cat

Fig. 14 Cinnamon cat

Fig. 15 Fawn cat

with eumelanic colours, but only the colours red and cream. The cream colour is obtained through the intervention of dilution genes.

There is a very special coat colour known as 'tortoiseshell'. In this coat, the colours red and black are perfectly mixed or are present as clearly separated and defined patches of colour. Cats displaying this coloration are usually female. As we know, the cat has 38 chromosomes: 36 autosomal and 2 sexual chromosomes. The male is **xy** and the female **xx**. Only chromosome **x** carries colour; **y** doesn't. This means the male can be **xOy** red or **xoy** non-red (i.e. black). The female can be:

- **xOxO** red (if both **x**'s carry orange)
- **xOxo** tortoiseshell (if one **x** carries orange and the other **x** doesn't)
- **xoxo** black (if both **x**'s do not carry orange)

The combination **xOxo** is the only instance (due to the presence of the double **x**) in which black and red can appear together.

Tortoiseshell female cats can be black (or another eumelanic colour) and red or can appear together with white (Fig. 16). When the white is present, the red and

Fig. 16 Diluted
tortoiseshell with white
kitten

black are confined into clearly defined and separate patches of colour. The females that have this particular coat are called tricolour or calico, the North American term. In presence of the dilution gene, black becomes blue and red becomes cream, giving rise to the blue-cream coat and, if white is present, to diluted tortoiseshell and white or diluted calico. In presence of Agouti (**A**), in the eumelanic areas the tabby pattern will appear. The term 'calico' derives from the Indian city of Calicut, in the Kerala region, that was a famous port in the sixteenth century thanks to the flourishing trade between Europe and India. In that city a raw cotton cloth called calico was also made, that was bleached and then dyed with vibrant colours. This term was later used in the United States to indicate multi-coloured objects. The distribution and percentage of colour is defined during the development stage of the embryo. Coats with grey or white underhair or with tabby stripes on the muzzle or in the red spots are not allowed. Alongside the classic black-red combination there is also the chocolate-cinnamon with red. All of these varieties are admitted with tabby markings (stripes). In the United States, the tortoiseshell or blue and cream tabby is also called patched tabby or torbie. Tortoiseshell cats are only females. Should this coat appear in a male specimen, the cat is almost always sterile.

Red

Red is the definition of the coat that has a magnificent golden-red, warm, pure coloration, evenly distributed from the root to the tip of the hair (Fig. 17). The standard defines it without any tabby markings (stripes) or lighter spotting. The

Fig. 17 Red and cream tabby cats

perfect red self (solid colour) is difficult to obtain because the non-agouti gene (**a**) that masks the stripes does not have a clean action on pheomelanic colours, as it does on eumelanic colours. For this reason, sometimes residual marking is visible on the head and legs, or, in the attempt to eliminate the stripes, coats of a much too light or washed-out red are obtained. Red self cats without stripes are obtained only by means of very careful selection. Nose leather and paw pads should be brick red.

Cream

In presence of dilution genes, red becomes a very soft and delicate pastel cream (Fig. 17). The colour is evenly distributed from the root to the tip of the hair and should be as light and homogeneous as possible, without any tabby markings (stripes), shadowing, light-coloured underhair or darker, pointed areas.

Blue-Cream

In this tortoiseshell cat, the dilution genes soften black into blue and red into cream (Fig. 16). The colours are perfectly mixed and well distributed, even on the extremities, to create a very light mixture with pastel hues. Like the tortoiseshell, the blue-cream cat is only female.

Siamese Pattern

In genetics, when speaking of the Siamese one speaks of coloured points. In fact, this particular coloration of the extremities is found in many breeds: Siamese, Thai, Persian (colourpoint) (Fig. 18), Sacred Cat of Birman, Ragdoll, Devon Rex and Cornish Rex (Si-rex). The gene involved is the one that regulates the intensity of body colour, i.e. gene **C** and its mutant alleles called albino alleles.

These alleles are all identified using letter **c** because they are found on the same locus (lower case because **c** is recessive to **C,** the gene responsible for coat colour intensity). They all have different suffixes that are the initials of the breeds in which their action plays a primary role.

Fig. 18 Colourpoint cat

- **C** full colour
- **cb** Burmese
- **cs** Siamese
- **ca** Blue-eyed Albino
- **c** Pink-eyed Albino

All of the alleles of this group are recessive to **C** but not among themselves; in fact, between **cb** (Burmese) and **cs** (Siamese) there is an incomplete dominance phenomenon. By crossing a Siamese (with light-coloured body and dark tips) with a Burmese (with shadows of colour, darker on the legs and lighter on the body), one obtains a Tonkinese (Fig. 19). The Tonkinese shows intermediate features: intensely coloured tips and body hair which is darker than the Siamese but lighter than the Burmese. Both (**cb** and **cs**), however, are dominant over **ca** (responsible for the Blue-eyed Albino), that in turn is dominant over **c** (Pink-eyed Albino) (**C > cs and cb cs > ca > c**).

cb Burmese
Albino alleles work by progressively decreasing the pigmentation of the eyes and hair. In the Burmese, because of the **cb** gene, black (**C**) becomes seal (sable or dark sepia) and the eyes, that are partially depigmented by the gene, tend towards yellow (Fig. 20).

cs Siamese
In the Siamese, the gene **cs** causes the seal colour to be limited only to the extremities (mask, ears, legs, feet and tail) while the rest of the body is coloured anywhere from beige to magnolia white. The eyes are deep blue.

Fig. 19 Tonkinese cat

Fig. 20 Burmese cat

ca Blue-Eyed Albino
In the Albino with blue eyes, the lack of pigmentation results in white hair and very pale blue eyes.

c Pink-Eyed Albino
In the Albino with pink eyes, caused by the presence of the **c** gene, in addition to the total lack of pigmentation, the eyes are pink because the iris is transparent and the retina's blood vessels become visible.

The Siamese gene causes the mutation of the structural gene of the tyrosinase enzyme, making it especially temperature-sensitive. An increase in temperature, in fact, deactivates it, and this is why on the body, that is warmer, there is a decrease in pigmentation (and therefore the coat is paler in colour). At the extremities, that are cooler, there is a larger deposit of pigment that creates the darker 'point' effect. Point kittens are born white because in the uterus the temperature is higher and more constant (38.5 °C), and their colourpoints start showing only a few days after birth. Climate changes can affect coat coloration too. In fact, cats living in warm climates are lighter coloured than those that live in cold climates. The colourpoint in Siamese cats regulates the colour of their body coat: the darker the points, the darker the rest of the coat (a seal point will have a darker coat compared to a red point). The coat also darkens with age. For this reason, Siamese cats have a rather short showing career, because the judges prefer more highly contrasting coats.

The Siamese pattern may appear in other cat breeds, such as the Sacred Cat of Birman, the Devon Rex, the Cornish Rex and others.

White Spotting

Coats with white spots or patches are quite common in nature. The white spots on a cat's coat are genetically determined by the **S** (white spotting) gene and are transmitted as independent entities. This explains why white spotting can be associated with any basic coat colour. Cats with white spotting are called bicolour and tricolour, and it is common to add 'and white' to the cat's colour name. In this way, a black cat with white spots becomes 'black and white', a red blotched tabby becomes a 'red blotched tabby and white', while the tortoiseshell and white cat is more simply called 'tricolour' or 'calico'. The **S** gene prevents coat coloration because it does not allow the melanin granules to settle in the follicles from which the hairs grow. As the **W** gene (responsible for the total depigmentation of the hair), **S** is a dominant epistatic gene but, unlike **W**, **S** does not affect the whole coat but just some patches of it, and its expression is enhanced by modifier polygenes that can amplify its action. The white spots show up more intensely if the **S** is homozygous (**SS**) compared to its heterozygous state (**Ss**). For this reason, it is quite common to see cats with just a few tufts of white hair on the chest and belly, or the extreme opposite whereby the coats are almost totally white with colour being limited to just a few areas of the head, back and tail. In fact, due to the variability of these coats, it is thought that there are various genes (or polygenes) conditioning the expression of the **S** gene. The formation of more extensive white areas confines colour into more distinct and visible spots. In tricolour coats, the spots of red and black are larger when the proportion of white is higher. Just like the **W** gene, the **S** gene seems to be linked to congenital deafness. It is possible for a white spotted cat to have deafness in the ear above the blue eye. White spotted coats are classified according to the percentage of white they contain.

Mitted

The small amount of white (1/4) is limited to the four feet. A white spot is preferred on the nose and/or between the eyes, while a white line should be present on the lower part of the body, starting from the throat and ending at the base of the tail. This coat is typically found in the Ragdoll.

Bicolour

This category includes coats that have a ratio of 2/3 colour to 1/3 white. Colour should be present on the muzzle (where an upturned 'V' is desirable), on the spine, head, tail and external area of the legs. White is desirable on the chest, belly and the inside area of the legs (Fig. 21). Also preferable is a white spot on the back, but its absence is not a penalty. This category allows for coats that show up to 50% white and 50% colour. Especially appreciated are specimens with a white 'flame' on their muzzle.

Harlequin

This coat has a much higher grade of white spotting compared to colour. Solid colour, in fact, covers only 1/6 of the coat and is limited to the top of the head, tail

Fig. 21 Bicolour cat

Fig. 22 Harlequin cat

and legs. Three or four clearly marked and separate colour spots are desirable on the back (Fig. 22). Colour spotting is random, but in any case, if located on the back, the spots cannot be fewer than four. A white flame on the muzzle is highly desirable.

Van

This category includes coats with colour markings on the head and tail. On the head, the colour is preferably confined to two large spots separated by a white line between the ears, while the tail should be evenly coloured down to the base. Not more than three colour spots on the body are accepted. Coats with more than three colour spots are considered to be Harlequin. The rest of the body is all-white. This name has been derived from the Turkish Van, the cat breed that features this coat.

Sacred Cat of Birman

This cat's coat is long-haired, colourpointed and features white 'gloves'. The white-gloved paws of this cat is the breed's signature feature, although this peculiar distribution of regular and symmetrical white spots on the four feet has divided genetists and scholars. Some authors claim that its genotype is similar to that of the colour-point (cscsll) but with the addition of the **S** gene (Piebald White Spotting gene). According to them, the expression of **S** is conditioned by modifier polygenes that allow white spots to follow a precise distribution on the four feet. Other authors instead describe the presence of **g** (gloves), a recessive autosomic gene that in double dose is capable of confining the white spotting to the extremities. The latter hypothesis now seems to be the most credited, although it is not precisely understood whether **g** can be considered a second gene totally different and independent from **S,** but capable of changing the expression of the latter (in this case, it would seem likely that there is a **Ssgg** genotype in which **S** codes for spotting and **g** codes for the white spotting's positioning on the feet) or whether it is an allele found on the same locus as **S**. Yet others believe that it is a dominant gene with incomplete penetrance and totally different from **S**.

Dominant White

White is not a colour but an absence of colour, coded by the dominant, epistatic gene **W**. This gene is responsible for the complete depigmentation of the hair (Fig. 23). It masks the expression of all other colours (epistasis), including white

Fig. 23 White cat

spotting and Siamese colourpoint, which means that a white cat can be defined as a cat of any colour painted white. The progeny of a homozygous white cat will be entirely white; on the contrary, a heterozygous white cat crossed with a non-white cat can generate coloured kittens as well. By examining the non-white progeny of such a cat, one can discover its hidden colour. If one crosses a white male cat with a red female cat, for example, and this cross produces a tortoiseshell kitten, then the hidden colour of the male cat will be black.

Sometimes the epistatic effect of this gene is not absolute. Often, a small spot remains visible on white kittens' heads, but disappears by adulthood. The W gene is unfortunately often associated with deafness, because it codes for a degeneration of the cochlea in the ear and atrophy of the organ of Corti. This genetic defect is congenital and irreversible. The white Oriental, also known as Foreign White, has the cs gene (Siamese) and the W gene (white), and therefore is a Siamese with the W gene (cscsW-) and not an albino Siamese from which it differs both by genotype and by phenotype.

Silver Coats

Cats with silver coats (smoke, shaded, Chinchilla and silver tabby) are perhaps the most striking and fascinating of them all. In all of these coats, only the tip of the hair is coloured, while the root section is white. All of these cats share the 'colour inhibitor' gene I that prevents the development of pigment in the hair and suppresses its yellow-greyish banding, which results in a silver coloration effect. Each single hair of these cats is coloured to varying degrees only at the tip, that can be of any colour: black, blue, red, tortoiseshell and so on, while the base, closest to the skin, is white. Even the skin remains normally pigmented. There are many theories regarding the genesis of silver coats. Until not long ago, the most widely accepted theory was based on a single gene responsible for this 'non-pigmentation', i.e. a mutation of the colour inhibitor gene I, a dominant autosomic gene that prevents pigment development in the hair (not to be confused with the Albino alleles) probably by limiting the quantity of pigment destined to the growing hair. The I gene suppresses the yellow-greyish bands of the tabby hair and at the same time codes for a pale silver colour at the base of the hair. In order to differentiate between silver tabby, Chinchilla and silver shaded, this theory envisaged the intervention of modifier polygenes capable of regulating the intensity of the I gene and therefore the different proportions between quantity of coloured hair and silver hair.

Other authors have proposed the presence of another gene, Ch, different and independent from I. This theory, called the 'two genes theory', is based on the assumption that the I gene erases the yellow-greyish bands and the Ch gene, dominant but independent from I, suppresses ticking and relegates it to the tip of the hair shaft (tipping).

The most recent theory proposes yet another solution to explain the many questions that arise when analysing the various coats. The presence of the I gene – colour inhibitor – is confirmed as regards the depigmentation and silver-colouring of the hair's base while, to justify the various widths of the underhair, several polygenes, called Widebanding Genes **Wb** (undercoat width genes), have been called into play. The polygenes are capable of regulating band width close to the base and act in varying degrees (low, medium, high) causing the widening of the pale band in the Agouti coat. To explain the tipping of the Chinchilla coat, another recessive gene, the superwide band gene (**swb**), is assumed to combine with **I** and with **Wb**.

This is the most widely accepted theory at the moment, but due to the complexity of the matter and to the many issues still to be cleared up, one continues to classify silver coats based on the varying silver to coloured hair ratio. Accordingly, based on the width of the coloured part of the hair shaft, called **tipping,** the following coats are distinguished.

Smoke

The Smoke is also known as the 'cat of contrast' because it has only a very small white band at the base of the hair in contrast with the very wide band of coloured tipping. The silver base should be evenly distributed all over the body, head legs and tail included (Fig. 24). The tipping (from the tip to midway down the hair shaft) is usually black but can also be blue, red or tortoiseshell. In longhair cats, the contrast is even more evident. A Smoke Persian, for example, looks entirely black but the contrast becomes clearly apparent as soon as it moves or when patted.

Shaded

This coat type has tipping (coloured part of the hair shaft) on about **1/3** of the hair (Fig. 25). The tipping can extend to the muzzle, legs and on the heel and results in a slightly darker colouring overall compared to the Chinchilla. The coat should not show tabby markings, dark spots or cream hues. The nose leather is brick red rimmed with a thin black line. The tips can be of various colours, the most common being black, although the variations of blue, chocolate, lilac and tortoiseshell are also admitted.

Fig. 24 Smoke coat

Fig. 25 Shaded cat

Fig. 26 Chinchilla cat

Chinchilla

The appearance of the Chinchilla (also called 'Shell') is that of a cat with only light silver tipping and a white underhair. The tipping involves about 1/8 of the hair (Fig. 26). The chin, chest, belly, inside of the thighs, the underside of the tail and the hock should be pure white. The head, ears, back, flanks, legs and tail are slightly shaded due to the presence of the tipping. Silver-shaded and Chinchilla coats are genetically identical, and can be found together in the same litter. When the polygenes have an intense action, the result is a Chinchilla, and when the action is bland the result is a silver shaded. Sometimes it is not easy to distinguish them. When in doubt, the colour of the heels will give the answer: silver heels mean silver shaded (Fig. 27), pure white heels mean Chinchilla (Fig. 28).

Fig. 27 Heels of
Chinchilla cat (lighter
color)

Silver Tabby

This is simply a tabby cat in which the **I** gene has erased the yellow-greyish bands, giving place to the white-silver hair to strikingly contrast with the striping above (Fig. 29).

Cameo

The base of the coat is silver and the tipping is red. Based on the length of the tipping, the cat coats are defined as Smoke Cameo, Shaded Cameo and Shell Cameo.

Golden

This is a special colouring of the coat having a warm apricot coloured underhair and black tipping. The Golden has the Agouti (**A**) gene, absence of the **I** gene (being **ii**) and the contemporary presence of wideband polygenes **Wb.** The genes described in this chapter are summarized in Table 1.

Fig. 28 Heels of silver
shaded cat (darker color)

Fig. 29 Silver tabby cat

The Vibrissae

The Vibrissae, commonly called 'whiskers', are very special hairs (Fig. 30). Almost three times larger and stiffer than normal hairs, they are located three times deeper into the dermis. They have a sheath of connective tissue rich in elastic fibres and they are served by a rich array of nerves and blood vessels. In addition to being found on the cat's cheeks on the sides of the mouth (12 on each side, in orderly rows), they are also located above the orbits as well as on the legs at the carpal level. Whiskers move constantly and stimulate the receptors of the nerve endings. For this reason, they constitute a powerful information system for monitoring the cat's immediate surroundings. These highly specialised receptors use afferent neurons to transmit signals to the trigeminal nerve ganglion, and from there to the part of cerebral cortex in charge of perceiving the somatic-sensorial stimuli, the minimal and almost imperceptible changes in the stimulated vibrissae, as well as all of the highly precise information regarding the extent, direction and duration of this change in status.

Fig. 30 Vibrissae

Fig. 31 White cat. The W gene (dominant white) masks the expression of all of the other genes responsible for coloration and colour distribution

Fig. 32 Tabby cat. All cats are genetically tabby and possess striping in their genotype

Fig. 33 Orange cats (O gene)

Fig. 34 Tabby kittens. Agouti is the background colour against which one sees the stripes of a tabby coat

Fig. 35 Siamese cat (cs gene)

The position of the vibrissae changes in function of the animal's activity and mood: when the cat attacks or is in a position of defence, it will point the vibrissae towards the rear. The vigil cat, focused on perceiving every single signal, points them forwards. When instead they are curved forwards and pointing down towards the ground, they are being used to recognize the ground and are ready to reveal any pit-holes or other types of unevenness.

The vibrissae pointed forwards to almost embrace the captured victim are being used to give the exact position of the prey and the direction of its fur or feathers so that the cat can understand from which end it must swallow it. Vibrissae also play an important role in defending a cat's eyes, because they function like eyelashes. It is enough to touch them lightly and the eyelids will close immediately. This proves highly useful when hunting, because in its predatory state the cat is concentrated on the prey, has its pupils totally dilated by adrenalin and therefore finds it difficult to focus on objects that are very close, such as small branches, bushes, grass or any other obstacle nearby. Because they just outpast its face, the vibrissae touch these obstacles first and cause the eyelids to close and defend the eyes.

The tactile functions of the vibrissae have been widely studied and debated. The vibrissae can act as airflow sensors. They are allegedly capable of detecting – and therefore of informing the cat about – the smallest vortices of air returned by the objects it encounters or the weaker air currents created when the air impacts an obstacle. This makes it easy for the cat to move and change position in the dark of night without bumping into objects. Only such a precise and perfect mechanism can explain the cat's incredible skill and precision when hunting at night: with its vibrissae, the cat can acquire an instantaneous and precise perception of its prey and capture it. This happens in blind cats too. A blind or partially sighted cat moves its head from one side to the other to perceive the ground's asperities and any obstacles with its vibrissae. The blind cat lacking vibrissae instead is highly deficient in this respect.

General References

1. Adalsteinnson S. Establishment of equilibrium for the dominant lethal gene for Manx taillessness in cats. Theor Appl Genet. 1980;58:49–53.
2. Affections héréditaires et congénitales des carnivores domestiques, Le point vétérinaire vol 28 N° spécial 1996.
3. Alhaidari Z, Von Tscharner C. Anatomie et physiologie du follicule, pileux chez les carnivores domestique. Prat Med Chir Anim Comp. 1997;32:181.
4. Alhaidari Z, Olivry T, Ortonne J. Melanocytogenesis and melanogenesis: genetic regulation and comparative clinical diseases. Vet Dermatol. 1999;7:10.
5. Anderson RE, et al. Plasma lipid abnormalities in the Abyssinian cat with a hereditary rod-cone degeneration. Exp Eye Res. 1991;53(3):415–7.
6. Baker HJ, Lindsey JR. Feline GM1 gangliosidosis. Am J Pathol. 1974;74:649–52.
7. Barnett KC, Gurger IH. Autosomal dominant progressive retinal atrophy in Abissinian cats. J Hered. 1985;76:168–70.
8. Bellhorn RW, Fischer CA. Feline central retinal degeneration. J Am Vet Med Assoc. 1970;157:842–9.

9. Bergsma DR, Brown KS. White fur, blue eyes and deafness in the domestic cat. J Hered. 1971;62:171–85.
10. Biller DS, et al. Polycystyc kidney disease in a family of Persian cats. J Am Vet Med Assoc. 1990;196:1288–90.
11. Bistner ST. Hereditary corneal distrophy in the Manx cat: a preliminary report. Investig Ophthalmol. 1976;15:15–26.
12. Bland van den Berg P, et al. A suspected lysosomal storage disease in Abyssinian cats. Genetic and clinical pathological aspects. J S Afr Vet Assoc. 1977;48:195–9.
13. Blaxter A, et al. Periodic muscle weakness in Burmese kittens. Vet Rec. 1986;118(22):619–20.
14. Bosher SK, Hallpike CS. Observations of the histopathological features, development and pathogenesis of the inner ear degeneration of deaf white cats. Proc R Soc Lond B Biol Sci. 1965;162:147–70.
15. Bosher SK, hallpike CS. Observations of the histogenesis of the inner ear degeneration of the deaf white cat. J Laryngol Otol. 1966;80:222–35.
16. Bourdeau P, et al. Alopecie hereditaire generalisee feline. Rec Med Vet. 1988;164:17.
17. Boyce JT, et al. Familial renal amyloidosis in Abyssinian cats. Vet Pathol. 1984;21(1):33–8.
18. Boyce JT, et al. Familial renal amyloidosis in Abyssinian cats. Vet Pathol. 1984;21:33–8.
19. Breton RR, Nancy CJ. Feline genetics. Net Pets; 1999.
20. Bridle KH, et al. Tail tip necrosis in two litters of Birman kittens. J Small Anim Pract. 1998;39(2):88–9.
21. Burditt LJ, et al. Biochemical studies on a case of feline mannosidosis. Biochem J. 1980;189:467–73.
22. Carlisle JL. Feline retinal atrophy. Vet Rec. 1981;108:311.
23. Casal M, et al. Congenital hypothricosis with thimic aplasia in nine Birman kittens. ACVIM abstracts N° 68, Washington, DC; 1993.
24. Centerwall WR, Benirschke K. Male tortoiseshell and calico cats. J Hered. 1973;64:272–8.
25. Chapman VA, Zeiner FN. The anatomy of polydactylism in cats with observations on genetic control. Anat Rec. 1961;141:205–17.
26. Chew DJ, et al. Renal amyloidosis in related Abyssinian cats. J Am Vet Med. Assoc. 1982;181:140–2.
27. Clark RD. Medical, genetic and behavioral aspects of purebred cats. Fairway: Forum publications Inc; 1992.
28. Collier LL, et al. Ocular manifestations of the Chédiak-Higashi syndrome in four species of animals. J Am Vet Med Assoc. 1979;175:587–90.
29. Collier LL, et al. A clinical description of dermatosparaxis in a Himalayan cat. Feline Pract. 1980;10(5):25–36.
30. Cooper ML, Pettigrew JD. The retinophthalamic pathways in Siamese cats. J Comp Neurol. 1979;187:313–48.
31. Cooper ML, Blasdel GG. Regional variation in the representation of the visual field in the visual cortex of the Siamese cat. J Comp Neurol. 1980;193:237–53.
32. Cork LC, et al. The pathology of feline GM2 gangliosidosis. Am J Pathol. 1978;90:723–34.
33. Cork LC, et al. GM2 ganglioside lysosomal storage disease in cats. Science. 1977;196:1014–7.
34. Cotter SM, et al. Hemofilia a in three unrelated cats. J Am Vet Med Assoc. 1978;172:166–8.
35. Counts DF, et al. Dermatosparaxis in a Himalayan cat. Biochemical studies of dermal collagen. J Invest Dermatol. 1980;74:96–9.
36. Creel D, et al. Abnormal retinal projections in cats with Chédiak-Higashi syndrome. Invest Ophthalmol Vis Sci. 1982;23:798–801.
37. Crowell WA, et al. Polycystic renal disease in related cats. J Am Vet Med Assoc. 1979;175:286–28.
38. Danforth CH. Hereditary of polydactyly in the cat. J Hered. 1947;38:107–12.
39. Davies M, Gill I. Congenital patellar luxation in the cat. Vet Rec. 1987;121:474–5.
40. De Maria R, et al. Beta-galactosidase deficiency in a Korat cat: a new form of feline GM1-gangliosidosis. Acta Neuropathol. 1998;96(3):307–14.

41. DeForest ME, Basrur PK. Malformations and the Manx syndrome in cats. Can Vet J. 1979;20:304–14.
42. Desnick RJ, et al. In: Desnick RJ, et al., editors. Animal models of inherited metabolic diseases. New York: Liss; 1982. p. 27–65.
43. Di Bartola SP, et al. Pedigree analysis of Abyssinian cats with familial amyloidosis. Am J Vet Res. 1986;47:2666–8.
44. Donovan A. Postnatal development of the cat retina. Exp Eye Res. 1966;5:249–54.
45. Ehinger B, et al. Photoreceptor degeneration and loss of immunoreactive GABA in the Abyssinian cat retina. Exp Eye Res. 1991;52(1):17–25.
46. Elverland HH, Mair IWS. Heredity deafness in the cat. An electron microscopic study of the spiral ganglion. Acta Otolaryngol. 1980;90:360–9.
47. Farrell DF, et al. Feline GM1 gangliosidosis: biochemical and ultrastructural comparisons with the disease in man. J Neuropathol Exp Neurol. 1973;32:1–18.
48. Flecknell PA, Gruffydd-Jones TJ. Congenital luxation of the patellae in the cat. Feline Pract. 1979;9(3):18–9.
49. Fraser AS. A note on the growth of the rex and angora cats. J Genet. 1953;51:237–42.
50. Freeman LJ. Ehlers-Danlos syndrome in dogs and cats. Semin Vet Med Surg. 1987;2:221.
51. French TW, et al. A bleeding disorder (von Willebrand's disease) in a Himalayan cat. J Am Vet Med Assoc. 1987;190:437–9.
52. Gorin MB, et al. Sequence analysis and exclusion of phosducin as the gene for the recessive retinal degeneration of the Abyssinian cat. Biochim Biophys Acta. 1995;1260(3):323–7.
53. Harpster NK. Cardiovascular diseases of the domestic cat. Adv Vet Sci Comp Med. 1977;21:39–74.
54. Haskins ME, et al. In: Desnick RH, editor. Animal models of inherited metabolic diseases. New York: Liss; 1982. p. 177–201.
55. Hearing JV. Biochemical control of melanogens and melanosomal organization. J Investig Dermatol Symp Proc. 1999;4:24–8.
56. Hendy-Ibbs PM. Hairless cats in Great Britain. J Hered. 1984;75:506–7.
57. Hendy-Ibbs PM. Familial feline epibulbar dermoids. Vet Rec. 1985;116:13–4.
58. Hirsch VM, Cunningham JA. Hereditary anomaly of neutrophil granulation in Birman cats. Am J Vet Res. 1984;45:2170–4.
59. Holbrook KA. Dermatosparaxis in a Himalayan cat. Ultrastructural studies of dermal collagen. J Invest Dermatol. 1980;74:100–4.
60. Hoskins JD. Congenital defects of the cat. In: Ettinger SJ, Feldman EC, editors. Textbook of veterinary internal medicine. Philadelphia: Saunders; 1995.
61. Howell JM, Siegel PB. Morphologic effects of the Manx factor incats. J Hered. 1966;57:100–4.
62. Jackson OF. Congenital bone lesions in cats with fold-ears. Bull Feline Advis Bur. 1975;14(4):2–4.
63. Jacobson SG, et al. Rhodopsin levels and rod-mediated function in abysinian cats with hereditary retinal degeneration. Exp Eye Res. 1989;49(5):843–52.
64. James CC, et al. Congenital anomalies of the lower spine and spinal cord in Manx cats. J Pathol. 1969;97:269–76.
65. Jezyk PF, et al. Alpha-mannosidosis in a persian cat. J Am Vet Med Assoc. 1986;189:1483–5.
66. Jones BR, et al. Preliminary studies on congenital hypothyroidism in a family of Abyssinian cats. Vet Rec. 1992;131(7):145–8.
67. Johnson CW. The Shaded American Shorthair, 1999 Cat Fanciers' Association Yearbook, CFA Inc, New Jersey.
68. Koch H, Walder E. A hereditary junctional mechanobullous disease in the cat. Proc World Congr Vet Dermatol. 1992;2:111.
69. Kramer JW, et al. The Chédiak-Higashi syndrome of cats. Lab Investig. 1977;36:554–62.
70. "La guide des chats" Selections du Reader's Digest, 1992.
71. Leipold HW. Congenital defects of the caudal vertebral column and spinal cord in Manx cats. J Am Vet Med Assoc. 1974;164:520–3.

72. Loxton H. The noble cat, aristocrat of the animal world. New York: Portland House; 1990.
73. Liu S-K. Pathology of feline heart disease. Vet Clin North Am. 1977;7(2):323–39.
74. Livingston ML. A possible hereditary influence in feline urolithiasis. Vet Med Small Anim Clin. 1965;60:705.
75. Loevy HT. Cytogenic analysis of Siamese cats with cleft palate. J Dent Res. 1974;53:453–6.
76. Loevy HT, Fenyes VL. Spontaneous cleft palate in a family of Siamese cats. Cleft Palate J. 1968;5:57–60.
77. Lomax TD, et al. Tabby pattern alleles of the domestic cat. J Hered. 1988;79(1):21–3.
78. Lorimer. The silver inhibitor gene. Cat Fanciers J.
79. Malik R. Osteochondrodysplasia in Scottish fold cats. Aust Vet J. 1999;77(2):85–92.
80. Martin AH. A congenital defect in the spinal cord of the Manx cat. Vet Pathol. 1971;8:232–9.
81. Mason K. A hereditary disease in the Burmese cats manifested as an episodic weakness with head nodding and neck ventroflexion. J Am Anim Hosp Assoc. 1988;24:147–51.
82. Muldoon LL, et al. Characterization of the molecular defect in a feline model for type-II GM2-gangliosidosis (Sandhoff's disease). Am J Pathol. 1994;144(5):1109–18.
83. Narfstrom K. Hereditary progressive retinal atrophy in the Abyssinian cat. J Hered. 1983;74:273–6.
84. Narfstrom K, et al. Retinal sensitivity in hereditary retinal degeneration in Abyssinian cats: electrophysiological similarities between man and cat. Br J Ophthalmol. 1989;73(7):516–21.
85. Neuwelt EA, et al. Characterization of a new model of GM2 gangliosidosis (Sandhoff's disease) in Korat cats. J Clin Invest. 1985;76(2):482–90.
86. Noden DM, et al. Inherited homeotic midfacial malformations in burmese cats. J Craniofac Genet Dev Biol Suppl. 1986;2:249–66.
87. Paasch H, Zook BC. The pathogenesis of endocardial fibroelastosis in Burmese cats. Lab Investig. 1980;42:197–204.
88. Paradis M, Scott DW. Hereditary primary seborrhea oleosa in Persian cats. Feline Pract. 1990;19:17.
89. Patterson DF, Minor RR. Hereditary fragility and hyperextensibility of the skin of cats. Lab Investig. 1977;37:170–9.
90. Pearson H, et al. Pyloric stenosis and oesophageal dysfunction in the cat. J Small Anim Pract. 1974;15:487–501.
91. Pedersen NC. Feline husbandry. Goleta: American Veterinary Publications Inc; 1991.
92. Prieur DJ, Collier LL. Morphologic basis of inherited coat color dilutions of cats. J Hered. 1981;72:178–82.
93. Prior JE. Luxating patellae in Devon rex cats. Vet Rec. 1985;117(7):154–5.
94. Robinson R. Devon rex: a third rexoid coat mutant in the cat. Genetica. 1969;40:597–9.
95. Robinson R. Expressivity of the Manx gene in cats. J Hered. 1993;84(3):170–2.
96. Robinson R. Genetics for cat breeders. 2nd ed. Oxford: Pergamon Press Ltd; 1987.
97. Robinson R. German rex: a rexoid coat mutant in the cat. Genetica. 1968;39:351–2.
98. Robinson R. The Canadian hairless or Sphinx cat. J Hered. 1973;64:47–8.
99. Robinson R. Oregon rex: a fourth rexoid coat mutant in the cat. Genetica. 1972;43:236–8.
100. Robinson R. The rex mutants of the domestic cat. Genetica. 1971;42:466–8.
101. Rubin LF. Hereditary cataract in Himalayan cats. Feline Pract. 1986;16(4):14–5.
102. Scott DW. Cutaneous asthenia in a cat. Vet Med (SAC). 1974;69:1256.
103. Searle AG, Jude AC. The rex type of coat in the domestic cat. J Genet. 1956;54:506–12.
104. Silson M, Robinson R. Hereditary hydrocephalus in the cat. Vet Rec. 1969;84:477.
105. Simpson J. The white spotting gene: new Zealand Cat Fancy Inc. (NZCF).
106. Sponenberg DP, Graf-Webster E. Hereditary meningoencephalocele in Burmese cats. J Hered. 1986;77:60.
107. Stebbins KE. Polycystyc disease of the kidney and liver in an adult Persian cat. J Comp Pathol. 1989;100(3):327–30.
108. Stephen G. Legacy of the cat. San Francisco: Cronicle Books; 1990.

109. Turner P, Robinson R. Melaninn inhibitor. A dominant gene in the domestic cat. J Hered. 1980;71:427–8.
110. der Linde V, Sipman JS, et al. Generalized AA-amyloidosis in Siamese and oriental cats. Vet Immunol Immunopathol. 1997;56(1–2):1–10.
111. Wilkinson GT, Kristensen TS. A hair abnormality in Abyssinian cats. J Small Anim Pract. 1989;30:27.
112. Wright M, Walter S. le livre du chat. Paris: Septimus editios; 1982.
113. Zook BC. The comparative pathology of primary endocardial fibroelastosis in Burmese cats. Virchow Arch (Pathol Anat). 1981;390:211–27.
114. Zook BC, et al. Encephalocele and other congenital craniofacial anomalies in burmese cats. Vet Med (SAC). 1983;78:695–701.

Approach to the Feline Patient: General and Dermatological Examination

Andrew H. Sparkes and Chiara Noli

Abstract

As a naturally solitary species that is both highly territorial and not naturally social, veterinary visits for the cat and the cat owner can be extremely challenging. The fact that the cat has been removed from its home territory (where it feels safe) and brought to the clinic (an unfamiliar environment) means that any cat will naturally experience anxiety, fear, and stress during the visit. For these reasons, it is important that any veterinary visit follows "cat-friendly" principles to ensure stress is minimized. This will help reduce the severity of stress-induced changes in laboratory parameters, facilitate an easier clinical examination, and ensure that owners will be willing to bring their cat back to the clinic when needed. This chapter will deal with how to perform a general and a dermatological examination, including the description of skin lesions and diagnostic procedures, in the feline patient.

Introduction

As a naturally solitary species that is both highly territorial and not naturally social, veterinary visits for the cat and the cat owner can be extremely challenging. The fact that the cat has been removed from its home territory (where it feels safe) and brought to the clinic (an unfamiliar environment) means that any cat will naturally experience anxiety, fear, and stress during the visit. For these reasons, it is important that any veterinary visit follows "cat-friendly" principles to ensure stress is minimized and anxieties are relieved rather than reinforced.

A. H. Sparkes (✉)
Simply Feline Veterinary Consultancy, Shaftesbury, UK

C. Noli
Servizi Dermatologici Veterinari, Peveragno, Italy

© Springer Nature Switzerland AG 2020
C. Noli, S. Colombo (eds.), *Feline Dermatology*,
https://doi.org/10.1007/978-3-030-29836-4_3

Using cat-friendly principles to minimize stress has numerous benefits. Not only does it improve the welfare of the feline patient, it also helps to reduce the severity of stress-induced changes in laboratory parameters, it facilitates an easier clinical examination, it reduces the risk of human injury from fear-induced feline aggression, and helps to ensure that owners will be willing to bring their cat back to the clinic when needed.

A more thorough discussion of cat-friendly principles can be found on International Cat Care's "Cat Friendly Clinic" web site (see www.catfriendlyclinic.org), but some of the important issues are covered here.

Before the Cat Arrives

For many owners, the process of taking a cat to the clinic is highly traumatic. They will have to catch the cat, confine it in a basket, take it away from its natural environment, often transport it in a car, and then bring it into the clinic. Understanding the implications of veterinary visits for cat owners, and what needs to be done to reduce the negative impact, will help enormously.

Advising owners on how best to bring the cat to the clinic and helping them remain calm and relaxed has a very positive effect, both on the owner and the cat. The cat will be exposed to many stressors such as:

- A strange cat basket
- An unfamiliar car journey
- An unfamiliar environment in the clinic
- Strange odors, sights, and noises on the journey and in the clinic
- Unfamiliar people and animals, which can be highly threatening
- Being handled and examined by unfamiliar people

Suitable cat carriers should be strong, escape proof, and allow easy access for both the cat, the owner and the clinic staff. Carriers with a large top opening are usually preferred as they allow easy and gentle lifting of the cat in and out of the carrier. The carrier should enable the cat to hide if possible, but if it is open on all sides (e.g., plastic coated wire baskets), then placing a blanket over the carrier to allow the cat to hide is helpful. Plastic carriers that allow the top half to be removed completely can be useful as some cats will feel safer remaining in their carrier during a consultation, and most of a clinical examination can be conducted with the cat in the carrier with the top removed.

Ideally, the carrier should be integrated as "part of the furniture" in the cat's home environment. If it is somewhere the cat choses to rest and sleep in on occasions, or if it is somewhere it is fed frequently, the cat will regard it as part of its territory rather than seeing it as a clue that a stressful journey is ahead, if it only comes out for veterinary visits. Ensuring that some of the cat's usual bedding is used in the carrier during the visit will also be reassuring to the cat as this will contain the odors the cat associates with its home territory. In addition, the use of synthetic feline facial pheromone spray or wipes in the carrier and/or on the bedding may be helpful. Asking the owner to bring extra bedding is also a good idea in case the cat soils the basket with feces or urine.

During a car journey, ensuring the carrier is restrained securely (eg, in a foot well) and will not move during the journey is important. Driving calmly will be helpful and if necessary cover the basket with a blanket to ensure the cat is able to hide.

With cats that are known to repeatedly become highly anxious and aroused during a veterinary visit and during the journey to the clinic, consideration can be given to the use of anxiolytic drugs such as gabapentin [1, 2]. While not recommended for routine use, there are undoubtedly some cats that will benefit from such an approach.

The Waiting Room

A well-designed waiting room with cat-friendly staff is important. The aim should be to create a calm and non-threatening environment for the cat to wait in so that anxiety is reduced rather than heightened. An atmosphere that reassures owners that the clinic is staffed by people who care about both them and their cats will also help to create a positive impression.

The waiting room should be designed and used in a way that minimizes the threats cats may feel (visual, aural, olfactory, etc.). Ideally, a clinic would have a separate waiting room for cats, but if this is not possible, consider physically separating the waiting room into two different areas for dogs and cats. Appropriate walls or barriers should be used to ensure visual contact is avoided between dogs and cats (Fig. 1), and measures should be taken to avoid having barking or noisy dogs in the waiting room (e.g., getting noisy dogs to wait outside).

Fig. 1 Having a separate waiting area for cats and cat owners that is quiet, and where cats cannot see dogs helps to reduce stress during veterinary visits

   The location and size of the cat waiting area should be appropriate for the clinic, and thought should be given to the route that cats take into and out of it. The cats should encounter minimal human and animal traffic while in the waiting area. The value of a feline-only area is greatly compromised if cats have to pass through a noisy area or pass by dogs to get to the consulting room. Having the cat waiting area adjacent to a cat consulting room can help overcome some of these problems.

   Other important considerations for the feline waiting area include:

- Having a low reception desk, or a wide shelf in front of the reception desk where owners are encouraged to place cat baskets (above the head height of most dogs). This helps to reduce anxiety as cats feel more threatened when at floor level;
- Prevent or reduce any noises from the consultation rooms reaching the waiting area;
- Ensure dogs are kept away from cat carriers, and reinforce this by asking dog clients to be considerate of cats in the waiting area;
- Try to ensure cats are not left to wait for excessive periods of time in the waiting room, but are able to move to the consulting room as quickly as possible;
- Direct visual contact with other cats can also be very threatening. This can be overcome in many ways such as erecting small partitions between seats to separate cats in the waiting area, or providing clean blankets or towels to cover the cat's carrier;
- Cats feel insecure if placed at floor level – having shelves, tables or chairs to place cat carriers on so they are raised up is very useful (Fig. 2). These should ideally be about 1.20 m from the ground and have partitions (or use covers) so that cats are not confronted with each other;
- Using a plug-in synthetic feline facial pheromone diffuser (Feliway, Ceva Animal Health) may also be of benefit in the environment.

Fig. 2 Having raised tables or shelves above floor level for owners to place the cat basket on while in the waiting room is another good way to provide reassurance to cats and help reduce anxiety

The Consultation Room

Where possible, a clinic should use a dedicated feline consulting room, free from the odor of dogs and other animals. For dermatologic examinations the room should be well lit, but with the ability to darken the room if needed (for example for evaluation with a Wood's Lamp). There should also be access to an illuminated medical magnifier.

Consideration should be given to allowing the cat to wander freely in the consulting room if it chooses, and so it is important there are no cupboards or furniture in the room that the cat could hide under, or small gaps that would make it difficult to retrieve the cat. The consulting room table should also have a clean non-slip surface so that the cat is able to grip well – this can be achieved with a rubber mat or perhaps a clean thick towel or blanket.

The use of synthetic feline facial pheromone sprays and diffusers in the consulting room may help to encourage a more relaxed atmosphere, but this is not a substitute for good empathetic handling techniques.

The Consultation Process

The aim of the consultation process should be to obtain a full history, undertake a full physical examination, and consider what further actions or investigations may be required in conjunction with the owner, while ensuring the cat remains as stress-free as possible. Irrespective of the suspected cause of the skin disease, a full history and full clinical examination should never be overlooked as there may be concurrent disease present and/or a systemic cause of the skin condition. A properly conducted dermatologic consultation, including ancillary tests, usually requires 45–60 minutes. About 20 minutes are dedicated to collecting and recording the signalment and history, examination of the patient requires about 10 minutes and the ancillary tests and discussions with the owner each require about 15 minutes. The times are indicative only and vary depending on the nature of the problem and the communicative ability of the owner.

The principles of "cat-friendly" handling should be adhered to at all times – see the AAFP/ISFM Feline-Friendly Handling Guidelines [3] – and the cat should be given time to acclimatize to this unfamiliar environment.

History Taking

Collecting and reviewing information on the medical and surgical history of the cat is a part of the routine healthcare examination. The history should be collected, as far as possible, in a systematic way – using a clinical history form is a valuable way of obtaining standardized data for all patients (Figs. 3, 4 and 5).

Clinical History

Date:	Cat's name:	Owner:	Clinician:

Background

Age: Sex: Breed: Time with owner:

Acquired from: ☐ Breeder   ☐ Rescue centre   ☐ Friend   ☐ Other:

Other cats: ☐ No   ☐ Yes ...... How many?:   Any problems?:

Habitat

Environment: ☐ Indoor   ☐ Indoor/Outdoor   ☐ Restricted outdoors   ☐ In at night   ☐ Oudoor only

Litter tray?   ☐ No   ☐ Yes ....   Type of litter?:

Contact with other cats?:   ☐ No   ☐ Yes ....   Describe:

Cat fights?   ☐ No   ☐ Yes ....   Describe:

Hunting?   ☐ No   ☐ Yes ....   Describe:

Access to poisons: ☐ No   ☐ Yes ....   Describe:

Nutrition

Diet type: ☐ Dry food   ☐ Wet food   ☐ Both   ☐ Other:

Type/brand usually fed:

Time last fed:

Routine Preventive Healthcare

Vaccination: ☐ FPV   ☐ FCV/FHV   ☐ Rabies   ☐ FeLV   ☐ Chlamydia   ☐ Other:

Last vaccine given: ☐ <12m   ☐ <36m   ☐ >36m   ☐ Never   ☐ Unknown

Flea/tick treatment (what and when):

Worming (what and when):

Heartworm (what and when):

Retrovirus status: ☐ Unknown   ☐ FeLV+   ☐ FeLV-   ☐ FIV+   ☐ FIV-   When tested:

Previous problems

Current problems

A lifelong partnership of care for the health and wellbeing of your cat • www.catcare4life.org

Fig. 3 Example of a feline clinical history form. Copies of this form are freely available from www.catcare4life.org

Nutritional Assessment Form

Date:	Cats name:		Owner:
Age:	Sex:		Breed:

How do you feed your cat:

☐ Bowl ☐ Puzzle feeder ☐ Floor ☐ Other _____

What do you currently feed your cat - please list all the things you offer including commercial foods, raw foods, and home prepared foods

Type of food offered List manufacturer, brand and flavour for commercial foods, and list type for any others	Form of food offered Dry, canned or sachet for commercial foods. Raw or cooked for others	How often do you give this (number of times offered daily or weekly)	Approximately how long have you fed this food	Approximately how much do you feed each time

Please list all the treats and snacks you feed your cat, including commercial treats, human foods, table scraps and any other treats.

Type of treat offered List manufacturer, brand and flavour for commercial foods, and list type for any others	Form of treat offered Dry, canned or sachet for commercial treats. Raw or cooked for others	How often do you give this (number of times offered daily or weekly)	Approximately how long have you fed this as a treat	Approximately how much do you give each time

What do you offer your cat to drink?:

☐ Water: ☐ Cow's Milk ☐ Commercial cat milk ☐ Other _____

Does your cat do any hinting (catch and eat wild animals), if so what?

☐No ☐ Yes .... If yes: ☐ Mice ☐ Rats ☐ Voles ☐ Birds ☐ Other _____

Have you noticed any recent change in:

1. Appetite: ☐ No change ☐ Increased ☐ Decreased
2. Weight ☐ No change ☐ Increased ☐ Decreased
3. Thirst ☐ No change ☐ Increased ☐ Decreased

CatCare *for* **Life** from **cat care** international in partnership with IDEXX LABORATORIES ROYAL CANIN Boehringer Ingelheim
www.icatcare.org

A lifelong partnership of care for the health and wellbeing of your cat • www.catcare4life.org

Fig. 4 Example of a feline nutritional history form. Copies of this form are freely available from www.catcare4life.org

Physical Examination

Date:	Cat's name:	Owner:	Clinician:

1. TPR, weight & condition

Temperature: Weight (kg):
Respiratory rate: BCS:
Pulse: MCS:

2. Attitude

☐ Bright & alert   ☐ Quiet   ☐ Lethargic   ☐ Dull
☐ Hyperactive   ☐ Other:

3. Hydration

☐ Normal   ☐ Other:

4. Face

☐ Normal   ☐ Other:

5. Eyes

☐ Normal   ☐ Other:

6. Ears

☐ Normal   ☐ Other:

7. Nose

☐ Normal   ☐ Other:

8. Mouth & pharynx

☐ Normal   ☐ Other:

Calculus: ☐ Mild   ☐ Moderate   ☐ Severe
Gingivitis: ☐ Mild   ☐ Moderate   ☐ Severe
Stomatitis: ☐ Mild   ☐ Moderate   ☐ Severe

9. Mucous membranes

☐ Normal   ☐ Pale   ☐ Icteric   ☐ Other:

10. Musculoskeletal system

☐ Normal   ☐ Other:

11. Rib spring

☐ Normal   ☐ Other:

12. Heart

☐ Normal
☐ Murmur   Grade: __/VI   ☐ Gallop
☐ Dysrrhythmia   ☐ Pulse deficit   CRT: ___

13. Lungs and breathing

Breathing: ☐ Normal ☐ Other:
Auscultation: ☐ Normal ☐ Other:
Percussion: ☐ Normal ☐ Other:

14. Abdomen

☐ Normal   ☐ Other:

15. Gastrointestinal tract

☐ Normal   ☐ Other:

16. Urogenital system

☐ Normal   ☐ Other:

17. Lymph nodes / tonsils

☐ Normal   ☐ Other:

18. Thyroid goitre

☐ None ☐ Left side ☐ Right side ☐ Bilateral

19. Nervous system

☐ Normal   ☐ Other:

20. Coat and skin

☐ Normal   ☐ Other:

21. Pain assessment

☐ Absent ☐ Unsure ☐ Mild ☐ Moderate ☐ Severe

Additional observations and plan:

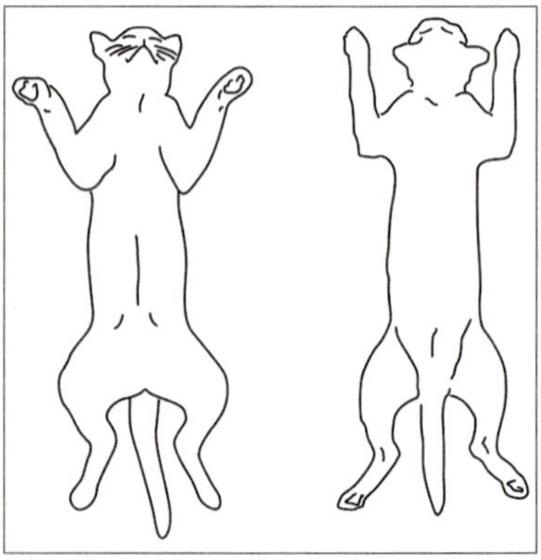

Fig. 5 Example of a feline physical examination form. Copies of this form are freely available from www.catcare4life.org

Clinical history and/or health questionnaires (e.g., including behavior, mobility, routine prophylactic therapy, and general health) can be given to owners to fill out, as far as possible, before bringing their cat to the clinic or while in the waiting room before the consultation. The assistance of a nurse or member of support staff may be valuable, but collecting such information before the consultation itself helps to streamline the process and gather all relevant information.

In particular for dermatological consultations it is highly desirable to record all the data collected on a special dermatological clinical record form. This form should be divided into sections for the signalment, history, clinical examination, list of differential diagnoses, ancillary tests, definitive diagnosis, therapy and follow-up (Fig. 6).

Many of the issues may be obvious, and may be a part of the existing medical record if the cat is a long-standing patient at the clinic. However, it is important to remember that some owners take their cat to more than one veterinary clinic, so other relevant problems the cat may have suffered should not be overlooked. Even when an accurate history is known, it is still important to consider:

- Any current medications (prescribed by the clinic or obtained elsewhere)
- Any non-prescription medications the owner may be using (e.g., nutritional supplements, parasiticides, alternative medications, etc.)
- Lifestyle (indoors, outdoors, other animals in the house, etc.)

In particular, questions regarding the skin disease for which the cat is presented should include:

- age of onset/duration of the problem;
- seasonality;
- initial site and lesion type and its modification during the course of the disease;
- severity and localization of pruritus, if present (Tables 1 and 2).

Reviewing the history during the clinical examination is an ideal opportunity to open the cat carrier and allow the cat time to come out voluntarily and explore the room. This helps acclimatize the cat to the environment and helps reduce stress during the subsequent examination.

History-taking should always include the use of open-ended questions such as:

- "How has Fluffy been doing since the last visit?"
- "Have you noticed any change in his appetite recently?"
- "Has there been any change in Fluffy's stool consistency?"

These are always better than leading questions such as:

- "Have you seen any diarrhea?"
- "Has he been eating more recently?"

Presenting complaint: _____

Past history: Owned since/origin_____

Prior diseases:

_____

_____

Recent history: Diet:

_____

Environment: inside   outside   hunts and eats prays?

Other animals in household _____

Appetite_____ Water intake _____ Urination _____ Feces _____

Preventative therapy: Vaccinations _____   Anthelmintics _____

Flea control _____

Cutaneous problem: Age of onset: _____

Initial localization and lesion type _____

_____

Current localization and lesion type _____

_____

Pruritus: absent   moderate   severe   season_____

Localization _____

Lesions on other animals or humans: _____

Prior therapy:
Drug   Date and Duration   Effect

_____ _____ _____

_____ _____ _____

_____ _____ _____

_____ _____ _____

_____ _____ _____

Fig. 6 Example of a feline dermatological examination history form

Localization of the lesions (draw):

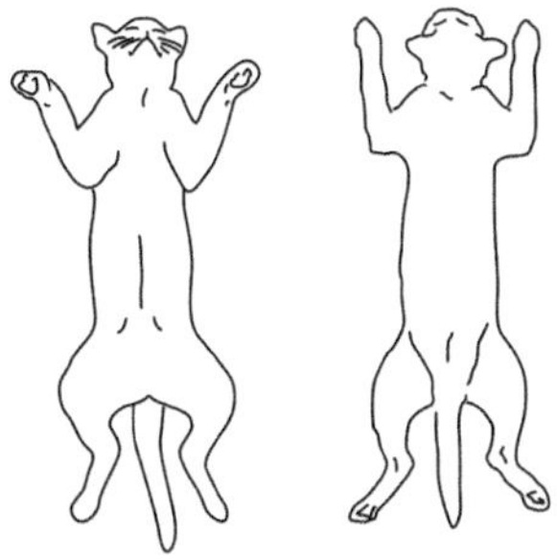

Description of the lesions (circle):

macule	papule	pustule
vesicle/bulla	collarette	wheal plaque
alopecia	scale	crust/excoriation
eschar	ulcer	hyperkeratosis
comedones	nodule	cellulitis
other _____		

nails and nail beds: _____coat: _____other:_____

Description of the clinical picture: _____

_____

_____

List of problems and differential diagnoses: _____

_____

_____

Ancillary tests and results: Scraping/Trichoscopy _____

Cytology _____

Biopsy   _____

Blood/urine _____

Culture _____

Other _____

Diagnosis: _____

Treatment: _____

_____

Next Appointment: _____

Fig.6 (continued)

Table 1 Feline skin diseases and potential severity of pruritus

Potential severity of pruritus
Absent
Non-inflammatory alopecia
Demodicosis (*D. cati*, uncomplicated)
Dermatophytosis (uncomplicated)
Moderate
Feline atopic syndrome (uncomplicated)
Adverse reaction to food (moderate severity)
Bacterial infection
Malassezia infection
Demodicosis by *D. gatoi* or by *D. cati*
Cheyletiellosis
Severe
Severe food allergy
Severe *Malassezia* infection
Notoedric mange

Table 2 Feline skin diseases and most frequent localization of pruritus

Most frequent localization of pruritus
Dorsum
Flea bite allergy
Cheyletiellosis
Psychogenic (licking)
Other allergies
Head
Otodectic mange
Adverse reaction to food
Notoedric mange
Neck
Flea bite allergy
Adverse reaction to food
Idiopathic neck lesion (consider welfare issues)
Abdomen (self-induced alopecia in cats from licking)
Feline atopic syndrome
Flea bite allergy
Adverse reaction to food
Flea infestation
Cheyletiellosis
Psychogenic

Importantly, a full history should also include a good nutritional assessment (Fig. 4), evaluating the cat's diet, lifestyle, feeding habits, etc. It is important that this is as comprehensive as possible and covers everything the cat has access to.

The cat's behavior and environment should not be overlooked. This will include whether the cat has free access outdoors, what other animals it regularly has contact with and whether the cat is known to hunt. It is also important to consider the potential interplay between many medical and behavioral issues, including dermatoses (e.g., psychogenic dermatoses).

Physical Examination

Patience, gentleness, and empathy are vital characteristics with cats in the consultation room. Even with the best environment and approach, some cats will remain very anxious and a full physical examination may not always be possible at the first attempt. Be prepared to take additional time if needed, and in some cases consider scheduling another appointment, or hospitalizing the cat if necessary.

As with the history taking, using a standardized form for physical examination and additional forms for special investigations such as dermatological examinations will be highly valuable (Figs. 5 and 6). Using a standardized form will ensure that the physical examination is performed systematically and that nothing is missed. This can be particularly important in cats as the order of events during the examination may have to be flexible and adapted to the needs of the individual cat (see below).

Important considerations during the physical examination include:

- Don't ever rush when examining a cat – taking a little extra time to do things slowly and at the cat's pace will be much more rewarding and less stressful.
- Always try to let the cat come out of its carrier by itself.
- Be flexible and let the cat choose – allowing the cat some control through exercising choice is a key method to reduce anxiety. The key is to find out and understand what makes the cat more relaxed and adapt the physical examination to suit the individual cat. Some cats may be happier on their owner's lap, others on the floor. Some may enjoy looking out of a window, while others prefer to stay sitting in their carriers or even hiding under a blanket. Try to be as adaptable as possible, be gentle, and take your time.
- Give the cat plenty of fuss and attention if that is what it likes, talk gently and aim to complete the majority of the physical examination without the cat realizing you are doing anything more than just stroking it.
- Providing some treats, if the cat will eat them, may also help to distract the cat.
- Sitting with the cat on the floor often helps and can make handling much easier.
- Some cats prefer to lie down, while others prefer to stand – try to do as much as possible with the cat in its preferred position.
- Always use the minimal amount of restraint necessary – *any* form of overt or heavy restraint will signal danger to the cat and escalate anxiety.
- Where needed, split the examination into short sections, and in between allow the cat to rest, change position, or wander round the room – give the cat a short break as soon as it starts to get restless.
- As sustained eye contact with a cat can be perceived as threatening by the cat, avoid direct eye contact where possible, and perform as much of the examination as possible with the cat facing away from you (Fig. 7).
- Be aware that older cats often suffer from osteoarthritis, which may make handling uncomfortable or painful.
- Perform more invasive examinations (such as taking the cat's temperature, where necessary) to last.

Fig. 7 The physical examination should be conducted gently and empathetically. Conducting much of the examination from behind the cat avoids direct eye contact which cats often perceive as threatening

In particular, for a dermatological examination, the following should be evaluated before any manipulation:

- Nutritional status
- Coat luster
- Thickness of the coat
- Any odors
- Localization of any obvious lesions

When possible the coat and skin should be inspected systematically. The authors normally follow a precise sequence so as not to forget any part of the body.

1. Rear of the animal
   - The area at the base of the tail is examined, extending forward along the dorsum to the neck.
   - The areas under the tail, anus, and perianal and perivulvar (females) skin are inspected.
   - The hind legs are examined, checking for any linear granulomas.
   - The hind feet are checked, examining all the interdigital spaces from underneath and on top. The nail beds are examined and all the nails are exposed and checked.
2. In recumbency
   - The medial aspect of the hind limbs and the inguinal and abdominal areas are examined. The external genitalia are inspected, including exteriorizing the penis and opening the vulva.
   - The sternal area, axillae, and medial aspect of the front limbs are examined.
3. Side of the animal
   - The lateral thorax and neck are examined and then the front leg and foot.
   - Examination of the appearance and odor of the external ear.
   - Repeat the examination from the other side.
4. Finally, the patient is examined from the front
   - The head is inspected, including opening the mouth and examination of the conjunctivae.

The whole process should be performed gently and quickly, in order not to stress the patient.

In exceptional cases, some cats are so fearful that a full examination is not achievable even with the most patient of handling. This is rare, but heavy restraint (scruffing the cat and pinning it to the table) will only make the experience much worse for the cat. In such cases, consider using chemical restraint to facilitate the examination.

Cats should be weighed at every clinic visit, and at least once or twice yearly. The percentage weight change should be calculated at each visit and trends noted. Human pediatric or feline-specific accurate electronic scales should be used if possible for optimum accuracy.

Skin lesions and their localization and distribution should be recorded for future reference. These include:

Macule A non-raised area of a color different to that of the surrounding skin. Hyperpigmented macules on the skin and mucosae of orange cats represent lentigo simplex (Fig. 8). Erythematous macules may be derived from peripheral vasodilation (as it occurs in many inflammatory skin diseases) or from hemorrhage (petecchiae). An extensive area of erythema is called erythroderma. Depigmented macules are typical of vitiligo in Siamese cats.

Papule A small, raised, erythematous lesion, it represents accumulation of inflammatory cells within the skin. Papules are typical, e.g., of the initial phases of eosinophilic granulomas. Papules are also a feature of parasitic skin diseases (mosquito-bite hypersensitivity, Fig. 9) and xanthomas.

Pustule An accumulation of inflammatory cells (pus) within or just under the epidermis. In cats, pustules are very rare and most frequently seen with pemphigus foliaceus (Fig. 10).

Fig. 8 Brown maculae (lentigo simplex) on the oral mucosae of a red cat

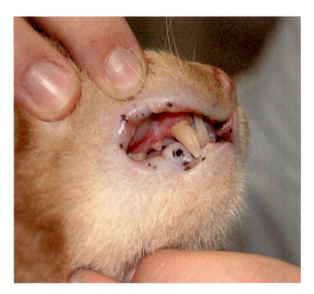

Fig. 9 Small papules and
erosions on the pinna of a
cat with mosquito bite
hypersensitivity

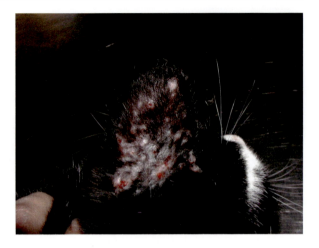

Fig. 10 A pustule on the
footpad of a cat with
pemphigus foliaceus

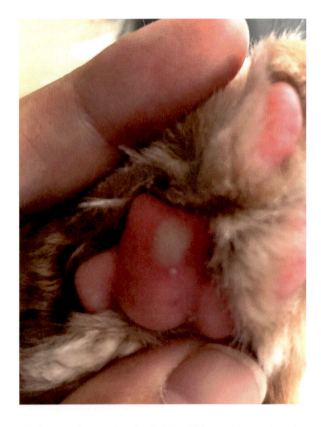

Vesicle An accumulation of clear or hemorrhagic fluid within or just under the
epidermis, a rare lesion often caused by autoimmune skin diseases.

Cysts Non-neoplastic well-circumscribed accumulations of liquid or keratin.
Multiple apocrine cysts, containing clear fluid, are observed in Persian cats (Fig. 11).

Fig. 11 Apocrine cysts on the muzzle of a Persian cat

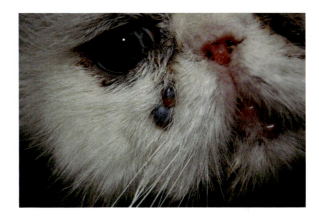

Fig. 12 Numerous nodules on the head of a cat with feline progressive histiocytosis

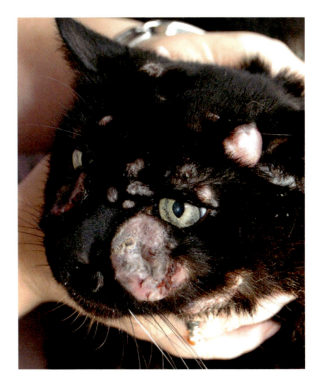

Nodule A raised protuberance caused by the infiltration or proliferation of cells and/or excessive connective stroma. Nodules are seen in bacterial disease (e.g., abscess), fungal infection (e.g., deep mycosis or dermatophyte mycetoma), sterile reactions (injection site granulomas, feline progressive histiocytosis, Fig. 12), or neoplasia.

Plaque A firm, raised area with a flattened surface, e.g., eosinophilic plaque (Fig. 13).

Wheal Raised well-circumscribed lesion consisting of edema of the superficial dermis. Wheals have an acute onset (a few hours) and tend to resolve quickly (over

Fig. 13 Typical aspect of a eosinophylic plaque in an allergic cat

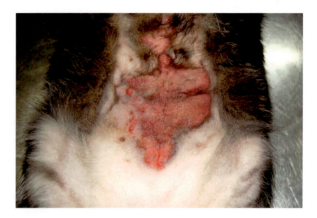

Fig. 14 Angioedema of the head of a cat

a few hours or 1 day). Wheals are a manifestation of a type 1 hypersensitivity reaction (immediate or anaphylactic) and are seen in reaction to intradermal allergen tests. Angioedema is edema extending to the deeper tissue and involving a larger area of the body (esp. the head, Fig. 14).

Comedones Commonly referred to as "blackheads", they represent an accumulation of keratin in the infundibulum of the hair follicle. Comedones in cats are seen on the chin in feline acne (Fig. 15).

Crust The accumulation of dried exudate (Fig. 16) or blood. The color depends on the material from which they were formed (blood = brown, pus = yellow). The eschar (Fig. 17) is a particular type of crust that contains dermal collagen fibers and is thus strongly anchored to the body (i.e., it cannot be easily pulled). Eschars are typically seen in cats in case of idiopathic ulcerative neck lesion and feline perforating dermatitis.

Fig. 15 Comedones and furuncolosis on the chin of a cat affected with acne

Fig. 16 Several yellow crusts (dry pus) on the pinna of a cat affected with pemphigus foliaceus

Fig. 17 An eschar on the neck of a cat affected with feline idiopathic ulcerative dermatitis

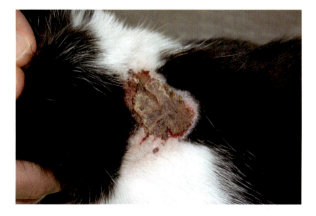

Fig. 18 Dry large flaques of exfoliation in a cat affected with paraneoplastic thymoma-associated exfoliative dermatitis

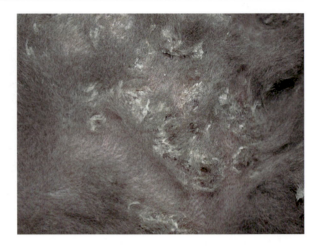

Fig. 19 Self-inflicted excoriations and ulcerations in a cat with adverse food reactions

Scale Dry accumulations of layers of the *stratum corneum* commonly called dandruff (Fig. 18). The presence of scales in cats is usually associated with dermatophytosis, sebaceous adenitis, or paraneoplastic diseases (feline exfoliative dermatitis due to thymoma).

Excoriation A self-induced lesion including ulceration and crusts, resulting from scratching and/or biting (Fig. 19).

Erosion A loss of epidermis down to the level of the basement membrane but leaving the dermis intact. Erosions are seen in some autoimmune diseases (e.g., pemphigus complex and diseases inducing dermo-epidermal separation) and in early cases of eosinophilic plaque (because of the abrasive action of the feline tongue).

Ulcer Tissue loss involving the epidermis and underlying tissues (dermis, less frequently subcutis). Examples of ulcers in cats are deep bacterial (e.g., atypical myco-

Fig. 20 Lesion of lip (indolent) ulcer

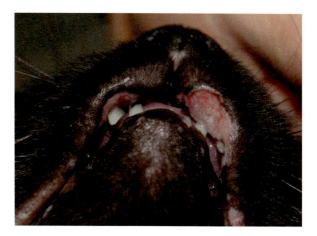

Fig. 21 Self-inflicted alopecia on the abdomen due to licking

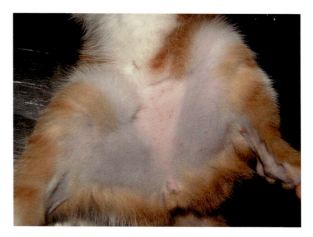

bacteria) or fungal (e.g., sporotrichosis) infections, for squamous cell carcinoma, idiopathic neck ulcer and lip (indolent) ulcer (Fig. 20).

Draining tract An opening in the tissue releasing exudate produced by a deeper inflammatory process (dermis or subcutis). Fistulization of abscesses or other inflammatory foci (sterile panniculitis, foreign body granuloma, etc.) permits the drainage of pus and the eventual expulsion of etiological agents, foreign bodies, or necrotic material.

Alopecia Can be used to describe both the complete loss of hair over one or more areas of the body and hypotrichosis, meaning thinning of the hair coat. It is important to differentiate alopecia caused by the loss of the hair together with the root from the loss of part of the hair shaft only. In cases of the loss of the hair root, for example, in endocrine or paraneoplastic alopecia, the hair at the periphery of the lesion can be easily epilated with traction. In the case of broken hair (e.g., self-induced alopecia, Fig. 21), the small remaining ends of the hair can be felt or seen with a magnifying lens as they leave the follicular ostia. The hair surrounding the areas of alopecia resists epilation.

Diagnostic Investigations

There are a number of diagnostic tests that, although mostly very straightforward, are extremely valuable in the diagnosis of various dermatoses. Again, it is important to consider appropriate chemical (sedation) for the cat in order to facilitate these tests wherever necessary. It is far less stressful to use appropriate chemical restraint than to struggle with heavy physical restraint of an anxious or fearful cat.

Trichogram

A trichogram involves microscopic examination of hairs (tip, shaft, and root) – ideally around 20–30 hairs are plucked and then examined. The preferred instrument to remove the hair is a pair of mosquito hemostats (preferably Klemmer) covered by small rubber or plastic tubing (to obtain even pressure and avoid causing artifacts by damaging the hair). In this way, samples from all stages in the growth cycle will be obtained rather than only resting-phase hairs. The hairs should be plucked in the direction of the hair growth to avoid fracturing them at the base. The hairs can be secured on a microscope slide either by placing them in mineral oil with a coverslip placed on top or using adhesive acetate tape. Examination is performed under 40× and 100× magnification.

The hair tips can be examined to determine whether there is pruritus (traumatic epilation) or spontaneous epilation. With traumatic hair loss the tips of the hairs will be broken and the usual slender tapering tips will be lost.

The hair roots can be examined to determine if the hairs are in anagen or telogen and see if there is normal hair cycling. Most should be in telogen (rough, spear-shaped bulb), with fewer in anagen (expanded bulb, may appear fringed, often pigmented, may appear club-shaped). In shorthair cats around 90% of hairs will be in telogen.

Hair shafts should also be examined for abnormalities including the presence of ectoparasites (*Demodex cati*), ectoparasite eggs (*Felicola subrostratus, Chyeletiella* spp.), and/or dermatophytes. Large adhesions of keratin on the hair shafts, called follicular casts, are seen in sebaceous adenitis.

Skin Scraping

A small window of skin is clipped if necessary. A small amount of mineral oil can be put on the skin surface to facilitate the skin scraping and a blunted scalpel blade or a Volkmann spoon (diameter 5–6 mm) is held perpendicular to the skin surface, which is scraped in the direction of the hair growth using moderate pressure.

For superficial parasites such as *Notoedres cati* and *Demodex gatoi*, scraping should not be so deep to cause capillary oozing of blood. The collected material can be smeared on a slide for examination.

Deep skin scraping involves repeated scraping at the same site until there is capillary oozing of blood. Pinching the skin prior to scraping to "squeeze" contents out from the follicles may also facilitate collection of follicular material and follicular

Demodex mites. In untreated patients with demodicosis, the number of mites detected is usually very high and hence it is rarely necessary to collect more than 2 or 3 samples and trichoscopy may be preferable over skin scrapings. When monitoring therapeutic success the mite numbers are low and numerous deep skin scrapings are necessary.

Examination of skin scrapes is performed under 40× to 400×, but initial examination should always be done at 40×. If there is heavy keratinous debris, taking "dry" skin scrapes and suspending the collected material in 10–20% potassium hydroxide which is then left for 20–30 minutes before examination under a coverslip may enhance visualization through "clearing" of the keratinous debris.

Tape Strip Test ("Scotch Tape" or "Acetate Tape" Test)

This test allows collection of superficial skin parasites, hairs, and yeasts. A 5–8 cm strip of clear sticky tape is repeatedly applied to the lesion or area of skin of interest. The skin can be clipped prior to performing the test if necessary. The tape is then applied (stuck) to a microscope slide. The free ends can be wrapped around the slide to help anchor it.

The preparation can be stained if necessary (e.g., looking for *Malassezia*) by applying a drop of suitable stain (e.g., the "blue" Diff-Quik stain) to the slide before applying the sticky tape. The slide is examined under 40× to 400× magnification.

Coat Brushings

Coat brushings are particularly helpful to look for fleas, but may also provide evidence of other superficial parasites. The cat is placed on a large sheet of white paper and is brushed vigorously both with and against the growth of the hair. Scale and debris are collected, are examined macroscopically, and can be placed on a slide and examined in mineral oil or a stain such as lactophenol cotton blue. Combing with a flea comb may also be a useful part of the procedure.

Wood's Lamp Illumination

Wood's lamp illumination is used to examine the hair coat (or collected hairs) under ultraviolet light to look for spontaneous fluorescence that is often associated with *Microsporum canis* infection. For optimal results, it is important to use the Wood's lamp in a dark room, and to allow 30–60 minutes for the eyes to adapt to the low light conditions. More information regarding this technique is presented in the chapter dealing with dermatophytosis (Chapter, Dermatophytosis).

Cytology

Several techniques have been developed to obtain material for a cytological examination.

Fine Needle Aspiration This technique is used for raised lesions, nodules, or accessible lymph nodes. A 21G (gray) needle is inserted into the center of the nodule and connected to a 5 or 10 ml syringe. While the needle is inside the mass several 1–2 ml aspirates are made, changing the needle position (angulation) without withdrawing the needle from the lesion. The suction is completely released before withdrawing the needle. If this has been done correctly, the plunger should return to 0 ml and the cells will be located in the needle lumen. If the plunger fails to return to zero, air has entered the syringe and the procedure should be repeated because the cells are in the cone of the syringe and difficult to remove. The needle is then removed from the cone and the syringe filled with air, reconnected to the needle and the cells are "sprayed" onto a microscope slide. If the sample is liquid, a smear is made, similar to that used for blood. If the material is solid, the material is spread by placing a second slight over the sample, applying little or no pressure.

Fine Needle Insertion A 24G needle is inserted into the mass and its angle changed, without connection to a syringe. The needle is then removed from the lesion, connected to a syringe filled with air and the samples are sprayed onto a slide and spread as described above. This technique is particularly indicated for lymph nodes, for very small lesions or when excessive blood is obtained with aspiration.

Impression Smear Impression smears are used for exudative lesions, superficial oily accumulations, pustules, crusts, or biopsy specimens cut in half. The slide is placed several times lightly onto the lesion or oily area. To sample a pustule or a crust, the lesion is opened with a 24G needle and a slide is applied to the drop of pus that comes out. Impression smears have the advantage of not deforming the cells but often result in a sample that is too thick. In such cases, search around the margins of the slide for a monolayer of cells.

Superficial Skin Scrapings As previously described, *Malassezia* can be demonstrated with a very superficial skin scraping of seborrheic skin by using a number 10 or 20 scalpel blade. The material is spread onto a microscope slide using the blade, fixed using a flame and stained with a standard stain.

Sampling with a Cotton Bud This technique is useful for collecting samples from draining tracts, interdigital spaces, claw folds, and external ear canals. The sample is applied to a slide by gently rolling the cotton bud.

Skin Biopsy

Skin biopsies are sometimes needed for investigation and diagnosis of dermatoses.

If the condition of the patient permits, it is better to perform the biopsy after 1–2 weeks of antibiotic therapy. This removes any secondary infection that can complicate the interpretation of the biopsy. Preferred antibiotics include cephalexin (20–30 mg/kg orally twice daily), cephadroxil (20–30 mg orally twice daily), or amoxicillin-clavulanate (20–25 mg orally twice daily). To avoid secondary infection and scarring, the antibiotic can be continued for 1 week after the biopsy. If the patient has been receiving glucocorticoid therapy and the condition of the patient permits, then the biopsy should be delayed until 15–20 days after treatment discontinuation or longer if long-acting depo-injections have been used.

Local anesthesia would be preferable, given that the procedure is minor and rapid and requires only one or two sutures, however in cats this is possible only if they are extremely quiet and when biopsies are taken from the trunk. If using local anesthesia, one should remember that no more than 1 ml of 2% lidocaine should be injected in cats, due to risk of cardiac toxicity. If multiple biopsies are necessary, lidocaine can be diluted 1:1 with saline, so that 2 ml of 1% lidocaine are obtained and can be used for up to 4 biopsies. In the majority of cases, however, general anesthesia is used.

In general, collecting several biopsies from representative lesions will facilitate the diagnosis. Wherever possible, early lesions, such as papules and pustules should be biopsied, and later evolution of these, such as ulcers and crusts should be avoided, however, if there is a range of lesions, biopysing all is prudent.

Prior to biopsy, lesions may be gently clipped, but it is preferable not to clean the skin as this may remove valuable diagnostic material. Disposable punch biopsies are usually the preferred method of biopsy collection. Biopsies from the edge of lesions, including adjacent apparently normal-looking skin, should be obtained with the elliptical excision biopsy technique.

The sample should be put in 10% fresh formalin and accompanied by a full clinical history. The pathologist should be informed of the signalment (age and breed), clinical signs, description and site of the lesions, and duration and evolution of the disease. Any treatment/medication, its duration and period of suspension should be included. Biopsies from different sites should be submitted in separate, numbered containers with a description in the history indicating the site and type of lesion for each biopsy.

References

1. Pankratz KE, Ferris KK, Griffith EH, et al. Use of single-dose oral gabapentin to attenuate fear responses in cage-trap confined community cats: a double- blind, placebo-controlled field trial. J Feline Med Surg. 2018;20:535–43.
2. Van Haaften KA, Eichstadt Forsythe LR, Stelow EA, Bain MJ. Effects of a single pre-appointment dose of gabapentin on signs of stress in cats during transportation and veterinary examination. J Am Vet Med Assoc. 2017;251:1175–81.
3. Rodin I, Sundhal E, Carney H, et al. AAFP and ISFM Feline-Friendly Handling Guidelines. J Feline Med Surg. 2011;13:364–75.

Part II

Problem Oriented Approach to…:

Alopecia

Silvia Colombo

Abstract

Alopecia, either spontaneous or self-induced, is a common presenting sign in cats. Definitions of alopecia and hypotrichosis and the clinical features of alopecia are given at the beginning of this chapter, followed by pathogenesis of the different types of alopecia. Clinical presentations of alopecia and its preferential localization in selected feline diseases are described, together with useful diagnostic hints coming from signalment and history. The diagnostic approach to alopecia implies the correct differentiation of the pathogenetic mechanisms underlying the clinical signs, which can be obtained by collecting history, examining the cat, and performing a microscopic examination of the hair. This is, in cats, a very important diagnostic test, which should always be performed at the beginning of the consultation, in order to differentiate spontaneous from self-induced alopecia. Dermatophytosis is very common in cats, and diagnostic tests to diagnose or rule out this disease should be carried out in all cases presenting with alopecia.

Definitions

Alopecia simply means hair loss. The word alopecia is derived from the ancient Greek word ἀλώπηξ (alṓpēx), which means fox. The term "alopecia" was used at that time to describe fox mange.

Hypotrichosis means that there is less than normal amount of hair (from the ancient Greek words υπο, below, and θριξ, hair), and this term is sometimes used as a synonym of partial alopecia. Although the exact meaning of these two terms is very similar, if not identical, the term hypotrichosis is preferred when there is a

S. Colombo (✉)
Servizi Dermatologici Veterinari, Legnano, Italy

© Springer Nature Switzerland AG 2020
C. Noli, S. Colombo (eds.), *Feline Dermatology*,
https://doi.org/10.1007/978-3-030-29836-4_4

Fig. 1 Congenital hypotrichosis in a domestic short-haired kitten

congenital deficiency of hair in both human and veterinary dermatology publications (Fig. 1) [1]. Strictly speaking, hypotrichosis should be used as a synonym of congenital alopecia.

Alopecia can be classified depending on severity (partial or complete), distribution (focal, multifocal, generalized, symmetrical), localization, and pathogenesis. Partial alopecia means that there is less than normal amount of hair, while complete alopecia describes absence of hair. Focal alopecia, occasionally also called localized alopecia, refers to a single patch of alopecia anywhere on the body (Fig. 2). If more patches are present, alopecia is defined as multifocal. Focal or multifocal alopecia in cats is a clinical presentation commonly observed in cases of dermatophytosis (Fig. 3). When a whole region of the body is involved, alopecia is described as diffuse or generalized. Diffuse alopecia may be symmetrical, when both sides of the body are equally affected. Generalized alopecia is normal in hypotrichotic breeds, such as the Sphynx cat [2].

Pathogenesis

In cats, the most useful classification of alopecia from a diagnostic point of view is the one based on pathogenesis. Alopecia can be spontaneous, when the hair falls off, or self-induced, when the cat actively removes the hair by continuous licking.

Spontaneous alopecia occurs as a consequence of two main pathogenetic mechanisms. When inflammation or infection targets the hair follicle and/or the hair shaft, the latter undergoes damage and falls off (Fig. 4). Hair may also be absent because the hair follicle is dysplastic or atrophic, thus not able to produce a normal hair shaft (Table 1).

Fig. 2 Focal alopecia on the front limb in a kitten affected by dermatophytosis

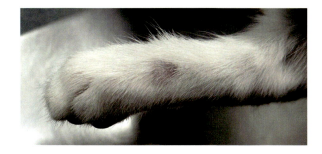

Fig. 3 Multifocal alopecia in a kitten affected by dermatophytosis

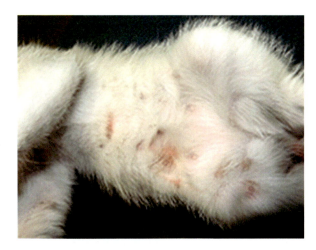

Fig. 4 Spontaneous alopecia following an adverse reaction to flea collar in an adult cat

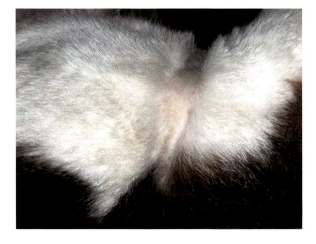

Self-induced alopecia is caused by the cat itself by excessive licking and, less commonly, by chewing, plucking hair, or scratching (Fig. 5). Licking in the feline species is a major component of grooming, a normal, genetically programmed feline behavior. Cats groom to remove dead hair, ectoparasites and dirt and to control body temperature. One study suggested that a healthy cat grooms for approximately 1 hour per day [3]. An increased frequency and/or intensity of this behavior is called

Table 1 Selected causes of spontaneous alopecia

Inflammation/infection of the hair follicle	Pyoderma
	Dermatophytosis
	Demodicosis (*Demodex cati*)
	Pemphigus foliaceus
	Pseudopelade
	Lymphocytic mural folliculitis
	Sebaceous adenitis
Dysplasia/atrophy of the hair follicle	Topical/injectable glucocorticoid administration
	Topical/systemic adverse drug reaction
	Telogen effluvium
	Spontaneous/iatrogenic hyperadrenocorticism
	Paraneoplastic alopecia
	Post-traumatic alopecia
	Alopecic breeds/congenital hypotrichoses
	Pili torti
	Cicatricial alopecia (scar)

Fig. 5 Self-induced alopecia in an allergic cat

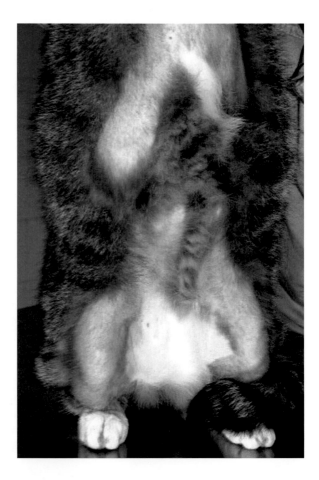

Table 2 Selected causes of self-induced alopecia

Pruritus	Pyoderma
	Dermatophytosis
	Malassezia overgrowth
	Flea infestation
	Cheyletiellosis
	Otodectic mange (erratic)
	Demodicosis (*Demodex gatoi*)
	Lynxacarus infestation
	Flea-bite hypersensitivity
	Adverse reaction to food
	Feline atopic syndrome
	Allergic contact dermatitis
	Feline lymphocytosis
Pain/neurologic	Feline hyperesthesia syndrome
	Irritant contact dermatitis
	Feline idiopathic cystitis
	Trauma
Behavioral	Psychogenic alopecia

overgrooming and may be the expression of pruritus, pain or behavioral problems (Table 2). Being the increased expression of a physiological behavior, overgrooming is often not recognized by the owner or not interpreted as a sign of pruritus or pain. Moreover, cats tend to express their discomfort by hiding away from the owners, who may not be aware of their pet's overgrooming.

Finally, one must remember that some diseases may cause both spontaneous alopecia due to damage to the hair follicle and self-induced alopecia due to pruritus. For example, some cases of dermatophytosis or demodicosis may be associated with pruritus.

Diagnostic Approach

Signalment and History

Infectious and ectoparasitic diseases such as dermatophytosis, demodicosis, flea infestation, or cheyletiellosis are commonly observed in kittens or in environmental conditions of crowding, such as breeding colonies or pet shops. Paraneoplastic syndromes and neoplasia are typically seen in older cats. Breed may be a relevant information in the diagnostic approach: Persian cats are predisposed to dermatophytosis (Fig. 6); congenital hypotrichosis has been recently reported in Birman cats [1]. A good knowledge of feline phenotypes is important, especially for breeds such as the Devon rex cat, which may have an extremely variable amount of hair on the trunk and is physiologically alopecic on the lateral and ventral neck.

History is also very relevant for the diagnosis. A scar may be easily diagnosed based on history, while previous trauma from a car accident or a fall may point towards a post-traumatic alopecia [4]. A detailed pharmacological history is

Fig. 6 Alopecia and scaling on the tail of a Persian cat with dermatophytosis

Fig. 7 Spontaneous alopecia in an old cat affected by hyperadrenocorticism and demodicosis

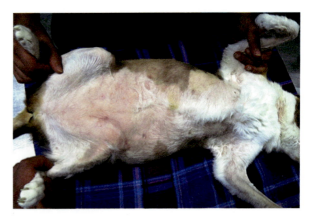

important when an adverse drug reaction is suspected. Sudden onset of alopecia in a queen that recently gave birth may suggest telogen effluvium. Seasonality of self-induced alopecia may orientate towards feline atopic dermatitis. Concurrent systemic clinical signs such as polyuria and polydipsia in an old, diabetic cat developing spontaneous alopecia should prompt testing for hyperadrenocorticism (Fig. 7), while self-induced alopecia of the abdomen and groin may be caused by feline idiopathic cystitis [5].

Clinical Presentation

Spontaneous alopecia may be partial or complete and, in general, hair can be easily epilated from the whole alopecic area, from the center of the lesion or from its periphery. The skin looks glabrous and smooth, and few short fragments of hair can be seen emerging from follicular ostia in selected diseases such as dermatophytosis.

Fig. 8 Close-up image of the abdominal skin of a cat with self-induced alopecia

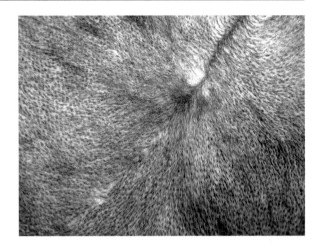

Self-induced alopecia is characterized by the presence of very short fragments of hair which can be observed by looking closely at the skin or with the help of a magnifying lens (Fig. 8). Hair cannot be easily epilated. Self-induced alopecia is often complete and may be symmetrical. The alopecic area usually has very well-defined margins, with abrupt change to normal hair.

Both spontaneous and self-induced alopecia in cats may be focal, multifocal or generalized, and may be associated with other skin lesions. The presence/absence and type of lesion accompanying alopecia is extremely useful to orient the diagnostic process (Table 3).

Focal alopecia and thickening of the affected skin, together with history of previous trauma, may allow the clinician to identify a scar. The skin may also be hypo- or hyperpigmented. Mild erythema and exfoliation associated with focal or multifocal alopecia in cats may suggest dermatophytosis. Pruritus may vary from absent to moderate and for this reason dermatophytosis should also be considered in the list of differential diagnoses of self-induced alopecia. A focal area of non-inflammatory alopecia with very thin skin, visible blood vessels and bruising suggests a reaction to one or repeated glucocorticoid injections in that site (Fig. 9). Generalized, predominantly ventral alopecia with shiny skin in an old cat is suggestive of paraneoplastic alopecia (Fig. 10) [6].

Focal or multifocal alopecia and erythema, mild scaling and occasionally comedones, associated with mild or no pruritus may be indicative of demodicosis due to *Demodex cati*, a follicular mite that usually causes disease in immunocompromised animals. Severe pruritus and self-induced alopecia with erythema and scaling raises the suspicion of demodicosis due to *Demodex gatoi*, a contagious, short-bodied mite living in the stratum corneum. Demodicosis is uncommon in the cat [7]. Severe scaling and self-induced alopecia, often with a dorsal distribution, may indicate cheyletiellosis. Self-induced alopecia, particularly if miliary dermatitis and/or eosinophilic plaques are concurrently observed, may be very suggestive of an allergic disease.

Table 3 Examples of lesions observed concurrently to alopecia in feline skin diseases

	Lesions	Disease
Spontaneous alopecia	Erythema, scaling, follicular casts	Dermatophytosis
	Erythema, scaling, comedones, follicular casts	Demodicosis
	Papules, crusting, scaling	Superficial pyoderma
	Pustules, yellow crusting	Pemphigus foliaceus
	Onychomadesis, onychorrhexis	Pseudopelade
	Scaling, hyperpigmentation	Lymphocytic mural folliculitis
	Scaling, crusting, follicular casts	Sebaceous adenitis
	Focal thinning, visible blood vessels, bruising	Topical/systemic glucocorticoid administration
	None	Telogen effluvium
	Thin skin, bruising, tears, scaling, comedones	Spontaneous/iatrogenic hyperadrenocorticism
	Shiny skin	Paraneoplastic alopecia
	Erythema, shiny skin, erosions/ulcers	Post-traumatic alopecia
	Absence of whiskers, claws, tongue papillae	Congenital hypotrichoses
	Thickening, hypo-/hyperpigmentation	Scar
Self-induced alopecia	Scaling	Cheyletiellosis
	Ceruminous otitis externa	Otodectic mange (erratic)
	Miliary dermatitis, eosinophilic plaque	Allergic diseases
	Papules, crusting, scaling	Superficial pyoderma
	Erythema, scaling, ceruminous otitis externa, paronychia, chin acne	*Malassezia* overgrowth
	Erythema, erosions/ulcers, plaques	Feline lymphocytosis
	Skin rolling	Feline hyperesthesia syndrome
	Erosions/ulcers	Trauma
	None	Feline idiopathic cystitis
	None	Psychogenic alopecia

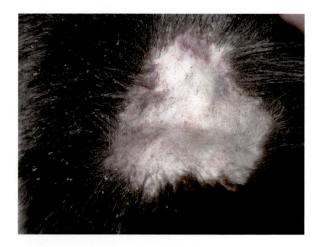

Fig. 9 Spontaneous, focal alopecia in a cat treated with repeated injections of a glucocorticoid

Fig. 10 Diffuse alopecia
and shiny skin on the
abdomen of a cat with
paraneoplastic alopecia

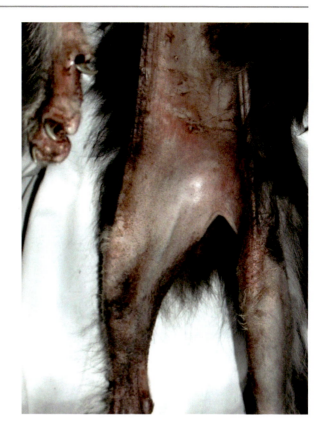

Together with the correct distinction between spontaneous and self-induced alopecia and identification of concurrent lesions, preferential localization of the clinical signs may help in listing the differential diagnoses (Table 4).

Diagnostic Algorithm

This section is illustrated in Fig. 11. Red squares with numbers represent the steps of the diagnostic process, as explained below.

1 Perform microscopic hair examination

Microscopic examination of hair shafts is the first test to perform in any case of alopecia in cats since it may yield useful information, beyond the distinction between spontaneous and self-induced alopecia.

First, the hair tips must be evaluated: broken tips indicate self-induced alopecia, while intact tips may indicate spontaneous hair loss, with the exception of dermatophytosis. Second, the hair shaft in its entire length should be carefully observed. Congenital abnormalities, such as *pili torti*, present with flattened hair shafts that twist on their own axis by 180 degrees at irregular intervals

Table 4 Common locations of alopecia in selected feline skin diseases

Spontaneous alopecia	
Distribution	Disease
Head, pinnae, paws, tail, generalized	Dermatophytosis
Head, neck, ear canal, generalized	Demodicosis
Head, pinnae, claw folds, abdomen	Pemphigus foliaceus
Head, abdomen, legs, paws	Pseudopelade
Head, pinnae neck, generalized	Sebaceous adenitis
Site of application/injection	Topical/systemic glucocorticoid administration
Trunk	Spontaneous/iatrogenic hyperadrenocorticism
Abdomen, ventral trunk, medial legs	Paraneoplastic alopecia
Rump	Post-traumatic alopecia
Generalized	Alopecic breeds/congenital hypotrichoses
Site of previous trauma	Cicatricial alopecia (scar)
Self-induced alopecia	
Rump	Flea infestation
Dorsum	Cheyletiellosis
Neck, rump, tail, ear canal	Otodectic mange (erratic)
Thorax, abdomen	Demodicosis
Rump	Flea-bite hypersensitivity
Abdomen, medial thighs, head, neck	Other allergic diseases
Chin, claw folds, face, ear canal, generalized	*Malassezia* overgrowth
Thorax, legs, pinnae, neck	Feline lymphocytosis
Dorsum	Feline hyperesthesia syndrome
Abdomen, groin	Feline idiopathic cystitis
Site of previous trauma	Post-traumatic alopecia

(Fig. 12) [8]. Moving towards the root, spores of dermatophytes arranged around the hair shaft or *Demodex* mites free or embedded in keratin casts may be identified. However, a negative result of hair examination does not rule out dermatophytosis and demodicosis.

2. Rule out non-dermatological causes of self-induced alopecia

When the microscopic examination of hair shafts indicates self-induced alopecia, non-dermatological causes must be carefully considered and history should specifically investigate for concurrent non-dermatological signs. If the alopecia occurs on the abdomen and groin only, urinalysis and bacterial culture and sensitivity testing should be performed to investigate feline idiopathic cystitis, urolithiasis and/or lower urinary tract infections. Ultrasound examination may help investigate other causes of abdominal pain. If the alopecia is focal and located, for example, on a single limb or on the dorsal spine, an x-ray examination may identify a previous trauma which may explain the cat's continuous licking at that site. When abnormal behavior such as rippling or rolling the skin along the lumbar spine is reported by the owner to occur frequently, a neurological examination should be recommended [9].

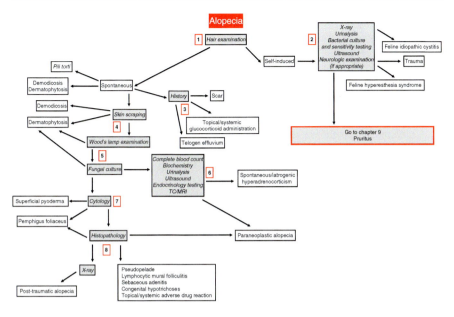

Fig. 11 Diagnostic algorithm of alopecia

Fig. 12 Microscopic examination of the hair shaft of a cat with *pili torti* (10X)

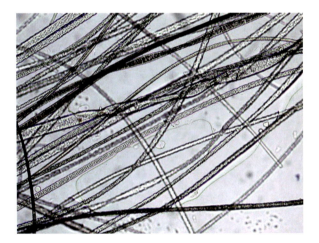

Finally, if all these potential causes of alopecia do not comply with the history and clinical presentation or have been ruled out, self-induced alopecia should be further investigated following the diagnostic approach to pruritus (Chapter, Pruritus).

3 Consider the patient's history

Spontaneous focal alopecia associated with variations of skin thickness and history of a wound in that site points towards a diagnosis of scar. If a topical glucocorticoid has been applied or a glucocorticoid injection has been given where alopecia developed and the skin appears thin, with bruising and visible vessels, the diagnosis is straightforward and may be supported by the observation of mostly telogen hair roots on microscopic examination. Sudden onset of diffuse alopecia in a queen who recently gave birth, for example, may suggest telogen effluvium. In this case, the remaining hair is easily epilated and microscopic examination of hair shows telogen roots only.

4 Perform skin scrapings

Skin scrapings are diagnostic for demodicosis and, together with microscopic examination of hair, may be strongly suggestive of dermatophytosis. In fact, the correct identification of dermatophyte spores surrounding hair shafts may be easier on skin scrapings than microscopic examination of hair shafts, because scraping the surface of the alopecic area is likely to collect more broken, infected hair [10].

5 Perform Wood's lamp examination and fungal culture

These two diagnostic tests, taken together, are diagnostic for dermatophytosis or, if negative results are obtained, are helpful to rule it out. Since dermatophytosis is the most common cause of alopecia in cats, a fungal culture is appropriate in all cases presenting with alopecia.

6 Consider non-dermatological clinical signs

In an old cat presenting with alopecia and systemic signs such as polyuria/polydipsia, polyphagia, vomiting or weight loss, one must consider the possibility of alopecia being caused by a systemic disease. If the alopecic skin appears thin, with bruising and/or tears developing after minimal traction, the cat should be investigated for hyperadrenocorticism. History may suggest iatrogenic hyperadrenocorticism, if glucocorticoids have been administered for a long time, or spontaneous hyperadrenocorticism if there is no history of glucocorticoids administration or the cat is diabetic. Ventrally distributed alopecia with shiny skin in a cat presenting with concurrent weight loss, depression, vomiting, and/or diarrhea may point towards a diagnosis of paraneoplastic alopecia and should prompt to perform an abdominal ultrasound examination.

7 Perform cytology

Cytology should be performed if other lesions, such as pustules, crusts or erosions/ulcers are present together with alopecia. The observation of large numbers of degenerate neutrophils with intracellular and extracellular bacteria indicates superficial pyoderma. If the neutrophils appear "healthy" and many acantholytic keratinocytes are seen, the results of cytological examination are suggestive of pemphigus foliaceus. If large numbers of eosinophils are seen, it is more likely that the cat is pruritic and alopecia should be further investigated following the diagnostic approach to pruritus (Chapter, Pruritus).

8 Take biopsies for histopathological examination

Histopathological examination may be useful to confirm paraneoplastic alopecia and should always be performed if pemphigus foliaceus is suspected. Some diseases presenting with alopecia can only be diagnosed with histopathology; examples are pseudopelade, sebaceous adenitis, congenital hypotrichosis, and adverse drug reactions. If the histopathological examination is suggestive of post-traumatic alopecia, radiological examination of the pelvis should be carried out to confirm the diagnosis.

References

1. Abitbol M, Bossé P, Thomas A, Tiret L. A deletion in FOXN1 is associated with a syndrome characterized by congenital hypotrichosis and short life expectancy in Birman cats. PLoS One. 2015;10:1–12.
2. Genovese DW, Johnson TL, Lamb KE, Gram WD. Histological and dermatoscopic description of sphynx cat skin. Vet Dermatol. 2014;25:523–e90.
3. Eckstein RA, Hart BL. The organization and control of grooming in cats. Appl Anim Behav Sci. 2000;68:131–40.
4. Declerq J. Alopecia and dermatopathy of the lower back following pelvic fractures in three cats. Vet Dermatol. 2004;15:42–6.
5. Amat M, Camps T, Manteca X. Stress in owned cats: behavioural changes and welfare implications. J Feline Med Surg. 2016;18:1–10.
6. Turek MM. Cutaneous paraneoplastic syndromes in dogs and cats: a review of the literature. Vet Dermatol. 2003;14:279–96.
7. Beale K. Feline Demodicosis. A consideration in the itchy or overgrooming cat. J Feline Med Surg. 2012;14:209–13.
8. Maina E, Colombo S, Abramo F, Pasquinelli G. A case of pili torti in a young adult domestic short-haired cat. Vet Dermatol. 2013;24:289–e68.
9. Ciribassi J. Feline hyperesthesia syndrome. Compend Contin Educ Vet. 2009;31:116–22.
10. Colombo S, Cornegliani L, Beccati M, Albanese F. Comparison of two sampling methods for microscopic examination of hair shafts in feline and canine dermatophytosis. Vet Dermatol. 2008;19(Suppl. 1):36.

General References

For definitions: Merriam-Webster Medical Dictionary. http://merriam-webster.com Accessed 10 May 2018.
Albanese F. Canine and feline skin cytology. Cham: Springer International Publishing; 2017.
Goldsmith LA, Katz SI, Gilchrest BA, Paller AS, Leffell DJ, Wolff K. Fitzpatrick's dermatology in general medicine. 8th ed. New York: The McGraw-Hill Companies; 2012.
Mecklenburg L. An overview on congenital alopecia in domestic animals. Vet Dermatol. 2006;17:393–410.
Miller WH, Griffin CE, Campbell KL. Muller & Kirk's small animal dermatology. 7th ed. St. Louis: Elsevier; 2013.
Noli C, Toma S. Dermatologia del cane e del gatto. 2nd ed. Vermezzo: Poletto Editore; 2011.

Papules, Pustules, Furuncles and Crusts

Silvia Colombo

Abstract

Papules, pustules, furuncles, abscesses and crusts are common lesions in cats. With the exception of abscesses, they are often observed in combinations, representing different stages of the same disease evolving into one another. In general, these lesions are the expression of inflammatory diseases, with infectious, parasitic, allergic or autoimmune pathogenesis. Clinical presentations of papules, pustules, furuncles, abscesses and crusts and their preferential localization in selected feline diseases are described, together with useful diagnostic hints coming from signalment and history. A feline-specific clinical presentation called miliary dermatitis is characterized by multiple, small crusted papules and pruritus. The diagnostic approach to papules, pustules, furuncles, abscesses and crusts requires performing the diagnostic tests in a systematic way. Dermatophytosis is very common in cats, and diagnostic tests to diagnose or rule out this disease should be carried out in all cases presenting with papules, pustules, crusts or as miliary dermatitis.

Definitions

A papule is a solid, erythematous, elevated skin lesion of less than 1 cm diameter [1]. Many papules close to each other may coalesce to form a plaque (Chapter, Plaques, Nodules and Eosinophilic Granuloma Complex Lesions).

A pustule is an elevated, circumscribed, hollow lesion containing pus and covered by epidermis. It may be centered around a hair follicle or may be interfollicular

S. Colombo (✉)
Servizi Dermatologici Veterinari, Legnano, Italy

© Springer Nature Switzerland AG 2020
C. Noli, S. Colombo (eds.), *Feline Dermatology*,
https://doi.org/10.1007/978-3-030-29836-4_5

in location. Pustules usually contain neutrophils, with or without bacteria, or less commonly eosinophils. They are fragile and often transient lesions, uncommonly observed in cats.

A furuncle is similar to a pustule, but it is larger in size and deeper in location, because it results from the complete destruction of the hair follicle. The wall of a furuncle is thicker than the roof of a pustule, and its content comprises pus, blood (in this case, it is also called hemorrhagic bulla) or a mixture. It is usually a very inflamed and painful lesion, centered around a hair follicle. The furuncle may open and drain pus, blood or a hemopurulent exudate.

An abscess is a circumscribed, fluctuant, dermal or subcutaneous collection of pus. It may open and drain on the skin surface, forming a draining tract.

A crust is an accumulation of dried exudate. The crust is yellowish when the dried material is pus, or brownish if dried blood is its main component (hemorrhagic crust). It may also contain microorganisms and epidermal cells, such as acantholytic keratinocytes or corneocytes and, if the crust encloses a tuft of hair, its removal results in focal alopecia.

Pathogenesis

Papules, pustules and crusts represent collections of inflammatory cells in the epidermis (pustule), dermis (papule) or on the skin surface (crust) as dead remnants of these cells. The inflammatory cells are attracted towards the superficial layers of the skin by infectious agents, parasites or allergens, or may be the expression of an autoimmune disease such as pemphigus foliaceus (Fig. 1).

The furuncle is a deeper lesion which results from the complete destruction of the hair follicle. The hair follicle is destroyed by severe inflammation, which is, in the feline species, most commonly induced by a bacterial infection as in complicated

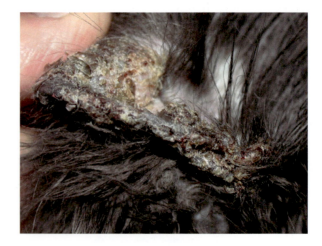

Fig. 1 Severe crusting due to drying of purulent exudate on the pinna of a cat with pemphigus foliaceus

Fig. 2 Furuncles on the chin of a cat affected by complicated chin acne

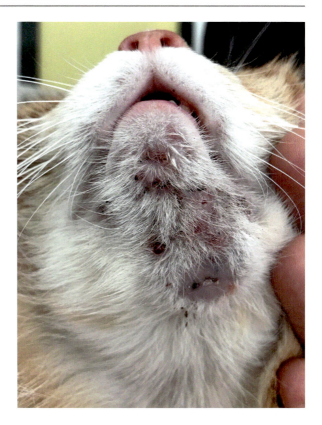

chin acne (Fig. 2) [2]. The hair shaft may be free in the dermis together with bacteria and other debris, and attracts more inflammatory cells behaving as a foreign body.

The abscess usually occurs following bite or claw wounds, with implantation of bacteria in the deep dermis and subcutis. The presence of bacteria attracts large numbers of neutrophils and other inflammatory cells at the infection site, until a large collection of pus is formed (Fig. 3).

Papules, pustules, furuncles and crusts may represent different stages of the same disease and can evolve into one another. A papule may develop into a pustule, which ruptures and becomes a small crust. Very uncommonly in the cat, a circular rim of scales may form when the crust comes off: this lesion is called epidermal collarette. A pustule may become a furuncle if the infection deepens and extends to involve and destroy the whole hair follicle. If the furuncle opens and drains exudate, a crust may form. When the crust comes off, an area of focal alopecia is the final result. Crusts may also cover other lesions, such as erosions and ulcers (Chapter, Excoriations, Erosions and Ulcers) (Fig. 4). This is important to keep in mind when examining the animal, because we may be able to identify different lesions which represent evolving stages of the disease or we may only find the final result of

Fig. 3 Retroauricular abscess in a stray cat

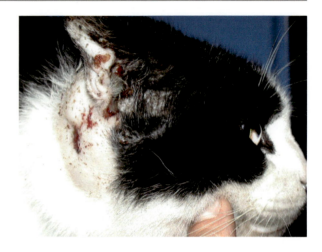

Fig. 4 Hemorrhagic crust covering an erosion/ulcer on the nose of a cat with herpesvirus infection

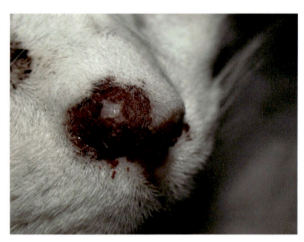

this process, which is the crust. Table 1 lists selected causes of papules, pustules, abscesses, crusts and furuncles in cats.

Diagnostic Approach

Signalment and History

Contagious diseases such as notoedric mange and dermatophytosis are most commonly observed in kittens, while neoplasia is typically seen in older cats. Cutaneous abscesses occur more often in intact male cats, as a consequence of fighting. Breed may be a relevant point in the diagnostic approach: Persian cats are predisposed

Table 1 Selected causes of papules, pustules, furuncles, abscesses and crusts

Papules	Notoedric mange
	Dermatophytosis
	Mosquito-bite hypersensitivity
	Allergic diseases
	Urticaria pigmentosa-like dermatitis
	Xanthomas
	Mast cell tumor
Pustules	Pemphigus foliaceus
Furuncles	Complicated chin acne
Abscess	Bacterial infections
Crusts	Trauma (including self-inflicted)
	Pyoderma
	Notoedric mange
	Dermatophytosis
	Subcutaneous and systemic fungal infections
	Herpesvirus dermatitis
	Poxvirus infection
	Allergic diseases
	Mosquito-bite hypersensitivity
	Adverse drug reactions
	Pemphigus foliaceus
	Complicated chin acne
	Perforating dermatitis
	Idiopathic facial dermatitis of Persian and Himalayan cats
	Squamous cell carcinoma
	Idiopathic/behavioral ulcerative dermatitis

to dermatophytosis and idiopathic facial dermatitis [3]. Urticaria pigmentosa-like dermatitis has been described in Devon rex and Sphynx cats [1, 4].

History is obviously of paramount importance for the diagnosis when previous trauma (including self-induced) is suspected in a cat examined for a crusting lesion. Especially in kittens, detailed information on where the pet was acquired must always be collected. Being found as a stray or adopted from a cattery may represent a predisposing factor for dermatophytosis, notoedric mange and herpesvirus dermatitis. Lifestyle is also relevant, because outdoor cats may be affected by mosquito-bite hypersensitivity and development of abscesses more commonly than indoor cats. Regularly hunting mice and voles is a predisposing factor for Poxvirus infection. Contagion of in-contact pets or people should prompt investigation for dermatophytes and ectoparasites.

Last but not least, one very important question to ask when taking the history is whether the cat is pruritic or not, and if pruritus is continuously present or seems to occur at a specific time of the year. Notoedric mange is a severely pruritic disease, and seasonal pruritus may suggest flea-bite or mosquito-bite hypersensitivity, and feline atopic syndrome.

Clinical Presentation

Papules and pustules are in most cases multiple lesions, sometimes with a grouped configuration. In feline urticaria pigmentosa-like dermatitis, papules may have a linear configuration [1]. A single or many furuncles may be observed in chin acne. The distribution of papules, pustules, furuncles and crusts may be localized or generalized. The abscess is usually a single lesion.

Papules and pustules are primary skin lesions; however, in most diseases, they represent one step in a pathological *continuum* of lesions. For example, although papules are the primary lesions in notoedric mange, they may not be visible, because they are covered by very thick crusts. Multiple, erythematous small papules covered by crusts, especially on the dorsum, may develop representing a feline-specific clinical presentation called miliary dermatitis [5, 6] (see later) (Fig. 5). The location of the lesions on the cat's body may be helpful in developing a correct list of differential diagnoses (Table 2).

An erythematous to hyperpigmented papular eruption, which may have a linear distribution on the ventrolateral chest and abdomen, is often pruritic and occurs in a Devon rex or Sphynx cats, is consistent with urticaria pigmentosa-like dermatitis (Fig. 6) [1, 4]. Small, erythematous, and crusted papules may suggest mosquito-bite hypersensitivity, when distributed on the dorsal nose, pinnae and footpads (Fig. 7) [7]. Pustules may be difficult to observe because they are transient, fragile lesions, but, when observed on the face, inner pinnae and abdomen, close to the nipples and on footpads should prompt investigation for pemphigus foliaceus (Fig. 8) [8].

Furuncles in cats are usually observed on the chin, where they develop when chin acne becomes complicated by secondary bacterial infection. A soft, fluctuant swelling, occasionally with a draining tract from which purulent exudate comes out, located on the face, neck, or tail base, most likely represents an abscess.

Crusts are extremely common lesions, as they are the final result of the pathological *continuum* of lesions described in this chapter, as well as of traumatic lesions.

Fig. 5 Miliary dermatitis in an allergic cat

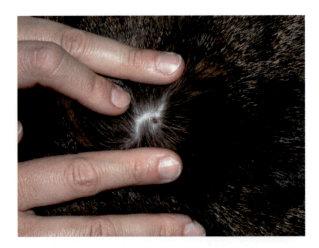

Table 2 Common locations of papules, pustules, abscesses, crusts and furuncles in selected feline skin diseases

Papules	
Distribution	Disease
Head, pinnae, neck, paws, perineum	Notoedric mange
Head, pinnae, paws, tail, generalized	Dermatophytosis
Head, pinnae, paws	Mosquito-bite hypersensitivity
Rump	Flea-bite hypersensitivity
Head, paws, bony prominence	Xanthomas
Pustules	
Head, pinnae, claw folds, abdomen	Pemphigus foliaceus
Furuncles	
Chin	Complicated chin acne
Abscesses	
Neck, shoulders, tail base	Bacterial infections
Crusts	
Site of previous trauma	Trauma
Face	Herpesvirus dermatitis
Face, ear canals	Idiopathic facial dermatitis of Persian and Himalayan cats
Head, pinnae	Squamous cell carcinoma

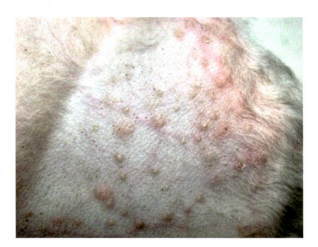

Fig. 6 Multiple erythematous papules in a Devon rex cat with urticaria pigmentosa-like dermatitis

One helpful clinical hint, when crusts are observed, is their color. If the crusts are dark brown, they are composed by dried blood and the lesion was most likely caused by a deep skin disease (ulcer) or (self-)trauma. If they are yellow, they represent dried purulent material and intact pustules should be carefully searched for. Very thick and dry light-colored crusts on the head, margins of pinnae, neck, paws,

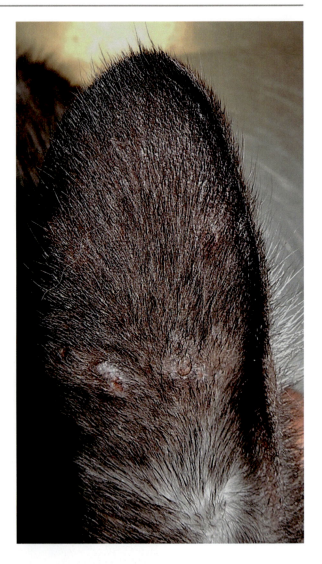

and perineum, associated with severe pruritus, are the predominantly observed
lesions in notoedric mange. Multiple, conical, very dry and thick crusted lesions
(eschars) developing at sites of previous trauma may indicate a rare feline disease
called acquired reactive perforating collagenosis or perforating dermatitis (Fig. 9)
[9]. These lesions are difficult to remove and usually cover an ulcerated, hemorrhagic area. Pruritus and adherent, black, variably dried exudate covering areas of
erythema or erosions distributed around the eyes, mouth, and chin are typical of
idiopathic facial dermatitis of Persian and Himalayan cats, also called dirty face
disease [3].

Fig. 8 Pustules and crusting on the inner pinna of a cat with pemphigus foliaceus

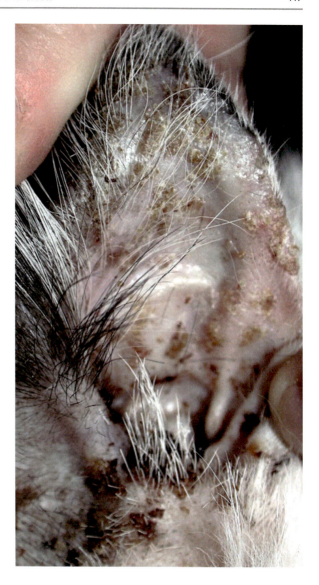

Miliary Dermatitis

Miliary dermatitis is a peculiar clinical presentation observed only in the cat. It is characterized by small, crusted papules "resembling millet seeds", hence the name, which are more easily felt by touching through the haircoat then seen. Miliary dermatitis mainly affects the trunk and neck and is often associated with pruritus and self-induced alopecia (Fig. 10) [5, 6]. Differential diagnoses of miliary dermatitis are listed in Table 3. Miliary dermatitis should be investigated following the diagnostic approach to pruritus (Chapter, Pruritus).

Fig. 9 Dry, thick, adherent yellow crust on the inner pinna of a young cat with perforating dermatitis

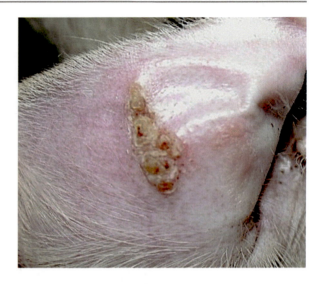

Fig. 10 Alopecia and miliary dermatitis on the dorsum of a cat affected by flea-bite hypersensitivity

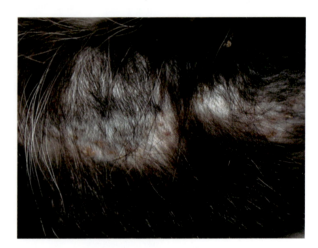

Table 3 Differential diagnoses of miliary dermatitis

Miliary dermatitis	Cheyletiellosis
	Other ectoparasites (*Lynxacarus radowski*)
	Dermatophytosis
	Flea-bite hypersensitivity
	Adverse reaction to food
	Feline atopic syndrome
	Adverse drug reaction
	Pemphigus foliaceus

Diagnostic Algorithm

This section is illustrated in Fig. 11. Red squares with numbers represent the steps of the diagnostic process, explained below.

1 Consider signalment, history and physical examination

Signalment, history, and physical examination may give the clinician extremely useful information for the diagnostic process. In a primarily outdoor intact male cat presenting with a fluctuant mass on the neck, for example, the most likely diagnosis is an abscess. When the main presenting signs are furuncles on the chin of a cat that suffers from chin acne, it is very likely that acne has become secondarily complicated by a bacterial infection. If physical examination reveals papules, pustules or crusts, a standardized sequence of diagnostic tests is usually required to make the diagnosis.

2 Perform skin scrapings

Skin scrapings must be performed whenever papules, pustules, crusts or furuncles are observed. Skin scrapings are diagnostic for notoedric mange and may identify *Demodex cati* mites in cases of chin furunculosis [10].

3 Perform Wood's lamp examination and fungal culture

These two diagnostic tests, taken together, are diagnostic for dermatophytosis or, if negative results are obtained, are helpful to rule it out. Since dermatophytosis may present with papules, pustules, miliary dermatitis and crusts in cats, a fungal culture is appropriate in all cases presenting with these lesions (Fig. 12).

4 Perform cytology

When the physical examination reveals the presence of an abscess, cytology from the purulent exudate should always be performed to support the diagnostic hypothesis. Usually, large numbers of degenerate neutrophils are visible, admixed with bacteria and variable numbers of macrophages, lymphocytes and plasma cells. To identify the bacteria species causing the abscess, bacterial culture and sensitivity testing should be performed. It is also advisable testing cats with abscesses for FIV and FeLV. Cytological examination of exudate draining from furuncles on the chin usually shows pyogranulomatous inflammation with bacteria. Bacterial culture and sensitivity testing for aerobes and anaerobes may be required to identify the causative microorganism and choosing the most effective antibiotic for treatment, if needed [2].

Papules, pustules and crusts should always be investigated by cytological examination, a simple test that often gives very useful information. The observation of large numbers of non-degenerate neutrophils admixed with many acantholytic keratinocytes suggests pemphigus foliaceus. Eosinophilic inflammation is very common in cats. If eosinophils are present in large numbers within a mixed inflammatory infiltrate in samples obtained from crusted, papular lesions on the bridge of the nose, mosquito-bite hypersensitivity is a likely diagnosis [7].

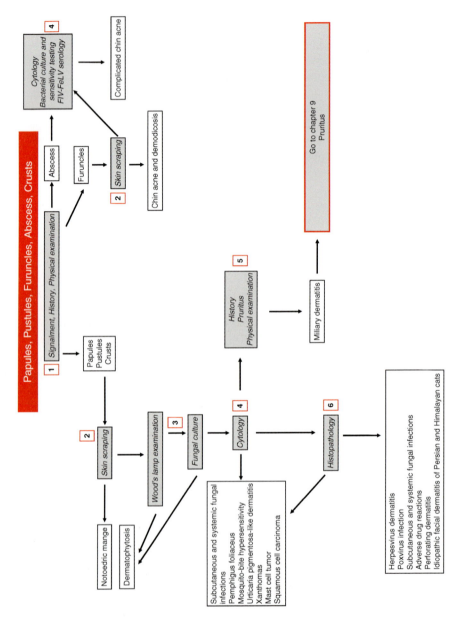

Fig. 11 Diagnostic algorithm to papules, pustules, furuncles, abscesses and crusts

Neutrophils, eosinophils and occasionally mast cells observed in samples from erythematous, hyperpigmented papules on the skin of a Devon rex or Sphynx cat are suggestive of urticaria pigmentosa-like dermatitis [4]. Finally, cytological examination may show a monomorphic population of well-differentiated mast cells in mast cell tumors or epithelial cells in small aggregates or as single cells, with aspects of squamous differentiation, often admixed with neutrophils and other inflammatory cells, in squamous cell carcinomas (Fig. 13). Cytological findings obtained from crusting, papular or pustular lesions must always be confirmed by biopsy and histopathological examination.

5 | Consider history, pruritus and physical examination

When skin scrapings, Wood's lamp examination and fungal culture yield negative results and cytological findings are nonspecific (e.g., neutrophilic inflammation), one must carefully re-consider the history and clinical findings. In a continuously or seasonally pruritic cat, presenting with crusted papules on

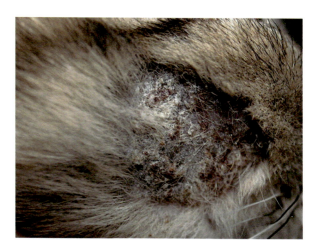

Fig. 12 Alopecia and crusting on the face of a cat with dermatophytosis

Fig. 13 Hemorrhagic crusting on the nose of a cat with squamous cell carcinoma

the dorsum, or, less commonly, with generalized distribution, miliary dermatitis should be further investigated following the diagnostic approach to pruritus (Chapter, Pruritus).

6 Take biopsies for histopathological examination

Histopathological examination should always be performed if pemphigus foliaceus, mosquito-bite hypersensitivity, urticaria pigmentosa-like dermatitis, or infectious, metabolic and neoplastic diseases are suspected, based on cytological findings. Other diseases with nonspecific cytological findings and requiring histopathological examination for the diagnosis are, for example, viral diseases, perforating dermatitis, idiopathic facial dermatitis of Persian and Himalayan cats and adverse drug reactions.

References

1. Vitale C, Ihrke PJ, Olivry T, Stannard AA. Feline urticaria pigmentosa in three related Sphinx cats. Vet Dermatol. 1996;7:227–33.
2. Jazic E, Coyner KS, Loeffler DG, Lewis TP. An evaluation of the clinical, cytological, infectious and histopathological features of feline acne. Vet Dermatol. 2006;17:134–40.
3. Bond R, Curtis CF, Ferguson EA, Mason IS, Rest J. An idiopathic facial dermatitis of Persian cats. Vet Dermatol. 2000;11:35–41.
4. Noli C, Colombo S, Abramo F, Scarampella F. Papular eosinophilic/mastocytic dermatitis (feline urticaria pigmentosa) in Devon rex cats: a distinct disease entity or a histopathological reaction pattern? Vet Dermatol. 2004;15:253–9.
5. Hobi S, Linek M, Marignac G, Olivry T, Beco L, Nett C, et al. Clinical characteristics and causes of pruritus in cats: a multicentre study on feline hypersensitivity-associated dermatoses. Vet Dermatol. 2011;22:406–13.
6. Diesel A. Cutaneous hypersensitivity dermatoses in the feline patient: a review of allergic skin disease in cats. Vet Sci. 2017:25. https://doi.org/10.3390/vetsci4020025.
7. Nagata M, Ishida T. Cutaneous reactivity to mosquito bites and its antigens in cats. Vet Dermatol. 1997;8:19–26.
8. Olivry T. A review of autoimmune skin diseases in domestic animals: I – superficial pemphigus. Vet Dermatol. 2006;17:291–305.
9. Albanese F, Tieghi C, De Rosa L, Colombo S, Abramo F. Feline perforating dermatitis resembling human reactive perforating collagenosis: clinicopathological findings and outcome in four cases. Vet Dermatol. 2009;20:273–80.
10. Beale K. Feline demodicosis: a consideration in the itchy or overgrooming cat. J Feline Med Surg. 2012;14:209–13.

General References

For definitions: Merriam-Webster Medical Dictionary. http://merriam-webster.com Accessed 10 May 2018.

Albanese F. Canine and feline skin cytology. Cham: Springer International Publishing; 2017.

Goldsmith LA, Katz SI, Gilchrest BA, Paller AS, Leffell DJ, Wolff K. Fitzpatrick's dermatology in general medicine. 8th ed. New York: The McGraw-Hill Companies; 2012.

Miller WH, Griffin CE, Muller CKL. Kirk's small animal dermatology. 7th ed. St. Louis: Elsevier; 2013.

Noli C, Foster A, Rosenkrantz W. Veterinary allergy. Chichester: Wiley Blackwell; 2014.

Noli C, Toma S. Dermatologia del cane e del gatto. 2nd ed. Vermezzo: Poletto Editore; 2011.

Plaques, Nodules and Eosinophilic Granuloma Complex Lesions

Silvia Colombo and Alessandra Fondati

Abstract

Plaques and nodules, including the lesions belonging to the eosinophilic granuloma complex (EGC), are common in cats. Plaques and nodules are caused in most cases by infectious, allergic, metabolic or neoplastic diseases. Clinical presentations of plaques and nodules and their preferential localization in selected feline diseases are described, together with useful hints coming from signalment and history. A feline-specific group of plaques or nodules, known as the EGC, and its specific features are also addressed in this chapter. EGC traditionally comprises eosinophilic plaque (EP), eosinophilic granuloma (EG) and lip (indolent) ulcer (LU). The diagnostic approach to plaques and nodules starts with the cytological examination, which may help the clinician to differentiate between the neoplastic and the inflammatory nature of the lesion. Histopathological examination is required to make or to confirm the diagnosis, and further testing is usually suggested by the histopathological diagnosis.

Definitions

A plaque is a flat elevation of the skin greater than 1 cm of diameter, and its size is, by definition, larger than its height. Plaques often form from a papule increasing in size or by coalescence of multiple papules.

S. Colombo (✉)
Servizi Dermatologici Veterinari, Legnano, Italy

A. Fondati
Veterinaria Trastevere - Veterinaria Cetego, Roma, RM, Italy

Clinica Veterinaria Colombo, Camaiore, LU, Italy

© Springer Nature Switzerland AG 2020
C. Noli, S. Colombo (eds.), *Feline Dermatology*,
https://doi.org/10.1007/978-3-030-29836-4_6

A nodule is a solid, palpable and circumscribed skin lesion greater than 1 cm of diameter. Nodules can be further characterized by their deepness, as epidermal, dermal or subcutaneous nodules. Nodules may open toward the skin surface and a draining tract may develop, with exudate of variable aspect and consistency coming out of the lesion. A peculiar type of nodule is the cyst, which is a cavity containing fluid or semisolid material lined by an epithelial wall.

Both nodules and plaques may be described by adding features such as number, size, shape, color, consistency (e.g., hard or soft), surface changes (e.g., alopecic, eroded, ulcerated) and relationship with the surrounding tissues (e.g., fixed, movable). A soft, fluctuant, circumscribed nodule containing a collection of pus is called abscess, and is described in Chapter, Papules, Pustules, Furuncles and Crusts. Other relevant descriptors are whether the lesion is pruritic or non-pruritic and whether it is painful or painless.

Plaques and nodules are common in cats and are the primary lesions of two of the clinical presentations of the eosinophilic granuloma complex (EGC). EGC traditionally comprises eosinophilic plaque (EP), eosinophilic granuloma (EG), and lip (indolent) ulcer (LU). These lesions affect the skin, lips and oral cavity of cats and have been initially grouped together because they were observed simultaneously on the same cat, therefore suggesting a common underlying cause. The EGC can be considered a "complex" in all respects, indeed, because EP, EG, and LU share clinical and histopathological aspects and a common etiopathogenesis, in which eosinophils play a pivotal role.

Pathogenesis

A plaque is a flat, solid lesion due to infiltration of inflammatory or neoplastic cells in the skin. In cats, it is most commonly associated with allergic or neoplastic diseases. It may develop because a papule increases in size or because many papules coalesce. In feline dermatology, the term plaque is most often used to describe the EP, a specific lesion belonging to the EGC (Fig. 1).

Fig. 1 Eosinophilic plaque in a flea-allergic cat

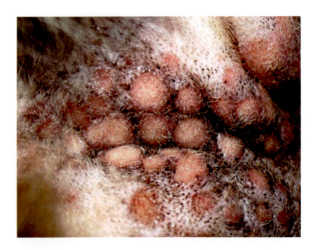

EGC is not a definitive diagnosis. It should be rather considered a cutaneous reaction pattern most likely incited by underlying allergic causes, including hypersensitivity reactions to flea and, less commonly, environmental and food allergens. Occasionally, stings or bites of arthropods other than fleas might be considered as triggering factors for cutaneous eosinophil recruitment. However, in some cases, no external inciting stimuli can be identified and EGC lesions remain idiopathic. However, it must be taken into account that the reliability of current available diagnostic procedures does not always allow to definitely confirm/exclude hypersensitivity reactions towards environmental allergens in the cat.

Based on observations of EGC in family-related cats, a genetic, inheritable "dysregulation" of the eosinophil response has been suggested to predispose to the development of EGC in the absence of detectable underlying causes, particularly in kittens.

A combined genetic and allergic etiopathogenesis has been also suggested for EGC [1]. A genetic predisposition to develop intense eosinophil responses might help to explain why only a few cats develop EGC lesions, whereas the hypothesized underlying allergic stimuli are so largely distributed and more commonly associated with different reaction patterns, such as head and neck pruritus, self-induced alopecia or miliary dermatitis. On the other hand, a genetically based "abnormal" eosinophil response would not fit with clustering of cases in unrelated in-contact cats or with the lack of predisposition to develop extra-cutaneous eosinophilic diseases in cats suffering from EGC.

Nodules also develop because of infiltration of inflammatory or neoplastic cells, however they are usually not flat and may extend deeper in the dermis and subcutaneous tissue. Non-neoplastic nodules may be induced by infectious agents, such as bacteria or fungi, or may be sterile, as it happens in EG or sterile nodular panniculitis. Uncommon causes of nodules or, rarely, plaques in cats are foreign bodies and deposition of calcium or lipids in the skin (Fig. 2) [2].

Cysts may be caused by congenital defects of development of different skin components or by obstruction of a sebaceous/apocrine duct (Fig. 3) [3]. Table 1 lists the most common causes of plaques and nodules in cats.

Fig. 2 Nodule of *calcinosis cutis* on the chin of a cat with chronic kidney disease

Fig. 3 Multiple cysts on
the muzzle of a Persian cat
with feline cystomatosis.
(Courtesy of Dr. Stefano
Borio)

Table 1 Selected causes
of plaques and nodules

Plaques	Eosinophilic plaque/granuloma
	Lip ulcer
	Papillomavirus infections
	Xanthomas
	Bowenoid in situ carcinoma
	Cutaneous lymphocytosis
	Mast cell tumor
	Progressive feline histiocytosis
Nodules	Botryomycosis
	Leprosy
	Rapidly growing mycobacterial infections
	Nocardiosis
	Dermatophytic mycetoma
	Eumycotic mycetomas
	Pheohyphomycosis
	Sporotrichosis
	Cryptococcosis
	Leishmaniosis
	Eosinophilic granuloma
	Calcinosis cutis
	Xanthomas
	Sterile granuloma/pyogranuloma syndrome
	Feline progressive histiocytosis
	Sterile nodular panniculitis
	Plasma cell pododermatitis
	Squamous cell carcinoma
	Basal cell tumors
	Follicular tumors
	Hemangiosarcoma
	Lymphangiosarcoma
	Mast cell tumor
	Sarcoid
	Vaccine-site fibrosarcoma
	Epitheliotropic/non-epitheliotropic cutaneous lymphoma
	Melanocytoma/melanoma
	Ceruminous cystomatosis

Diagnostic Approach

Signalment and History

Plaques and nodules are usually observed in adult or older cats and are due to infectious, allergic, metabolic or neoplastic diseases in the majority of cases. Nodular lesions with breed predisposition are dermatophytic mycetoma (Fig. 4) and apocrine cystomatosis, occurring more commonly in Persian cats, and mast cell tumor, being more often diagnosed in Siamese cats [4].

History should investigate the cat's lifestyle, since most bacterial and fungal infections presenting with nodules require a penetrating wound to develop. For this reason, these diseases are more likely to occur in cats allowed to go outdoors. More specifically, contact with pigeon droppings has been advocated in cryptococcosis and contact with decaying plant material in sporotrichosis, while leprosy syndrome is usually reported in hunting or fighting cats [5, 6]. History of travelling or living in endemic areas may suggest diseases such as leishmaniosis, which occurs in specific geographic locations [7].

When a cat is presented for a nodular lesion, neoplastic diseases should always be included in the list of differential diagnoses. Useful information may be gathered by enquiring about the age and time of lesion development, changes in its appearance and size and concurrent systemic signs presented by the cat. Vaccinal history is also very relevant, because cats are predisposed to vaccination-site fibrosarcoma (Fig. 5) [8]. History of prolonged sun exposure in a white cat may suggest squamous cell carcinoma (Fig. 6).

EGC lesions may be observed in cats of any breed, sex, and age; however, they frequently occur in young cats and occasionally appear in few month-old kittens. Lesions onset varies from acute (a few days) in EP to slow in LU and EG. Pruritus varies from intense in EP to variable in EG and absent in LU. If pruritus is absent and lesions are not clearly visible, as in selected cases of linear EG on caudal thighs,

Fig. 4 Large nodule on the leg of a Persian cat affected by dermatophytic mycetoma

Fig. 5 Relapse of vaccination-site fibrosarcoma in a cat

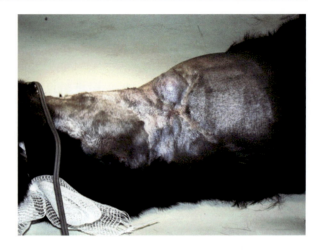

Fig. 6 Large nodules diagnosed as squamous cell carcinoma on the dorsum of a congenitally alopecic cat

lesions are usually identified by the owner when touching the cat (Fig. 7). Normally, EGC lesions are chronically persistent or recurrent, but, especially in kittens, EG may spontaneously regress with no further relapses.

Clinical Presentation

Plaques are, in most cases, lesions belonging to the EGC and may be single or more commonly multiple. Clinical features of the EGC, including EP, EG, and LU, have been well delineated and are considered quite distinctive [9]. The EP appears as intensely pruritic, oozing, eroded, firm, coalescing papules and plaques affecting sites accessible to being licked, such as ventral abdomen and inner thighs. Secondary bacterial infection and regional lymphadenopathy are common [10].

Fig. 7 Linear granuloma on the hind limb of a cat

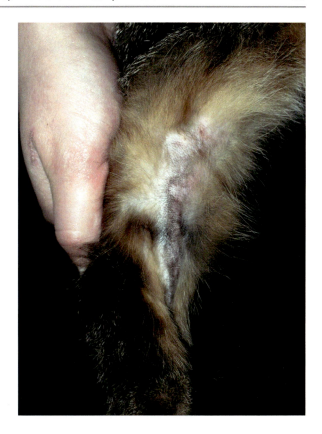

EG classically occurs as firm, yellowish, variably pruritic, alopecic, erythematous and crusting papules and plaques with a striking linear configuration when affecting the caudal thigh. EG may also appear as single, yellowish papulo-nodular lesions located anywhere on the body, including paws, mid-lower lip/chin, lip commissure and oral cavity. Pedal EG lesions are frequently ulcerated and crusted whereas mucosal lesions appear as irregularly surfaced yellowish nodules, frequently located on the tongue and the palate.

LU refers to an apparently non-pruritic and non-painful, reddish-brown to yellowish, glistening, non-bleeding, well-circumscribed, frequently concave ulcer with raised margins and the aspect of an ulcerated plaque rather than a true ulcer. The LU occurs most commonly on the midline of the upper lip, at the philtrum or adjacent to the upper canine tooth, mono- or bilaterally (Fig. 8).

EGC lesions with overlapping features of more than one form are commonly observed and lesion definition can be difficult, as is the case of solitary or linearly grouped ulcerated EG resembling LU or EP. Lesions might be therefore described as papules, plaques and nodules belonging to the EGC, with no further clinical distinction. This observation raises the question on the adequacy of the currently

Fig. 8 Bilateral ulcer on
the upper lips

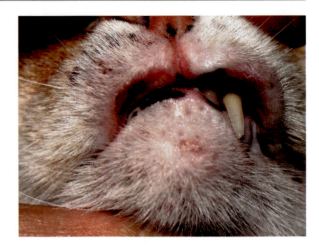

Fig. 9 Single,
erythematous, and
exfoliative nodule on the
front limb of a cat with
epitheliotropic cutaneous
lymphoma

adopted nomenclature that represents a mixture of clinical (plaque and ulcer) and
histologic (eosinophilic and granuloma) terms.

Considering that the striking clinical phenotype of EGC consists of firm, raised
papules, plaques and nodules and of sharply demarcated ulcers, the main clinical
differential diagnoses include deep bacterial, including mycobacterial, or fungal
infections and neoplasia. Specifically, the main differentials to be taken into account
are squamous cell carcinoma for LU and mast cell tumor, cutaneous lymphocytosis,
and cutaneous infiltration of mammary adenocarcinoma for EP.

In xanthomas, plaques may be whitish-yellow in color, occasionally ulcerated
and occurring on the head and extremities, while they can be hyperkeratotic and
hyperpigmented in papillomas or Bowenoid in situ carcinoma [2, 11]. Erythematous,
eroded, round plaques or nodules clinically indistinguishable from eosinophilic
plaques may be observed in cutaneous lymphocytosis or epitheliotropic cutaneous
lymphoma (Fig. 9) [12, 13].

Nodules may be single or multiple. In terms of usefulness for the diagnosis, relevant clinical features of nodules are location (Table 2), consistency and presence or absence of draining tracts. Soft, fluctuant nodules draining exudate on the trunk may represent sterile nodular panniculitis or mycobacterial infection (Fig. 10). A nodule affecting the bridge of the nose and deforming the cat's profile (roman nose) may suggest cryptococcosis or nasal lymphoma. Swelling of one or more footpads may indicate plasma cell pododermatitis (Fig. 11) [14]. Occasionally, nodules may drain an exudate containing macroscopically visible granules (grains). The grains are usually white in bacterial botryomycosis, yellow in dermatophytic mycetomas and of variable colors in eumycotic mycetomas [5]. A nodule in the

Table 2 Common locations of plaques and nodules in selected feline skin diseases

Plaques	
Distribution	Disease
Head, extremities	Xanthomas
Abdomen, groin, axillae	Eosinophilic plaque
Nodules	
Abdomen, groin, rump	Rapidly growing mycobacterial infections
Abdomen	Nocardiosis
Head, extremities, tail base	Sporotrichosis
Dorsal nose	Cryptococcosis
Caudal thighs, chin, oral cavity, paws	Eosinophilic granuloma
Paws	Calcinosis cutis
Footpads	Plasma cell pododermatitis
Trunk	Sterile nodular panniculitis
Pinnae, eyelids, nasal planum	Squamous cell carcinoma
Abdomen	Lymphangiosarcoma
Interscapular, trunk	Vaccine-site fibrosarcoma
Ear canals, pinnae	Ceruminous cystomatosis

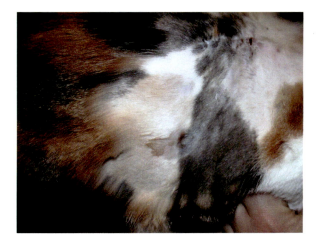

Fig. 10 Fluctuant nodules with small ulcers and draining tract on the flank and rump due to mycobacterial infection (*M. smegmatis*)

Fig. 11 Plasma cell
pododermatitis with
ulceration of the central
metacarpal footpad

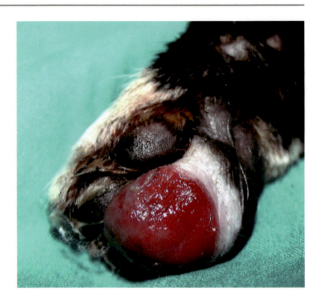

interscapular region or dorsolateral thorax should raise suspicion of vaccine-site fibrosarcoma [8]. Multiple, grey-bluish nodules affecting the face and/or the ear canals and inner aspect of the pinnae may indicate ceruminous cystomatosis, particularly in Persian cats [3].

Diagnostic Algorithm

This section is illustrated in Figs. 12a, b. Red squares with numbers represent the steps of the diagnostic process, explained below.

1 Perform cytology

When the lesion is a nodule or a plaque, cytology is the first diagnostic test to perform during the consultation. Techniques useful to obtain samples for cytological examination from these lesions are fine needle insertion or aspiration and impression smears, if the nodule is ulcerated or if there is a draining tract. However, impression smears may be difficult to interpret in "open" lesions due to potential sample contamination. Cytology allows the clinician to differentiate between inflammatory and neoplastic infiltrates in most cases, and to select the most appropriate diagnostic tests to perform thereafter. When a monomorphous cell population with few or no inflammatory cells is observed, neoplasia should be suspected. Cytological examination allows to further characterize the cell population as composed by epithelial, mesenchymal, or round cells, and, in some cases, it may be diagnostic of a specific neoplasia (e.g., well-differentiated mast cell tumor). In the majority of cases, however, the lesion must be biopsied or excised to perform histopathological examination and properly "name" the tumor.

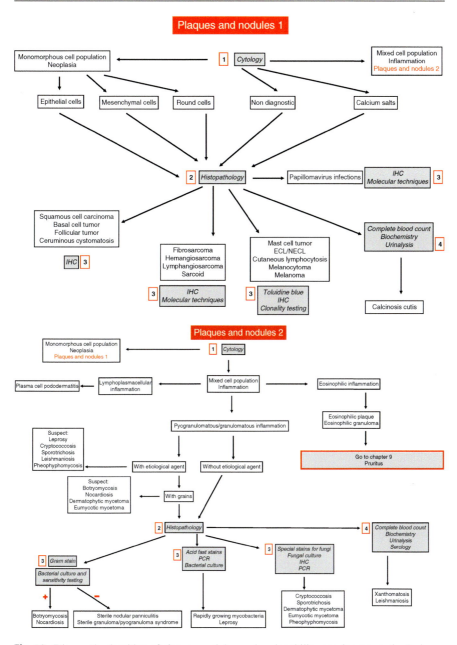

Fig. 12 Diagnostic algorithm of plaques, nodules, and eosinophilic granuloma complex lesions

A mixed cell population observed on cytology indicates inflammation. Inflammatory cells most commonly identified include neutrophils, eosinophils, macrophages, lymphocytes, plasma cells and mast cells, often accompanied by a variable amount of red blood cells. The relative percentage of one cell type in respect to other inflammatory cells is used in cytology to define the different types of inflammation, such as pyogranulomatous (neutrophils and macrophages in variable proportions, epithelioid macrophages and giant histiocytic cells), granulomatous (same as before, with very few or no neutrophils), eosinophilic and lymphoplasmacellular inflammation. Etiological agents such as bacteria, fungi, and parasites can also be detected, as well as calcium salts in *calcinosis cutis* due to chronic kidney disease. Depending on the type of inflammation and the microorganism(s) observed, a diagnosis can be made in some cases. For example, amastigotes of the genus *Leishmania* in the cytoplasm of macrophages indicate leishmaniosis, or yeasts of the genus *Cryptococcus* within pyogranulomatous inflammation are suggestive of cryptococcosis. When the same type of inflammation is observed with unstained, rod-shaped bacteria within macrophages, mycobacterial diseases should be suspected. Whenever the exudate contains grains, cytology from a squashed grain may be useful: filamentous bacteria may suggest nocardiosis, while cocci or rods may point toward a diagnosis of bacterial botryomycosis. If grains appear amorphous and hyphae are detected at the periphery of a grain, dermatophytic mycetoma or eumycotic mycetoma are likely diagnoses. All these diagnoses should be confirmed by histopathological examination and cultures from biopsy samples. Histopathology and culture are also mandatory in all the cases in which cytology reveals granulomatous or pyogranulomatous inflammation without evidence of etiological agents. A predominantly eosinophilic inflammation together with characteristic clinical findings points toward a lesion of the EGC, while lymphoplasmacellular inflammation suggests plasma cell pododermatitis. In case of EGC lesions, the diagnostic workup is irrespective of the clinical form, the distribution of the lesions, and the presence or absence of pruritus (Fig. 9b). Histopathological examination may be performed to confirm the diagnosis.

Cytological examination may also be non-diagnostic, because too few cells are obtained or the sample is heavily contaminated by blood. In ceruminous cystomatosis, for example, a clear fluid containing variable numbers of macrophages can be obtained. In these cases, histopathological examination must be carried out.

2 Take biopsies for histopathological examination

Histopathological examination is mandatory whenever a neoplastic disease is suspected. However, a non-neoplastic nodule or plaque also requires, in the majority of cases, histopathological examination to make or confirm the diagnosis and suggest further diagnostic tests. It must be remembered that the histological appearance of EGC lesions does not always reflect the clinical form and the eosinophilic infiltrate density is quite variable. LU, for instance, is commonly reported as a neutrophilic fibrosing dermatitis rather than an eosinophilic dermatitis. A progression of histological lesions has been described from a dermal eosinophilic infiltrate to fibrosis and neutrophilic ulceration, in a few months, in the LU of the upper lip. These findings might help to explain why LU is infrequently described as an eosinophil-rich dermatitis. Being clinicians reluctant to biopsy the cat's lip, the majority of LU lesions might be present for months at the time they are histologically examined. In fact, LU are mostly biopsied to rule out neoplasia rather than to confirm the diagnosis of EGC. When collecting biopsy samples, some fresh tissue, preferably from the deep portion of the samples, should be stored in a sterile tube and frozen for possible microbial culture, molecular studies or both.

3 Histopathological examination may be diagnostic for neoplasia or, in difficult cases, additional testing may be needed. Depending on the type of tumor identified or suspected on histopathological examination, special stains (e.g., toluidine blue or Giemsa for mast cell tumor), immunohistochemistry, or clonality testing (to differentiate cutaneous lymphocytosis from epitheliotropic cutaneous lymphoma) may be suggested by the pathologist to make the diagnosis.

Special staining, immunohistochemistry and molecular techniques such as polymerase chain reaction (PCR) may also be useful to identify or characterize infectious agents which are difficult to see on "standard" histopathology or to grow on culture. Gram stain is useful to identify bacteria, while acid fast stains such as Ziehl-Neelsen may be necessary to visualize mycobacteria. Periodic-acid of Schiff (PAS) stain is commonly used to identify fungi in tissues. Immunohistochemistry, PCR and/or other molecular techniques may be applied to diagnose papillomavirus infections, mycobacterial diseases and some uncommon fungal infections (pheohyphomycosis). If a deep bacterial or fungal infection is suspected, tissue cultures are recommended to identify the causative microorganisms. Cultures should be preferably performed in specialized Veterinary Labs, and clinicians should inform of the clinical suspicion. In selected cases, sensitivity testing can help to choose the correct antimicrobial treatment. A negative result, together with compatible clinical and histopathological findings, confirms the diagnosis in sterile diseases such as sterile nodular panniculitis and sterile granuloma/pyogranuloma syndrome.

4 Perform complete blood count, biochemistry, urinalysis, and serology

Complete blood count, biochemistry, and urinalysis are useful when a metabolic disease such as xanthomas or *calcinosis cutis* due to renal failure is suspected, based on the results of histopathological examination. If results of cytology and/or histopathology suggest a diagnosis of leishmaniosis, serology should also be performed. FIV and FeLV serology should be also carried out, especially in cats affected by infectious diseases.

References

1. Colombini S, Clay Hodgin E, Foil CS, Hosgood G, Foil LD. Induction of feline flea allergy dermatitis and the incidence and histopathological characteristics of concurrent lip ulcers. Vet Dermatol. 2001;12:155–61.

2. Vogelnest LJ. Skin as a marker of general feline health: cutaneous manifestations of systemic disease. J Feline Med Surg. 2017;19:948–60.
3. Chaitman J, Van der Voerdt A, Bartick TE. Multiple eyelid cysts resembling apocrine hidrocystomas in three Persian cats and one Himalayan cat. Vet Pathol. 1999;36:474–6.
4. Moriello KA, Coyner K, Paterson S, Mignon B. Diagnosis and treatment of dermatophytosis in dogs and cats. Clinical consensus guidelines of the world association for veterinary dermatology. Vet Dermatol. 2017;28:266–e68.
5. Backel K, Cain C. Skin as a marker of general feline health: cutaneous manifestations of infectious disease. J Feline Med Surg. 2017;19:1149–65.
6. Gremiao IDF, Menezes RC, Schubach TMP, Figueiredo ABF, Cavalcanti MCH, Pereira SA. Feline sporotrichosis: epidemiological and clinical aspects. Med Mycol. 2015;53:15–21.
7. Pennisi MG, Cardoso L, Baneth G, Bourdeau P, Koutinas A, Miró G, Oliva G, Solano-Gallego L. LeishVet update and recommendations on feline leishmaniosis. Parasit Vectors. 2015;8:302–20.
8. Hartmann K, Day MJ, Thiry E, Lloret A, Frymus T, Addie D, Boucraut-Baralon C, Egberink H, Gruffydd-Jones T, Horzinek MC, Hosie MJ, Lutz H, Marsilio F, Pennisi MG, Radford AD, Truyen U, Möstl K. Feline injection-site sarcoma: ABCD guidelines on prevention and management. J Feline Med Surg. 2015;17:606–13.
9. Buckley L, Nuttall T. Feline eosinophilic granuloma complex(ities) some clinical clarification. J Feline Med Surg. 2012;14:471–81.
10. Wildermuth BE, Griffin CE, Rosenkrantz WS. Response of feline eosinophilic plaques and lip ulcers to amoxicillin trihydrate–clavulanate potassium therapy: a randomized, double-blind placebo-controlled prospective study. Vet Dermatol. 2011;23:110–e25.
11. Munday JS. Papillomaviruses in felids. Vet J. 2014;199:340–7.
12. Gilbert S, Affolter VK, Gross TL, Moore PF, Ihrke PJ. Clinical, morphological and immunohistochemical characterization of cutaneous lymphocytosis in 23 cats. Vet Dermatol. 2004;15:3–12.
13. Fontaine J, Heimann M, Day MJ. Cutaneous epitheliotropic T-cell lymphoma in the cat: a review of the literature and five new cases. Vet Dermatol. 2011;22:454–61.
14. Dias Pereira P, Faustino AMR. Feline plasma cell pododermatitis: a study of 8 cases. Vet Dermatol. 2003;14:333–7.

Further Readings

"Plaque, nodule". Merriam-Webster Medical Dictionary. http://merriam-webster.com Accessed 31 Jan 2018.
Albanese F. Canine and feline skin cytology. Cham: Springer International Publishing; 2017.
Goldsmith LA, Katz SI, Gilchrest BA, Paller AS, Leffell DJ, Wolff K. Fitzpatrick's dermatology in general medicine. 8th ed. New York: The McGraw-Hill Companies; 2012.
Gross TL, Ihrke PJ, Walder EJ, Affolter VK. Skin diseases of the dog and cat. Clinical and histopathologic diagnosis. 2nd ed. Oxford: Blackwell Publishing; 2005.
Miller WH, Griffin CE, Muller CKL. Kirk's small animal dermatology. 7th ed. St. Louis: Elsevier; 2013.
Noli C, Toma S. Dermatologia del cane e del gatto. 2nd ed. Vermezzo: Poletto Editore; 2011.

Excoriations, Erosions and Ulcers

Silvia Colombo

Abstract

Excoriations, erosions and ulcers are relatively common lesions in the cat, and in general, they are quite non-specific. Excoriations are, by definition, self-induced lesions due to scratching, while erosions and ulcers develop spontaneously. Full thickness skin wounds are very suggestive of cutaneous asthenia or acquired skin fragility syndrome, depending on the cat's age. Erosions and ulcers are often secondarily infected and may appear more severe because of pruritus due to infection. A peculiar feline clinical presentation, common in allergic diseases, is "head and neck pruritus," with excoriations and ulcers being self-induced. This presentation is usually investigated following the diagnostic approach to pruritus. The most relevant clinical feature of erosions and ulcers is their location, which may be helpful for the diagnosis. In general, histopathology is the most important diagnostic test to make a specific diagnosis in erosive/ulcerative feline skin diseases.

Definitions

An excoriation is a superficial abrasion of the epidermis that results from scratching, or, less commonly, from licking or biting. It is a self-induced lesion and may show a linear pattern, directly reflecting its pathogenesis.

An erosion is a superficial, moist, circumscribed lesion that results from loss of a part or all of the epidermis and does not involve the dermis. An erosion does not bleed and heals without scarring.

An ulcer is a circumscribed skin defect in which the epidermis and at least the superficial dermis have been lost, and it is deeper than the erosion. The ulcer involves

S. Colombo (✉)
Servizi Dermatologici Veterinari, Legnano, Italy

© Springer Nature Switzerland AG 2020   137
C. Noli, S. Colombo (eds.), *Feline Dermatology*,
https://doi.org/10.1007/978-3-030-29836-4_7

also the adnexa and may heal with scarring. Further features used to describe an ulcer relate to its margins, surface and presence of exudate eventually covering its bottom. The margins, for example, may be thickened, regular or irregular, and the bottom may be clean, hemorrhagic or necrotic. There may be a crust or purulent exudate covering the ulcerated area.

Erosion and ulcer are difficult to be differentiated clinically, because the depth of a skin defect can only be defined with certainty by histopathological examination. For this reason, when describing a typical lesion or a disease, the words erosive and ulcerative are always used together.

Pathogenesis

The pathogenetic mechanisms underlying the formation of erosions and ulcers vary from external trauma, to congenital defects causing reduced skin resistance, to direct infectious or autoimmune damage to the skin. Erosions and ulcers are, in the vast majority of cases, complicated by self-induced trauma due to pruritus and/or by secondary infections. The clinical appearance of any erosive and ulcerative disease may therefore evolve, and lesions may become deeper and more severe.

A peculiar feline clinical presentation, common in allergic diseases, is ulceration due to "head and neck pruritus" (Fig. 1) [1]. Cats scratch using their hind paws and claws and may cause severe and extensive ulcers to these locations.

Despite its name, the feline-specific lesion known as "indolent ulcer" or "lip ulcer" (Fig. 2) is an ulcerated plaque and is described in Chapter, Plaques, Nodules and Eosinophilic Granuloma Complex Lesions [2].

Fig. 1 Head and neck pruritus in an allergic cat

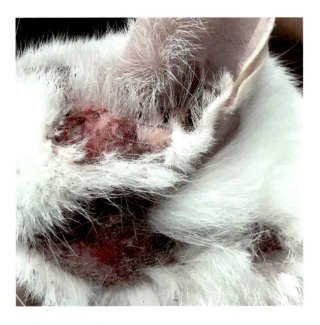

Fig. 2 Severe, bilateral
indolent ulcer with necrotic
material in the center

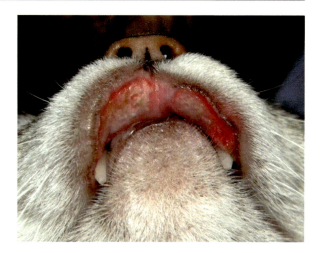

In diseases such as cutaneous asthenia or acquired skin fragility syndrome, full thickness lacerations and skin detachment occur following minor trauma, and the lesions should be better described as wounds. Table 1 lists selected causes of excoriations, erosions and ulcers in cats.

Diagnostic Approach

Signalment and History

Erosive/ulcerative skin diseases such as cutaneous asthenia and dystrophic or junctional epidermolysis bullosa are congenital and present at birth or shortly thereafter [3, 4]. In other cases, the disease has a delayed onset but is clinically apparent in young adult cats of specific breeds (idiopathic facial dermatitis of Persian and Himalayan cats, ulcerative nasal dermatitis of Bengal cats) [5, 6]. Full thickness wounds following minor trauma may occur in senior to geriatric cats in acquired skin fragility syndrome, which may be caused by hyperadrenocorticism (Fig. 3) or other diseases [7, 8]. Traumatic excoriations or wounds may be seen more often in tomcats, while neoplastic diseases are more common in older cats.

History is very relevant for the diagnosis of feline erosive/ulcerative diseases. Previous or concurrent respiratory clinical signs may suggest herpesvirus dermatitis, while an outdoor lifestyle may predispose the cat to trauma, deep bacterial, mycobacterial or fungal infections, or, in white cats, to squamous cell carcinoma (Fig. 4) [9, 10]. Previous or concurrent drug administration should prompt the clinician to include adverse drug reactions and toxic epidermal necrolysis among the differential diagnoses, particularly if the lesions have a sudden onset [11]. Finally, the presence of pruritus may be a relevant information as it may be typical of some

Table 1 Selected causes of excoriations, erosions and ulcers

Excoriations	Self-trauma
Erosions/ulcers	Herpesvirus dermatitis
	Leprosy
	Rapidly growing mycobacterial infections
	Subcutaneous fungal infections
	Systemic fungal infections
	Myiasis
	Leishmaniosis
	Head and neck pruritus[a] (Table 3)
	Adverse drug reactions
	Pemphigus foliaceus
	Pemphigus vulgaris
	Vesicular diseases of the dermo-epidermal junction
	Erythema multiforme
	Toxic epidermal necrolysis
	Vasculitis
	Hyperadrenocorticism/acquired skin fragility syndrome
	Idiopathic facial dermatitis of Persian and Himalayan cats
	Ulcerative nasal dermatitis of Bengal cats
	Junctional/dystrophic epidermolysis bullosa
	Cutaneous asthenia
	Trauma
	Indolent ulcer[b]
	Idiopathic/behavioral ulcerative dermatitis
	Plasma cell pododermatitis
	Squamous cell carcinoma

[a]Chapter, Pruritus
[b]Chapter, Plaques, Nodules and Eosinophilic Granuloma Complex Lesions

Fig. 3 Full thickness skin wound in a geriatric cat with hyperadrenocorticism

Fig. 4 Squamous cell carcinoma involving the lower eyelid and nose of a white cat

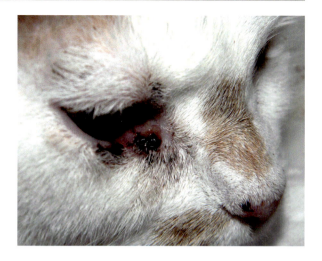

diseases such as idiopathic/behavioral ulcerative dermatitis and of the clinical pattern described as "head and neck pruritus" [1, 12]. However, one must remember that pruritus may also be due to secondary infections of the erosion/ulcer.

Clinical Presentation

Erosions and ulcers are relatively common lesions in the cat and are quite nonspecific. Accompanying primary or secondary lesions are uncommon, with the exception of crusts covering the erosions/ulcers. Nodules and plaques may have an eroded or ulcerated surface, as it happens in indolent ulcer and eosinophilic plaque. On the other hand, full thickness wounds are very specific once a traumatic etiology has been ruled out. Skin tears with minimal or no bleeding, occurring following minor traction, suggest cutaneous asthenia or acquired skin fragility syndrome, depending on the patient's age [3, 7, 8, 13]. The concurrent presence of skin hyperextensibility is a feature of cutaneous asthenia, while thin, irregular scars represent resolved lesions and may be observed in both conditions.

The most useful clinical features of erosions/ulcers are the lesions' location (Table 2) and the presence or absence of pruritus. The face is the most common site for erosions/ulcers due to herpesvirus (Fig. 5) or calicivirus infections, occasionally with oral cavity involvement [9, 10]. Idiopathic facial dermatitis of Persian and Himalayan cats is initially characterized by an accumulation of adherent black material around the eyes, nose and mouth, and inflamed, eroded/ulcerated skin lesions underneath the exudate develop with time [5]. These lesions may be severely pruritic and secondary infections are common. Multiple erosions/ulcers, covered by crusts, on the tip of the pinnae, eyelids and/or nasal planum of a white cat should prompt the clinician to investigate squamous cell carcinoma. A hyperkeratotic, scaly, occasionally ulcerated nasal planum in Bengal cats has been reported and is thought to be a congenital disease [6]. Application of spot-on products to prevent

Table 2 Common locations of erosions/ulcers in selected feline skin diseases

Distribution	Disease
Oral cavity	Herpesvirus dermatitis Pemphigus vulgaris Vesicular diseases of the dermo-epidermal junction
Abdomen, groin	Rapidly growing mycobacterial infections Eosinophilic plaques
Upper lip	Indolent ulcer
Dorsal neck	Adverse drug reaction (spot-on, injection) Idiopathic/behavioral ulcerative dermatitis "Head and neck pruritus"
Footpads	Plasma cell pododermatitis
Trunk	Hyperadrenocorticism Acquired skin fragility syndrome
Nasal planum	Ulcerative nasal dermatitis of Bengal cats Squamous cell carcinoma Pemphigus foliaceus
Muzzle	Idiopathic facial dermatitis of Persian and Himalayan cats Pemphigus foliaceus Herpesvirus dermatitis "Head and neck pruritus"
Pinnae	Squamous cell carcinoma Pemphigus foliaceus
Eyelids	Squamous cell carcinoma Pemphigus vulgaris Vesicular diseases of the dermo-epidermal junction

Fig. 5 Large erosion/ulcer on the muzzle of a cat with herpesvirus infection

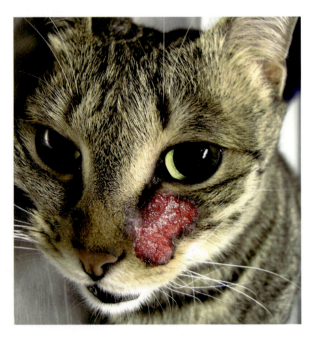

ectoparasites may cause erosions/ulcers on the dorsal neck. Ulcerative lesions and draining tracts discharging exudate on the abdomen may be observed in rapidly growing mycobacterial infections [10]. Severely swollen, ulcerated metacarpal and/ or metatarsal footpads are suggestive of plasma cell pododermatitis [14]. In erythema multiforme, maculopapular lesions evolving to erosions/ulcers and crusts usually present with a generalized distribution [11].

Head and Neck Pruritus

Pruritus and self-induced erosions/ulcers involving the head, pinnae and neck are commonly observed and represent a peculiar clinical presentation in the feline species [1]. Variably sized erosions and ulcers are caused by the cat's scratching with the hind paws, and the owner is usually well aware of the cat's pruritus. These lesions may be very severe, secondary infections are often present, and their depth may reach the subcutis (Fig. 6). Other feline presentations typically observed in pruritic skin diseases, such as miliary dermatitis and self-induced alopecia, may be concurrently observed. Differential diagnoses of head and neck pruritus are listed in Table 3. Head and neck pruritus should be investigated following the diagnostic approach to pruritus (Chapter, Pruritus).

Idiopathic/behavioral ulcerative dermatitis (Fig. 7), which presents as a very severe and extremely pruritic, usually single crusted ulceration affecting the dorsal neck, deserves a specific comment, since its etiopathogenesis is controversial. Suggested causes of idiopathic ulcerative dermatitis involve allergic diseases,

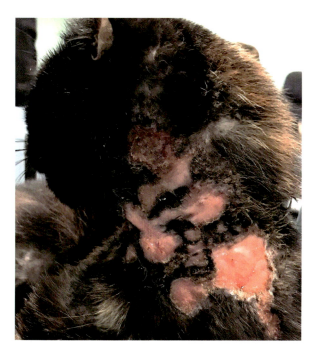

Fig. 6 Very severe erosions/ulcers in a food allergic cat

Table 3 Differential diagnoses of head and neck pruritus

Disease
Herpesvirus dermatitis
Dermatophytosis
Notoedric mange
Otodectic mange
Demodicosis (*Demodex gatoi*)
Trombiculiasis
Lynxacarus infestation
Flea-bite hypersensitivity
Adverse reaction to food
Feline atopic syndrome
Mosquito-bite hypersensitivity
Adverse drug reaction
Pemphigus foliaceus
Idiopathic facial dermatitis of Persian and Himalayan cats
Idiopathic/behavioral ulcerative dermatitis

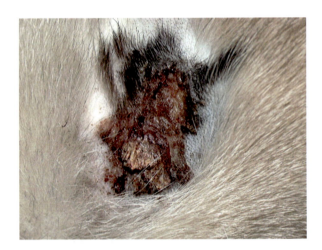

Fig. 7 Idiopathic/behavioral ulcerative dermatitis on the dorsal neck

secondary infections, neurological diseases and a behavioral disorder, although most cases, as the name suggests, are idiopathic [12].

Diagnostic Algorithm

This section is illustrated in Fig. 8. Red squares with numbers represent the steps of the diagnostic process, as explained below.

1 Consider history and clinical examination.

When examining a cat with erosions/ulcers, the first step is to separate these lesions from full thickness wounds or skin lacerations following minor trauma,

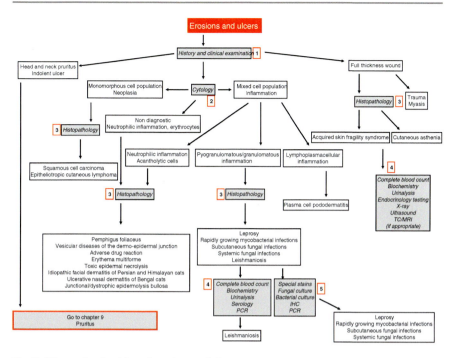

Fig. 8 Diagnostic algorithm of erosions and ulcers

such as manual traction. Extreme fragility of the skin only occurs in two conditions, in cats. The first one is cutaneous asthenia, clinically apparent in kittens or young cats, while the second one occurs in old felines and gathers different diseases under the name "acquired skin fragility syndrome" [3, 7, 8, 13]. History would also help us, in most cases, to decide if the wound occurred following a major trauma such as a car accident. The presence of insect larvae in the wound indicates myiasis. If the cat is pruritic on the head and neck, or the main lesion is the "indolent" lip ulcer or an eroded plaque (Chapter, Plaques, Nodules and Eosinophilic Granuloma Complex Lesions), particularly when associated with self-induced alopecia, one should follow the diagnostic approach to pruritus, described in Chapter, Pruritus.

2 Perform cytology.

Cytologic examination of erosions/ulcers is often disappointing, because in most cases one can only observe non-specific findings, such as red blood cells and neutrophils. When neutrophils are present admixed with acantholytic cells, the main clinical suspicion is pemphigus foliaceus (Fig. 9). If a mixed cell population comprising large numbers of plasma cells and lymphocytes are observed and the cytology sample comes from a footpad, the diagnosis is plasma cell pododermatitis [14]. Eosinophils may be observed on samples taken from the tiny erosions underneath the crusts in miliary dermatitis, or from the eroded surface of an eosinophilic plaque. Pyogranulomatous inflammation is also a non-specific

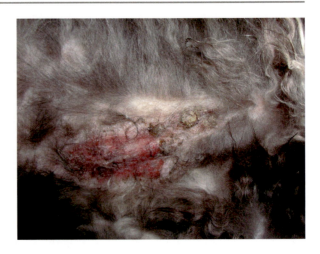

Fig. 9 Pemphigus foliaceus in a domestic short-haired cat

cytological picture; however, it is more commonly observed in infectious diseases such as mycobacterial or fungal infections and leishmaniosis. Occasionally, a monomorphous cell population is seen on cytology, and this finding may suggest a neoplastic disease.

3   Perform histopathology.

Histopathology is of paramount importance in erosive/ulcerative feline skin diseases. First of all, it can confirm the diagnosis of neoplasia and acquired skin fragility syndrome. In cutaneous asthenia, histopathological comparison with a skin sample obtained from a cat of the same age and from the same site may be required, as well as special stains and electron microscopy. The majority of autoimmune, immune-mediated and idiopathic erosive/ulcerative skin diseases can be diagnosed by histopathology. In case of infectious diseases, a standard histopathological examination (hematoxylin-eosin, H&E) is in most cases only indicative, due to the difficulty of identifying the etiological agent without further diagnostic procedures, such as special stains or immunohistochemistry.

4   Perform blood testing, serology, urinalysis and diagnostic imaging.

In an old cat presenting with full thickness wounds and a histopathological diagnosis of acquired skin fragility syndrome, it is necessary to identify the causative disease in order to attempt a treatment. Skin fragility syndrome is often caused by hyperadrenocorticism; however, severe cachexia, diabetes mellitus, hepatic lipidosis or inflammatory and neoplastic diseases affecting the liver, nephrosis and some infectious diseases have all been reported [8, 13]. When history, clinical examination and histopathology suggest leishmaniosis, the diagnostic process should be completed by complete blood count, biochemistry, urinalysis, serology and/or PCR [10].

5   As said before, standard histopathology (H&E staining) may in some cases be suggestive of an infectious disease, without confirming a specific diagnosis. In these cases, further testing is mandatory and special stains (PAS for fungi, Ziehl-Neelsen for acid-fast bacteria, immunohistochemistry for leishmania and viruses), cultures and PCR should be requested depending on the case.

References

1. Hobi S, Linek M, Marignac G, Olivry T, et al. Clinical characteristics and causes of pruritus in cats: a multicentre study on feline hypersensitivity-associated dermatoses. Vet Dermatol. 2011;22:406–13.
2. Buckley L, Nuttall T. Feline Eosinophilic Granuloma Complex(ITIES): some clinical clarification. J Fel Med Surg. 2012;14:471–81.
3. Hansen N, Foster SF, Burrows AK, Mackie J, Malik R. Cutaneous asthenia (Ehlers–Danlos-like syndrome) of Burmese cats. J Feline Med Surg. 2015;17:954–63.
4. Medeiros GX, Riet-Correa F. Epidermolysis bullosa in animals: a review. Vet Dermatol. 2015;26:3–e2.
5. Bond R, Curtis CF, Ferguson EA, Mason IS, Rest J. An idiopathic facial dermatitis of Persian cats. Vet Dermatol. 2000;11:35–41.
6. Bergvall K. A novel ulcerative nasal dermatitis of Bengal cats. Vet Dermatol. 2004;15:28.
7. Boland LA, Barrs VR. Peculiarities of feline hyperadrenocorticism: update on diagnosis and treatment. J Feline Med Surg. 2017;19:933–47.
8. Furiani N, Porcellato I, Brachelente C. Reversible and cachexia-associated feline skin fragility syndrome in three cats. Vet Dermatol. 2017;28:508–e121.
9. Hargis AM, Ginn PE. Feline herpesvirus 1-associated facial and nasal dermatitis and stomatitis in domestic cats. Vet Clin North Am Small Anim Pract. 1999;29(6):1281–90.
10. Backel K, Cain C. Skin as a marker of general feline health: cutaneous manifestations of infectious disease. J Feline Med Surg. 2017;19:1149–65.
11. Yager JA. Erythema multiforme, Stevens–Johnson syndrome and toxic epidermal necrolysis: a comparative review. Vet Dermatol. 2014;25:406–e64.
12. Titeux E, Gilbert C, Briand A, Cochet-Faivre N. From feline idiopathic ulcerative dermatitis to feline behavioural ulcerative dermatitis: grooming repetitive behaviors indicators of poor welfare in cats. Front Vet Sci. 2018; https://doi.org/10.3389/fvets.2018.00081.
13. Vogelnest LJ. Skin as a marker of general feline health: cutaneous manifestations of systemic disease. J Feline Med Surg. 2017;19:948–60.
14. Dias Pereira P, Faustino AMR. Feline plasma cell pododermatitis: a study of 8 cases. Vet Dermatol. 2003;14:333–7.

General References

For definitions: Merriam-Webster Medical Dictionary. http://merriam-webster.com Accessed 10 May 2018.

Albanese F. Canine and feline skin cytology. Cham: Springer International Publishing; 2017.

Goldsmith LA, Katz SI, Gilchrest BA, Paller AS, Leffell DJ, Wolff K. Fitzpatrick's dermatology in general medicine. 8th ed. New York: The McGraw-Hill Companies; 2012.

Miller WH, Griffin CE, Campbell KL. Muller & Kirk's Small Animal Dermatology. 7th ed. St. Louis: Elsevier; 2013.

Noli C, Toma S. Dermatologia del cane e del gatto. 2nd ed. Vermezzo: Poletto Editore; 2011.

Scaling

Silvia Colombo

Abstract

Exfoliative diseases in cats are clinically characterized by dry or greasy scaling and, less commonly, by follicular casts. In normal skin, there is a continuous turnover of cells, with new keratinocytes being produced in the basal layer and migrating upward to become non-nucleated corneocytes in the stratum corneum. Corneocytes are shed in the environment and are not visible to the naked eye. When this process is abnormal, scales become macroscopically visible. The most common cause of scaling in cats is poor grooming, usually associated with older age, obesity or concurrent systemic diseases. Greasy scaling is often associated with *Malassezia* overgrowth, while follicular casts are rare in the feline species. The diagnostic approach involves ruling out ectoparasitic diseases and dermatophytosis, evaluating the presence or absence of *Malassezia* spp. by cytology, and assessing the cat's general health status, especially in older patients. Histopathology is usually required to make the diagnosis of the majority of exfoliative dermatoses.

Definitions

A scale is a small, thin, dry piece of cornified layer detaching from the skin, and scaling is the process of shedding scales. In English, scale and squame as well as scaling, desquamation and exfoliation are synonyms and are used indifferently. The informal term used to describe scaling is dandruff. In normal conditions, exfoliation occurs continuously, without the formation of visible scaling. Desquamation becomes visible when it occurs in increased amount, because the epidermal differentiation is abnormal.

S. Colombo (✉)
Servizi Dermatologici Veterinari, Legnano, Italy

© Springer Nature Switzerland AG 2020
C. Noli, S. Colombo (eds.), *Feline Dermatology*,
https://doi.org/10.1007/978-3-030-29836-4_8

149

Scales may be further characterized as dry or greasy, and their color may be white, silver, yellow, brown or grey, depending on the causative disease. Dry scaling is common in cats, while greasy scaling is observed only in a few skin diseases. Scales are also often described as pityriasiform, which means small, thin, whitish and similar to oat bran, or psoriasiform, a term used to describe larger, thicker and often silvery scales. Scales arranged in a circle are described as an epidermal collarette and are rarely observed in cats. The epidermal collarette is the final evolutive stage of a papule or a pustule (Chapter, Papules, Pustules, Furuncles and Crusts).

A follicular cast is an accumulation of keratin and follicular content which adheres to the hair shaft, protruding from the follicular ostium. This material often glues together a tuft of hair or may accumulate around a single hair shaft. Follicular casts are very uncommon in cats but may represent a useful clinical hint toward the diagnosis.

Pathogenesis

In the normal skin, there is a continuous turnover of cells, with new keratinocytes being produced in the basal layer, maturing in the spinous layer, and dying to become corneocytes in the horny layer. Corneocytes are shed in the environment and are not visible to the naked eye. In abnormal situations, scaling becomes obvious because the corneocytes detach in larger clusters. This may be due to increased production or reduced shedding of the horny layer or to abnormalities of the superficial lipid film, that covers and protects the skin surface. An increased production of the horny layer may occur in congenital diseases such as ichthyosis or primary seborrhea; however, these conditions are extremely rare in cats [1]. More commonly, the increased thickness of the horny layer is a response to an external insult, such as sunrays damage in solar dermatitis (Fig. 1), which may evolve into actinic keratosis and squamous cell carcinoma, or to ectoparasites feeding on the skin surface in cheyletiellosis. Another pathogenetic mechanism underlying scaling dermatoses is the infiltration of inflammatory or neoplastic cells in the skin, as it happens in erythema multiforme (Fig. 2), exfoliative dermatitis with/without thymoma or epitheliotropic cutaneous lymphoma [2–4].

In cats, reduced shedding of horny layer is usually due to poor grooming, more commonly in older or obese cats or in cats affected by systemic diseases such as diabetes mellitus or hyperthyroidism. The lipid film that protects the skin is, at least in part, produced by the sebaceous glands. Diseases affecting and destroying these glands, such as sebaceous adenitis or leishmaniosis, may present with scaling [3, 5].

Although very uncommonly, scaling in cats may also be greasy. This may occur due to excessive production of glandular secretions in primary seborrhea, a very rare disease in cats, and in the more common tail gland hyperplasia, also known as

Fig. 1 Scaling, erythema, and mild crusting on the pinna of a white cat with solar dermatitis

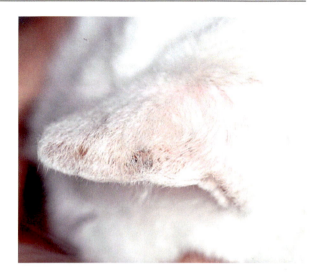

Fig. 2 Scaling on the footpads of a cat affected by erythema multiforme

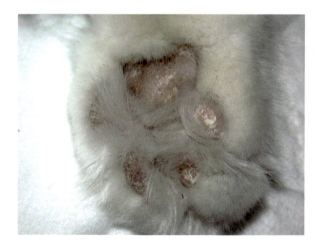

stud tail (Fig. 3). *Malassezia* overgrowth may be related to sebaceous gland hyperplasia and abnormalities of the superficial lipid film, and consequently presents with greasy scaling [6, 7].

Follicular casting is uncommon in the feline species. It may be the clinical expression of follicular damage, as in demodicosis and dermatophytosis, or of destruction of the sebaceous glands in sebaceous adenitis [3]. A rare, recently reported congenital disease called sebaceous gland dysplasia occurs in kittens and is clinically characterized by generalized hypotrichosis, scaling and follicular casts [8]. Selected feline diseases presenting with scaling are listed in Table 1.

Fig. 3 Greasy seborrhea
on the dorsal tail of a
Persian cat with tail gland
hyperplasia

Diagnostic Approach

Signalment and History

Exfoliative diseases such as dermatophytosis (Fig. 4) or cheyletiellosis are commonly observed in kittens or in environmental conditions of crowding, such as breeding colonies or pet shops. Congenital diseases presenting with dry or greasy scaling and/or follicular casting are observed in kittens, although primary seborrhea and sebaceous gland dysplasia are extremely rare diseases [2, 9]. The most common cause of dry scaling in senior and geriatric cats is poor grooming, which may be due to obesity (Fig. 5) or concurrent systemic diseases such as chronic renal insufficiency, hyperthyroidism or diabetes mellitus. Less commonly, aged

Table 1 Selected diseases presenting with dry or greasy scaling and follicular casts in cats

Poor grooming due to obesity or systemic disease
Cheyletiellosis
Demodicosis
Dermatophytosis
Malassezia overgrowth
Leishmaniosis
Adverse drug reaction
Erythema multiforme
Sebaceous gland dysplasia
Primary seborrhea
Sebaceous adenitis
Plasma cell pododermatitis
Solar dermatitis
Tail gland hyperplasia
Exfoliative dermatitis (thymoma-associated or not)
Epitheliotropic cutaneous lymphoma

Fig. 4 Scaling and erythema on the margin of the pinna of a kitten affected by dermatophytosis

cats may be affected by neoplastic diseases or paraneoplastic syndromes [4, 9]. Dermatophytosis should always be considered in a Persian cat presenting with scaling and alopecia, regardless of age and lifestyle (Fig. 6). White cats and cats with white ears and/or muzzle are predisposed to solar dermatitis, if they are allowed outdoors and like laying in the sun.

History in adult to older cats should always include concurrently or previously administered drugs which may cause adverse drug reactions. Finally, in a cat presenting with scaling, FIV and FeLV statuses should be evaluated. A FeLV-associated giant cell dermatosis has been reported to cause generalized, severe exfoliation, and both viral infections may predispose the cat to other infectious diseases [10].

Fig. 5 Mild generalized
scaling in a geriatric and
obese cat

Fig. 6 Focal alopecia and
scaling in a Persian cat
with dermatophytosis

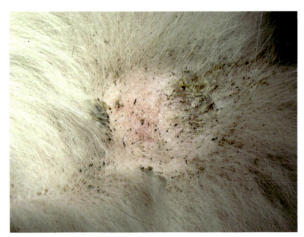

Clinical Presentation

Dry scaling of variable severity is common in cats, and further clinical features
should be considered to help listing the differential diagnoses. Generalized or
dorsally distributed pityriasiform scaling in an older cat may be simply due to
poor grooming, while in a recently acquired kitten, it may suggest cheyletiellosis,
especially if pruritus is also reported. Exfoliative erythroderma, a clinical presentation characterized by scaling, erythema, and often alopecia, has been reported in
senior or geriatric cats affected by epitheliotropic cutaneous lymphoma, although
the disease is rare in the cat [4]. Exfoliative dermatoses in cats are often generalized, with a few exceptions. When scaling is associated to focal or multifocal
alopecia, dermatophytosis is a possible diagnosis. Mild scaling and erythema on
the pinnae of a white cat should prompt the clinician to include solar dermatitis
in the differentials.

Fig. 7 Severe psoriasiform scaling in thymoma-associated exfoliative dermatitis

Psoriasiform scaling is uncommon in cats. In middle-age to older cats, non-pruritic, severe, generalized, psoriasiform scaling with a history of starting on the head and neck and associated alopecia and erythema may be suggestive of thymoma or non-thymoma-associated exfoliative dermatitis (Fig. 7) [9, 11]. In thymoma-associated cases, coughing, dyspnea, depression, anorexia and weight loss are usually observed after the skin lesions. Psoriasiform scaling, follicular casting and alopecia with a generalized distribution, associated with deposition of dark-colored debris on the eyelids, may be consistent with sebaceous adenitis (Fig. 8), an extremely rare disease in the cat [3]. Scaling limited to the footpads may be a clinical feature of plasma cell pododermatitis (Fig. 9) [12].

Localized or generalized greasy scaling, erythema, pruritus and rancid smell may suggest *Malassezia* overgrowth (Fig. 10). This disease may be observed both in young, allergic cats and in older felines with severe systemic diseases, neoplasia or paraneoplastic syndromes [6, 7]. Greasy scaling on the dorsal aspect of the tail is the clinical presentation of tail gland hyperplasia, also known as stud tail.

Diagnostic Algorithm

This section is illustrated in Fig. 11. Red squares with numbers represent the steps of the diagnostic process, as explained below.

1 Perform skin scrapings and microscopic examination of scales and skin debris.
   The diagnostic approach to scaling in cats begins with simple tests to diagnose or rule out ectoparasites. Multiple skin scrapings should be performed to
diagnose or rule out demodicosis, and fungal spores can also be seen surrounding and invading fragments of hair shafts in dermatophytosis. Cheyletiellosis
may also be diagnosed with skin scrapings, although the most commonly used
test is microscopic examination of acetate tape strips, after collecting scales

Fig. 8 Severe exfoliation and alopecia on the ventral trunk of a cat affected by sebaceous adenitis

Fig. 9 Scaling on the footpads in a mild case of plasma cell pododermatitis

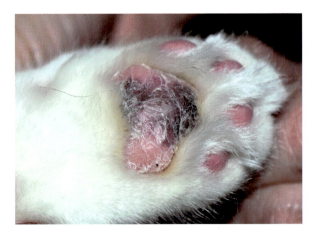

Fig. 10 Greasy, brown scaling in the interdigital spaces of an allergic Devon Rex cat with *Malassezia* overgrowth

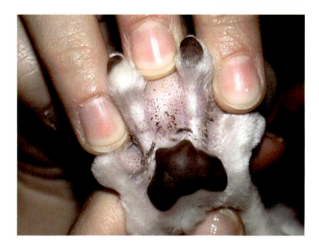

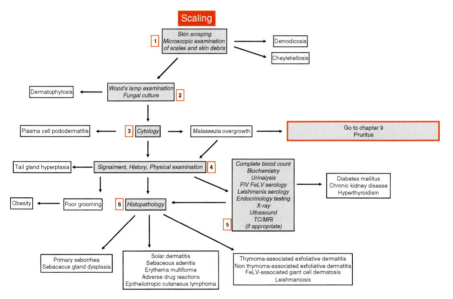

Fig. 11 Algorithm of the approach to feline scaling

directly from the cat's coat or from material collected from the examination table after vigorous stroking of the coat.

2 Perform Wood's lamp examination and fungal culture.

These two diagnostic tests, taken together, are diagnostic for dermatophytosis or, if negative results are obtained, are helpful to rule it out. Since dermatophytosis is common in cats, a fungal culture is appropriate in all cases presenting with scaling.

3 Perform cytology.

Cytology is particularly useful when greasy scaling is a presenting sign. Samples may be taken by impression smear, using a cotton swab or a piece of acetate tape to look for *Malassezia* yeasts. Since these yeasts may be identified in both young, allergic cats and older felines with systemic diseases, neoplasia or paraneoplastic syndromes, further investigations should always be carried out based on signalment, history and clinical examination. When a scaly footpad has to be sampled because plasma cell pododermatitis is suspected, fine needle capillary suction and aspiration are the preferred techniques.

4 Consider the patient's signalment, history and physical examination.

In a young to adult cat presenting with greasy scaling on the dorsal tail, after ruling out ectoparasitic and fungal diseases, the diagnosis of tail gland hyperplasia is straightforward. When the patient is a senior or geriatric cat and presents with generalized dry scaling, poor grooming is a major differential. An aged cat may groom with difficulty because it is obese or because it suffers from a metabolic disease. Depending on other clinical signs, when identified on general physical examination, a variety of diagnostic tests may be appropriate.

5 Perform blood testing, urinalysis and diagnostic imaging.

In an older cat, basic information should always be obtained by taking a blood sample and a urine sample for complete blood count, biochemistry, urinalysis and serum total thyroxine (T4) concentrations. This will be useful also if sedation or general anesthesia is planned for biopsies. FIV and FeLV serology must be carried out if a cutaneous disease linked to one of these viruses is suspected, although this may become obvious only after histopathological examination. The same applies to serology for leishmaniosis, a rare disease in cats. Thoracic radiography and/or CT/MRI may be diagnostic for thymoma, which is often associated with exfoliative dermatitis.

6 Take biopsies for histopathological examination.

Histopathological examination usually confirms the diagnosis, whether the clinician is facing a congenital disease or an acquired one. Figure 11 summarizes the most important exfoliative disorders which require biopsies for the diagnosis.

References

1. Paradis M, Scott DW. Hereditary primary seborrhea oleosa in Persian cats. Feline Pract. 1990;18:17–20.
2. Yager JA. Erythema multiforme, Stevens-Johnson syndrome and toxic epidermal necrolysis: a comparative review. Vet Dermatol. 2014;25:406–e64.
3. Noli C, Toma S. Case report three cases of immune-mediated adnexal skin disease treated with cyclosporin. Vet Dermatol. 2006;17(1):85–92.
4. Fontaine J, Heimann M, Day MJ. Cutaneous epitheliotropic T-cell lymphoma in the cat : a review of the literature and five new cases. Vet Dermatol. 2011;22(5):454–61.

5. Pennisi MG, Cardoso L, Baneth G, Bourdeau P, Koutinas A, Miró G, et al. LeishVet update and recommendations on feline leishmaniosis. Parasit Vectors. 2015;8:1–18.
6. Mauldin EA, Morris DO, Goldschmidt MH. Retrospective study: the presence of Malassezia in feline skin biopsies. A clinicopathological study. Vet Dermatol. 2002;13:7–14.
7. Ordeix L, Galeotti F, Scarampella F, Dedola C, Bardagi M, Romano E, Fondati A. Malassezia spp. overgrowth in allergic cats. Vet Dermatol. 2007;18:316–23.
8. Yager JA, Gross TL, Shearer D, Rothstein E, Power H, Sinke JD, Kraus H, Gram D, Cowper E, Foster A, Welle M. Abnormal sebaceous gland differentiation in 10 kittens ('sebaceous gland dysplasia') associated with generalized hypotrichosis and scaling. Vet Dermatol. 2012;23:136–e30.
9. Turek MM. Cutaneous paraneoplastic syndromes in dogs and cats : a review of the literature. Vet Dermatol. 2003;14:279–96.
10. Gross TL, Clark EG, Hargis AM, Head LL, Hainesh DM. Giant cell dermatosis in FeLV-positive cats. Vet Dermatol. 1993;4:117–22.
11. Brachelente C, vonTscharner C, Favrot C, Linek M, Silvia R, Wilhelm S, et al. Non thymoma-associated exfoliative dermatitis in 18 cats. Vet Dermatol. 2015;26:40–e13.
12. Dias Pereira P, Faustino AMR. Feline plasma cell pododermatitis: a study of 8 cases. Vet Dermatol. 2003;14:333–7.

General References

For definitions: Merriam-Webster Medical Dictionary. http://merriam-webster.com Accessed 10 May 2018.
Albanese F. Canine and feline skin cytology. Cham: Springer International Publishing; 2017.
Goldsmith LA, Katz SI, Gilchrest BA, Paller AS, Leffell DJ, Wolff K. Fitzpatrick's dermatology in general medicine. 8th ed. New York: The McGraw-Hill Companies; 2012.
Gross TL, Ihrke PJ, Walder EJ, Affolter VK. Skin diseases of the dog and cat. Clinical and histopathologic diagnosis. 2nd ed. Oxford: Blackwell Publishing; 2005.
Miller WH, Griffin CE, Campbell KL. Muller & Kirk's small animal dermatology. 7th ed. St. Louis: Elsevier; 2013.
Noli C, Toma S. Dermatologia del cane e del gatto. 2nd ed. Vermezzo: Poletto Editore; 2011.

Pruritus

Silvia Colombo

Abstract

Pruritus, also called itching, is an irritating sensation in the upper surface of the skin, thought to result from stimulation of sensory nerve endings. Pruritus is common in cats and can be further classified based on its distribution (localized or generalized), location on the animal's body and severity (mild, moderate, or severe). From a clinical point of view, pruritus in cats is most commonly caused by ectoparasitic, allergic, infectious or immune-mediated diseases. Cats manifest pruritus by overgrooming, which makes it particularly difficult to recognize and evaluate, and to be differentiated from pain or a behavioral problem. In a very young cat, ectoparasites and dermatophytosis are common, while in an adult cat, allergic and immune-mediated skin diseases should also be considered. History is relevant for concurrent drug administration or systemic disease and for severity and seasonality of pruritus. Most pruritic cats present with one (or more) of four clinical patterns, namely, head and neck pruritus, miliary dermatitis, self-induced alopecia and the eosinophilic granuloma complex. The diagnostic approach to pruritus should always be carefully followed in each of its steps in order to make a correct diagnosis.

Definitions

Pruritus, also called itching, is defined as an unpleasant feeling that causes the desire to scratch. In the vast majority of cases, the irritating sensation develops in the skin and is thought to result from stimulation of sensory nerve endings. In rare cases, pruritus may originate in the central nervous system. Pruritus is extremely common in veterinary dermatology and may be due to a wide variety of diseases.

S. Colombo (✉)
Servizi Dermatologici Veterinari, Legnano, Italy

© Springer Nature Switzerland AG 2020
C. Noli, S. Colombo (eds.), *Feline Dermatology*,
https://doi.org/10.1007/978-3-030-29836-4_9

It is an obvious clinical sign in dogs, while it can be very subtle in cats because it can be expressed as excessive grooming, which is a normal feline behavior, or because cats often hide from owners when they feel the desire to scratch. Pruritus is further classified based on its distribution (localized or generalized), location on the animal's body and severity (mild, moderate, or severe).

Pathogenesis

The vast majority of the information about mechanisms, pathways, and mediators of pruritus comes from human or laboratory animal studies and has been reviewed elsewhere [1, 2]. From a clinical point of view, pruritus in cats is usually caused by ectoparasitic, allergic, infectious or immune-mediated diseases and can be worsened by concurrent factors such as stress, boredom, dry skin or high environmental temperature (Table 1). Although pruritus may be interpreted, in some cases, as a defense mechanism (scratching or licking to remove ectoparasites), skin lesions often occur as a consequence of behaviors carried out by the cat to relieve it.

Cats manifest pruritus by overgrooming, in other words by increasing the frequency and intensity of a normal, programmed feline behavior. Cats groom to keep their skin and hair coat clean and healthy, to remove ectoparasites and dirt, to control their body temperature and to relieve tension or stress [3, 4]. Grooming in cats

Table 1 Selected causes of pruritus in cats	Pruritus	Herpesvirus infection
		Superficial pyoderma
		Complicated chin acne
		Flea infestation
		Cheyletiellosis
		Notoedric mange
		Otodectic mange
		Demodicosis (*Demodex gatoi*)
		Trombiculiasis
		Dermatophytosis
		Malassezia overgrowth
		Flea-bite hypersensitivity
		Adverse reaction to food
		Feline atopic syndrome
		Mosquito-bite hypersensitivity
		Allergic/irritant contact dermatitis
		Adverse drug reaction
		Hyperthyroidism
		Pemphigus foliaceus
		Lymphocytic mural folliculitis
		Familial pedal eosinophilic dermatosis
		Urticaria pigmentosa-like dermatitis
		Idiopathic facial dermatitis of Persian and Himalayan cats

include oral grooming, which is stroking the tongue through the pelage and nibbling with the incisor teeth, and scratch grooming, which is scratching with the hind paws [5]. According to one study, indoor, ectoparasite-free adult cats spend 50% of their time sleeping or resting. Of the time spent awake, oral grooming accounts for about 1 hour per day and scratch grooming for about 1 minute per day. Ninety-one percent of oral grooming is directed to multiple body regions, while scratch grooming is always directed to single regions [5].

Being the increased expression of a physiological behavior, overgrooming is often not recognized by the owner or not interpreted as a sign of pruritus, pain or stress. Moreover, cats tend to express their discomfort by hiding away from owners, who may not be aware of their pet's overgrooming. For all these reasons, pruritus can be particularly difficult to recognize and evaluate in cats and to be differentiated from pain (e.g., licking the abdomen due to cystitis) or a behavioral problem (causing licking, scratching, or hair pulling).

Idiopathic ulcerative dermatitis presents as a very severe and extremely pruritic, usually single, crusted ulceration affecting the dorsal neck (Fig. 1) in which pruritus, neuropathic itch and behavioral disorder have all been considered relevant in the disease pathogenesis. Idiopathic ulcerative dermatitis is diagnosed by exclusion of diseases which may induce pruritus to the dorsal neck, such as allergies and

Fig. 1 Idiopathic/ behavioral ulcerative dermatitis on the dorsal neck

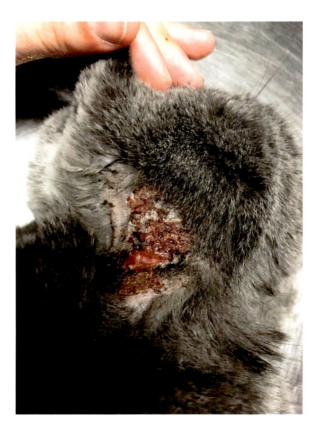

Table 2 Examples of non-dermatological diseases to be differentiated from pruritic skin diseases

| Feline idiopathic cystitis |
| Psychogenic alopecia |
| Feline idiopathic/behavioral ulcerative dermatitis |
| Feline orofacial pain syndrome |
| Feline hyperesthesia syndrome |
| Localized neuropathies |

ectoparasites. A recent case report proposed that idiopathic ulcerative dermatitis may be a neuropathic itch syndrome, and the cat responded completely to topiramate, an anti-epileptic drug [6]. However, the same disease has also been investigated from a behavioral point of view. In 13 affected cats, in an open, uncontrolled study, environmental enrichment and improvement of overall welfare led to resolution of skin lesions, and psychotropic drugs were employed only in one case. The authors of this study proposed to change the disease name to feline behavioral ulcerative dermatitis [7].

Finally, an orofacial pain syndrome has been reported in cats. This syndrome occurs more commonly, although not exclusively, in Burmese cats and is clinically characterized by self-trauma to the face and oral cavity and occasionally by mutilation of the tongue. The disease may be associated with teeth eruption, dental disease, and stress and is suspected to be a neuropathic disorder, which should be considered in cats presenting with severe facial excoriations or ulcers [8].

In conclusion, overgrooming, which includes excessive licking and excessive scratching, may be the expression of non-dermatological diseases, which should always be considered when listing the differential diagnoses in an apparently "dermatological" case presenting for "pruritus" (Table 2).

Diagnostic Approach

Signalment and History

Depending on the cat's age, some diseases may be more likely than others. In a very young cat, ectoparasites and dermatophytosis are common, especially if the kitten has been found as a stray or adopted from a cattery, where crowding plays an important role too. Some diseases, such as dermatophytosis, cheyletiellosis and notoedric mange, are very contagious. These diseases may affect in contact animals as well as people, and questions about the presence of skin lesions on other pets or family members are mandatory. In an adult cat, allergic and immune-mediated skin diseases should also be considered, while in an older cat, hyperthyroidism may occur and explain the excessive grooming. Older cats may present with pruritus due to *Malassezia* overgrowth, which can be the marker of an underlying systemic disease or paraneoplastic syndrome (Fig. 2) [9].

Fig. 2 Alopecia and
brown greasy material
typical of *Malassezia*
overgrowth in an old cat
with pancreatic
paraneoplastic alopecia

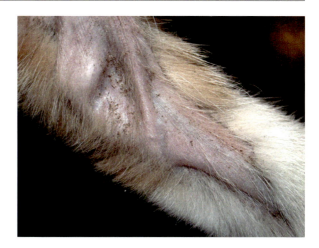

Fig. 3 Alopecia and
excoriations on the head
and pinnae of a cat affected
by seasonal feline atopic
syndrome

History should include drugs administered for other diseases, which may cause adverse drug reactions, and ectoparasite prevention. Immunosuppressive therapy or systemic disease may predispose the cat to dermatophytosis if it gets exposed, for example, because the owner adopts a new kitten. Seasonality of pruritus may be useful to limit the list of differentials: ectoparasites and seasonal feline atopic syndrome are more likely in a cat scratching in spring and summer (Fig. 3). Severity of pruritus should also be analyzed in depth because some diseases are characterized by extremely severe pruritus (notoedric mange) while in others pruritus may be very mild (cheyletiellosis, dermatophytosis).

Persian cats of any age are predisposed to dermatophytosis. An older Persian cat may be affected by dermatophytosis, if it is an asymptomatic carrier, without the need for contact with a diseased animal [10].

Clinical Presentation

In cats, pruritus is expressed by overgrooming; however, only increased scratching is easily recognized by the owner. Since they scratch with their hind paws, excoriations usually involve areas that the cat can reach, such as the face, ears, head and neck. The so called "head and neck pruritus" is a common clinical presentation in pruritic cats (Fig. 4) [11]. Variably sized excoriations, erosions and ulcers in these locations may be very severe and deep and are often secondarily infected. This clinical presentation is specifically addressed in Chapter, Excoriations, Erosions and Ulcers.

Less obviously, alopecia may be caused by an overgrooming, pruritic cat [11]. Self-induced alopecia is characterized by the presence of very short hair fragments which can be observed by looking closely at the skin or with the help of a magnifying lens. Hair cannot be easily epilated. The alopecic area usually has very well-defined margins, with abrupt change to normal hair, and involves parts of the body that can be reached by the tongue (Fig. 5). Self-induced alopecia is described in Chapter, Alopecia.

Miliary dermatitis is a peculiar feline clinical presentation also associated with pruritus [11]. It is characterized by small, crusted papules "resembling millet seeds," hence the name, which are more easily felt by touching through the haircoat than

Fig. 4 Excoriations on the head of an allergic cat

Fig. 5 Self-induced alopecia on the abdomen of a cat with flea-bite hypersensitivity

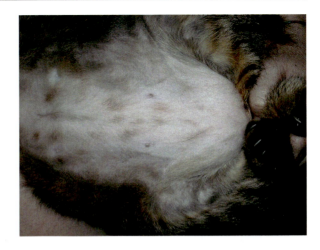

Fig. 6 Small, crusted papules typical of miliary dermatitis

seen (Fig. 6). Miliary dermatitis is often associated with self-induced alopecia and is addressed in Chapter, Papules, Pustules, Furuncles and Crusts.

Another clinical pattern associated with pruritus is a group of lesions named eosinophilic granuloma complex or eosinophilic dermatitides (Figs. 7 and 8) [12]. These conditions, or clinical presentations, are often caused by allergic diseases and are discussed in Chapter, Plaques, Nodules and Eosinophilic Granuloma Complex Lesions.

Many pruritic diseases, in cats, can be associated with one or more of the four previously described clinical patterns and/or with recurrent otitis (Chapter, Otitis). However, each disease has its own preferential distribution of pruritus and lesions on the animal body (Table 3). Some other unusual presentations are also associated with pruritus and may be caused by hypersensitivity reactions to food or environmental allergens, at least in some cases. Lymphocytic mural folliculitis, for example, is a histopathological reaction pattern occasionally identified in allergic cats

Fig. 7 Eosinophilic
plaques on the abdomen

Fig. 8 Bilateral indolent
ulcer in a domestic
short-haired cat

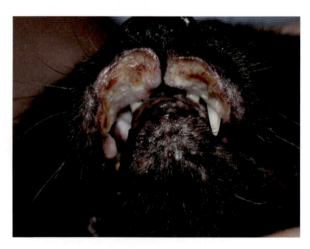

presenting with pruritus, localized or generalized, partial or complete alopecia and scaling (Fig. 9) [13]. Urticaria pigmentosa-like dermatitis occurs in Devon Rex or Sphynx cats and is clinically characterized by an erythematous to hyperpigmented papular eruption, which is often pruritic (Fig. 10) [14, 15].

Pruritic and non-pruritic dermatoses may be secondarily infected by bacteria or yeasts. Although this occurs in cats much less frequently than in dogs, one should always keep into consideration and diagnose/rule out these diseases when examining a pruritic cat [9, 16].

Diagnostic Algorithm

This section is illustrated in Fig. 11. Red squares with numbers represent the steps of the diagnostic process, as explained below.

Table 3 Common locations of selected feline skin diseases associated with pruritus

Locations	Disease
Face	Herpesvirus infection
Chin	Complicated chin acne
Rump	Flea infestation
Thorax, abdomen	Demodicosis (*Demodex gatoi*)
Dorsum	Cheyletiellosis
Ear canal	Otodectic mange
Pinnae, paws, abdomen	Trombiculiasis
Pinnae, face, neck, paws, perineum	Notoedric mange
Head, pinnae, paws, tail, generalized	Dermatophytosis
Chin, claw folds, face, ear canal, generalized	*Malassezia* overgrowth
Rump	Flea-bite hypersensitivity
Dorsal nose, pinnae, paws	Mosquito-bite hypersensitivity
Abdomen, medial thighs, head, neck	Other allergic diseases
Head, pinnae, claw folds, abdomen	Pemphigus foliaceus
Paws	Familial pedal eosinophilic dermatosis
Face	Idiopathic facial dermatitis of Persian and Himalayan cats

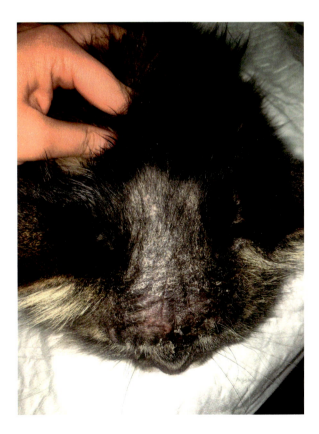

Fig. 9 Alopecia, scaling and hyperpigmentation on the head of a cat with lymphocytic mural folliculitis

Fig. 10 Coalescing,
crusted, and non-crusted
papules in a Sphynx cat
with urticaria pigmentosa-
like dermatitis

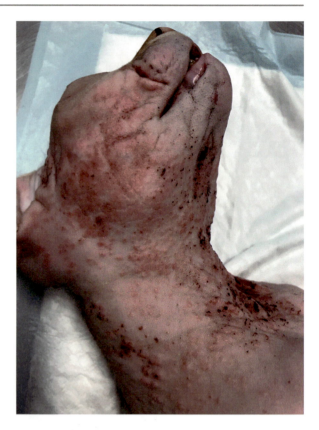

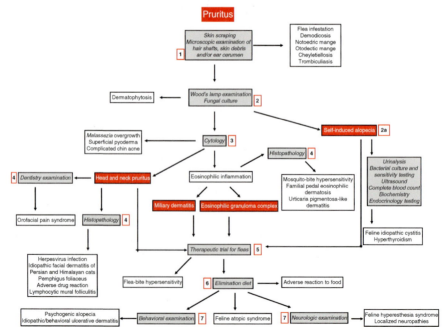

Fig. 11 Diagnostic algorithm of pruritus

1 Perform skin scrapings and microscopic examination of hair, skin debris and/or ear cerumen.

In the diagnostic approach to pruritus, it is mandatory to begin with simple tests to diagnose or rule out ectoparasites. Multiple skin scrapings are useful for notoedric mange and demodicosis, and fungal spores can be seen surrounding and invading fragments of hair shafts in dermatophytosis. Cheyletiellosis and trombiculiasis may be diagnosed by microscopic examination of acetate tape strips, after collecting samples directly from the cat's coat or, for *Cheyletiella* spp., from the material collected from the examination table after vigorous stroking of the coat. This latter way of collecting specimens may also be used to find flea dirt, together with coat combing. If pruritus is mainly affecting the ears, microscopic examination of ear cerumen is required to diagnose otodectic mange.

2 Perform Wood's lamp examination and fungal culture.

The second step is to rule out or diagnose dermatophytosis, which may have already been suspected after microscopic examination of hair shafts. Wood's lamp examination may support the diagnostic hypothesis and fungal culture is required to confirm dermatophytosis. If negative results are obtained, these tests are helpful to rule it out. Since dermatophytosis is common in cats, a fungal culture is appropriate in all cases, although pruritus can be of variable severity.

2a If the clinical presentation is self-induced alopecia involving the groin and abdomen, urinalysis, bacterial culture and sensitivity testing, and abdominal ultrasound should be performed to investigate feline idiopathic cystitis or other urinary tract diseases. Self-induced alopecia in an old cat may also be caused by hyperthyroidism, and hematology, biochemistry and endocrine testing should be carried out in this specific situation.

3 Perform cytology.

Cytology is the easiest and quickest diagnostic test to support the clinical suspicion of diseases characterized by eosinophilic inflammation, which are numerous and very common in cats. Eosinophilic plaque and granuloma and miliary dermatitis are often clinical patterns of allergy, characterized by eosinophilic inflammation, and the diagnostic process should continue to identify the primary disease. On the other hand, familial pedal eosinophilic dermatosis, mosquito-bite hypersensitivity and urticaria pigmentosa-like dermatitis show eosinophilic inflammation on cytology and are specific diseases which should be confirmed by histopathological examination. Cytology is also important because secondary bacterial or yeast infections may complicate the primary disease and increase the severity of pruritus, although this occurs less frequently in cats compared to dogs. Samples may be taken by impression smear, using a cotton swab or a piece of acetate tape to look for *Malassezia* yeasts, bacteria and inflammatory cells. Finally, identification of acantholytic cells admixed with neutrophils may suggest pemphigus foliaceus.

⌐4⌐ Perform histopathology.

As anticipated, histopathology is the confirmatory diagnostic test for many feline diseases cytologically characterized by eosinophilic inflammation. When pruritus affects mainly the face, histopathological examination is required to diagnose idiopathic facial dermatitis of Persian and Himalayan cats and herpesvirus infection, although in this latter disease immunohistochemistry may be necessary to confirm the etiology. In cases with clinical manifestation of severe self-trauma to the face and oral cavity, a dental examination may be required to investigate orofacial pain syndrome. Histopathological examination is useful to diagnose pemphigus foliaceus and, together with history, adverse drug reactions.

⌐5⌐ Perform a therapeutic trial for fleas.

In the majority of cases presenting for pruritus, ectoparasites and dermatophytosis can be ruled out at the beginning of the diagnostic approach, and cytological examination only shows secondary infections or eosinophilic inflammation, which is neither specific nor particularly useful. These cases usually present with one of the four clinical patterns typical of pruritus and should be investigated in a systematic way. The first step is a therapeutic trial for fleas, which may have not been identified during the initial investigations for ectoparasites. A positive response to the trial suggests flea-bite hypersensitivity.

⌐6⌐ Perform an elimination diet.

If the therapeutic trial for fleas is unsuccessful, the second step is performing an elimination diet with novel protein sources or a hydrolyzed diet, to be carried out for at least 8 weeks. If the cat improves on the diet, challenge with the previous food is required to diagnose an adverse reaction to food.

⌐7⌐ After ruling out food as the cause of pruritus, the clinician is left with a possible diagnosis of feline atopic syndrome. There are different treatment options for environmental allergy in cats and the diagnosis is confirmed by response to treatment.

Depending on history and clinical presentation, in some cases, a behavioral problem can be suspected, especially if the cat presents with self-induced alopecia or ulcerative dermatitis affecting the dorsal neck. In other cases, a neurologic problem such as feline hyperesthesia syndrome may be considered and needs to be investigated. These conditions are usually addressed only when all the other differentials have been ruled out, and the cat does not respond to treatment for feline atopic syndrome.

References

1. Metz M, Grundmann S, Stander S. Pruritus: an overview of current concepts. Vet Dermatol. 2011;22:121–31.
2. Gnirs K, Prelaud P. Cutaneous manifestations of neurological diseases: review of neuropathophysiology and diseases causing pruritus. Vet Dermatol. 2005;16:137–46.
3. Beaver BV. Feline behavior. A guide for veterinarians. Second edition. St. Louis: WB Saunders; 2003.

4. Bowen J, Heath S. Behaviour problems in small animals. Practical advice for the veterinary team. Philadelphia: Elsevier Saunders; 2005.
5. Eckstein RA, Hart BL. The organization and control of grooming in cats. Appl Anim Behav Sci. 2000;68:131–40.
6. Grant D, Rusbridge C. Topiramate in the management of feline idiopathic ulcerative dermatitis in a two-year-old cat. Vet Dermatol. 2014;25:226–e60.
7. Titeux E, Gilbert C, Briand A, Cochet-Faivre N. From feline idiopathic ulcerative dermatitis to feline behavioral ulcerative dermatitis: grooming repetitive behavior indicators of poor welfare in cats. Front Vet Sci. 2018; https://doi.org/10.3389/fvets.2018.00081.
8. Rusbridge C, Heath S, Gunn-Moore D, Knowler SP, Johnston N, McFadyen AK. Feline orofacial pain syndrome (FOPS): a retrospective study of 113 cases. J Feline Med Surg. 2010;12:498–508.
9. Mauldin EA, Morris DO, Goldschmidt MH. Retrospective study: the presence of *Malassezia* in feline skin biopsies. A clinicopathological study. Vet Dermatol. 2002;13:7–14.
10. Moriello KA, Coyner K, Paterson S, Mignon B. Diagnosis and treatment of dermatophytosis in dogs and cats.: clinical consensus guidelines of the world Association for Veterinary Dermatology. Vet Dermatol. 2017;28(3):266–8.
11. Hobi S, Linek M, Marignac G, et al. Clinical characteristics and causes of pruritus in cats: a multicentre study on feline hypersensitivity-associated dermatoses. Vet Dermatol. 2011;22:406–13.
12. Buckley L, Nuttall T. Feline eosinophilic granuloma complex(ITIES): some clinical clarification. J Feline Med Surg. 2012;14:471–81.
13. Rosenberg AS, Scott DW, Erb HN, McDonough SP. Infiltrative lymphocytic mural folliculitis: a histopathological reaction pattern in skin-biopsy specimens from cats with allergic skin disease. J Feline Med Surg. 2010;12:80–5.
14. Noli C, Colombo S, Abramo F, Scarampella F. Papular eosinophilic/mastocytic dermatitis (feline urticaria pigmentosa) in Devon rex cats: a distinct disease entity or a histopathological reaction pattern? Vet Dermatol. 2004;15:253–9.
15. Ngo J, Morren MA, Bodemer C, Heimann M, Fontaine J. Feline maculopapular cutaneous mastocytosis: a retrospective study of 13 cases and proposal for a new classification. J Feline Med Surg. https://doi.org/10.1177/1098612X18776141.
16. Yu HW, Vogelnest L. Feline superficial pyoderma: a retrospective study of 52 cases (2001-2011). Vet Dermatol. 2012;23:448–e86.

General References

For definitions: Merriam-Webster Medical Dictionary. http://merriam-webster.com Accessed 10 May 2018.
Albanese F. Canine and feline skin cytology. Cham: Springer International Publishing; 2017.
Goldsmith LA, Katz SI, Gilchrest BA, Paller AS, Leffell DJ, Wolff K. Fitzpatrick's Dermatology in General Medicine. 8th ed. New York: The McGraw-Hill Companies; 2012.
Miller WH, Griffin CE, Muller CKL. Kirk's small animal dermatology. 7th ed. St. Louis: Elsevier; 2013.
Noli C, Toma S. Dermatologia del cane e del gatto. 2nd ed. Vermezzo: Poletto Editore; 2011.

Otitis

Tim Nuttall

Abstract

Otitis externa and media are common in cats, although almost all infections are secondary. The underlying conditions must be diagnosed and managed for resolution. The approach to feline otitis is different from that in dogs. There are important differences in the ear anatomy of dogs and cats, although there is less breed variation among cats. The role of primary, secondary, predisposing and perpetuating (PSPP) factors is less clear in feline otitis, with fewer predisposing and perpetuating problems. The primary aetiology of otitis is different from dogs with less of a role for hypersensitivity dermatoses. There are a variety of cat-specific conditions, including inflammatory polyps, cystoadenomatosis, and proliferative and necrotising otitis. This chapter will describe the anatomy and physiology of feline ears, how to use clinical examination, cytology, culture and imaging in diagnosis, ear cleaning, treatment of otitis externa and otitis media, and the diagnosis and management of specific ear conditions in cats.

Introduction

Feline otitis requires a different approach to diagnosis and treatment compared to canine otitis. The aetiology is different and many conditions are specific to cats. Otitis is less common and less well associated with common skin diseases in cats than in dogs. For example, otitis has been reported in 16% [1] to 20% [2] of cats with hypersensitivity dermatitis. In contrast, up to 80% of dogs with atopic dermatitis may suffer from recurrent otitis externa [3]. In addition, the PSPP (primary, secondary, predisposing and perpetuating) approach is less useful in cats compared

T. Nuttall (✉)
Royal (Dick) School of Veterinary Studies, University of Edinburgh, Roslin, UK
e-mail: tim.nuttall@ed.ac.uk

© Springer Nature Switzerland AG 2020
C. Noli, S. Colombo (eds.), *Feline Dermatology*,
https://doi.org/10.1007/978-3-030-29836-4_10

to dogs. While it is true that ear infections are invariably secondary and that there are a number of defined primary causes of otitis in cats, the role of predisposing factors and perpetuating problems in initiating otitis and the progression to chronic disease is less clear. Finally, cats can be more sensitive to ototoxicity than dogs, and topical treatments and ear cleaners must be selected and used with care.

It is very important to recognise that cats with recurrent ear infections have an underlying problem – they do not have an antimicrobial deficiency! Overuse of antimicrobials can mask the primary condition (which can get more severe and difficult to manage) and select for antimicrobial resistance (which can complicate future treatment). Successful management requires diagnosis and appropriate treatment of the primary cause.

Anatomy and Physiology

Feline ear canal anatomy is similar to that in dogs, although with much less variation among breeds and individuals [4, 5] (Chapter, Structure and Function of the Skin).

The Pinnae

With the exception of some breeds such as the Scottish Fold, cats have an upright pinna. The skin of the pinnae and ear canals is continuous with the skin over the rest of the body. The dorsal surface is covered with densely haired skin that is loosely attached to the underlying cartilage. The skin on the ventral surface is tightly attached to the cartilage. Hairs arising on the rostral margin of the pinna fold back across the pinnal surface and ear canal opening and probably limit entry of foreign material (Fig. 1). Touching these hairs can elicit ear flicking or head shaking. They can be extensive in long-haired breeds. The ventral surface of the pinna is otherwise hairless. There is a complex array of cartilage folds at the base of the pinna. The most important to recognise is the tragus (Fig. 1), which forms the lateral margin of the vertical canal. The opening to the vertical ear canal is located behind the tragus.

Ear Canals and Cerumen

The skin of the ear canals is continuous with that of the ventral pinna. It is thin, hairless and tightly opposed to the underlying ear canal cartilages. The auricular cartilage is continuous with the pinna and forms the vertical ear canal (Fig. 2). It is loosely embedded within connective tissue and is reasonably mobile. The auricular cartilage is connected to the annular cartilage by fibrous tissue, which gives some flexibility and mobility. The junction is visible as a dorsal ridge extending into the ear canal lumen (Fig. 3). The annular cartilage forms the short horizontal ear canal

Fig. 1 The inner surface of the pinna (rostral is to the bottom left; caudal to the upper right). Blue arrow – hairs arising from the rostral pinna margin and extending caudally across the pinna; black arrow – the tragus

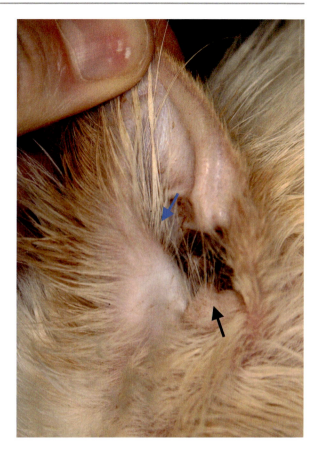

and is connected to the bony external auditory meatus by fibrous tissue. This gives some flexibility but the horizontal ear canal is much less mobile than the vertical ear canal and pinna. The horizontal ear canal is typically 6–9 mm in diameter, which can limit access of otoscope cones.

Cerumen is typically sparse in healthy cat ears and has a film-like creamy consistency compared to dogs. Outward, lateral migration of the stratum corneum moves the cerumen, desquamated cells and debris to the opening of the ear canals. Here, the cerumen dries, detaches and is removed through normal grooming.

Tympanic Membrane

The tympanic membrane (Fig. 4) separates the external ear canals from the middle ear. It is housed within the bony external auditory meatus (Fig. 2) facing horizontally at an angle from dorso-lateral to ventro-medial, although it can be near vertical in some cats. The dorsal pars flaccida is narrow and much less prominent than in

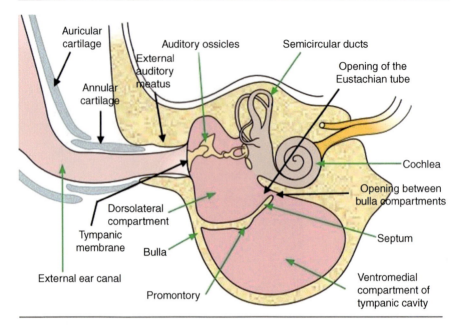

Fig. 2 Schematic diagram of the external ear canals, middle ear and inner ear

Fig. 3 View of the base of the vertical ear canal. The arrow indicates the ridge or shelf formed at the junction of the vertical and horizontal ear canals

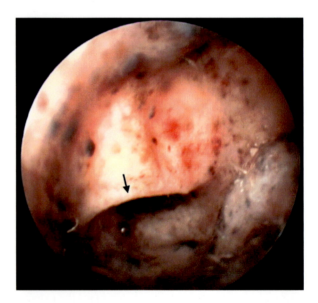

dogs. The pars tensa forms a thin, grey-white, translucent membrane with prominent striations radiating from the manubrium of the malleus. The malleus forms a white straight or slightly curved structure running ventrally from the rostro-dorsal edge of the tympanic membrane. The concavity of the curve faces rostrally, but this is less marked than in dogs. The malleus is surrounded by a ring of blood vessels.

Fig. 4 Normal tympanic membrane in a healthy cat. A – attachment of the malleus; B – *pars flaccida*; C – *pars tensa*

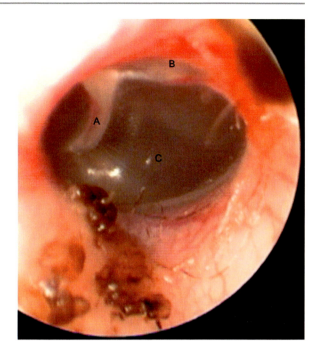

Middle Ear

The middle ear (Fig. 2) is divided by a bony shelf into ventro-medial (pars endotympanica) and dorso-lateral (pars tympanica) compartments. The dorso-lateral compartment is bounded laterally by the tympanic membrane, dorsally by the epitympanic recess and medially by the lateral wall of the cochlea. The epitympanic recess contains the auditory ossicles and openings to the cochlear (auditory window) and vestibular systems (vestibular window). The Eustachian tube opens in the medial wall of the middle ear near the origin of the bony shelf and connects the middle ear with the pharynx. A sympathetic nerve trunk runs close to the opening of the Eustachian tube. It is quite superficial in cats and vulnerable when removing polyps, flushing the middle ear and/or treating otitis (which can result in Horner's syndrome). The middle ear is lined with a mucous epithelium continuous with that of the Eustachian tube and pharynx. The Eustachian tube allows air pressure to equalise across the tympanic membrane and mucus from the middle ear to drain into the pharynx. A small opening in the bony shelf allows mucus from the ventral bulla compartment to drain through the Eustachian tube.

General Approach to Otitis

1. Identify and treat the infection.
   - Use ear cytology to identify *Malassezia*, bacteria and inflammatory cells.
   - When necessary, perform culture to identify the microorganisms and their antimicrobial susceptibility.

2. Identify and manage the primary cause.
   - Perform a thorough history and full clinical examination.
   - Examine the ear canals for clinical lesions, type of discharge, mites, foreign bodies, inflammatory polyps and tumours.
   - Examine the tympanic membrane for signs of rupture and otitis media.
3. Identify and manage any predisposing and perpetuating causes.
   - Clinically assess the extent and severity of chronic inflammatory changes.
   - Consider radiographs, CT scans or MRI scans.

Diagnostic Procedures

Examining the Ear

The ears should be carefully examined for any abnormalities. Certain clinical signs are often highly specific for primary conditions in feline otitis (see below). Healthy ears should be freely mobile, pliable, non-painful and non-pruritic and have little to no discharge. Very firm, immobile ear canals are often irreversibly fibrosed and/or mineralised. The skin should be pale, thin and smooth. The ear canals should be open with a thin, smooth and pale lining, little to no discharge and a normal tympanic membrane. However, the narrow diameter of cats' ear canals can make otoscopic examination of the horizontal ear canal difficult. The tympanic membrane may not be easily visible in diseased ears. This should not preclude treatment, although the possibility of a ruptured membrane should be considered. Full examination may therefore need sedation, anaesthesia, removal of any discharge and/or treatment to open the ear canal lumen. Despite this, careful examination in a conscious cat can help identify the cause of the otitis as well as the extent and severity of secondary changes without otoscopic examination. The nature of the discharge can indicate the likely problem and/or infection, but the appearance of the dried discharge at the opening of the ears can be misleading and fresh material from the ear canals should be evaluated.

Cytology

Cytology is mandatory in all cases of otitis where it can identify the most likely organisms. This is particularly useful in mixed infections, where culture may identify several organisms with different susceptibility patterns. Samples should be collected from the ear canal with swabs or curettes. Mites can be found in material collected in mineral oil under x40 magnification. Air-dried or heat-fixed material stained with a modified Wright–Giemsa stain can be examined under high magnification (x400 or x1000 oil immersion) to see cells and microorganisms.

It is important to recognise biofilms, which form thick, dark and slimy exudates. They appear as variably thick veil-like material (Fig. 5a, b) on cytology that may obscure bacteria and cells. Biofilms are becoming increasingly common in otitis. Many microorganisms can produce biofilms, which facilitate adherence to the ear canal epidermis, middle ear lining and surrounding hairs. They also inhibit antimicrobials by providing physical protection and by altering metabolic activity. The net result is that much higher antimicrobial concentrations predicted by *in vitro* testing are required – in effect the minimum inhibitory concentration (MIC) is increased. Specific anti-biofilm measures (see below) should be used in all cases where they are present.

The numbers of yeast, cocci, rods, neutrophils and epithelial cells should be quantified. Staphylococci (Fig. 5b) and *Malassezia* spp. (Fig. 6) are straightforward to identify, and a good estimate of their probable sensitivity can be made based on knowledge of local resistance patterns and previous treatment. Gram-negative bacteria (Fig. 7), however, are harder to differentiate on cytology alone.

Fig. 5 (**a**) The typical dark, thick and slimy appearance of biofilm from the ear canals; (**b**) – a biofilm-associated staphylococcal infection showing coccoid bacteria embedded in a veil-like matrix (Rapi-Diff II® stain; ×400 magnification). Staphylococci typically form pairs, groups of 4 and irregular clumps

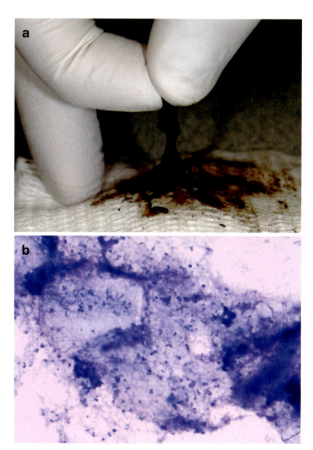

Fig. 6 *Malassezia* otitis
with large numbers of
budding yeasts (Rapi-Diff
II® stain; ×400
magnification)

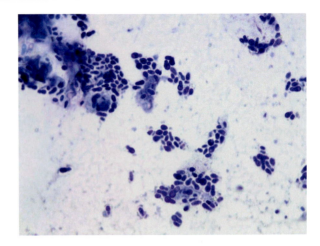

Fig. 7 *Pseudomonas* otitis
with large numbers of rod
bacteria (A) and
neutrophils (B) embedded
in a slimy matrix (C)
consistent with a biofilm.
All bacteria that stain with
modified Wright–Giemsa
stains will stain dark
blue – their Gram-negative
identity can only be
inferred (Rapi-Diff II®
stain; ×400 magnification)

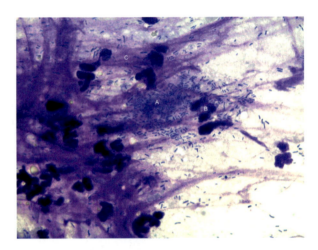

Bacterial Culture and Antimicrobial Susceptibility Testing

Bacterial culture and susceptibility testing is not necessary in most cases of otitis externa if topical therapy is used, as the antibiotic concentration in these products greatly exceeds the minimum inhibitory concentration (MIC) for the bacteria.

Great care must be taken in interpreting antibiotic susceptibility and resistance results because the susceptibility–resistance breakpoints are based on tissue concentrations after systemic treatment. This does not necessarily mean that the bacteria are resistant to the antimicrobial because sufficiently high antibiotic levels, as achieved with topical therapy, may still exceed the MIC. Sensitivity data are therefore very poorly predictive of the response to topical drugs because concentrations in the ear canal are much higher. The response to treatment is best assessed using clinical signs and cytology.

Bacterial cultures can help identify the bacteria involved in the infection. This can be useful for less common organisms that are hard to differentiate on cytology and/or where the antimicrobial susceptibility patterns are less predictable.

Antibiotic sensitivity data should be used to predict the efficacy of systemic drugs in case these are used (e.g. in case of otitis media), although the concentration in the ear tissues may be low and high doses are needed. In addition, biofilms that inhibit antimicrobial penetration and efficacy effectively increase the *in vivo* MIC, meaning that *in vitro* tests over-estimate antimicrobial susceptibility.

Diagnostic Imaging

Diagnostic imaging techniques include radiography, CT and MRI. Radiography (Fig. 8) is the most widely available but is the least sensitive. A full series should include dorso-ventral, lateral, right and left lateral oblique, and, where necessary, rostro-caudal open mouth views [6]. CT scans (Fig. 9) are less widely available, but are fast, can be done under sedation, and are highly accurate for bony and soft-tissue changes. Post-contrast evaluation of bone- and soft-tissue weighted views can reveal the extent and severity of inflammation as well differentiate tissue density

Fig. 8 Rostro-caudal open mouth radiograph of a cat with unilateral otitis media. The left middle ear compartments have a normal dark air-filled appearance. The right middle ear is filled with opaque material consistent with soft-tissue or fluid

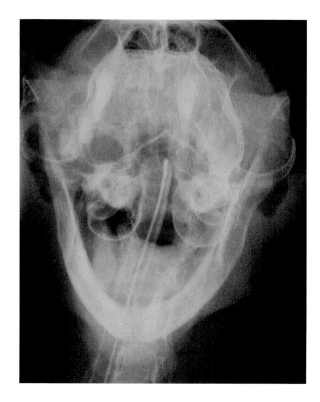

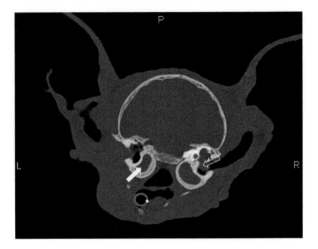

Fig. 9 CT scan of a cat with bilateral otitis media. There is some thickening of the tympanic bulla wall consistent with chronic inflammation (compare to the bulla walls in Fig. 8). The ventral compartments are filled with soft tissue, which density analysis revealed to be fluid. The right dorsal compartment is also filled with soft tissue that was shown to be solid. The cat had an inflammatory polyp in the right ear and mucoid congestion in the ventral bullae in both ears

(e.g. solid tissue, fat and fluid). MRI is best for evaluating the soft tissues and nerves around the ears, but will not image bony structures adequately.

Ear Cleaning and Ear Flushing

Ear cleaning removes debris and microbes from the canal [4, 5, 7]. Some ear cleaners have broad-spectrum antimicrobial activity [8]. Very waxy or exudative ears should be cleaned daily during treatment, but this isn't necessary if there is less debris. It is important to demonstrate effective ear-cleaning techniques to owners.

Ear Cleaners

Ceruminolytic (lift debris off the epidermis) and ceruminosolvent (soften cerumen) cleaners (i.e. propylene glycol, lanolin, glycerine, squalane, butylated hydroxytoluene, cocamidopropyl betaine and mineral oils) are useful for softening and removing dry waxy debris and/or wax plugs. Surfactant-based ear cleaners (i.e. docusate sodium, calcium sulfosuccinate and similar detergents) are better in more seborrhoeic ears and purulent ears. Tris-EDTA has very little ceruminolytic or detergent activity, but is soothing in ulcerated purulent ears and is safer if the tympanic membrane is ruptured. Astringents (i.e. isopropyl alcohol, boric acid, benzoic acid, salicylic acid, sulphur, aluminium acetate, acetic acid and silicon dioxide) can help prevent maceration of the epithelial lining of the canal. Antimicrobials (e.g. p-chlorometaxylenol [PCMX], chlorhexidine and ketoconazole) can help treat and prevent infections. Tris-EDTA has little antimicrobial activity by itself, but high concentrations can potentiate the effect of antibiotics and chlorhexidine [9, 10]. Ear cleaners should be used with some caution in cats, and some ingredients (e.g. detergents, acids and alcohols) can irritate the ears and/or trigger ototoxicity.

Ear Flushing

Thorough ear flushing under general anaesthetic is the only way to clean the deeper ear canal and middle ear [4, 5, 7]. A tomcat catheter, urinary catheter or feeding tube can be inserted into the ear canals and, if necessary, middle ear under direct visualisation through an operating head or, preferably, video otoscope. The longer length, narrower diameter, magnification and visual clarity of a video otoscope make the procedure much easier, more accurate and safer. Ear flushing should be done with saline or water to minimise the risk of ototoxicity or Horner's syndrome. This is flushed and aspirated into the ear canals and/or middle ear until clean. It may be necessary to use a ceruminolytic initially to help soften and loosen stubborn debris, but this should be carefully and thoroughly flushed out afterwards.

Myringotomy

A myringotomy (deliberate rupture of the tympanic membrane) should be considered if the ear drum is intact but there is evidence of middle ear disease (e.g. clinical signs, abnormal tympanic membrane and/or diagnostic imaging findings) [4, 5, 7]. A catheter, stylet, spinal needle or curette can be used to puncture the ventro-lateral portion of the tympanic membrane (Fig. 4). This avoids the important structures in the epitympanic recess. It's only possible to get access to the dorso-lateral tympanic bulla through the tympanic membrane as the bony shelf (while allowing some communication) prevents direct entry into the ventro-medial compartment.

Diseases of the Pinna

Diseases of the pinna (Table 1) are relatively common in cats. They are normally associated with more generalised skin conditions that are covered in detail elsewhere in this book. In contrast to the more specific causes of otitis, these conditions rarely affect the ear canals.

Table 1 Diseases of the pinna

Alopecia	Hyperadrenocorticism Hypothyroidism (very rare) Dermatophytosis *Demodex* spp. (rare) Follicular dysplasia (Devon rex cats) Lymphocytic and other mural folliculitis Alopecia mucinosa
Pruritic and eosinophilic dermatitis	Mosquito-bite hypersensitivity Other biting insects (including rabbit fleas) Head and neck dermatitis Eosinophilic granuloma complex *Notoedres cati* and *Sarcoptes scabiei* (rare)
Pustules and crusting	Pemphigus foliaceus
Necrosis and ulceration	Drug reactions Vasculitis Cold agglutinin disease Frostbite
Thickening, scaling and pigmentation (with or without distortion)	Actinic keratosis Cutaneous horn Multicentric squamous cell carcinoma *in situ* (Bowen's disease) Auricular chondritis
Nodules and ulcers	Squamous cell carcinoma
Nodules, ulcers and sinus tracts	Cat bite abscess or cellulitis Aural haematoma (rare) *Cryptococcus* spp. and other deep fungal infections Deep bacterial infections (e.g. *Actinomyces* and *Nocardia* spp.) Mycobacterial infections

Otitis Externa

Clinical Signs

Clinical signs of otitis externa include pruritus, head shaking, inflammation and discharge. The division into erythroceruminous and suppurative otitis is less clear than in dogs. Waxy ceruminous to seborrheic discharges may be sterile or involve *Malassezia* or (less commonly) a bacterial overgrowth (i.e. there are no neutrophils or other inflammatory exudates). Purulent otitis is relatively more common than in dogs (where erythroceruminous otitis predominates). However, severe ulcerative *Pseudomonas* otitis is uncommon in cats.

Acute, unilateral otitis externa can be seen with foreign bodies, whereas chronic unilateral otitis suggests neoplasia or an inflammatory polyp. Bilateral otitis externa in cats is most commonly associated with otodectic mange, but chronic cases can be associated with adverse food reactions or other hypersensitivity diseases.

Table 2 Primary, predisposing and perpetuating factors in otitis externa

Primary (the actual cause of the ear disease)	*Otodectes cynotis* *Demodex* spp. (rare) Foreign bodies (rare) Adverse food reactions Feline atopic syndrome (feline atopic dermatitis) Inflammatory polyps Cystoadenomatosis Proliferative and necrotising otitis Ceruminous gland neoplasia Seborrheic otitis
Predisposing (make otitis more likely or more likely to be severe)	Conformation (e.g. pendulous, hairy, narrow ears and ceruminous ears; rare in cats) Swimming (very rare in cats) Maceration or irritation of canal epithelium with cleaning or medication
Perpetuating (prevent resolution)	Chronic pathological changes (e.g. decreased epithelial migration, sebaceous and ceruminous hyperplasia, increased discharge, oedema, fibrosis, thickening and stenosis, and calcification) Otitis media

The PSPP Approach

Ear infections are almost always secondary to primary, predisposing and perpetuating factors (the PSPP approach – Table 2) [4, 5]. The primary cause of the otitis must be diagnosed and treated. Predisposing factors may not be easily countered or managed, but should alert clinicians to animals that will be more likely than others to have recurrent otitis. However, with the exception of excessive cleaning or medication, these are less important in cats compared to dogs. Failure to address all the perpetuating causes of otitis externa commonly results in relapsing chronic otitis. Perpetuating causes can change over the course of chronic otitis and will eventually lead to irreversible changes that require a total ear canal ablation.

Bacterial and *Malassezia* Infections

Staphylococcal bacteria (e.g. *Staphylococcus pseudintermedius* and *S. felis*) are most common, but other organisms can include streptococci, *Pasteurella multocida*, *E. coli*, *Klebsiella pneumoniae*, and/or *Proteus*, *Pseudomonas*, *Corynebacteria* and *Actinomyces* spp. [4, 5]. Mixed bacterial infections are frequently seen. Many staphylococci and Gram-negative strains can produce biofilms [11]. These inhibit cleaning, prevent penetration and activity of antimicrobials (effectively increasing the MIC) and provide a protected reservoir of bacteria (Figs. 5 and 7). They may also enhance the development of antimicrobial resistance, especially in Gram-negative bacteria that acquire stepwise resistance mutations to concentration-dependent antibiotics.

Table 3 First-line topical antimicrobials

Fusidic acid	Gram-positive only Effective against MRSA and MRSP Synergistic with framycetin against Gram-positive bacteria
Florfenicol	Broad-spectrum but not effective against *Pseudomonas* spp.
Polymyxin B	Broad-spectrum and effective against *Pseudomonas* spp. Inactivated by organic debris and needs a clean ear canal Synergistic with miconazole against Gram-negative bacteria
Gentamicin	Broad-spectrum and effective against *Pseudomonas* spp.
Neomycin	Broad-spectrum but limited efficacy against *Pseudomonas* spp.
Fluoroquinolones	Broad-spectrum and effective against *Pseudomonas* spp. Additive activity with silver sulfadiazine against *Pseudomonas* spp.
Nystatin Terbinafine Clotrimazole Miconazole Posaconazole	Broad-spectrum antifungals

MRSA methicillin-resistant *Staphylococcus aureus*, *MRSP* methicillin-resistant *S. pseudintermedius*

First-Line Antimicrobial Treatment

In general, topical antimicrobials are more effective than oral antibiotics for resolving otitis externa (Table 3). High antimicrobial concentrations (usually mg/ml) can overcome apparent antibiotic resistance. It is important to use an adequate volume to penetrate into the ear canals – 0.5–1 ml is sufficient for most cats, but this may be too much in very small animals.

The efficacy of concentration-dependent drugs (e.g. fluoroquinolones and aminoglycosides) depends on delivering concentrations of at least 10x MIC once daily. Time-dependent drugs (penicillins and cephalosporins) require concentrations above MIC for at least 70% of the dosing interval. This is readily obtained with topical therapy, which achieves high local concentrations that probably persist in the absence of systemic metabolism. Most topical medication should be effective with once daily dosing, although some are licensed for twice daily administration. Products licensed for use in dogs should be used with care in cats as ototoxicity is possible.

Topical Antibiotic Treatment of Multi-drug-Resistant Bacteria

If bacteria persist on cytology despite 1–2 weeks of appropriate treatment, then antibiotic resistance should be suspected. Other reasons for treatment failure include polyps, neoplasia, foreign bodies and other underlying conditions; debris, biofilms and failure to clean the ears; stenosis, otitis media and other perpetuating factors; and poor compliance. *Pseudomonas* are inherently resistant to many antibiotics, and they readily develop further resistance if treatment is ineffective. There are a variety of approaches to multi-drug-resistant infections (Table 4), although none of these are licensed in cats and they must be used with care.

Table 4 Antibiotics useful in multi-drug-resistant infections

Antibiotic	Treatment regimes
Ciprofloxacin	0.2% sol. 0.15–0.3 ml/ear q 24 h
Enrofloxacin	2.5% injectable sol. diluted 1:4 with TrizEDTA, saline or Epi-Otic® topically q 24h; 22.7 mg/ml sol. 0.15–0.3 ml/ear q 24 h
Marbofloxacin	1% injectable sol. diluted 1:4 with saline or TrizEDTA topically q 24h; 2 mg/ml in TrizEDTA topically q 24h; 20 mg/ml sol. 0.15–0.3 ml/ear q 24h
Clavulanate–ticarcillin[a]	16 mg/ml in TrizEDTA topically q 24h; reconstituted injectable sol. 0.15–0.3 ml/ear q 12h; 160 mg/ml sol. 0.15–0.3 ml/ear q 12h; potentially ototoxic
Ceftazidime[a]	10 mg/ml in TrizEDTA topically q 24h; 100 mg/ml 0.15–0.3 ml/ear q 12h
Silver sulfadiazine	Dilute to 0.1–0.5% in saline; additive activity with gentamicin and fluoroquinolones
Amikacin	2 mg/ml in TrizEDTA topically q 24h; 50 mg/ml 0.15–0.3 ml/ear q 24h; susceptibility maintained if there is resistance to other aminoglycosides; potentially ototoxic
Gentamicin	3.2 mg/ml in TrizEDTA topically q 24; ototoxicity possible but uncommon

[a]Reconstituted sol. Stable for up to 7 days at 4 °C or for 1 month frozen

Tris-EDTA

Tris-EDTA damages bacterial cell walls and increases antibiotic efficacy, which can overcome partial resistance. It is best given 20–30 minutes before the antibiotic but can be co-administered. It is well tolerated and non-ototoxic. Tris-EDTA shows additive activity with chlorhexidine, gentamicin and fluoroquinolones at high concentrations [9, 10, 12].

Treatment of Biofilms and Mucus

Biofilms can be physically broken up and removed by thorough flushing and aspiration. Topical Tris-EDTA and N-acetylcysteine (NAC) can disrupt biofilms, facilitating their removal, and enhancing penetration of antimicrobials. NAC, however, is potentially irritating (particularly in concentrations above 2%). Systemic NAC (600 mg orally twice daily) can help dissolve biofilms in the middle ear. Systemic NAC and bromhexine (1–2 mg/kg orally q12h) can liquefy mucus, facilitating drainage in otitis media due to chronic mucosal inflammation of inflammatory polyps (see below).

Systemic Antimicrobial Therapy

Systemic therapy may be less effective in otitis externa because bacteria are present only in the external ear canal and cerumen, there is no inflammatory discharge and penetration to the lumen is poor. Systemic treatment is indicated when the ear canal cannot be treated topically (e.g. stenosis or compliance problems or if topical adverse reactions are suspected) and in otitis media. High doses of drugs with good tissue penetration (e.g. clindamycin or fluoroquinolones) should be considered. Oral itraconazole (5 mg/kg orally once daily) can be administered if systemic antifungal therapy is indicated.

Otitis Media

Aetiology and Pathogenesis

Otitis media may be primary or secondary. Inflammatory polyps (see below) are the most common cause of primary otitis media in cats. Chronic otitis externa can result in maceration and rupture of the tympanic membrane (Fig. 10), especially with stenosis of the horizontal ear canal and Gram-negative bacterial infections. Chronic upper respiratory infections lead to inflammation and increased bacterial colonisation of the nasopharynx, which may ascend up the Eustachian tube. Less commonly, infections may spread to the middle ear from retrobulbar or para-aural abscesses, or other severe local or systemic infections. Middle-ear infections may rupture through the tympanic membrane into the ear canal or spread to the para-aural tissues and/or central nervous system.

Clinical Signs

Clinical signs associated with otitis media can be chronic, mild and vague until neurological deficits become evident (Table 5) (Figs. 11 and 12).

Fig. 10 Ruptured ear drum in cat with chronic upper respiratory tract and middle ear infections. The middle ear mucosa (A) is visible behind the malleus (B)

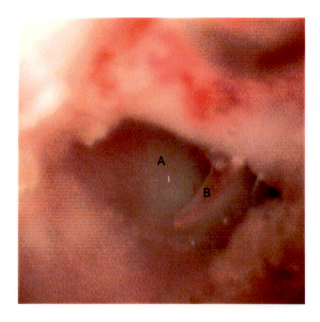

Table 5 Clinical signs associated with otitis media in cats

Otitis media	Neurological deficits
Head shaking, rubbing or scratching Dullness and pain; may avoid hard foods and handling around the head Reduced ability to localise sound (unilateral) Deafness (bilateral)	Horner's syndrome (miosis, partial ptosis and apparent enophthalmos); sympathetic trunk Ataxia and nystagmus; peripheral vestibular syndrome Facial paresis; facial nerve

Fig. 11 Head tilt in a cat with otitis media associated with a *Pasteurella* infection

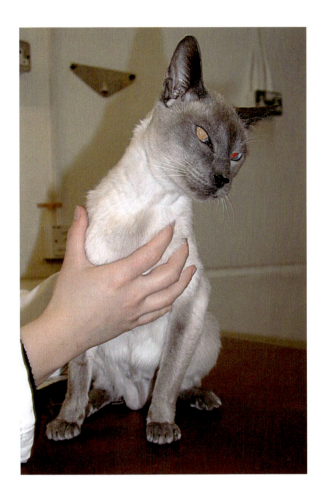

Diagnosis

Using otoscopy and myringotomy to demonstrate an abnormal tympanic membrane and/or fluid or debris in the tympanic bulla is diagnostic. Imaging such as radiography, CT or MRI is useful when stenosis limits otoscopic examination and will reveal the extent of the otitis media and lytic, proliferative and/or expansive changes in the tympanic bulla.

Fig. 12 Horner's
syndrome in a cat with
otitis media caused by an
inflammatory polyp

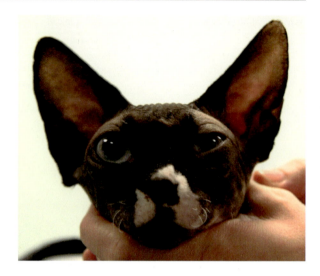

Treatment

Many cases of otitis media will resolve with medical therapy, but most cases will also require flushing of the middle ear. A myringotomy will be necessary if the tympanic membrane is intact.

Systemic antibiotics should be selected based on bacterial culture of the middle ear, taking into account the ability of the antibiotic to penetrate the middle ear. Penetration of antibiotics into chronically inflamed ears could be poor, and high doses of drugs with a high volume of distribution (e.g. fluoroquinolones) should be considered [13]. Systemic therapy can be challenging if susceptibility is limited to topical and/or parenteral drugs. Topical antibiotics and glucocorticoids in saline or Tris-EDTA can be directly administered into the middle ear; gentamicin, fluoroquinolones, ceftazidime and dexamethasone do not appear to be specifically ototoxic when administered this way [5, 14]. It is unclear how long these drugs persist in the middle ear, but, as this is essentially a blind-ended sac, drugs are likely to be active in the middle ear for a few days. Repeated instillation of mg/ml solutions every 5–7 days could therefore be useful in multidrug-resistant infections that are refractory to oral medication. Treatment should continue until clinical resolution, which may take 6–8 weeks in severe cases. Oral or topical glucocorticoids and/or mucolytics should also be administered in appropriate cases. The tympanic membrane generally heals within 2–3 weeks if infection and inflammation are controlled.

Chronic otitis media, osteomyelitis of the tympanic bulla, cholesteatoma and para-aural abscesses may be refractory to medical treatment. A total ear canal ablation/lateral bulla osteotomy is required in these cases.

Foreign Bodies

Clinical Signs

Potential foreign bodies include grass awns, hair and other organic debris, cotton wool, cotton bud tips and (occasionally) broken pieces from catheters or forceps. These cause variable, acute and sometimes extreme pain and pruritus. Some cases may present with chronic otitis externa that is poorly responsive to treatment. Most cases are unilateral but bilateral foreign bodies can be seen. Foreign bodies may penetrate the tympanic membrane and cause otitis media.

Diagnosis

Otoscopic examination will reveal variable amounts of inflammation and exudates. The nature of the exudate will depend on the secondary infection. Large foreign bodies are usually visible, but the cat may need to be sedated or anaesthetised to allow ear cleaning and complete visualisation of the ear canal. Advanced imaging may be needed to detect a foreign body that has penetrated fully into the middle ear.

Treatment

The foreign body must be removed using forceps and/or ear flushing. Care should be taken to check that it has been completely removed – tiny fragments left behind can be enough to perpetuate the problem. The inflammation and secondary infection should be treated appropriately (see above), but they normally quickly resolve once the foreign body has been removed.

Inflammatory Polyps

Aetiology and Pathogenesis

Inflammatory polyps are a common cause of otitis media and externa in cats [15]. They are most common in young cats but can be seen in older individuals. Most are unilateral but they can be bilateral. Their aetiology is unknown but may involve an abnormal inflammatory reaction to the commensal nasopharyngeal microflora or to respiratory virus infections (although virus isolation has been negative) [16].

Clinical Signs

The polyps usually arise in or near to the opening of the Eustachian tube [15]. They may extend down the Eustachian tube and into the nasopharynx, causing snoring, an altered voice, sneezing, coughing, gagging and/or retching. Polyps within the ear usually present with otitis media (if the tympanic membrane is intact; Fig. 12) and/ or otitis externa (if the tympanic membrane is ruptured and the polyp extends into the ear canals). This is usually associated with a secondary bacterial infection and a purulent discharge in the ear canals (Fig. 13). It is unusual for the polyp to extend in the ventro-medial compartment of the tympanic bulla. However, this is often full of mucus arising from obstruction of the Eustachian tube and draining foramen by the polyp and inflamed mucosa (Fig. 9). This can become stagnant, inspissated and/ or infected.

Diagnosis

A history of unilateral otitis, purulent discharge, Horner's syndrome and/or otitis media is highly suggestive of an inflammatory polyp [15]. The polyp may be visible in the ear canal (after cleaning any discharge; Fig. 14) or in the nasopharynx (which usually requires sedation and pulling the soft palate rostrally). Diagnostic imaging can be used to confirm the extent and severity of the polyp and secondary infection. Computerized tomography (CT) has much better sensitivity and specificity than radiography. Density analysis of the pre- and post-contrast CT images weighted for bone and soft tissue will differentiate solid polyp tissue from fluid (mucus and/ or pus) and show areas of active inflammation or infection (Fig. 9). This allows

Fig. 13 Waxy and suppurative debris in the ear canal of a cat with an inflammatory polyp. The polyp could not be seen until the debris was flushed away. Repeated courses of antibiotics had led to an MRSA infection

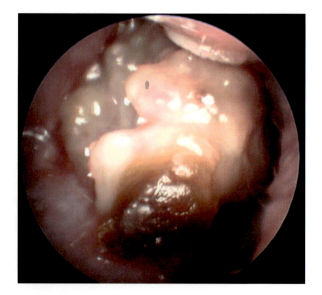

Fig. 14 Inflammatory
polyp in the horizontal ear
canal

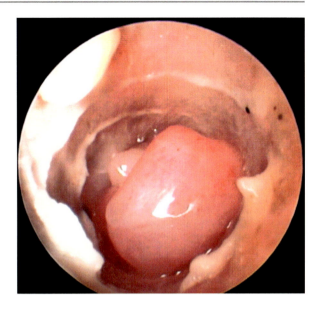

accurate assessment of the extent of the polyp; for example, while radiographs can suggest that all compartments of the middle ear are affected, a CT scan can show the true extent of the solid polyp, mucus build-up and involvement of the Eustachian tube. CT scans will also more accurately assess changes to the bony structures of the tympanic bulla and middle ear (e.g. osteomyelitis, sclerosis, proliferation and/ or lysis). This can be critical for planning treatment; for example, in most cases, CT scans show that solid polyps don't extend into the ventral compartment, making a ventral bulla osteotomy unnecessary. Other types of tumours should be suspected if a polyp-like mass develops in older cats, and histopathology can be used to confirm the diagnosis.

Treatment

Polyps in the nasopharynx, dorsolateral middle ear and external ear canal can be removed via traction with forceps under anaesthetic [17, 18]. The polyp is grasped firmly with a set of forceps and gradual continuous traction applied to pull the polyp from the middle ear. They can usually be removed in one piece, but sometimes multiple attempts are needed (Fig. 15). Twisting the polyp stalk before or during traction can help limit bleeding. Small polyps in the horizontal ear canal can be removed using alligator forceps through an operating head or video otoscope. Flatter nodules that are difficult to grasp with forceps can be ablated with a laser (Fig. 16). Solid polyps in the ventro-medial bulla compartment have to be removed via a ventral bulla osteotomy, although this is uncommon.

Otitis externa and media are often present, and material from the ear canals and middle ear should be collected for cytology and culture. The polyps themselves are invariably sterile. The ear canals and middle ear should be flushed and treated appropriately (see treatment of otitis externa and otitis media above).

Systemic glucocorticoids (e.g. 2 mg/kg prednisolone or 0.2 mg/kg of dexamethasone daily to resolution and then slowly tapered) will reduce post-traction inflammation, help open the bulla foramen and Eustachian tube and may help to reduce the recurrence rate [19]. N-acetylcysteine (600 mg orally q12h) or bromhexine (2 mg/kg q12h) can help liquefy mucus and facilitate drainage from the middle ear.

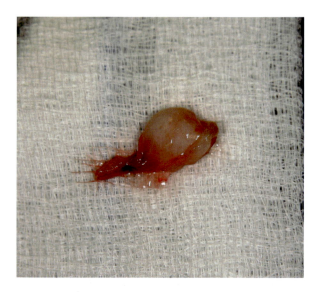

Fig. 15 Inflammatory polyp (see Fig. 14) after removal by traction. The intact stalk indicates that it has been successfully removed intact

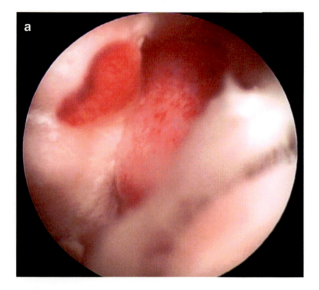

Fig. 16 (**a**) Sessile inflammatory polyps arising in the horizontal ear canal; (**b**) the horizontal ear canal following laser ablation of the polyps and ear flushing

Fig. 16 (continued)

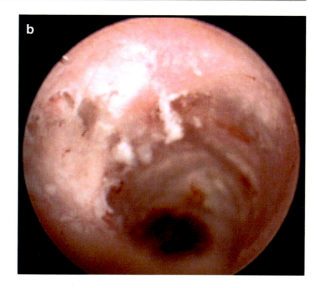

   Potential complications include Horner's syndrome, vestibular syndrome and facial nerve paralysis. Horner's syndrome is particularly common as a sympathetic nerve trunk is very close to the most common origin of polyps near the opening of the Eustachian tube (Fig. 12). These problems are usually temporary, and permanent deficits are rare.

Cystoadenomatosis (Cystomatosis)

Aetiology and Pathogenesis

Cystoadenomatosis results in multiple, pigmented, ceruminous papules, nodules and cysts in the ventral pinna and external ear canal [4, 5, 20]. The cause is unknown, but may involve genetic predisposition and inflammatory triggers. Persian cats may be predisposed. The cysts eventually block the ear canal resulting in otitis externa and secondary infections.

Clinical Signs

The clinical appearance is highly suggestive. Early cases present with multiple blue-black comedones and papules around the opening of the ear canal (Fig. 17). These gradually increase in number and size, forming nodules and cysts that may spread onto the pinna and into the vertical and (less commonly) horizontal ear canals (Fig. 18). Ruptured cysts release a brown-black fluid. Cysto adenomatosis is

Fig. 17 Multiple
blue-black comedones and
cysts on the pinna of a cat
with cysto adenomatosis

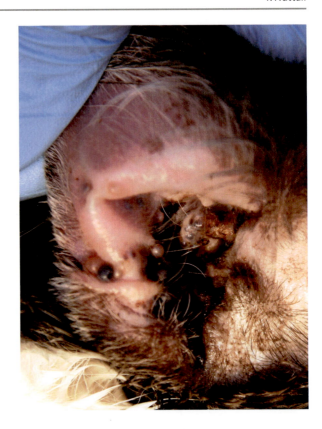

Fig. 18 Multiple cysts
obstructing the ear
canals in a cat with
cysto adenomatosis

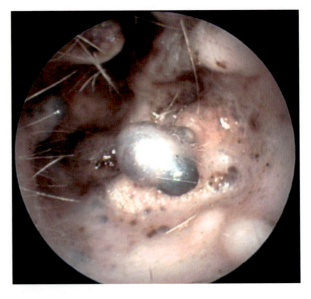

not usually pruritic or painful unless secondary infections are present. Ceruminous adenomas and adenocarcinomas are more commonly seen in older animals and form solitary or small numbers of discreet tumours.

Diagnosis

The clinical appearance is largely pathognomonic. If necessary, histopathology will differentiate adenocarcinoma, adenoma and cysto adenomatosis. Ear cytology can be used to identify secondary *Malassezia* and/or bacterial infections. Advanced imaging can be performed if an otitis media is suspected.

Treatment

Medical treatment is suitable for early and/or mild cases. Any secondary otitis should be treated appropriately. Systemic and/or topical steroids can decrease swelling and ceruminous hyperplasia in the ears. Regular maintenance therapy with topical steroids can be used to maintain remission.

More extensive nodules and cysts can be ablated with a CO_2, diode or other laser, or electrocautery (Fig. 19). Topical antibiotics/steroid combinations should be used postoperatively to reduce inflammation and prevent infection. Laser ablation is very effective with prolonged remission. Regular topical steroids may reduce the recurrence rate.

Fig. 19 Same cat as in Fig. 18; here the cysts have been ablated using a laser

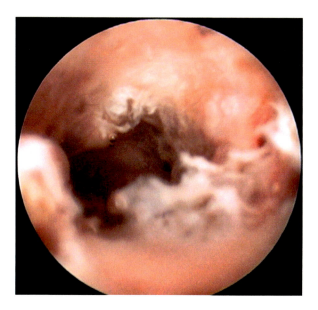

Total ear canal ablation with lateral bulla osteotomy is necessary in cases where medical therapy or laser ablation isn't appropriate or available. Surgery is curative, although at the expense of the ear.

Otodectes and Other Parasites

Aetiology and Pathogenesis

Otodectes cynotis mites are the most common cause of parasitic otitis, although others can include trombiculid mites (mainly on the pinnae), *Demodex* species (mainly on the pinnae and only rarely in the ear canals) and *Otobius megnini* (the spinose ear tick; very rare in cats). *Otodectes* are highly contagious between cats and other species, and are common in multi-cat situations (particularly with young animals and/or where there is a high turnover). Most *Demodex* are not contagious, although *Demodex gatoi* may be contagious between cats in a household.

Clinical Signs

Otodectes cynotis infestation is characterised by large amounts of dry, brown, waxy debris with variable erythema and pruritus (Fig. 20). The pruritus may be severe and extend onto the head and neck, making *Otodectes* a differential diagnosis in head and neck dermatitis. Cats can develop hypersensitivity reactions to *Otodectes*, and small numbers of mites can still be associated with clinical signs. Some cats are apparently asymptomatic carriers of *Otodectes*. *Demodex* species cause similar clinical signs. *Otobius* can result in a variably severe inflamed and painful otitis externa.

Diagnosis

The history and clinical signs are highly suggestive of *Otodectes*. Careful otoscopic examination often reveals the mites as they move (Fig. 21). *Otodectes* and *Demodex* can be seen on microscopic examination of the waxy debris broken up in liquid paraffin (Fig. 22). Treatment trials for *Otodectes* (see below) are warranted if mites aren't found as small numbers can be missed in sensitised cats. *Demodex* may be found on hair plucks, tape strips or skin scrapes of the pinnae or elsewhere. *Otobius* ticks should be obvious on otoscopic examination.

Fig. 20 Characteristic dry waxy discharge at the opening of the ear canal in a Persian cat with *Otodectes*

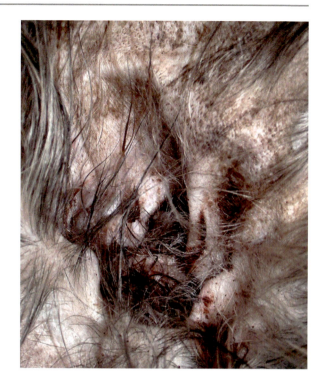

Fig. 21 *Otodectes* seen in the ear canal using a video otoscope

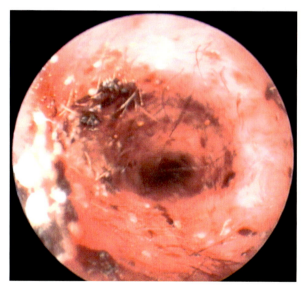

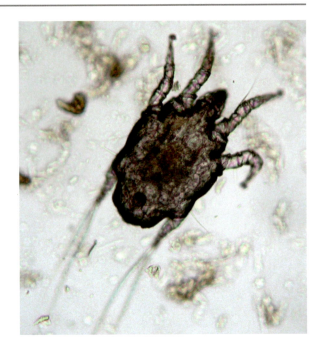

Fig. 22 Adult *Otodectes* mite found by collecting the ear canal discharge into liquid paraffin (x100 magnification)

Treatment

Most anti-mite products are effective against *Otodectes*, including topical fipronil, selamectin and imidacloprid/moxidectin. Isoxazoline drugs seem to be highly effective against *Otodectes* and *Demodex*. All potentially exposed cats and dogs should be treated. *Otobius* ticks can be killed with an appropriate product before careful removal using forceps. Any secondary otitis usually resolves rapidly but can treated if necessary.

Neoplasia

Aetiology and Pathogenesis

Ear canal neoplasia is fairly common in older cats. Most tumours are ceruminous adenomas, but malignant adenocarcinomas comprise up to 50% of ceruminous tumours [4, 5, 20]. Other tumours in the ear canals are rare. The obstruction usually results in a secondary otitis externa. Malignant tumours can result in local destruction and invasion and potentially spread to local lymph nodes and distal organs. Tumours arising from external tissues can occasionally invade the middle ear and/or ear canals.

Clinical Signs

Most tumours arise at the base of the pinna and upper vertical canal, although they can be seen at any depth in the ear canals. The nodules can be obscured by discharge if there is a secondary infection. Swelling of the surrounding tissues and/or local lymph nodes is suggestive of local spread and/or metastasis.

Diagnosis

The clinical presentation is usually obvious, although single lesions deeper in the ear canal should be distinguished from inflammatory polyps and multiple tumours from cysto adenomatosis. Cytology and/or histopathology can help differentiate neoplasia from polyps and benign from more malignant tumours (Fig. 23). Aspirates should be taken from local lymph nodes if there is any suspicion of malignancy or spread. Imaging (especially CT scans) can be used to determine the extent and severity of local invasion, lymph node involvement and metastasis to internal organs such as the lungs.

Treatment

Accessible benign tumours can be surgically excised. Lasers can be used to excise and ablate tumours deeper in the ear canals where surgical excision isn't possible. Vertical ear canal surgery or total ear canal ablation can be used to remove tumours if lasers aren't available or wider surgical margins are required (Fig. 24). The prognosis is good unless there has been metastatic spread.

Fig. 23 Fine needle aspiration cytology from a ceruminous adenoma; there are numerous epithelial cells forming a well-defined and differentiated sheet with minimal pleomorphism

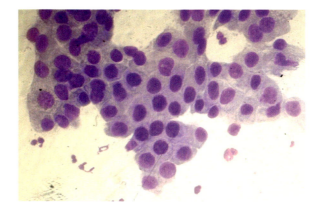

Fig. 24 Benign
ceruminous adenoma in
the horizontal ear canal of
a cat following a total ear
canal ablation. A laser
could have been used to
ablate the tumour *in situ*
and preserve the ear canal

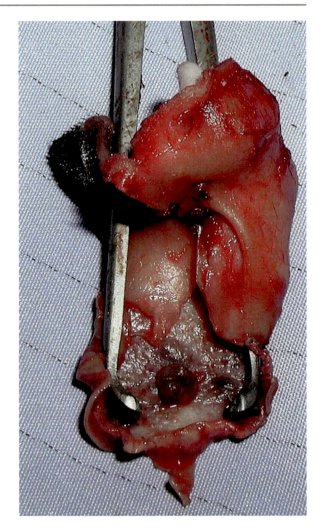

Seborrheic Otitis

Seborrheic otitis is a common problem of uncertain aetiology and clinical significance [4, 5]. Dark waxy to greasy scales build up on the inside of the pinna and around the opening of the ear canal (Fig. 25). The lower vertical and horizontal ear canals are usually normal. The cats may have no other clinical signs but can shake their heads, flick their ears or scratch at their ears.

This may be secondary to inflammation, and affected cats should be carefully evaluated for other clinical signs consistent with primary causes of otitis, which should be managed appropriately. Seborrheic otitis may be seen in Persian and related cats with idiopathic facial dermatitis. Sphynx and Devon rex cats can have

Fig. 25 Build-up of sebaceous/ceruminous material in the vertical ear canal. The focal accumulation of the material is suggestive of glandular hyperplasia or hypersecretion

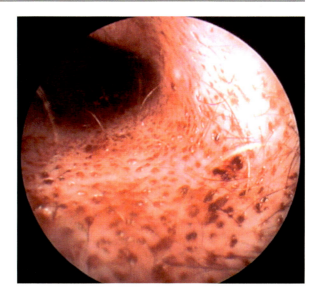

an asymptomatic build-up of cerumen on the inner pinnae. Cytology should be used to determine whether there is a bacterial or *Malassezia* infection. However, the waxy build-up may simply be material accumulating on the pinna after epidermal migration out of the ear canals. If there are no other clinical signs, the cats don't need treatment. If necessary, gentle wiping with a ceruminosolvent/ceruminolytic ear cleaner can reduce the build-up.

Proliferative and Necrotising Otitis

Aetiology and Pathogenesis

The aetiology of this condition is unknown, but it is probably immune-mediated [21, 22]. The lesions show T-cell-mediated keratinocyte apoptosis similar to erythema multiforme [22]. It was first recognised in kittens but has now been reported in adult and older cats, although most cases are seen in cats less than 4 years old [4, 5].

Clinical Signs

The clinical signs are very suggestive. The lesions are usually symmetrical and most commonly affect the base of the pinna, opening of the ears and the vertical ear canal. Occasionally, the lips, face, periocular skin or remote sites may become involved. Affected cats develop erythematous and hyperkeratotic plaques with tightly adherent crusts. More severe cases may have erosions, ulceration and haemorrhage (Fig. 26). Secondary bacterial or *Malassezia* ear infections may obscure the clinical lesions.

Fig. 26 Proliferative and
necrotising otitis in a cat
with erythematous plaques,
erosions and crusts

Diagnosis

The diagnosis is usually based on the history and clinical signs. Cytology should
be used to identify secondary infection, which should be treated appropriately [21].
Where necessary, cytology and histopathology can be used to confirm the diagnosis and eliminate differential diagnoses affecting the pinna and ear canals such as
pemphigus foliaceus, eosinophilic granuloma syndrome and thiamazole-associated
drug reactions.

Treatment

The prognosis is generally good [21]. Many cases, especially in kittens or young
cats, may spontaneously resolve. There is usually a good response to topical 0.1%
tacrolimus or systemic ciclosporin. Some cats may not need further treatment after
resolution, but some may need long-term therapy to maintain remission.

Para-aural Abscessation

Para-aural abscesses are uncommon in cats. Causes include severely infected otitis externa and/or media that extends into the surrounding soft tissues, bite wounds, deep bacterial, mycobacterial and/or fungal infections, complications from ear canal and/or middle ear surgery, and traumatic ear canal avulsion (usually following a road traffic accident) (Fig. 27). Advanced imaging (especially contrast-enhanced CT scans) can reveal the extent and severity of the infection and inflammation, including sinus tracts into adjacent structures and tissues (which can include the central nervous system). Samples for cytology and culture should be taken from the affected tissues (Fig. 28), as secondary bacterial infections on the surface may obscure the causative organisms. This may require surgical exploration.

Treatment depends on the primary cause and may include surgical exploration, debridement and flushing, vertical ear canal surgery or a total ear canal ablation/lateral bulla osteotomy. Surgical sites, sinus tracts and debrided tissues should be thoroughly flushed. Drains may be necessary with deep abscesses or sinus tracts. Antimicrobials should be selected using culture and administered until clinical cure. The course of treatment will depend on the extent and severity of infection

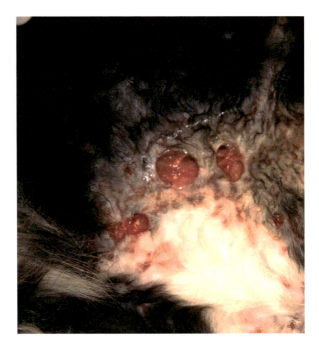

Fig. 27 Extensive para-aural abscesses, ulcers and sinus tracts in a cat with an *Actinomyces* infection following a bite wound

Fig. 28 Deep tissue cytology from the cat in Fig. 27. There is pyogranulomatous inflammation with a prominent multinucleated giant cell. This has several beaded filamentous organisms characteristic of *Actinomyces*

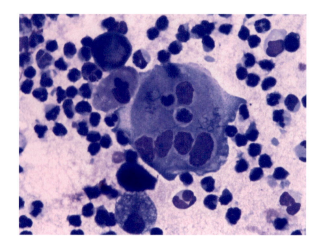

and inflammation. Prolonged courses aren't necessary after surgery provided that the site is clean and closed well, but longer courses (4–6 weeks or longer) will be necessary with deep infections (especially with slow growing organisms such as *Nocardia*, *Actinomyces*, mycobacteria and fungi).

References

1. Ravens PA, Xu BJ, Vogelnest LJ. Feline atopic dermatitis: a retrospective study of 45 cases (2001–2012). Vet Dermatol. 2014;25:95.
2. Hobi S, Linek M, Marignac G, Olivry T, Beco L, Nett C, et al. Clinical characteristics and causes of pruritus in cats: a multicentre study on feline hypersensitivity-associated dermatoses. Vet Dermatol. 2011;22:406–13.
3. Hensel P, Santoro D, Favrot C, Hill P, Griffin C. Canine atopic dermatitis: detailed guidelines for diagnosis and allergen identification. BMC Vet Res. 2015;11:196.
4. Miller WH, Griffin CE, Campbell KL. Diseases of eyelids, claws, anal sacs, and ears. In: Muller and Kirk's small animal dermatology. 7th ed. St Louis: Elsevier-Mosby; 2013. p. 723–73.
5. Harvey RG, Paterson S. Otitis externa: an essential guide to diagnosis and treatment. Boca Raton: CRC Press; 2014.
6. Hammond GJC, Sullivan M, Weinrauch S, King AM. A comparison of the rostrocaudal open mouth and rostro 10 degrees ventro-caudodorsal oblique radiographic views for imaging fluid in the feline tympanic bulla. Vet Radiol Ultrasound. 2005;46:205–9.
7. Nuttall TJ, Cole LK. Ear cleaning: the UK and US perspective. Vet Dermatol. 2004;15:127–36.
8. Swinney A, Fazakerley J, McEwan N, Nuttall T. Comparative *in vitro* antimicrobial efficacy of commercial ear cleaners. Vet Dermatol. 2008;19:373–9.
9. Buckley LM, McEwan NA, Nuttall T. Tris-EDTA significantly enhances antibiotic efficacy against multidrug-resistant *Pseudomonas aeruginosa in vitro*. Vet Dermatol. 2013;24:519.
10. Clark SM, Loeffler A, Schmidt VM, Chang Y-M, Wilson A, Timofte D, et al. Interaction of chlorhexidine with trisEDTA or miconazole *in vitro* against canine meticillin-resistant and -susceptible *Staphylococcus pseudintermedius* isolates from two UK regions. Vet Dermatol. 2016;27:340–e84.

11. Pye CC, Yu AA, Weese JS. Evaluation of biofilm production by *Pseudomonas aeruginosa* from canine ears and the impact of biofilm on antimicrobial susceptibility *in vitro*. Vet Dermatol. 2013;24:446–E99.
12. Pye CC, Singh A, Weese JS. Evaluation of the impact of tromethamine edetate disodium dihydrate on antimicrobial susceptibility of *Pseudomonas aeruginosa* in biofilm *in vitro*. Vet Dermatol. 2014;25:120.
13. Cole LK, Papich MG, Kwochka KW, Hillier A, Smeak DD, Lehman AM. Plasma and ear tissue concentrations of enrofloxacin and its metabolite ciprofloxacin in dogs with chronic end-stage otitis externa after intravenous administration of enrofloxacin. Vet Dermatol. 2009;20:51–9.
14. Paterson S. Brainstem auditory evoked responses in 37 dogs with otitis media before and after topical therapy. J Small Anim Pract. 2018;59:10–5.
15. Greci V, Mortellaro CM. Management of Otic and Nasopharyngeal, and nasal polyps in cats and dogs. Vet Clin North Am Small Anim Pract. 2016;46:643.
16. Veir JK, Lappin MR, Foley JE, Getzy DM. Feline inflammatory polyps: historical, clinical, and PCR findings for feline calici virus and feline herpes virus-1 in 28 cases. J Feline Med Surg. 2002;4:195–9.
17. Greci V, Vernia E, Mortellaro CM. Per-endoscopic trans-tympanic traction for the management of feline aural inflammatory polyps: a case review of 37 cats. J Feline Med Surg. 2014;16:645–50.
18. Janssens SDS, Haagsman AN, Ter Haar G. Middle ear polyps: results of traction avulsion after a lateral approach to the ear canal in 62 cats (2004–2014). J Feline Med Surg. 2017;19:803–8.
19. Anderson DM, Robinson RK, White RAS. Management of inflammatory polyps in 37 cats. Vet Rec. 2000;147:684–7.
20. Sula MJM. Tumors and tumorlike lesions of dog and cat ears. Vet Clin North Am Small Anim Pract. 2012;42:1161.
21. Mauldin EA, Ness TA, Goldschmidt MH. Proliferative and necrotizing otitis externa in four cats. Vet Dermatol. 2007;18:370–7.
22. Videmont E, Pin D. Proliferative and necrotising otitis in a kitten: first demonstration of T-cell-mediated apoptosis. J Small Anim Pract. 2010;51:599–603.

Part III

Feline Skin Diseases by Etiology

Bacterial Diseases

Linda Jean Vogelnest

Abstract

Accurate diagnosis of feline bacterial skin diseases is important for both patient well-being and appropriate use of antibiotics in times of increasing antimicrobial resistance. This chapter reviews knowledge of clinical lesions and historical features associated with feline bacterial infections, skin diagnostics relevant to efficient and accurate diagnosis, and current treatment recommendations. Deep infections including nocardiosis and mycobacteriosis (Chapter, Mycobacterial Diseases) are well-reported, and although accurate diagnosis is important, and treatment may be lengthy and challenging, they do occur only rarely. In contrast, superficial bacterial pyoderma (SBP) is a more common feline presentation that may be under-recognised, most typically complicating underlying allergic skin disease, but also associated with a range of underlying diseases and factors. SBP is reviewed in this chapter, along with deeper infections including deep bacterial pyoderma, cellulitis and wound abscessation, dermatophilosis, necrotizing fasciitis and environmental saprophytic bacterial infections including nocardiosis. Confirmation of bacterial skin disease in cats is readily achievable in a general practice setting. Cytology is often the most valuable tool, used in conjunction with clues from the history and physical examination and supplemented with skin surface or tissue culture and/or histopathology when indicated. Cytology methods relevant to bacterial infections in the cat are detailed in this chapter. Treatment principles are also discussed, including the potential role of methicillin-resistant staphylococci in feline pyoderma, with a focus on current worldwide recommendations that may supersede some outdated clinic protocols.

L. J. Vogelnest (✉)
University of Sydney, Sydney, NSW, Australia

Small Animal Specialist Hospital, North Ryde, NSW, Australia
e-mail: lvogelnest@sashvets.com

© Springer Nature Switzerland AG 2020
C. Noli, S. Colombo (eds.), *Feline Dermatology*,
https://doi.org/10.1007/978-3-030-29836-4_11

Introduction

Bacterial dermatoses in the cat occur in two broad presentations reflecting the depth of skin invasion. Superficial infections, involving the epidermis and follicular epithelium, are most common and primarily associated with multiplication of resident skin microbiota secondary to reduced local and/or systemic host defences. Deep bacterial infections, involving the dermis and/or subcutaneous tissues, may be extensions of superficial infection or associated with traumatic implantation of a range of environmental or commensal bacterial species. Some rare but life-threatening deep bacterial infections have a propensity for body dissemination.

Normal Feline Cutaneous and Mucosal Bacterial Microbiota

There is limited knowledge about normal commensal bacteria in cats, with most studies culture-based and focused on staphylococcal isolates. The mouth, followed by the perineum, appears to be the most consistent staphylococcal carriage site [1]. Fifteen species of staphylococci were identified by MALDI-TOF testing of isolates from the oropharynx of healthy cats in Brazil, with *S. aureus* the only coagulase-positive staphylococcus (CoPS) species, with a range of coagulase-negative staphylococci (CoNS) [2]. However, α-haemolytic streptococci were more frequently isolated than staphylococci from healthy mouths of free-roaming cats in Spain, followed by two Proteobacteria (*Neisseria* spp. and *Pasteurella* spp.) [3].

Staphylococci have also been less frequently identified as resident skin bacteria in normal cats, with *Micrococcus* spp., *Acinetobacter* spp. and *Streptococcus* spp. most common [4]. Of staphylococci isolated, CoNS including *S. felis*, *S. xylosus* and *S. simulans* have been more frequent than CoPS [4–6], with *S. felis* potentially misidentified as *S. simulans* in some studies [5, 7]. Either *S. intermedius* (reclassified as *S. pseudintermedius* in 2005) [1, 8] or *S. aureus* [5, 9, 10] are variably reported as the more frequent CoPS isolates. *Escherichia coli*, *Proteus mirabilis*, *Pseudomonas* spp., *Alcaligenes* spp. and *Bacillus* spp. are less frequent isolates from normal feline skin [4, 5].

More recent genomic DNA studies in healthy cats ($n = 11$) identified a greater diversity and number of bacteria on normal feline skin than culture-based studies. Haired skin had the greatest diversity of species, the pre-aural space the greatest richness and evenness of species, and mucosal surfaces (nostril, conjunctiva, reproductive) and the ear canal (contrasting to dogs) the lowest species diversity. As for culture-based studies, *Staphylococcus* spp. did not dominate, with Proteobacteria (*Pasteurellaceae*, *Pseudomonadaceae*, *Moraxellaceae* [e.g. *Acinetobacter* spp.]) most frequent, followed by Bacteroides (*Porphyromonadaceae*), Firmicutes (*Alicyclobacillaceae*, *Staphylococcaceae*, *Streptococcaceae*), Actinobacteria (*Corynebacteriaceae*, *Micrococcus* spp.) and Fusobacteria. It is acknowledged that some species including *Propionibacterium* spp. may have been under-recognised in this study [11].

Bacterial residents vary between individuals [4, 11] and may also vary between healthy and diseased states. Carriage of staphylococci is known to increase in humans and dogs with atopic dermatitis. Similarly, *Staphylococcus* spp. were more frequently detected in allergic cats (*n* = 10) compared to normal healthy cats, with more dominance at some anatomic sites (e.g. ear canal) [11]. *Staphylococcus* spp. were also more prevalent in diseased mouths compared to normal mouths [3]. In contrast, there was no statistical difference in isolation of *Staphylococcus* spp. in another study (*n* = 98) from healthy skin compared to inflamed skin [9].

In summary, the feline studies to date suggest, in contrast to dogs, that Proteobacteria including *Acinetobacter* spp., *Pasteurella* spp. and *Pseudomonas* spp. are more common on normal feline skin than *Staphylococcus* spp., and amongst staphylococci, that CoNS appear to dominate. It is uncertain if staphylococci in general, and CoPS or CoNS in particular, multiply more readily on diseased skin.

Superficial Bacterial Pyoderma

Feline superficial bacterial pyoderma (SBP) is increasingly recognised and reported in 10–20% of cats presenting to dermatology referral [12–14]. As in other species, SBP in cats is a secondary disease, most commonly reported with hypersensitivities [12–14]; 10% of cats presenting to referral in the USA [14] and 60% in Australia had confirmed underlying allergy, most commonly atopic dermatitis [13]. Recurrent pyoderma is also commonly reported [13, 15].

Bacterial Species

Although *Staphylococcus* spp. are considered the likely pathogens [1, 2, 9, 12], weaker adherence of *S. pseudintermedius* and *S. aureus* to normal feline corneocytes in contrast to canine and human corneocytes has been documented [16], and the casual bacterial species in feline SBP have only been confirmed in a small number of cats. *S. aureus* was isolated in pure culture from papules and crusts of one cat, with concurrent neutrophils on skin cytology, and complete resolution of lesions by 10 days of antibiotic therapy [17]. *S. felis* was isolated from the nostrils and skin lesions (excoriations) of another cat with suspected underlying flea bite hypersensitivity, with concurrent neutrophils and intracellular cocci on cytology, and complete resolution of lesions by 14 days of antibiotic therapy and flea control [5]. Eosinophilic granuloma complex lesions may also be complicated by secondary pyoderma, and the most common isolates from surface swabs and/or tissue biopsies from eosinophilic plaques or lip ulcers (*n* = 9), with concurrent neutrophils and intracellular cocci on cytology, were *S. pseudintermedius* and *S. aureus*. Other isolates detected in this study included CoNS, *Pasteurella multocida*, *Streptococcus canis* and *Pseudomonas aeruginosa* [12].

A number of other bacterial culture studies, predominantly on laboratory isolates from a range of skin lesions unconfirmed as pyoderma, have focused on
staphylococci; whether isolates were pathogenic or incidental is uncertain, and non-
staphylococcal isolates are rarely reported [4, 7, 9, 17–19]. CoNS are the most common isolates in a number of studies, accounting for 96% of isolates from 'inflamed
skin' ($n = 24$) [9], the second most frequent isolate (*S. simulans*) from abscesses,
miliary dermatitis, excoriations, exfoliative dermatitis or eosinophilic plaques
($n = 45$) [17] and the most frequent isolates (*S. felis* followed by *S. epidermidis*)
from unspecified 'dermatitis' [7]. Less common CoNS isolates include *S. hyicus*, S.
xylosus and *S. schleiferi* subsp. *schleiferi* [9, 17].

CoPS have been more prevalent in some studies on diseased feline skin [4], with
S. aureus ($n = 69$) [9, 17] or *S. intermedius* ($n = 9$ [5]; $n = 30$ [20]) the most frequent
isolates, and *Streptococcus* spp. (10%), *Proteus* spp. (10%), *Pasteurella* spp. and
Bacillus spp. (10%) also reported [20].

The relative importance of staphylococci in general, and CoNS and CoPS in
particular, to feline pyoderma and whether there is one predominant causal species
as for bacterial pyoderma in humans (*S. aureus*) and dogs (*S. pseudintermedius*) is
currently uncertain.

Clinical Presentation

A median age of onset of 2 years is documented for feline SBP, although a wide
range is reported (6 months to 16.5 years), with older cats also frequently affected
(first presentation at >9 years of age in 23% of cats) [13]. Pruritus is common,
particularly with underlying hypersensitivity, reported in 92% of cats with SBP in
Australia and often severe (56%) [13]. Lesions associated with feline SBP often
reflect self-trauma, consisting most typically of multifocal, crusted, alopecic,
excoriated and erosive to ulcerative lesions (Figs. 1, 2, 3 and 4). Eroded papules,
eosinophilic plaques, eosinophilic granulomas and rare pustules are also reported.
The most frequent lesional sites are the face, neck, limbs and ventral abdomen
[12, 13, 21].

Diagnosis

Although some clinical lesions have been recognised as useful diagnostic clues
for bacterial pyoderma in dogs [22, 23], SBP lesions in cats are less characteristic, with many non-specific presentations (e.g. erosions, crusting). Diagnostic tests
are thus important to confirm a diagnosis of feline pyoderma (see later section on
"Cytology", Table 1) and are strongly encouraged prior to consideration of treatment with systemic antimicrobials [22–24].

Cytology has been considered the most useful single test, with the presence
of neutrophils and intracellular or associated bacteria being diagnostic (Fig. 11a)

Fig. 1 Feline secondary bacterial pyoderma (SBP): exudative erosions and crusting

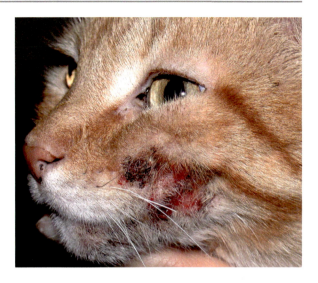

Fig. 2 Feline SBP: alopecia, erythema and focal crusting

[12, 13, 22, 25]. In canine pyoderma, cytology is considered mandatory when typical lesions (pustules) are not present or scant and is also essential to identify concurrent or alternate *Malassezia* dermatitis [23]. The morphology of bacteria on cytology (cocci and/or rods) will also guide valid empirical treatment choices and/or the need for bacterial culture. Adhesive tape impressions are applicable to all superficial skin lesions, in particular dry lesions and restricted body sites, while glass slide impressions are suitable for erosive to ulcerative lesions [22]. In canine SBP, it is reported that inflammatory cells and bacteria may be absent or scarce with concurrent immunosuppression from disease or drugs [23].

Histopathology is infrequently discussed in relation to diagnosis of SBP; however, it can provide further diagnostic confirmation, especially if samples are

Fig. 3 Feline SBP: erythematous eroded plaques

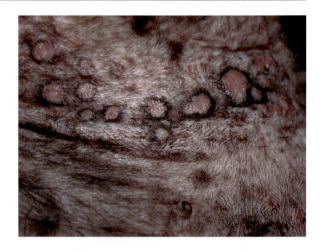

Fig. 4 Feline SBP: well-demarcated alopecia and erythema with focal crusting

collected without prior skin surface cleansing or disinfection as bacterial colonies are frequently observed within the crusts (Fig. 5) (see later section on Histopathology). Histopathology is also valuable to aid exclusion of other differentials for atypical presentations or when a diagnosis is uncertain [22].

Bacterial culture is not helpful for diagnosis of SBP, particularly when assessed independently of cytology, as isolation may simply reflect normal commensal species not involved in disease (see later section on Bacterial Culture) [6]. A heavy pure culture of one bacterial species is more likely associated with a pathogen than mixed-species isolation, but concurrent cytology remains essential [1]. Coagulase status of any staphylococci isolated is less helpful for feline pyoderma, as both CoNS and CoPS are potentially pathogenic. Despite a limited role diagnostically, culture and antibiotic susceptibility testing (C&S) can be important to guide appropriate antibiotic therapy, particularly when antimicrobial resistance is more likely.

Table 1 Differential diagnoses and valuable diagnostic tools for cutaneous lesions associated with bacterial infections in cats

Lesion	Common differentials	Less common differentials for lesion	Diagnostic tools
Papules	SBP, allergy[a] dermatophytosis	Ectoparasites (*Otodectes*, larval ticks, trombiculids); pemphigus foliaceus	History (parasiticides, exposure/contagion), cytology (tape impression), biopsy (histo)
Alopecia, erythema, scaling, crusting	SBP, dermatophytosis, allergy[a], actinic keratoses (non-pigmented skin)	Demodicosis (*D. gatoi, D. cati*), pemphigus foliaceus, ectoparasites (*Cheyletiella*, lice)	History (potential exposure/contagion, pruritus or lesions first), cytology (tape impression), biopsy (histo)
Erosion, ulceration, crusting	SBP, allergy[a], SCC (non-pigmented skin)	Herpes viral dermatitis, SCC in situ, cutaneous vasculitis	History (degree of pruritus, recurrent/seasonal), cytology (tape or slide impression), biopsy (histo)
Erythematous plaques	SBP, allergy[a]	Cutaneous xanthoma	Cytology (tape or slide impression), biopsy (histo)
Nodules (lip, chin, linear)	SBP, DBP, allergy[a]	Mycetoma, neoplasia (SCC)	Cytology (FNA), biopsy (histo)
Nodules (poorly demarcated)	Bacterial cellulitis/abscessation	Mycobacteria, *Nocardia*, sterile panniculitis	Cytology (FNA), biopsy (histo, C&S)
Nodules (discrete)	Neoplasia (variety), eosinophilic granuloma	Pseudomycetoma (bacterial, dermatophyte), mycetoma, histiocytosis, sterile pyogranuloma	Cytology (FNA), biopsy (histo, C&S)
Pustules (rare)	SBP, pemphigus foliaceus	Dermatophytosis	Cytology (impression after rupture), biopsy (histo)

[a]Atopic dermatitis, adverse food reactions and/or flea bite hypersensitivity

C&S culture and antibiotic susceptibility testing, *DBP* deep bacterial pyoderma, *FNA* fine needle aspirates, *histo* histopathology, *SBP* superficial bacterial pyoderma, *SCC* squamous cell carcinoma

Treatment

There are limited studies evaluating treatment of feline SBP, and most recommendations are anecdotal. However, recent guidelines stress the importance of confirming a diagnosis of SBP prior to considering systemic antibacterial therapy (see later section on Antibiotic Stewardship (Box 1)) [1, 22, 23]. Over-utilisation of antibiotics without confirmation of diagnosis is well-recognised, and the common practice of prescribing antibiotics 'just in case' is strongly discouraged [22–24, 26]. Topical antiseptic therapy is a more valid 'just in case' choice; however, prior cytology is always recommended [1].

Fig. 5 Bacterial colonies, usually cocci, are frequently observed in biopsies from feline cutaneous lesions with secondary bacterial infection (H&E, 400×). (Courtesy of Dr. Chiara Noli)

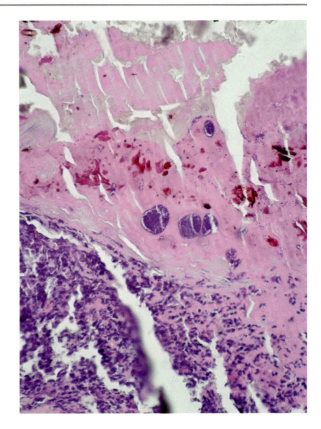

Topical Therapy

Although cats are often considered less tolerant of topical therapies, and even in dogs topical therapy is considered under-utilised [23], topical therapy has been recommended as the optimal sole antibacterial treatment for superficial infections whenever achievable for the pet and owner, particularly for localised or mild lesions. It is also recommended as the best option for pyoderma associated with methicillin-resistant staphylococci (MRS) [1]. Topical therapy has the advantage of more rapid lesion resolution, reduced duration of systemic antibiotics, physical removal of bacteria and debris from the skin surface and reduced impact on bystander commensals [1, 23]. The response in dogs with SBP to daily chlorhexidine spray (4%) for 4 weeks concurrently with twice weekly bathing with chlorhexidine shampoo was comparable to oral amoxicillin-clavulanic acid (amoxi-clav) [27]. Other small studies have similarly shown sole topical therapy to be effective [1].

Although a range of topical formulations are discussed for use in dogs, it is acknowledged there is limited evidence for efficacy and safety to guide optimal choices and protocols [23]. There is even less evidence in cats. However, the author has found a range of topical antiseptics and antibiotics helpful in the treatment of SBP in some cats, particularly for localised lesions. Chlorhexidine solution (2–3%

> **Box 1: Important Principles of Treatment for Cutaneous Bacterial Infections in Cats in Line with Good Antimicrobial Stewardship**
> - Have sufficient evidence to confirm a diagnosis of bacterial infection prior to instigating treatment (unless severe and life-threatening): Avoid 'just-in-case' usage.
>   - Cytology is essential; culture of bacteria from a skin surface swab does not confirm infection.
> - Choose antibiotics wisely, based on recommended treatment guidelines:
>   - Use first-line antibiotics for empirical use, assuming relevant options exist for the confirmed infection.
>   - Only use second-line antibiotics if adverse events limit use of first-line choices and if culture and sensitivity testing (C&S) supports efficacy.
>   - Do not use third-line antibiotics (e.g. cefovecin, fluoroquinolones) unless C&S indicates absence of other first- or second-line choices: Avoid justification due to 'ease of use' without actively discussing first-line oral alternatives.
> - Use correct dose and duration of treatment:
>   - Dose at the upper end of dose range as skin blood supply is comparatively poor, and weigh patients: slightly over-dose rather than under-dose.
>   - Follow duration guidelines for the confirmed infection, and re-evaluate clinical and cytological response prior to cessation of therapy.

once or twice daily), silver sulfadiazine 1% cream or mupirocin 2% ointment (twice daily) have apparent efficacy and safety [12, 13], and fusidic acid 1% viscous eye drops (Conoptal®; twice daily) may also be useful, particularly for facial/periocular lesions. Concern has been raised over the use of both mupirocin and fusidic acid in veterinary patients, potentially encouraging resistance in resident human staphylococci, and it has been recommended to restrict their use to cases without other practical choices [1, 23]. Shampoo therapy (chlorhexidine or piroctone olamine) once to twice weekly may be adjunctive for treatment or to inhibit recurrence of SBP, although it is poorly tolerated in many cats.

Excessive grooming and exacerbated self-trauma in response to topical therapies in cats, especially to ointments or creams, may sometimes limit their use. Body suits or conforming bandages may be helpful, particularly in cats with severe pruritus. Despite a common concern of owners that licking will remove topical medications, there is no evidence to confirm that grooming notably reduces efficacy of topical therapy, as lipophilic medications will be quickly absorbed after application.

Systemic Therapy
There is a lack of consensus on the most appropriate systemic antibiotics for treatment of SPB and some variation in recommendations with geographical region [23, 28]. First-line antibiotics are considered suitable choices for empirical therapy,

assuming a diagnosis is confirmed (e.g. intracellular cocci on cytology). Culture and antibiotic susceptibility testing (C&S) is important for cases that respond poorly to appropriate empirical therapy, or if there is higher risk of MRS (repeated antibiotic courses, other household pet carriers, some geographical regions) [1, 12].

Amoxi-clav and cephalexin are generally considered first-line choices for feline SBP (see later section on Antibiotic Stewardship) [12, 13]. Amoxi-clav was effective for eosinophilic plaques and partially effective for lip ulcers with concurrent bacterial infection [25]. Doxycycline is used in some countries for first-line therapy of SBP, but resistance in some geographical regions [29], and potential value for MRS and multidrug-resistant staphylococci in others [10], suggests it may be less appropriate for first-line use. There is also debate over the use of cefovecin as first-line treatment for feline SBP, and although it is commonly adopted, third-generation cephalosporins are considered critically important antibiotics in human medicine, reserved for life-threatening diseases [26, 30–32]. It has thus been recommended cefovecin is not appropriate for first-line treatment for feline SBP, unless, due to compliance issues, no other treatment is possible.

Second-line antibiotics may be considered if first-line antibiotics are not effective or serious side effects (real or potential due to previous history) limit the use of first-line choices. The major second-line choices for feline SBP are clindamycin or doxycycline, with preceding C&S optimal as efficacy is less predictable than for first-line choices (see later section on Antibiotic Stewardship). Lower sensitivity of staphylococcal isolates has been documented to clindamycin compared to amoxi-clav and cephalexin in South Africa [8] and to erythromycin in Malaysia [29]. Cefovecin is another potential second-line choice, when all avenues of oral administration of first-line and initial second-line choices have been exhausted. Second-generation fluoroquinolones (FQ) (enrofloxacin, marbofloxacin) are a final consideration, but restriction to cases with no other alternatives based on C&S is recommended. Ease of administration of FQ and low incidence of side effects are not justification for their use as first-line or early second-line options.

Third-line antibiotics are rarely indicated for feline SBP, with topical therapies, even requiring hospitalisation and/or sedation where necessary, preferable. They include third-generation FQ (orbifloxacin, pradofloxacin), aminoglycosides (amikacin, gentamicin) and rifampicin. Critical antibiotics, reserved for life-threatening infections in humans, with veterinary use discouraged, are not a consideration for treatment of SBP in any species (see later section on Antibiotic Stewardship).

Duration of Therapy

Although there is an absence of scientific evidence to confirm an optimal duration of therapy for SBP in either dogs or cats, current expert opinion recommends a 3-week therapy as most appropriate [1, 26]. Shorter courses may be considered, until clinical lesions and microbiological evidence of infection have resolved; however, re-evaluation of patients is essential to make this assessment [1, 28].

Treatment of the Primary Disease

It is well-recognised that the underlying primary cause of SBP must be managed to limit recurrence. However, there is less clarity on whether treatment of SBP and primary diseases need to occur concurrently or sequentially. As immunosuppressive therapy is contraindicated when treating infectious diseases, as a general rule, it is advised that SBP treatment be completed, prior to commencing any sustained glucocorticoid therapy (e.g. for primary allergy). In some cases of very active primary disease, resolution of SBP may not readily occur until the primary disease is more controlled. Management of primary atopic dermatitis in particular can be very challenging in some cats prone to secondary bacterial infections [13]. Ciclosporin therapy may be a more valid allergy treatment choice than glucocorticoids in this scenario, sparing innate immune responses (neutrophils, macrophages), albeit with slower onset of effect.

Deep Bacterial Infections

Chin Nodular Swelling: Secondary Deep Bacterial Pyoderma

Feline chin acne most typically presents with brown to black comedones and hair casts on the ventral chin and occasionally the margins of the lower or upper lips (Chapter, Idiopathic Miscellaneous Diseases). A proportion of affected cats develop notable swelling with draining tracts, often due to secondary deep bacterial infection. Of cats with feline acne presenting to referral hospitals in the USA, 42% had deep bacterial infection ($n = 72$) [33], and 45% had bacteria isolated from tissue cultures ($n = 22$), including all cats with evidence of folliculitis and furunculosis on histopathology. The most frequent bacteria isolated, typically in pure culture, were CoPS, followed by α-haemolytic streptococci, *Micrococcus* sp., *E. coli* and *Bacillus cereus*. Of note, *Pseudomonas aeruginosa* was isolated in heavy growth from the tissue biopsy of one healthy control cat [34].

Clinical Presentation

Deep pyoderma typically presents with large papules to nodular swelling with draining tracts (Fig. 6) and less commonly diffuse swelling. Lesions may be pruritic and/or painful, and regional lymph node enlargement can occur [3, 33, 34].

Diagnosis

Cytology from fine needle aspirates (FNA) or expressed discharge after initial surface cleansing may reveal intracellular bacteria within neutrophils and/or macrophages. Careful examination may be required as bacteria can be sparse in samples from nodular lesions despite marked inflammation.

Histopathology will typically reveal folliculitis, furunculosis and perifollicular to nodular pyogranulomatous inflammation (Fig. 7); bacteria present within follicular ostia or lumina in this setting, at least focally, confirm a diagnosis. Feline acne

Fig. 6 Feline chin acne: nodular swellings and drain tracts as a consequence of a deep bacterial infection. (Courtesy of Dr. Chiara Noli)

Fig. 7 Histopathological section from feline chin acne (H&E, 40×): multifocal nodular pyogranulomatous inflammation in the mid and deep dermis, mostly centred on the hair follicles, which appear completely destroyed. Haemorrhage is evident, which is reflected clinically by haemopurulent exudate. (Courtesy of Dr. Chiara Noli)

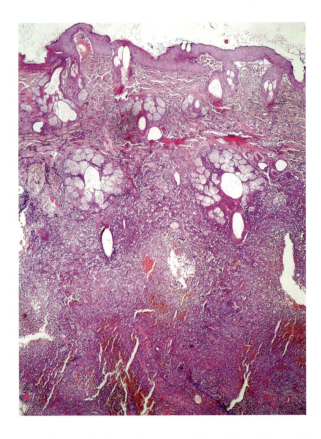

is associated with a spectrum of histopathology changes, with periglandular and/ or perifollicular inflammation usually dominating. Sebaceous gland ductal dilation and pyogranulomatous inflammation of sebaceous glands are also reported [34].

The presence of folliculitis and furunculosis without causal bacteria is suggestive of a role for secondary bacterial pyoderma, but exclusion of other causes including dermatophytosis is important, and special stains are warranted.

Bacterial culture of sterile tissue biopsies or FNA from affected regions is required to identify causal species and enable antibiotic susceptibility testing.

Treatment

Systemic antibiotics are indicated; if intracellular cocci are evident on cytology, empirical treatment with cephalexin or amoxi-clav is often considered suitable. If bacterial rods are present on cytology, or in geographical regions where MRSP is more common, C&S is recommended, optimally from tissue biopsies. The optimal duration of therapy for deep pyoderma is undetermined; however, a minimum of 4–6 weeks is often advised, continuing for at least 2 weeks beyond resolution or stasis of lesions [1, 26]. Comedones typically persist in feline acne following resolution of the bacterial infection, so further treatment of the underlying pathology is important to limit recurrent infection (Chapter, Idiopathic Miscellaneous Diseases) [33].

Discrete Nodules: Bacterial Pseudomycetoma

Some bacteria rarely cause localised discrete deep infections forming skin nodules that mimic fungal or neoplastic causes. Infections presumably occur following traumatic implantation of bacteria, which are most commonly *Staphylococcus* spp., but may be *Streptococcus* spp., *Pseudomonas* spp., *Proteus* spp. or *Actinobacillus* species.

Clinical Presentation

Single or multiple inflammatory nodules, with or without draining tracts, are typical. Discharge may contain small white grains or granules, composed of compact bacterial colonies [35]. A single case with less typical overlying thick crusting is reported in an FIV-positive cat, with concurrent SBP supported by cytology findings [36].

Diagnosis

Cytology of FNA from intact nodules or impression smears of freshly expressed exudate should reveal numerous bacteria, most typically cocci but dependent on causal species. Histopathology will reveal nodular to diffuse pyogranulomatous dermatitis and/or panniculitis with numerous macrophages, multinucleate giant cells and central aggregations of bacteria, often with a brightly eosinophilic amorphous periphery (Splendore-Hoeppli phenomenon) (Figs. 8 and 9) [35, 36].

Treatment

Surgical excision/drainage is important for resolution, as systemic antibiotics will often not penetrate into the central walled-off bacteria.

Fig. 8 Histopathological section from a lesion of bacterial pseudomycetoma (H&E 40×). There is a multifocal nodular pyogranulomatous inflammation with large bacterial colonies covered by bright red proteinaceous material, which appear clinically as white granules in the exudate. (Courtesy of Dr. Chiara Noli)

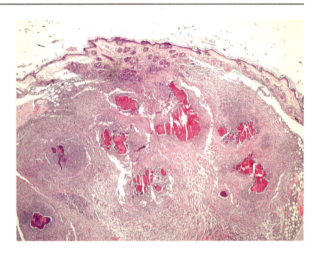

Fig. 9 A bacterial colony (dark blue in the centre, is surrounded by amorphous eosinophilic material (Splendore-Hoeppli phenomenon) (H&E, 400×). (Courtesy of Dr. Chiara Noli)

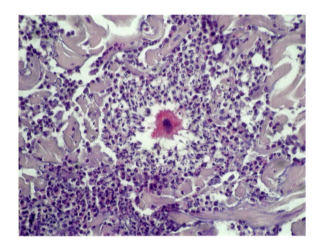

Subcutaneous Nodular Swellings with Abscessation: Anaerobic Bacteria

Painful rapidly progressing subcutaneous swellings are common in cats due to implantation of anaerobic bacteria, most typically associated with fight wounds although less commonly with other skin trauma including surgical wounds or catheterisation. Causal bacteria are often anaerobic or facultatively anaerobic oral commensals, including *Pasteurella multocida*, *Fusobacterium* spp., *Peptostreptococcus* spp., *Porphyromonas* spp. and gas-producing species such as *Clostridium* spp. and *Bacteroides* species [37].

Fig. 10 Swelling, ulceration, fistulisation and necrosis of the abdominal skin of a cat due to infection with anaerobic bacteria. (Courtesy of Dr. Chiara Noli)

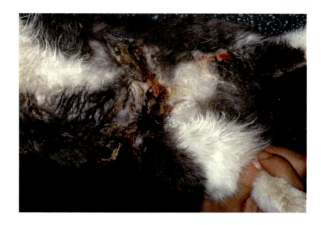

Clinical Presentation

Poorly demarcated areas of oedema and swelling are typical, which progress to abscessation (Fig. 10) and sometimes overlying skin necrosis. Lesions are often single, but may be multiple, and are usually painful. There is often associated pyrexia and malaise, especially with larger lesions or when bacteria produce toxins. Purulent abscess contents often have a putrid smell, and tissue crepitus may be apparent.

Diagnosis

The clinical presentation is usually diagnostic. Cytology of abscess contents, or FNA from oedematous areas in early lesions, should reveal intense neutrophilic inflammation, with bacterial rods and/or cocci often readily apparent. Mixed infections are not unusual. Culture is generally not required, but anaerobic sampling would be important to accurately identify most causal bacteria.

Treatment

Early lesions are usually managed successfully with systemic antibiotics, with most organisms sensitive to amoxi-clav or metronidazole. *Bacteroides* spp. may be resistant to ampicillin and clindamycin [30]. Surgical drainage of abscesses, with aeration and cleansing of infected tissue, is important to resolution.

Subcutaneous Nodular Swellings with Ulceration and Draining Tracts: *Nocardia, Rhodococcus* and *Streptomyces*

A number of bacterial species, many of which are ubiquitous environmental saprophytes, are rare causes of poorly demarcated nodular swellings with focal ulceration and draining tracts in cats. Infections are often locally invasive, and some species have a propensity to disseminate, particularly in immunocompromised cats. Most infections presumably occur following traumatic implantation.

Diagnostic tests are essential to accurately confirm the cause of this presentation. In addition to multiple potential bacteria, differential diagnoses include mycobacteria (Chapter, Mycobacterial Diseases), saprophytic fungi (Chapter, Deep Fungal Diseases) and sterile panniculitis (Chapter, Idiopathic Miscellaneous Diseases).

Cytology of FNA from oedematous tissue or fluid pockets or of smears from draining tracts (after initial skin surface cleansing) will typically reveal neutrophils and epithelioid macrophages, sometimes with multinucleate giant cells, regardless of the causal organism. Organisms will more often be detected within macrophages, with morphology varying with the causal species.

Histopathology of tissue biopsies will reveal nodular to diffuse pyogranuloma-tous dermatitis and/or panniculitis. Specials stains help elucidate the likely causal bacteria [38].

Bacterial culture from sterile fluid aspirates or tissue biopsies may be needed to confirm causal species and is optimal to determine antimicrobial susceptibility testing. It is important to alert the laboratory of the potential for unusual bacterial species with special culture requirements.

PCR testing can be useful for retrospectively identifying pathogens from formalin-fixed tissue samples if fresh samples are not available for bacterial culture [39].

Nocardiosis

Nocardia are ubiquitous soil and decaying vegetation saprophytes that may cause rare but potentially serious infection in cats, typically following implantation into skin wounds. Infection is more common in cats than dogs and may remain localised and indolent or be fulminant with wide dissemination; the latter course is more likely in immunocompromised hosts. *N. nova* is the most frequently identified causal species, but infections with *N. farcinica* or *N. cyriacigeorgica* also occur. Skin infections are most common, with occasional cases restricted to pulmonary or abdominal infection [40].

Clinical Presentation Progressive irregular nodules and punctate draining sinuses are typical (Fig. 11), often with concurrent malaise and respiratory signs. Skin infection may start with discrete abscesses that gradually extend into discharging, non-healing wounds. The extremities, ventral abdomen and inguinal areas are more often affected, and lymphadenopathy is common. Discharge may contain gritty granules (bacterial microcolonies) [40].

Diagnosis Filamentous bacteria that stain at least partially with acid-fast stains are typically prevalent on cytology and histopathology and appear branching or beaded (Fig. 12). Organisms may be found within clear lipid vacuoles [40]. Bacterial culture is slow; it is important to forewarn laboratories with potential cases.

Treatment Prompt early treatment of acute lesions, even in immunocompromised patients, can result in good outcomes. Surgical debridement and drainage to reduce residual organisms are optimal, and aggressive early excision, with potential later corrective surgery, is indicated. C&S is important to maximise treatment success. *N. nova* tends to have less resistance than other species and is often susceptible to sulphonamides,

Fig. 11 Localised
swelling, ulceration and
training tract in a cat
affected by cutaneous
nocardiosis. (Courtesy of
Dr. Carolyn O'Brien)

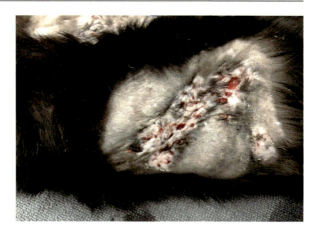

Fig. 12 Cytology of
nocardiosis: multiple
groups of bacteria (grains)
and slender and
filamentous *Nocardia
asteroides* microorganisms
are evident (MGG 1000×).
(Courtesy of Dr. Nicola
Colombo)

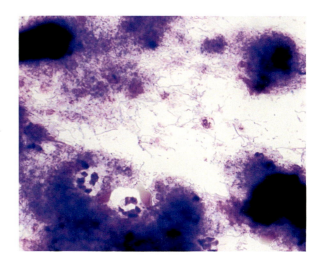

tetracyclines (minocycline, doxycycline), clarithromycin and ampicillin/amoxicil-
lin, but paradoxically not to amoxi-clav (clavulanic acid induces β-lactamase production in these species) nor to FQ. Amoxicillin (20 mg/kg twice daily) combined
with clarithromycin (62.5–125 mg/cat twice daily) and/or doxycycline (5–10 mg/kg
twice daily) is recommended over sulphonamides. Long-term therapy is generally
required (3–6 months), and recurrence is common with shorter treatment. *N. farcinica* is less commonly identified but is often multidrug resistant and highly pathogenic. Initial parenteral therapy with amikacin and/or imipenem combined with
trimethoprim-sulphonamides is a consideration [40].

Rhodococcosis

Rhodococcus equi is a ubiquitous soil-borne bacterium commonly pathogenic in
horses, where it produces a pyogranulomatous pneumonia and enteritis with high mortality in young foals. Infection is also increasingly documented in humans with immunocompromise and is reported in a small number of cats, involving skin (nodules with

focal ulceration and draining tracts, most frequently on the extremities), abdominal or thoracic cavities and/or the respiratory tract [41–43]. In one report, a pyogranulomatous skin disease and cellulitis (Fig. 13), different from usual presentations in cats, were described in a 2-year-old female domestic shorthaired cat [43]. Infection in local lymph nodes, presumably via lymphatic spread, is reported [41–43]. Implantation of organisms via skin wounds is proposed, with highest risk in cats with exposure to horses; infected foals shed copious bacteria into the environment via faeces [41].

Diagnosis Cytology of FNA samples and/or histopathology usually readily reveals gram-positive cocci to coccobacilli within macrophages (Fig. 14) [42, 43]. Bacterial culture is essential to confirm a diagnosis; the bacteria grow readily with aerobic culture within 48 hours, but organisms may be protected within macrophages in fluid samples, so tissue samples may be optimal [42].

Treatment C&S is important to guide potential therapy. *R. equi* infections are often refractory to conventional therapies in horses, and although a combination of rifampicin and erythromycin has been recommended, increasing resistance is rec-

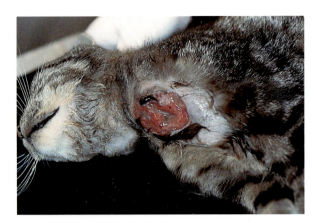

Fig. 13 Cutaneous *Rhodococcus equi* infection in a cat: pyogranulomatous dermatitis and cellulitis with superficial ulceration. (Courtesy of Dr. Anita Patel)

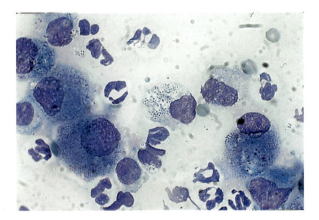

Fig. 14 Cytology of case in Figure 13: intracellular *Rhodococcus equi* organisms are evident in macrophages (MGG 1000×). (Courtesy of Dr. Anita Patel)

ognised [28]. In a confirmed feline case with a chronic limb lesion, *R. equi* displayed intermediate sensitivity to amoxi-clav, rifampicin and erythromycin and sensitivity to cephalexin and gentamicin, but the cat deteriorated despite initial cephalexin and later surgical debridement and gentamicin therapy and was euthanised [42]. In another case with sensitivity reported to doxycycline, enrofloxacin and cefuroxime, response to enrofloxacin and later doxycycline was poor [43]. However, doxycycline was reported effective in thee kittens with *R. equi* pneumonia, from two litters in a cattery in Australia where the source of the infection was undetermined [41].

Streptomycosis
Streptomyces spp. are ubiquitous environmental bacteria that very rarely cause irregular nodular lesions with draining tracts and dark tissue granules on the limbs and ventral abdomen of cats. One cat without skin lesions had mesenteric and lymph node infection. Two cats were FIV and/or FeLV positive and two cats had unknown viral status [38].

Diagnosis Gram-positive rods to coccobacilli were present on cytology and histopathology, and bacteria were identified by PCR testing [38].

Treatment All four cats failed to respond to surgical and/or multiple antibiotic therapy and were euthanised following 6–18 months of disease [38].

Dermatophilosis
Dermatophilosis is a contagious and potentially zoonotic disease caused by *Dermatophilus congolensis,* which most commonly affects cattle, sheep and horses in tropical and subtropical climates. The organism does not survive readily in the environment, and infected or carrier animals are the main source. Infection is reported very rarely in cats. Two presumptive cases presented with nodular swelling and draining tracts overlying infected lymph nodes, with associated skin surface crusting. Characteristic gram-positive branching filamentous bacteria were evident on histopathology, and both cats resided on farms in tropical northern Australia. Surgical excision was curative in one cat, and the other cat was euthanised prior to diagnostics [44]. *Dermatophilus congolensis* was isolated in pure culture from crusts in another cat presenting with crusting and exudation on the ventrolateral lip margins; it was reported sensitive to oxytetracycline and penicillin, but resistant to ampicillin, amoxicillin, gentamicin and cefoperazone [45]. Characteristic filamentous branching bacteria (Fig. 15) were present on cytology from a fourth cat with draining tracts on two lower limbs; bacterial culture was negative, but the cat responded completely to amoxicillin therapy for 10 days [46].

Streptococcal Infection
One case of extensive oedematous swelling with multifocal ulceration and draining tracts is reported on the hindlimb of a cat, associated with numerous clusters and chains of gram-positive cocci, identified by tissue PCR as *Streptococcus* spp., in skin

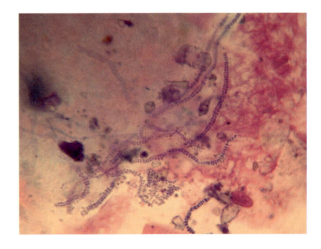

Fig. 15 Cytology of *Dermatophilus congolensis*: long colonies, like train tracks, are characteristic (Diff Quik, 1000x)

and underlying bone. Clusters of bacteria surrounded by eosinophilic amorphous material (Splendore-Hoeppli phenomenon) were present on histopathology [39].

Actinomycosis

Actinomyces spp. are oral saprophytes in a variety of animals including dogs and cats, which are most commonly associated with soft tissue and bone infections in the jaws of cattle. Rare cutaneous infections are reported in dogs, characterised by nodular swellings with discharge, typically on the extremities. Although abdominal infection with *Actinomyces* spp. is documented in one cat, and isolation of *Actinomyces* spp. is reported concurrently with other bacterial species, or from lesions without concurrent histopathological confirmation, there are no confirmed reports of *Actinomyces* spp. causing cutaneous infections in cats [47, 48].

Rapidly Progressive Oedematous Swelling to Necrosis and Septic Shock: Necrotizing Fasciitis

Necrotizing fasciitis is a rapidly progressive and frequently fatal syndrome caused by severe bacterial infection of subcutaneous tissue (fascia) and adjacent skin, typically associated with septic shock. *Streptococcus canis* is a recognised cause of fulminant disease in humans and dogs and has also been associated with an outbreak of fatal necrotizing fasciitis in shelter cats in southern California. Clonal bacteria were identified and spread via close physical contact was proposed. *S. canis* is a normal inhabitant of the urinary, reproductive and gastrointestinal tracts of dogs and cats, and although infections are rare and most typically associated with immunocompromise, necrotizing fasciitis can occur in immunocompetent hosts. In contrast to dogs where *S. canis* is mainly associated with skin infections, respiratory tract infections are more typical in cats [49]. One case associated with *S. canis* in a single cat following minor limb trauma is also reported [50].

Fig. 16 Large areas of necrosis and ulceration in a cat with necrotising fasciitis. (Courtesy of Dr. Susan McMillan)

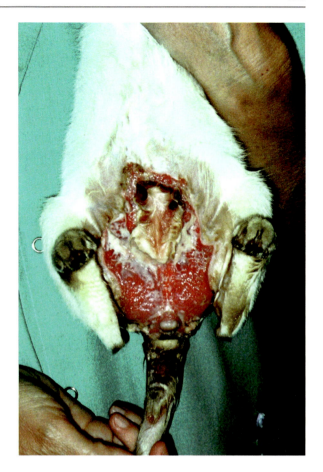

Another form of necrotizing fasciitis in people, occurring after minor skin trauma (catheterisation, hospitalisation), has been associated with multiple concurrent bacteria including *Staphylococcus* spp., *Streptococcus* spp., *Pseudomonas* spp. and *E coli*. Single case reports in cats are described due to *Acinetobacter baumannii* [51] and multiple bacteria (*E. coli*, *Enterococcus* sp. and *S. haemolyticus*; *E. coli*, *Enterococcus faecium* and *S. epidermidis*) [52, 53].

Clinical Presentation Poorly demarcated painful regions of oedema and erythema are typical, associated with rapid development of signs of septic shock (pyrexia, severe malaise, collapse). Skin lesions progress to large areas of skin necrosis (Fig. 16).

Diagnosis FNA of affected regions reveals neutrophilic inflammation, and causal bacteria are usually apparent intracellularly within neutrophils. Bacterial culture of sterilely collected fluid or tissue samples is required to confirm the causal species. It is important to interpret culture results in conjunction with bacterial morphology from cytology and/ or histopathology, as contaminant species may be cultured from exudative lesions.

Treatment Most cases reported in cats have been fatal. Urgent extensive surgical debridement, with removal of the bacterial nidus and all necrotic tissue to limit further extension along fascial planes, is recognised as crucial for suspected cases prior to availability of diagnostic test results, together with broad-spectrum intravenous antimicrobial treatment and critical care. Reconstructive surgery may be required after recovery [50].

Diagnostic Tools for Feline Cutaneous Bacterial Infections

Clinical Lesions and Historical Features

Prior to reaching for diagnostic tests, careful clinical examination and history taking for each case can focus the diagnostic possibilities and guide the most appropriate test choices. Knowledge of the more likely differentials for specific skin lesions, and the major differentials when bacterial infections are being considered, is helpful (see Table 1).

Cytology

Cytology is often the most useful initial test when considering bacterial dermatoses and may confirm a diagnosis. The most suitable technique will vary with the clinical lesions (see Table 1).

Adhesive tape impressions are suitable for all superficial skin lesions, including alopecia, scaling, crusting, excoriations, ulceration and papules. More exudative lesions can be gently blotted with a dry gauze swab prior to sampling. Good quality adhesive tape (clear, transparent, strongly adhesive; 18–20 mm width) is optimal for use on standard glass slides. Tape strips (~5–6 cm long) are pushed firmly onto lesional skin, squeezing gently on intact papules or plaques and repositioning repeatedly until adhesiveness reduces. Tapes are stained with a Romanowsky stain (e.g. Diff-Quik®) without initial fixation (dissolves the adhesive, reducing clarity). Use of the red stain is useful in cats to aid identification of eosinophils. Tapes can be dipped into stain pots, as for glass slides (Fig. 17).

Glass slide impressions are suitable for moist lesions, including erosions and ulcers, and for sampling pustules after rupture with a sterile needle. Slides are stained with a Romanowsky stain including the fixative. Heat fixing is not required.

Fine needle aspirates are suitable for deeper lesions, including larger papules and nodules. The skin surface should be gently disinfected with alcohol prior to sampling. Aspirated samples are quickly sprayed from the hub of the needle onto glass slides using an air-filled syringe. Slides are air-dried prior to routine staining with a Romanowsky stain or with gram and/or acid-fast stains for identification of less common bacterial species.

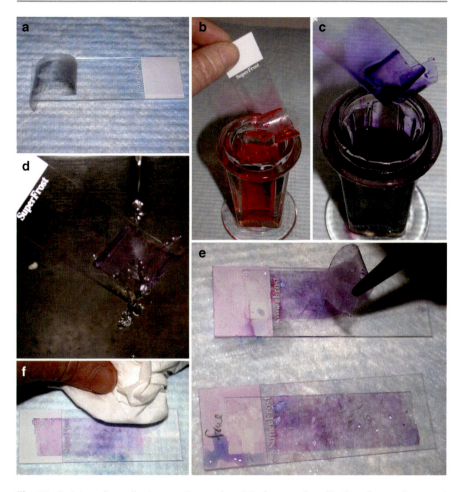

Fig. 17 Staining of an adhesive tape impression: (**a**) after sample collection, the tape is pressed firmly at one end, adhesive side down, onto a glass slide and curled into a slightly offset cylinder; (**b**) tape is dipped into red stain of Diff-Quik® (6 × 1 s dips); (**c**) tape is dipped into blue stain of Diff-Quik® (6 × 1 s dips); (**d**) stain is rinsed off tape under a gentle stream of water; (**e**) tape is uncurled by grasping free edge with forceps and laid flat on the glass slide; (**f**) tape is dried and flattened firmly onto the glass slide by wiping the surface firmly with a tissue

Interpretation of Cytology Samples Bacteria are very sparse in an oil immersion field (OIF: 1000x magnification). Oil immersion is required for accurate recognition of bacteria on normal skin surface samples despite being readily culturable from skin surface swabs (which sample thousands of OIF). The presence of increased numbers of bacteria clustered (colonising) on keratinocytes represents bacterial overgrowth (Fig. 18), while bacteria present intracellularly or closely associated within neutrophils (Figs. 19 and 20) and/or macrophages confirm infection. In deeper samples (e.g. FNA), bacteria should be absent if

Fig. 18 Numerous cocci clustered on keratinocytes on an adhesive tape impression confirm bacterial overgrowth, while cocci intracellularly within one intact neutrophil (multi-lobed nucleus) suggest concurrent focal bacterial infection (100x lens, oil immersion; stained with Diff-Quik® as per Fig. 17)

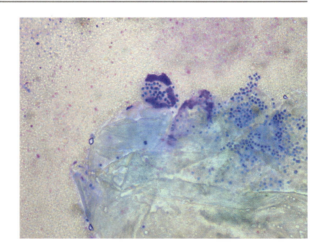

Fig. 19 Cocci intracellularly and associated with degenerate neutrophil remnants and nuclear streaming on an adhesive tape impression confirm bacterial infection (100x lens, oil immersion; stained with Diff-Quik® as per Fig. 17)

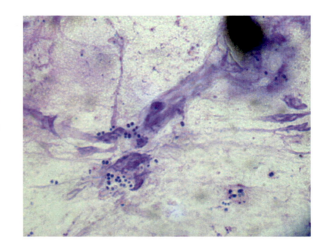

sterile technique was successfully employed; the presence of any bacteria is abnormal. Adhesive tape impressions require some experience for efficient and accurate examination. Keratinocytes typically dominate, staining pale to mid-blue and ranging from sheets of flat polyhedral cells to single or clustered shards (follicular cells). Inflammatory cells stain purple, with neutrophils most prevalent; they may be in small clusters or form peripheral rims around keratinocyte sheets. Eosinophils may also be present, particularly in cases with underlying hypersensitivity. Neutrophils should be plentiful in erosive or ulcerative samples but may be relatively sparse in drier lesions. Neutrophils degenerate quickly on the skin surface, often appearing as elongated strands of nuclear material (nuclear streaming). Tapes should be scanned under low power microscopy (4x lens) for areas of dense cells or neutrophil clusters to examine under higher power (see Fig. 20). Microscope oil is placed directly on the tape surface to examine under OIF.

Fig. 20 Keratinocytes distributed singly and in sheets on an adhesive tape impression with a central neutrophil cluster (4x lens; stained with Diff-Quik® as per Fig. 17)

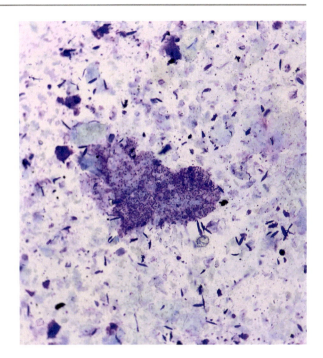

Bacterial Culture

Culture and antibiotic susceptibility testing (C&S) is vital for bacterial infections caused by species with unpredictable antimicrobial sensitivity profiles, such as rods and many of the environmental bacteria that cause sporadic deep infections. In contrast, empirical therapy based on cytology is considered appropriate for many cases of SBP [22]. C&S is indicated in severe life-threatening infections, if rod-shaped bacteria are evident on cytology (where sensitivity is less predictable), if empirical therapy does not resolve lesions or when antibiotic resistance is more likely in that geographical region or patient [1, 22]. There is no current evidence to support any negative influence of current antibiotic therapy on isolation of causative bacteria; thus, withdrawal of systemic or topical antibiotics is considered unnecessary [23].

Superficial Skin Sampling Collection of culture samples from primary lesions is optimal with pustules ruptured and papules incised with a needle prior to sampling with a culture swab, without preceding skin disinfection [22, 23]. Sterile tissue biopsy may be more reliable for papules [23]. In dogs with SBP, dry culture swabs were equally effective as moistened swabs or light scrapings for sampling a range of superficial lesions, including papules. Swabs were rubbed vigorously for 5–10 seconds on representative lesions, confirmed as SBP on cytology, without prior skin disinfection [54]. Culture swabs from the skin surface have also been well utilised for numerous feline skin culture studies sampling a range of skin lesions [5, 7, 9, 17, 19]. Swabs should be immediately placed in transport medium and optimally refrigerated prior to transit to limit overgrowth of contaminants, particularly in warm climates.

Multiple strains of *S. pseudintermedius*, with distinct antimicrobial resistance profiles, have recently been detected from single lesions in canine SBP, with pustules and, to a lesser extent, papules associated with less species and strain diversity than collarettes and crusts. Pustules and papules were swabbed after incision with the tip of a sterile needle. Crusts and collarettes were sampled by touching a culture swab to the edges of lesions [55]. These findings reinforce the value of sampling primary lesions whenever possible and raise the potential importance of collecting multiple samples from a range of primary lesions to aid identification of all potential pathogens collectively contributing to infection in a patient.

Deeper Skin Sampling FNA or tissue biopsies collected with sterile technique are appropriate for bacterial culture from nodular lesions, with tissue samples most reliable. The surface epidermis may be excised after sample collection to help avoid isolation of contaminants. Swabs of discharging tracts are not suitable, as a range of contaminant bacteria are readily isolated [22]. When an infectious cause remains uncertain, and a range of infectious agents with varying culture requirements are differentials, tissue culture samples can be held refrigerated in a sterile container on a sterile saline-moistened swab, pending histopathology.

Culture Techniques Minimum microbiology evaluation should include complete speciation of staphylococci, regardless of tube coagulase status, and an antibiogram for all cultured isolates [1]. In-house culturing can be clinically misleading, resulting in erroneous and ineffective treatments and is not recommended, particularly for superficial skin sampling [28].

Culture Interpretation Culture results should always be interpreted in light of concurrent cytology findings and the likely pathogens in that location. Growth of bacteria in the laboratory alone does not confirm a pathogenic role. The morphology of cultured isolates must be consistent with morphology of bacteria evident on cytology for isolates to be relevant. Even bacteria with alarming multidrug resistance profiles can be inadvertent contaminants or incidental commensals, without any role in the current skin disease [1, 22]. However, correctly discerning the relevance of cultured isolates is not always straightforward; although CoPS are proposed as the major skin pathogens, commensal CoNS and a variety of environmental saprophytes may be pathogenic at times, particularly with concurrent immunosuppression [1, 22].

Histopathology

Skin biopsies for histopathology are essential to confirm a diagnosis for many deep nodular lesions. Multiple excisional biopsies are optimal, sampling any smaller peripheral lesions in addition to large lesions and avoiding central areas of large lesions which may be necrotic. Larger lesions should be sectioned to ensure adequate formalin penetration. Biopsies for histopathology should be placed in formalin immediately after collection. Biopsy samples can also be retained frozen for potential PCR or other molecular testing.

Histopathology is less often indicated for superficial infections but may be important where cytology results are inconclusive or presentations are atypical for SBP. Punch biopsies are suitable for small lesions (pustules, papules) or uniform lesions (plaques, erythema, crusting). Elliptical samples are most useful for transitional areas and edges of ulcerative lesions.

PCR Testing

PCR testing can be helpful to identify species not readily culturable in the laboratory. It is ideally performed on fresh tissue biopsies, collected with sterile technique, but can also be performed on formalin-fixed samples, assuming fixation in formalin was for <24 hour. PCR detection from swab samples does not confirm any role as a pathogen for environmental bacteria (e.g. *Nocardia* spp.) as detection may simply reflect skin contaminants.

Treatment Principles for Feline Cutaneous Bacterial Infections and Antimicrobial Stewardship

Antimicrobial Resistance and Stewardship

Increasing development of antimicrobial resistance is of profound concern in recent years and has marked impact on human and animal health and related economics. It is undeniable that antimicrobial use can result in antimicrobial resistance in the species that is being treated and that some resistant pathogens or resistance mechanisms can be transmitted bi-directionally between animals and humans [1, 28, 56].

Methicillin resistance of *Staphylococcus* spp. relevant to veterinary medicine has been recognised as a serious problem worldwide since the late 1990s, with geographical variation in incidence, but rapid escalation of resistant *S. pseudintermedius* (MRSP), *S. aureus* (MRSA) and *S. schleiferi* species. Acquisition of methicillin resistance confers resistance to all β-lactam antibiotics, including cephalosporins. MRS isolates also frequently acquire co-resistance to other classes of antibiotics, especially FQ and macrolides [18, 19]. MRSP in particular is not uncommonly multidrug resistant (resistance to at least six antibiotic classes). As *S. pseudintermedius* is a major canine pathogen and a recognised feline pathogen, this has created significant new veterinary challenges [1].

Inappropriate use of antibiotics in the veterinary arena is considered an important factor promoting progression of resistance [1, 28, 56].

- **Cefovecin:** Despite being reported as the most frequently chosen antibiotic for use in cats in recent studies, and specifically the most frequently used for skin infections or abscesses, it is a third-generation cephalosporin, which is considered 'highest priority/critically important antimicrobials' in human medicine, reserved for life-threatening infections or when culture and susceptibility testing does not indicate alternate antibiotic choices [26, 31]. Reported use is often 'just in case', without any clinical and/or cytological evidence to confirm a role for bacterial infection [31]. Alarmingly, only 0.4% of prescriptions in >1000 cats had C&S testing performed at the time of use and none prior to use. In addition, nearly 23% had concurrent glucocorticoid treatment, with long-acting methyl-prednisolone acetate injections in 38% of these, although these drugs are contraindicated in the face of active infections [31]. Prescription of cefovecin due to convenience of administration is not a justification for valid use.
- **Fluoroquinolones:** There is evidence that FQ therapy can promote colonisation with bacteria carrying more resistance genes. FQ therapy was a significant risk factor for isolation of MRS, multidrug-resistant staphylococci, and FQ-resistant staphylococci from mucosal samples in dogs in a recent study in England [56]. Clindamycin and amoxi-clav therapy were not significantly associated with detection of antibiotic resistance, but cephalexin was, potentially due to longer treatment courses typically used in contrast to amoxi-clav. FQ maintained this effect at 1 month post-treatment and cephalexin until at least 3 months post-treatment [56]. FQ should not be used as first-line treatment options.

Feline MRS Infections There are increasing reports of MRSP and MRSA skin isolates from cats with skin lesions, although rarely with confirmed pyoderma, in multiple regions of the world [6, 8, 10, 57]. Variable co-resistance of isolates is documented, including MRSA with FQ resistance (11.8%) in Australia [10], MRSP with multidrug resistance in Thailand [57] and MRSP also resistant to TMS (30.8%), chloramphenicol (7.7%) or clindamycin (7.7%) in Australia [10]. MRSP isolates from cats are typically sensitive to rifampicin, FQ (second- or third-generation) and amikacin. CoNS that are more frequently isolated in cats are also often methicillin-resistant and multidrug resistant [6].

Risk factors increasing the likelihood of MRS infections in cats are currently unknown. Risk factors identified in dogs include prior antibiotic therapy, eating animal stools and contact with veterinary hospitals. Despite confirmed sharing of staphylococcal isolates including MRSP between pets, dogs from multidog households appear less likely to have mucosal MRS [56].

Antimicrobial Stewardship The appropriate use of antimicrobials to reduce promotion of further antimicrobial resistance is an important concept referred to as antimicrobial stewardship. The first important principle of appropriate antibiotic usage is to prescribe antibiotics only in patients with sufficient evidence to confirm a diagnosis of bacterial infection. Use of antibiotics 'just in case', especially without prior diagnostics or when diagnostics fail to confirm bacterial infection, is strongly discouraged [23, 24, 26, 30, 31].

The second important principle of appropriate antibiotic usage is wise choice of antibiotic, based on the likely causal bacteria and their likely sensitivity profiles. Empirical choice is appropriate for diseases where causal pathogens are fairly predictable and have fairly predictable antibiotic susceptibility profiles and first-line antibiotics (see later) are appropriate. Use of antibiotics that have greater value for some resistant bacteria (second- or third-line antibiotics) is not suitable without evidence from C&S that they are appropriate and first-line choices are not, unless facing life-threatening situations.

The final important principle of appropriate antibiotic usage is to use the correct dose and duration of the chosen antibiotic, taking care to weigh patients accurately prior to therapy and rounding doses up rather than under-dosing (see Table 2). Although sound evidence is lacking, it is generally recommended that treatment of superficial infections continues for 3 weeks and deep infections for at least 4 weeks (and sometimes many months for difficult pathogens). See specific diseases for further guidelines.

Antibiotic Choices

Antibiotic classes are divided into generations based on differences in their spectrum of activity [30], and they can also be divided into groups based on current prescribing guidelines. There is no clear consensus on optimal antibiotic choices for bacterial infections in either dogs or cats [1, 26, 28, 30, 31, 58], with a general paucity of scientific evidence to clarify. The following recommendations for feline cutaneous bacterial infections are based on a compilation of current expert opinion in both veterinary and human medicine.

First-line antibiotics are considered most appropriate for empirical therapy of diagnosed infections, as they are generally well-tolerated and have high efficacy against the expected causal bacteria [26]. Empirical therapy appears suitable for treatment of feline pyoderma. First-line choices for feline pyoderma are the following:

- *Amoxi-clav or cephalexin* – both reported with high levels of sensitivity to isolated *Staphylococcus* spp. [8] Even in regions where MRS are common in canine SBP, MRS infections in cats appear very rare, and most reports are of laboratory isolates [18, 19].

Second-line antibiotics should only be used when there is culture evidence that first-line drugs will not be effective or as initial empirical therapy for severe infections while awaiting C&S results if resistance to first-line drugs is more likely. This classification includes newer broad-spectrum antibiotics important to animal and human health, so reserving their use to necessary cases is prudent. Not all second-line choices are equal, with a hierarchical consideration recommended, guided by regional data [30]. Second-line antibiotics relevant to treatment of feline skin infections include the following:

Table 2 Systemic antibiotic choices for feline bacterial infections in line with antimicrobial stewardship guidelines* [26, 28, 30, 31, 58]

Diagnosis	First-line: potential empirical therapy (dose mg/kg, frequency)				Second-line: only when C&S supports use and first-line not suitable; or while C&S pending if resistance likely (dose mg/kg, frequency)				Third-line: only when C&S supports use and no other choices (dose mg/kg, frequency)			Critical: (no veterinary use)
	AMC (20–25 BID)	CX (20–25 BID)	DXY[a] (5 BID)	METR (10 BID)	CLI (5.5–11 BID)	FQ 2nd Marbo (2.7–5.5 SID), Enro (5 SID)	CHL (50 BID)	TMS (15 BID)	CFV[e] (8 q 14d)	FQ 3rd Prado (7.5 SID) Orbi (2.5–7.5 SID)	GNT, AMK, RIF	VAN, TEI, TEL, LIN
SBP/ DBP	S[b,c]	S[b]	M	A (DBP only)	Some MSSP/MSSA Some MRSP/MRSA				MSSP/MSSA only	Some MRSP/MRSA	MRSP/MRSA	
Abscess /cellulitis	S	M	M	S	S[d]	R				R		
Nocardia	R	R	M	R		R		M		R		
Rhodococcus	R	R	M	R							M	
Uncertain	No antibiotics "Just in case" use strongly discouraged[24,56]											
Side effects	GIT (mild)	GIT (more)	Oesophageal stricture (water-swallow)		GIT (mild)	Retinal degeneration (enro, higher doses)	Myelosuppression; aplastic anaemia in people handling	Blood dyscrasia		Retinal degeneration (orbi, higher doses)	Severe risk: renal, hepatic, ototoxic	

*Some regional variation acceptable: judicious antibiotic use requires consideration of local availability, veterinary licensing, recommendations for human use, and regional antimicrobial susceptibility data [1]

Antibiotic abbreviations: *AMC* amoxicillin–clavulanic acid, *AMK* amikacin, *CFV* cefovecin, *CHL* chloramphenicol, *CLI* clindamycin, *CX* cephalexin, cefadroxil; *d* day, *DXY* doxycycline, *Enro* enrofloxacin, *FQ 2nd* second generation fluoroquinolone, *FQ 3rd* third generation fluoroquinolone, *GNT* gentamicin, *LIN* linezolid, *Marbo* marbofloxacin, *Orbi* orbifloxacin, *Prado* pradofloxacin, *q* every, *RIF* rifampicin, *TEI* teicoplanin, *TEL* telavancin, *TMS* trimethoprim sulphonamide, *VAN* vancomycin

General abbreviations: A potential adjunctive value only: not as sole treatment, C&S culture and antibiotic susceptibility testing, DBP deep bacterial pyoderma, GIT gastro-intestinal tract, M some resistant isolates, at least in some geographical regions, MSSP methicillin-sensitive Staphylococcus pseudintermedius, MRSP methicillin-resistant Staphylococcus pseudintermedius, MSSA methicillin-sensitive Staphylococcus aureus, MRSA methicillin-resistant Staphylococcus aureus, MSSA methicillin-sensitive Staphylococcus aureus, R high levels of resistance for common causal bacteria, S typically high levels of sensitivity for causal bacteria, SBP superficial bacterial pyoderma

[a]May be best considered 2nd-line, particularly in regions where MRSP-isolates are more often susceptible to doxycycline; Minocyclin 8mg/kg once daily can be used if doxycycline unavailable/expensive

[b]Assuming intracellular cocci are present on cytology

[c]May be the choice when cocci and rods are present on cytology; C&S is indicated if rods are exclusively present on cytology

[d]Resistance occurs with some Bacteroides spp.. and most gram-negative bacteria

[e]Often considered second-line, or even first-line; however, third-generation cephalosporins are considered third-line in human medicine

- *Clindamycin* – registered for use in many countries for skin and soft-tissue infections. Although there is some debate in veterinary medicine, macrolide antibiotics are not first-line choices in human medicine [30]. Clindamycin has also been shown to have lower levels of sensitivity to staphylococcal isolates in some studies, and a bacterial culture and susceptibility test is recommended prior to its use [8].
- *Doxycycline* – considered first-line in some regions. However, it may be generally less suitable as a first-line choice considering that high levels of resistance are documented in staphylococcal isolates in some regions [10, 29], even though lower resistance in others [8]. Minocycline has a similar spectrum of action to doxycycline and is less expensive and more available in some countries but may be associated with more gastrointestinal irritation [30].
- *Cefovecin* – effective against some gram-negative and anaerobic bacteria in addition to gram-positive bacteria, providing a broader spectrum of activity than second-generation cephalosporins such as cephalexin. There is generally poor activity against *Pseudomonas* spp. and enterococci. Although typically considered first- or second-line in veterinary medicine, third-generation cephalosporins are considered critically important antibiotics in human medicine reserved for life-threatening diseases (third-line), so classification as second-line is questioned [30].
- *Second-generation FQ (enrofloxacin, difloxacin, marbofloxacin, ciprofloxacin)* – primarily target gram-negative bacteria, which are less frequent skin pathogens.
- *Trimethoprim-sulphonamides* – greater risk of side effects in cats and lower sensitivity of many bacteria compared to other choices reduce the suitability of this option; may be effective for some MRS.

Third-line antibiotics are very important to animal and human health, especially for treatment of multidrug-resistant bacteria, and their use should be only considered when C&S indicates a lack of other treatment choices. Many are not licensed for veterinary use [26, 30]. Their use for superficial infections is strongly discouraged. Third-line choices for cats with severe bacterial cutaneous infections include the following:

- *Third-generation FQ (pradofloxacin and orbifloxacin)* – have an increased gram-positive and anaerobic spectrum compared to second-generation FQ, in addition to good gram-negative coverage; considered unlikely to be effective for *Nocardia* spp. [30].
- *Aminoglycosides (gentamicin, amikacin)* – potential considerations only for life-threatening skin infections, but have considerable risk of severe renal side effects, requiring careful monitoring, concurrent fluid therapy and brief duration therapy

- *Other new and old antibiotics* (*chloramphenicol, clarithromycin, rifampicin, imipenem, piperacillin*) – potential use for MRS and multidrug-resistant bacteria, but considerable potential for moderate to severe side effects
- *Newest generation antibiotics* (e.g. *vancomycin, teicoplanin, telavancin, linezolid*) – deemed of critical importance to human health and strongly discouraged/unavailable for veterinary use [1, 26]

Management of Veterinary Patients with MRS Infection

Transmission of MRS between humans and various animal species including cats is documented [1, 28]. MRSA and methicillin-resistant CoNS, including *S. haemolyticus, S. epidermidis* and *S. fleurettii*, were co-isolated from multiple cats, horses and humans on one farm in Europe, with isolates sharing the same characteristics [59]. Concern is thus raised when MRS infections are documented in veterinary species, when greater bacterial numbers are likely to increase the risks of transmission.

It is currently recommended that pets with MRS infections have limited contact with other pets or humans until their infections are controlled and that good hand hygiene and heightened cleaning protocols are used in the home environment to reduce potential transmission. Veterinary hospitals are also recognised as potential sources of MRS transmission, and adherence to strict hand hygiene (proper washing/drying and use of alcohol-based hand sanitizers) between handling all patients and regular cleaning and disinfection protocols will reduce the risks of transmission, with MRS susceptible to commonly used disinfectants. Barrier nursing protocols for hospitalised patients with known MRS infections are recommended [1, 56].

Despite concerns over the potential challenges of treatment of MRS infection, resistant isolates are not more virulent or likely to cause infection than non-resistant isolates. There is no current evidence to support attempted decolonisation of patients colonised by MRS, and thus, screening of clinically normal animals for carriage of MRS is currently not recommended [1].

Conclusion

Feline cutaneous bacterial infections range from common secondary to rare but potentially life-threatening deep and disseminated infections. Causal pathogens include normal skin and mucosal commensals and a range of environmental saprophytes. Development of antimicrobial resistance, particularly methicillin resistance in staphylococci, poses increasing veterinary challenges. Accurate and efficient diagnosis is important to expedite appropriate treatment and to limit further promotion of antibiotic resistance by restricting use of antibiotics to patients with confirmed disease.

References

1. Morris DO, Loeffler A, Davis MF, Guardabassi L, Weese JS. Recommendations for approaches to methicillin-resistant staphylococcal infections of small animals: diagnosis, therapeutic considerations and preventative measures: Clinical Consensus Guidelines of the World Association for Veterinary Dermatology. Vet Dermatol. 2017;28:304–30.
2. Rossi CC, da Silva DI, Mansur Muniz I, Lilenbaum W, Giambiagi-deMarval M. The oral microbiota of domestic cats harbors a wide variety of *Staphylococcus* species with zoonotic potential. Vet Microbiol. 2017;201:136–40.
3. Weese JS. The canine and feline skin microbiome in health and disease. Vet Dermatol. 2013;24:137–45.
4. Patel A, Lloyd DH, Lamport AI. Antimicrobial resistance of feline staphylococci in South-Eastern England. Vet Dermatol. 1999;10:257–61.
5. Patel A, Lloyd DH, Howell SA, Noble WC. Investigation into the potential pathogenicity of *Staphylococcus felis* in a cat. Vet Rec. 2002;150:668–9.
6. Muniz IM, Penna B, Lilenbaum W. Methicillin-resistant commensal staphylococci in the oral cavity of healthy cats: a reservoir of methicillin resistance. Vet Rec. 2013;173:502.2. https://doi.org/10.1136/vr.101971.
7. Igimi SI, Atobe H, Tohya Y, Inoue A, Takahashi E, Knoishi S. Characterization of the most frequently encountered Staphylococcus sp. in cats. Vet Microbiol. 1994;39:255–60.
8. Qekwana DN, Sebola D, Oguttu JW, Odoi A. Antimicrobial resistance patterns of *Staphylococcus* species isolated from cats presented at a veterinary academic hospital in South Africa. BMC Vet Res. 2017;13:286. https://doi.org/10.1186/s12917-017-1204-3.
9. Abraham JK, Morris DO, Griffeth GC, Shofer FS, Rankin SC. Surveillance of healthy cats and cats with inflammatory skin disease for colonization of the skin by methicillin-resistant coagulase-positive staphylococci and *Staphylococcus schleiferi* ssp. *schleiferi*. Vet Dermatol. 2007;18:252–9.
10. Saputra S, Jordan D, Worthing KA, Norris JM, Wong HS, Abraham R, et al. Antimicrobial resistance in coagulase-positive staphylococci isolated from companion animals in Australia: a one year study. PLoS One. 2017;12:e0176379. https://doi.org/10.1371/0176379.
11. Older CE, Diesel A, Patterson AP, Meason-Smith C, Johnson TJ, Mansell J, Suchodolski J, Hoffmann AR. The feline skin microbiota: the bacteria inhabiting the skin of healthy and allergic cats. PLoS One. 2017;12:e0178555. https://doi.org/10.1371/vr.0178555.
12. Wildermuth BE, Griffin CE, Rosenkrantz WS. Feline pyoderma therapy. Clin Tech Small Anim Pract. 2006;21:150–6.
13. Scott DW, Miller WH, Erb HN. Feline dermatology at Cornell University: 1407 cases (1988–2003). J Fel Med Surg. 2013;15:307–16.
14. Yu HW, Vogelnest LJ. Feline superficial pyoderma: a retrospective study of 52 cases (2001–2011). Vet Dermatol. 2012;23:448–55.
15. Whyte A, Gracia A, Bonastre C, Tejedor MT, Whyte J, Monteagudo LV, Simon C. Oral disease and microbiota in free-roaming cats. Top Companion Anim Med. 2017;32:91–5.
16. Wooley KL, Kelly RF, Fazakerley J, Williams NJ, Nuttal TJ, McEwan NA. Reduced in vitro adherence of Staphylococcus spp. to feline corneocytes compared to canine and human corneocytes. Vet Dermatol. 2006;19:1–6.
17. Medleau L, Blue JL. Frequency and antimicrobial susceptibility of *Staphylococcus* spp isolated from feline skin lesions. J Am Vet Med Assoc. 1988;193:1080–1.

18. Morris DO, Rook KA, Shofer FS, Rankin SC. Screening of *Staphylococcus aureus*, *Staphylococcus intermedius*, and *Staphylococcus schleiferi* isolates obtained from small companion animals for antimicrobial resistance: a retrospective review of 749 isolates (2003–04). Vet Dermatol. 2006;17:332–7.

19. Morris DO, Maudlin EA, O'Shea K, Shofer FS, Rankin SC. Clinical, microbiological, and molecular characterization of methicillin-resistant Staphylococcus aureus infections of cats. Am J Vet Res. 2006;67:1421–5.

20. Selvaraj P, Senthil KK. Feline Pyoderma – a study of microbial population and its antibiogram. Intas Polivet. 2013;14(11):405–6.

21. White SD. Pyoderma in five cats. J Am Anim Hosp Assoc. 1991;27:141–6.

22. Beco L, Guaguere E, Lorente Mendez C, Noli C, Nuttall T, Vroom M. Suggested guidelines for using systemic antimicrobials in bacterial skin infections (1): diagnosis based on clinical presentation, cytology and culture. Vet Rec. 2013;172:72–8.

23. Hillier A, Lloyd DH, Weese JS, Blondeau JM, Boothe D, Breitschwerdt E, et al. Guidelines for the diagnosis and antimicrobial therapy of canine superficial bacterial folliculitis (Antimicrobial Guidelines Working Group of the International Society for Companion Animal Infectious Diseases). Vet Dermatol. 2014;25:163–74.

24. Singleton DA, Sanchez-Vizcaino F, Dawson S, Jones PH, Noble PJ, Pinchbeck GL, et al. Patterns of antimicrobial agent prescription in a sentinel population of canine and feline veterinary practices in the United Kingdom. The Vet J. 2017;224:18–24.

25. Wildermuth BE, Griffin CE, Rosenkrantz WS. Response of feline eosinophilic plaques and lip ulcers to amoxicillin trihydrate–clavulanate potassium therapy: a randomized, double-blind placebo-controlled prospective study. Vet Dermatol. 2011;23:110–8.

26. Beco L, Guaguere E, Lorente Mendez C, Noli C, Nuttall T, Vroom M. Suggested guidelines for using systemic antimicrobials in bacterial skin infections (2): antimicrobial choice, treatment regimens and compliance. Vet Rec. 2013;172:156–60.

27. Borio S, Colombo S, La Rosa G, De Lucia M, Dombord P, Guardabassi L. Effectiveness of a combined (4% chlorhexidine digluconate shampoo and solution) protocol in MRS and non-MRS canine superficial pyoderma: a randomized, blinded, antibiotic-controlled study. Vet Dermatol. 2015;26:339–44.

28. Weese JS, Giguere S, Guardabassi L, Morley PS, Papich M, Ricciuto DR, et al. ACVIM consensus statement on therapeutic antimicrobial use in animals and antimicrobial resistance. J Vet Intern Med. 2015;29:487–98.

29. Mohamed MA, Abdul-Aziz S, Dhaliwal GK, Bejo SK, Goni MD, Bitrus AA, et al. Antibiotic resistance profiles of *Staphylococcus pseudintermedius* isolated from dogs and cats. Malays J Microbiol. 2017;13:180–6.

30. Whitehouse W, Viviano K. Update in feline therapeutics: clinical use of 10 emerging therapies. J Feline Med Surg. 2015;17:220–34.

31. Burke S, Black V, Sanchez-Vizcaino F, Radford A, Hibbert A, Tasker S. Use of cefovecin in a UK population of cats attending first-opinion practices as recorded in electronic health records. J Feline Med Surg. 2017;19:687–92.

32. Hardefeldt LY, Holloway S, Trott DJ, Shipstone M, Barrs VR, Malik R, et al. Antimicrobial prescribing in dogs and cats in Australia: results of the Australasian Infectious Disease Advisory Panel Survey. J Vet Intern Med. 2017;31:1100–7.

33. Scott DW, Miller WH. Feline acne: a retrospective study of 74 cases (1988–2003). Jpn J Vet Dermatol. 2010;16:203–9.

34. Jazic E, Coyner KS, Loeffler DG, Lewis TP. An evaluation of the clinical, cytological, infectious and histopathological features of feline acne. Vet Dermatol. 2006;17:134–40.

35. Walton DK, Scott DW, Manning TO. Cutaneous bacterial granuloma (botryomycosis) in a dog and cat. J Am Anim Hosp Assoc. 1983;183(19):537–41.

36. Murai T, Yasuno K, Shirota K. Bacterial pseudomycetoma (Botryomycosis) in an FIV-positive cat. Jap J Vet Dermatol. 2010;16:61–5.

37. Norris JM, Love DN. The isolation and enumeration of three feline oral *Porphyromonas* species from subcutaneous abscessed in cats. Vet Microbiol. 1999;65:115–22.
38. Traslavina RP, Reilly CM, Vasireddy R, Samitz EM, Stepnik CT, Outerbridge C, et al. Laser capture microdissection of feline *Streptomyces spp* pyogranulomatous dermatitis and cellulitis. Vet Pathol. 2015;205(52):1172–5.
39. De Araujo FS, Braga JF, Moreira MV, Silva VC, Souza EF, Pereira LC, et al. Splendore-Hoeppli phenomenon in a cat with osteomyelitis caused by *Streptococcus* species. J Feline Med Surg. 2014;16:189–93.
40. Malik R, Krockenberger MB, O'Brien CR, White JD, Foster D, Tisdall PL, et al. Nocardia infections in cats: a retrospective multi-institutional study of 17 cases. Aust Vet J. 2006;84: 235–45.
41. Gunew MN. *Rhodococcus equi* infection in cats. Aust Vet Practit. 2002;32:2–5.
42. Farias MR, Takai S, Ribeiro MG, Fabris VE, Franco SR. Cutaneous pyogranuloma in a cat caused by virulent *Rhodococcus equi* containing an 87 kb type I plasmid. Aust Vet J. 2007;85:29–31.
43. Patel A. Pyogranulomatous skin disease and cellulitis in a cat caused by *Rhodococcus equi*. J Small Anim Pract. 2002;43:129–32.
44. Miller RI, Ladds PW, Mudie A, Hayes DP, Trueman KF. Probable dermatophilosis in 2 cats. Aust Vet J. 1983;60:155–6.
45. Kaya O, Kirkan S, Unal B. Isolation of *Dermatophilus congolensis* from a cat. J Veterinary Med Ser B. 2000;47:155–7.
46. Carakostas MC. Subcutaneous dermatophilosis in a cat. J Am Vet Med Assoc. 1984;185:675–6.
47. Sharman MJ, Goh CS, Kuipers RG, Hodgson JL. Intra-abdominal actinomycetoma in a cat. J Feline Med Surg. 2009;11:701–5.
48. Koenhemsi L, Sigirci BD, Bayrakal A, Metiner K, Gonul R, Ozgur NY. *Actinomyces viscosus* isolation from the skin of a cat. Isr J Vet Med. 2014;69:239–42.
49. Kruger EF, Byrne BA, Pesavento P, Hurley KF, Lindsay LL, Sykes JE. Relationship between clinical manifestations and pulsed-field gel profiles of *Streptococcus canis* isolates from dogs and cats. Vet Microbiol. 2010;146:167–71.
50. Nolff MC, Meyer-Lindenberg A. Necrotising fasciitis in a domestic shorthair cat – negative pressure wound therapy assisted debridement and reconstruction. J Small Anim Pract. 2015;56:281–4.
51. Brachelente C, Wiener D, Malik Y, Huessy D. A case of necrotizing fasciitis with septic shock in a cat caused by *Acinetobacter baumannii*. Vet Dermatol. 2007;18:432–8.
52. Plavec T, Zdovc I, Juntes P, Svara T, Ambrozic-Avgustin I, Suhadolc-Scholten S. Necrotising fasciitis, a potential threat following conservative treatment of a leucopenic cat: a case report. Vet Med (Praha). 2015;8:460–7.
53. Berube DE, Whelan MF, Tater KC, Bracker KE. Fournier's gangrene in a cat. J Vet Emerg Crit Care. 2010;20:148–4.
54. Ravens PA, Vogelnest LJ, Ewen E, Bosward KL, Norris JM. Canine superficial bacterial pyoderma: evaluation of skin surface sampling methods and antimicrobial susceptibility of causal *Staphylococcus* isolates. Aust Vet J. 2014;92:149–55.
55. Larsen RF, Boysen L, Jessen LR, Guardabassi L, Damborg P. Diversity of *Staphylococcus pseudintermedius* in carriage sites and skin lesions of dogs with superficial bacterial folliculitis: potential implications for diagnostic testing and therapy. Vet Dermatol. 2018;29:291–5.
56. Schmidt VM, Pinchbeck G, Nuttall T, Shaw S, McIntyre KM, McEwan N, et al. Impact of systemic antimicrobial therapy on mucosal staphylococci in a population of dogs in Northwest England. Vet Dermatol. 2018;29:192–202.
57. Kadlec K, Weiß S, Wendlandt S, Schwarz S, Tonpitak W. Characterization of canine and feline methicillin-resistant *Staphylococcus pseudintermedius* (MRSP) from Thailand. Vet Microbiol. 2016;194:93–7.

58. Lappin MR, Bondeau J, Boothe D, Breitschwerdt FB, Guardabassi L, Lloyd DH, et al. Antimicrobial use Guidelines for Treatment of Respiratory Tract Disease in Dogs and Cats: Antimicrobial Guidelines Working Group of the International Society for Companion Animal Infectious Diseases. J Vet Intern Med. 2017;31:279–94.
59. Loncaric I, Kunzel F, Klang A, Wagner R, Licka T, Grunert T, et al. Carriage of methicillin-resistant staphylococci between humans and animals on a small farm. Vet Dermatol. 2016;27:191–4.

Mycobacterial Diseases

Carolyn O'Brien

Abstract

Cats may be infected with a variety of both rapidly- and slowly-growing myco-bacterial species, which cause a variety of clinical syndromes in cats, from local-ized skin disease to disseminated and potentially fatal infections. Cutaneous disease is the most common manifestation for all causative species; however, some species may have internal involvement, with any organ system, skeletal or soft tissue structure potentially infected. Infections by rapidly-growing myco-bacteria generally result in fistulating panniculitis of the inguinal region or less commonly, axillae, flanks or dorsum, whereas those caused by members of the slow-growing taxons typically present with solitary or multiple nodular skin lesions and/or local lymphadenopathy, especially of the head, neck and/or limbs. Most affected cats do not appear to have an underlying immunosuppressive con-dition, and no association has been made with a positive retroviral status. Most cases occur in adult cats with unrestricted outdoor access. Depending on the causative species and the extent of disease when first diagnosed, these infections can be challenging to treat. Generally, localized cutaneous infection caused by all species has a relatively favorable prognosis if treated with an appropriate combination of drugs and surgery, if necessary. If the cat acquires systemic infec-tion, the prognosis becomes significantly worse. The commitment of the owner to the implementation of a potentially expensive and time-consuming schedule of multidrug therapy for many months may also influence the outcome. The zoo-notic potential of these organisms is generally low, however cat-to-human trans-fer of *Mycobacterium bovis* has been reported.

C. O'Brien (✉)
Melbourne Cat Vets, Fitzroy, Victoria, Australia
e-mail: cob@catvet.net.au

© Springer Nature Switzerland AG 2020
C. Noli, S. Colombo (eds.), *Feline Dermatology*,
https://doi.org/10.1007/978-3-030-29836-4_12

251

Mycobacteria are aerobic, nonmotile, Gram-positive, nonspore-forming bacilli in the phylum *Actinobacteria*. Of the more than 180 mycobacterial species identified [1], almost all are environmental saprophytes. However, a few, such as the *Mycobacterium tuberculosis* complex (MTB), *M. leprae* and its relatives, members of the *M. avium* complex (MAC), such as *M. avium* subsp. *paratuberculosis* and *M. lepraemurium*, appear to have evolved into obligate pathogens.

Mycobacterial species can be divided genetically and phenotypically into two main groups: rapidly growing (RGM) and slowly growing mycobacteria (SGM). The RGM are ancestral to the SGM, with the latter forming a distinct genetic sub-branch based on analysis of housekeeping genes and, more recently, whole-genome analysis [2]. The *M. abscessus/chelonae* complex appears to be the genetically oldest group identified, with *M. triviale* and also the closely related *M. terrae* group the likely evolutionary links between the RGM and the SGM [2].

Mycobacterial infections cause a variety of clinical syndromes in cats, from minor localized skin disease to potentially fatal disseminated infections. Cutaneous disease is the most common manifestation for all causative species; however, some species, particularly the MTB and MAC, may have internal involvement with any organ system, skeletal or soft tissue structure, potentially infected.

Few investigations have examined substantial cohorts of cats with mycobacteriosis and only some definitively identified the causative mycobacterial species via genetic analysis. These studies are typically limited to animals from a particular geographical region and may not be representative of the disease in cats domiciled elsewhere, especially with regard to incidence and causative species.

Typically, cats with mycobacterial infections do not appear to have an immunosuppression, and no association has been made with a positive retroviral status, unlike MAC infections in people with human immunodeficiency virus/acquired immunodeficiency syndrome. Regardless of the causative species, most cases occur in adult cats with unrestricted outdoor access, although MAC infections have been occasionally reported in exclusively indoor cats.

Rapidly Growing Mycobacteria

Etiology and Epidemiology

The RGM are environmental saprophytes widely distributed as free-living organisms in both terrestrial and aquatic biomes. The RGM are so named as they are able to grow on synthetic culture media within 7 days at 75–113° F (24–45° C).

RGM have low inherent pathogenicity and generally tend to cause opportunistic infections in cats, mostly through breaches in the integument, for example, via cat-scratch wounds. They have a low tendency to cause systemic disease unless the host is immunocompromised, although occasionally inhalation of organisms may lead to pneumonia in apparently immunocompetent individuals. The disease manifests in cats primarily as ventral abdominal panniculitis and tends to be caused by the *M. smegmatis*, *M. margaritense*, *M. fortuitum*, and *M. chelonae-abscessus* groups.

Cases are reported from the Americas (Brazil, southeastern and southwestern United States, Canada), Oceania (Australia and New Zealand), and Europe (Finland, the Netherlands, Germany, and the United Kingdom). The incidence of particular causative organisms varies between geographical regions. *M. smegmatis* and *M. margaritense*, followed by *M. fortuitum* groups, cause most infections in cats in eastern Australia, whereas, in the southwestern United States, *M. fortuitum* group followed by *M. chelonae* infections appear to be more common.

Cats with a prominent ventral abdominal fat pad appear to have a predisposition toward RGM infection. This is likely due to the preference of the organisms for tissues rich in lipid, which may provide triglycerides for growth and perhaps protection from the host immune response. In cats that do not have a significant amount of subcutaneous fat, the ability to establish experimental infections appears to be limited [3].

Clinical Features

Typically, lesions caused by RGM are located in the inguinal region or, less commonly, axillae, flanks, or dorsum. Initially, the infection appears as a circumscribed plaque or nodule of the skin and subcutis. Subsequently, the affected cat develops alopecic areas of thin epidermis which overlies and is adherent to diseased subcutaneous tissue; this results in a characteristic "pepper pot" appearance (Fig. 1). The characteristic focal purple depressions in the skin break down to become fistulae exuding a watery discharge that may become purulent with secondary infection. The lesions may eventually involve the entire ventral abdomen, flanks, perineum, and occasionally the limbs. Internal organs or lymph node involvement is not likely; however, the abdominal wall is rarely involved.

Most cats do not have signs of systemic illness unless the skin lesions become secondarily infected with *Staphylococcus* and *Streptococcus* spp., in which case the patient may display lethargy, pyrexia, anorexia, weight loss, and reluctance to move.

Diagnosis

Fine needle aspiration and cytology may establish the presence of pyogranulomatous inflammation, and subcutaneous exudate may be obtained with this technique to allow the culture of the organism, thus establishing the diagnosis.

RGM are not typically visible on either Romanowsky-stained cytological samples or hematoxylin and eosin-stained histopathology sections of biopsy tissue. Instead, they are visualized using acid-fast stains, such as Ziehl-Neelsen (ZN) or Fite's.

RGM may be few in number and difficult to visualize in acid-fast stained cytological material, and the diagnosis is not excluded if organisms are not visualized. The organisms may be lost during processing of cytologic and histopathologic

Fig. 1 The typical
appearance of dermatitis/
panniculitis caused by a
rapidly growing
mycobacterial species,
Mycobacterium smegmatis.
(Courtesy of Nicola
Colombo)

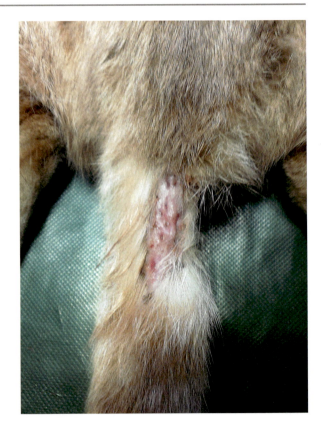

samples as they tend to exist extracellularly in fat vacuoles in tissues. Occasionally, positive results on mycobacterial culture or molecular methods such as polymerase chain reaction (PCR) may be obtained on samples that are "acid-fast bacilli (AFB)-negative" on cyto- or histopathological evaluation.

Punch biopsies of the skin are usually inadequate for obtaining representative tissue samples, and a deep subcutaneous tissue biopsy from the margin of the lesion is preferred. The histopathologic characteristics of RGM dermatitis/panniculitis include an ulcerated or acanthotic dermis overlying multifocal to diffuse pyogranulomatous inflammation, which tends to extend well into the subcutis. In the pyogranulomas, a rim of neutrophils often surrounds a clear, inner zone of degenerate adipocytes, which may contain scant AFB with an outer collection of epithelioid macrophages (Fig. 2). A mixed inflammatory response, predominantly comprising neutrophils and macrophages, but also containing lymphocytes and plasma cells, is found between each pyogranuloma. AFB may also occasionally be visualized within macrophages but can be very hard to find within tissue sections.

When attempting to culture mycobacteria from panniculitis lesions, material swabbed directly from cutaneous draining sinus tracts usually contains high numbers of contaminating skin bacteria, which outcompete the RGM on culture media.

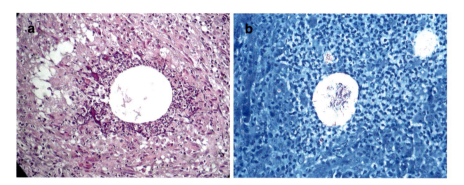

Fig. 2 (**a**) Histopathological aspect of rapidly growing mycobacterial infection: pyogranulomatous inflammation with a rim of neutrophils surrounding a clear, inner zone of degenerate adipocytes, which contain acid fast bacteria (H&E 400×); (**b**) Ziehl-Neelsen stain of the same sample: rod shaped bacteria are stained in red and can be easily recognised (400×). (Courtesy of Dr. Chiara Noli)

Fine-needle samples obtained through intact skin decontaminated with 70% ethanol or surgically collected subcutaneous tissue biopsies are therefore preferred. Uncontaminated samples of RGM grow readily on routine media such as blood [4] and MacConkey agar (without crystal violet), so there is usually no need for the clinician to request "mycobacterial media" culture for these organisms specifically.

Treatment and Prognosis

Depending on the causative species and the extent of disease when first diagnosed, these infections can be challenging to treat. They often have a high rate of recurrence, frequently require protracted courses of therapy, and may have a substantial incidence of inherent and/or acquired drug resistance.

Susceptibility data is especially useful for organisms that may have inherently variable drug susceptibility, such as *M. fortuitum*, or for recurrent or chronically persistent RGM infections, especially where the cat has undergone prior antibiotic treatment which may have induced acquired drug resistance. Ideally, treatment should begin with one or two oral antimicrobials (doxycycline, a fluoroquinolone, and/or clarithromycin). These are usually chosen empirically until results of culture and susceptibilities are known. In Australia, doxycycline and/or a fluoroquinolone – preferably pradofloxacin – are best, whereas, in the United States, clarithromycin is the drug of choice initially. *M. smegmatis* group tends to be inherently resistant to clarithromycin, and some isolates may be resistant to the enrofloxacin or ciprofloxacin, although this does not rule out susceptibility to pradofloxacin [5]. Members of the *M. fortuitum* group are typically susceptible to fluoroquinolones, however, demonstrate variable expression of the erythromycin-inducible methylase (*erm*) gene which confers macrolide resistance [6]. Approximately 50% of *M. fortuitum* isolates are susceptible to doxycycline [7]. *M. chelonae-abscessus* group isolates tend to be resistant to all drugs available for oral dosing apart from clarithromycin and linezolid. Where indicated by drug susceptibility data, refractory cases may be treated with clofazimine, amikacin, cefoxitin, or linezolid. It is recommended to commence treatment at standard dose rates increased slowly to the high end of the dose range, unless adverse effects are observed.

Treatment duration is variable, but it is recommended to continue therapy for 1–2 months past resolution of all clinical signs. Some animals with recalcitrant lesions benefit from en bloc resection of isolated areas of infection, often necessitating reconstructive surgery [8] or vacuum-assisted wound closure [9, 10].

Public Health Risks

Zoonotic transmission of RGM organisms from infected animals to humans is very unlikely. There is one report of *M. fortuitum* infection in an otherwise healthy middle-aged woman, after a cat bite to the forearm [11].

Slowly Growing Mycobacteria

The SGM taxon includes a large number of opportunistic environmental species: the obligate pathogens, *M. leprae* and *M. lepromatosis*, and the members of the *M. tuberculosis* complex. There are also a number of fastidious species included – traditionally classified as the causative species of "feline leprosy" – that are incapable of growing in axenic culture; thus, their epidemiological niche is unclear.

Tuberculous Mycobacteria

Cats are naturally resistant to *M. tuberculosis*, but occasional infections likely transmitted directly from humans are reported [12]. Disease in cats is most commonly caused by *M. bovis* and *M. microti* [13]. *M. bovis* has worldwide endemicity. However, much of Continental Europe, parts of the Caribbean, and Australia are free of the disease due to widespread surveillance, slaughter of test-positive cattle, the pasteurization of milk, and the absence of a wildlife host. *M. microti* is endemic to Europe and the United Kingdom (UK). Its main reservoir appears to be voles, shrews, wood mice, and other small rodents [14].

The exact route of transmission of these MTB species to cats is unclear. Numerous potential rodent prey species collected from areas of southwest England were found to be infected with *M. bovis* [15]. Suspected nosocomial contamination of surgical wounds has been reported [16].

MAC and Other Slowly Growing Saprophytes

Disease in cats is caused by several saprophytic slowly growing mycobacterial species, mostly members of the MAC, which are found worldwide in water sources and soil. Certain slowly growing species are more common in some environmental niches or particular geographical areas, for example, *M. malmoense* or biofilms with *M. intracellulare* in the UK and Sweden. Some have highly restricted, focal areas of endemicity, for example, *M. ulcerans* infection.

As with MTB complex, the clinical picture is determined by the route of infection. Cats likely acquire skin lesions via transcutaneous inoculation of contaminated environmental material. Most cats with slow-growing mycobacterial infections have unrestricted outdoor access, and almost all of these cases had no overt predisposing conditions.

Fastidious Mycobacteria

"Feline leprosy" has been diagnosed in New Zealand, Australia, western Canada, the UK, southwestern United States, continental Europe, New Caledonia, the Greek islands, and Japan. Historically, New Zealand and Australia have reported the highest number of cases worldwide.

Genetic studies have identified the involvement of several "non-culturable" species of mycobacteria: *M. lepraemurium*, *Candidatus* "M. tarwinense,' [17, 18] *Candidatus* "M. lepraefelis,' [19] and *M. visibilis*, although the latter has not been reported for many years [20]. *M. lepraemurium* tends to cause disease in young male cats, whereas *Candidatus* "M. tarwinense" and *Candidatus* "M. lepraefelis" are more likely to cause disease in middle-aged to older cats. There is no gender preponderance for *Candidatus* "M. tarwinense" infection, whereas *Candidatus* "M. lepraefelis" is slightly more likely to cause disease in males.

Clinical Features

The majority of cats with SGM infection have solitary or multiple nodular skin lesions and/or local lymphadenopathy, especially of the head, neck, and/or limbs (Fig. 3). Ulceration of cutaneous lesions and the skin overlying affected lymph nodes may be

Fig. 3 Large, ulcerated nodules on the lateral thigh of a young male cat with *Mycobacterium lepraemurium* infection. Despite the widespread nature of the cutaneous lesions, this cat was cured with multidrug therapy including rifampicin and clofazimine. (Courtesy of Dr. Mei Sae Zhong)

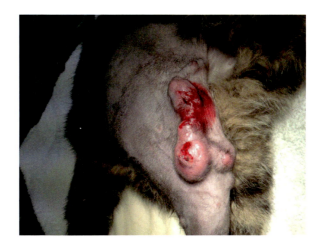

observed, and infection may occasionally involve contiguous muscle and bone, which is more often the case with MTB complex species than other causative agents. In some cases, the dermal lesions may be widespread, involving many cutaneous sites. Host factors (age, concurrent illness, immunological status), the causal species, or the route and size of the inoculum may influence the nature of the disease.

If systemic disease is detected, the most common causative agents are either the MTB complex mycobacteria (especially in the UK and New Zealand) or members of the MAC. Rarely, systemic infections by other mycobacterial species, including other slowly growing saprophytes and *Candidatus* "M. lepraefelis," have been documented.

Diagnosis

Differential diagnoses of nodular skin and subcutaneous lesions include *Nocardia* and *Rhodococcus* spp. (which may also be acid-fast), fungi, or algal infections and primary or metastatic neoplasia. There are no pathognomonic clinical features that differentiate mycobacterial infections from other etiologies, and collection of representative tissue samples for cytology or histopathology and microbiology is necessary for the diagnosis.

It is vital in areas endemic for the MTB complex that the diagnosis is not based simply on cytologic or histopathologic findings. An attempt to identify the causative agent should be made in every case, ideally via a mycobacterium reference laboratory or equivalent, especially where mandatory reporting of such cases may result in compulsory euthanasia.

The diagnosis of cutaneous infections caused by SGM is often relatively simple, provided there is a high index of suspicion. Personal protective equipment should be worn during any procedure which involves handling of discharging or ulcerated lesions and/or surgical or necropsy tissues, when members of the MTB complex are a possible cause of disease.

Ideally, at the time of biopsy sampling for histopathology, a piece of fresh tissue wrapped in sterile saline-moistened gauze swabs placed in a sterile container should be collected, if microbiological processing is needed. The pathology laboratory should ideally be notified before submission, as SGM culture and identification requires specialized expertise.

Romanowsky-stained cytological samples of cutaneous nodules will demonstrate granulomatous to pyogranulomatous inflammation, and mycobacteria are recognized by their characteristic "negatively staining" appearance (Fig. 4), usually located within macrophages. As with the RGM, SGM are not typically visible on Romanowsky-stained cytological or hematoxylin and eosin-stained histopathology sections, except *M. visibile* and *Candidatus* "M. lepraefelis." Instead, Ziehl-Neelsen (ZN) staining (Fig. 5) or similar (e.g. Fite's) is required. Depending on mycobacterial species and host immune response, bacterial numbers may be variable.

MTB complex organisms produce characteristic solitary to coalescing granulomas ("tubercules"). Granulation tissue surrounds a layer of mixed inflammatory cells, consisting of macrophages, neutrophils, lymphocytes, and plasma cells. The

Fig. 4 Numerous macrophages with cytoplasm filled with many achromatic rod-shaped areas (Diff Quick, 1000×). (Courtesy of Dr. Francesco Albanese)

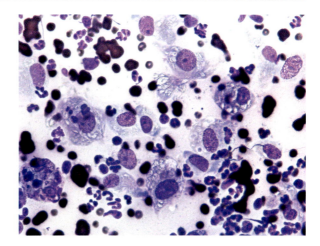

Fig. 5 Many brightly red and rod-shaped Mycobacteria are well recognizable with Ziehl-Neelsen staining (1000×). (Courtesy of Dr. Francesco Albanese)

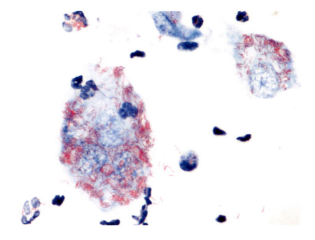

center of the granuloma contains epithelioid macrophages and some neutrophils, with variable but usually low numbers of AFB, with or without necrotic tissue.

Cutaneous MAC infections cause pyogranulomatous or granulomatous inflammation with a variable fibroblastic response. The fibroblastic reaction may, on occasion, be so pronounced as to make it difficult to differentiate the disease from an inflamed fibrosarcoma (so-called "mycobacterial pseudotumor") [21]. AFB found both within macrophages and spindle cells identify the underlying etiology in these cases (Fig. 6). In the absence of a prominent fibroblastic response, lesions may resemble lepromatous leprosy.

The pathological picture of feline leprosy is subdivided into multi-bacillary (lepromatous) and pauci-bacillary (tuberculoid) forms [22]. "Multi-bacillary" leprosy is thought to correspond with a weak cell-mediated immune (CMI) response. Typically, many foamy or multinucleate macrophages, containing huge numbers of mycobacteria, are observed. There is no necrosis, and lesions contain virtually no lymphocytes and plasma cells. "Pauci-bacillary" leprosy, in which moderate to few

Fig. 6 Histopathological appearance of MTB complex infection: granulomas consisting of macrophages, neutrophils, lymphocytes and plasma cells. The center of the granuloma contains epithelioid macrophage (H&E 400×) (Courtesy of Dr. Chiara Noli)

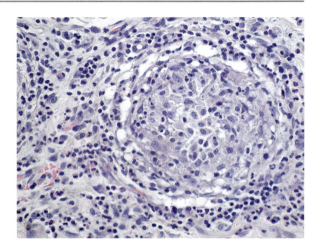

observable AFB are found within pyogranulomatous inflammation dominated by epithelioid histiocytes, is thought to occur with a more effective CMI response. Moderate numbers of lymphocytes and plasma cells are also observed, with multifocal to coalescing necrosis. Involvement of peripheral nerves, a feature of human leprosy, is not seen in cats.

Except where samples have become contaminated with environmental mycobacteria, molecular methods, such as PCR and sequencing, can provide a highly accurate diagnosis on fresh or frozen tissue, formalin-fixed paraffin-embedded tissue sections, and Romanowsky-stained cytology slides [23]. It should be remembered that for samples in which no AFB are visualized microscopically, mycobacterial infections cannot be excluded with a negative PCR result.

A feline IFN-γ ELISPOT test is currently commercially available [24]. This test utilizes both bovine tuberculin and ESTAT6/CFP10 for the identification of cats infected with either *M. bovis* or *M. microti* and is able to differentiate the two mycobacteria. It is reported as having a sensitivity of 90% for detecting feline *M. bovis* infections, 83.3% sensitivity for detecting feline *M. microti* infections, and 100% specificity for both.

Serum antibody tests (multi-antigen print immune-assay (MAPIA), TB STAT-PAK, and Rapid DPP VetTB) have been evaluated in cats with TB [25]. Overall sensitivity was 90% for detection of *M. bovis* infection and greater than 40% for *M. microti*, with a specificity of 100%.

It is important to remember that these tests do not explicitly differentiate active from latent infection or prior exposure. Culture of organisms from clinical samples obtained from cats with appropriate signs and diagnostic findings remains the gold standard for the diagnosis of active TB.

Treatment

Controlled studies of feline mycobacteriosis treatment are lacking, and the existing literature consists of a few retrospective observational case series and case reports.

Table 1 Drugs typically chosen to treat feline mycobacterial infections

Drug	Dose	Side effects/comments
Clofazimine	25 mg/cat PO q 24 h or 50 mg/cat q 48 h	Skin and body fluid discoloration (pink-brown), photosensitization, pitting corneal lesions, nausea, vomiting, and abdominal pain Possible hepatotoxicity Monitor serum hepatic enzymes[a]
Clarithromycin	62.5 mg/cat PO q 12 h	Cutaneous erythema and edema, hepatotoxicity, diarrhea, and/or vomiting, neutropenia, thrombocytopenia
Azithromycin	5–15 mg/kg PO q 24 h	Vomiting, diarrhea, abdominal pain, hepatotoxicity
Rifampicin	10 mg/kg PO q 24 h	Hepatotoxicity and/or inappetence, cutaneous erythema/pruritus, anaphylaxis Monitor serum hepatic enzymes[a]
Doxycycline	5–10 mg/kg PO q 12 h	Hydrochloride or hyclate formulations may cause esophageal irritation and possibly stricture
Enrofloxacin Marbofloxacin Orbifloxacin	5 mg/kg PO q 24 h 2 mg/kg PO q 24 h 7.5 mg/kg PO q 24 h	Enrofloxacin may cause retinal toxicity in cats; marbofloxacin or orbifloxacin are preferred if available Most *M. avium* complex organisms are resistant to second-generation fluoroquinolones
Pradofloxacin	7.5 mg/kg PO q 24 h	Give without food unless gastrointestinal side effects occur
Moxifloxacin	10 mg/kg PO q 24 h	Vomiting and anorexia; dose can be divided 12 hourly and/or administered with food

[a]Alanine transferase and alkaline phosphatase

There have been occasional reports of spontaneous resolution of *M. lepraemurium* infection; [26, 27] however, the vast majority of SGM infections require treatment to achieve a cure. Table 1 lists the drugs and doses typically chosen to treat feline mycobacteriosis.

The initiation of empirical treatment is required in almost all cases of SGM infection, as identification of the causative mycobacterium may take weeks to months (or may not be available at all). The choice of initial treatment will depend on [1] the suspected etiological agent, [2] owner factors such as finances and ability/willingness to medicate the cat orally for an extended period, and [3] the presence of comorbidities that may restrict the use of certain drugs, for example, hepatic disease when using rifampicin.

Therapy should include at least rifampicin, clarithromycin (or azithromycin), and/or pradofloxacin (or moxifloxacin). In areas where infection with the fastidious organisms is common the inclusion of clofazimine, if available, would also be a reasonable choice. Ethambutol and isoniazid have been used to treat feline TB, although toxicity tends to limit their use. They tend only to be prescribed if there is drug resistance to the more commonly utilized agents. If the infection is restricted to a localized cutaneous site, surgical excision may be a beneficial adjunct to antibiotic therapy.

Medical treatment can be subsequently modified depending on identification of the mycobacterial species involved, response to treatment, and/or, if available, the results of drug susceptibility testing. Therapy should extend for at least 2 months post-surgical resection or beyond resolution of clinical signs. Unless diagnosed with MTB complex infection, quarantine of the cat is not necessary. Some of the drugs, especially clofazimine, induce photosensitivity; it is recommended that owners keep the cat indoors in the summer months.

Prognosis

Localized cutaneous infection caused by all slowly growing species has a good prognosis if treated promptly with a combination of appropriate antibiotics, and if possible surgical resection. If cutaneous disease progresses to systemic infection, the prognosis becomes significantly worse. Treatment is potentially expensive and time-consuming. Cats can be notoriously tricky to medicate, and the provision of multidrug therapy for many months may also affect the outcome.

Public Health Risks

The only SGM that appears to carry a definite risk of cat-to-human transfer is *M. bovis*, although this risk seems to be low. A report from the UK details the infection in four people (two clinically and two sub-clinically affected) associated with an infected pet cat [28]. A laboratory worker seroconverted after exposure to research cats that were infected after accidentally being fed infected meat [29]. At this time, instances of cat-to-human transmission of *M. microti* infection have not been reported.

There is one report of a person contracting *M. marinum* secondary to a cat scratch [30]. However, this likely represented mechanical inoculation, rather than true zoonotic transfer. Likewise, there appears to be almost no risk of humans acquiring infections from any of the fastidious organisms from cats; however, as the ecology and transmission of these mycobacterial species are not understood, it is difficult to determine their potential for zoonotic transfer completely.

The Advisory Board on Cat Diseases (based in Europe) recommends that all people in contact with an infected cat should be made aware of the potential but low risk of zoonotic transfer of feline mycobacteriosis [31]. As a minimal precaution, the use of gloves is recommended when treating these animals. This is especially important for anyone in contact with the cat who is immunocompromised. Veterinary staff should utilize personal protective equipment when handling cats with cutaneous lesions, collecting biopsies, or performing necropsy studies.

References

1. Gupta RS, Lo B, Son J. Phylogenomics and comparative genomic studies robustly support division of the genus Mycobacterium into an emended genus Mycobacterium and four novel genera. Front Microbiol. 2018;9:67.

2. Fedrizzi T, Meehan CJ, Grottola A, Giacobazzi E, Serpini GF, Tagliazucchi S, et al. Genomic characterization of nontuberculous mycobacteria. Sci Rep. 2017;7:45258.
3. Lewis DT, Hodgin EC, Foil S, Cox HU, Roy AF, Lewis DD. Experimental reproduction of feline Mycobacterium fortuitum panniculitis. Vet Dermatol. 1994;5(4):189–95.
4. Drancourt M, Raoult D. Cost-effectiveness of blood agar for isolation of mycobacteria. PLoS Negl Trop Dis. 2007;1(2):e83.
5. Govendir M, Hansen T, Kimble B, Norris JM, Baral RM, Wigney DI, et al. Susceptibility of rapidly growing mycobacteria isolated from cats and dogs, to ciprofloxacin, enrofloxacin and moxifloxacin. Vet Microbiol. 2011;147(1–2):113–8.
6. Nash KA, Andini N, Zhang Y, Brown-Elliott BA, Wallace RJ. Intrinsic macrolide resistance in rapidly growing mycobacteria. Antimicrob Agents Chemother. 2006;50(10):3476–8.
7. Brown-Elliott B, Philley J. Rapidly growing mycobacteria. Microbiol Spectr. 2017;5:TNMI7-0027-2016.
8. Malik R, Wigney DI, Dawson D, Martin P, Hunt GB, Love DN. Infection of the subcutis and skin of cats with rapidly growing mycobacteria: a review of microbiological and clinical findings. J Feline Med Surg. 2000;2(1):35–48.
9. Guille AE, Tseng LW, Orsher RJ. Use of vacuum-assisted closure for management of a large skin wound in a cat. J Am Vet Med Assoc. 2007;230(11):1669–73.
10. Vishkautsan P, Reagan KL, Keel MK, Sykes JE. Mycobacterial panniculitis caused by Mycobacterium thermoresistibile in a cat. JFMS Open Rep. 2016;2(2):2055116916672786.
11. Ngan N, Morris A, de Chalain T. Mycobacterium fortuitum infection caused by a cat bite. N Z Med J. 2005;118(1211):U1354.
12. Alves DM, da Motta SP, Zamboni R, Marcolongo-Pereira C, Bonel J, Raffi MB, et al. Tuberculosis in domestic cats (Felis catus) in southern Rio Grande do Sul. Pesquisa Veterinária Brasileira. 2017;37(7):725–8.
13. Gunn-Moore DA, McFarland SE, Brewer JI, Crawshaw TR, Clifton-Hadley RS, Kovalik M, et al. Mycobacterial disease in cats in Great Britain: I. Culture results, geographical distribution and clinical presentation of 339 cases. J Feline Med Surg. 2011;13(12):934–44.
14. Cavanagh R, Begon M, Bennett M, Ergon T, Graham IM, De Haas PE, et al. Mycobacterium microti infection (vole tuberculosis) in wild rodent populations. J Clin Microbiol. 2002;40(9):3281–5.
15. Delahay RJ, Smith GC, Barlow AM, Walker N, Harris A, Clifton-Hadley RS, et al. Bovine tuberculosis infection in wild mammals in the south-west region of England: a survey of prevalence and a semi-quantitative assessment of the relative risks to cattle. Vet J. 2007;173(2):287–301.
16. Murray A, Dineen A, Kelly P, McGoey K, Madigan G, NiGhallchoir E, et al. Nosocomial spread of mycobacterium bovis in domestic cats. J Feline Med Surg. 2015;17(2):173–80.
17. Fyfe JA, McCowan C, O'Brien CR, Globan M, Birch C, Revill P, et al. Molecular characterization of a novel fastidious mycobacterium causing lepromatous lesions of the skin, subcutis, cornea, and conjunctiva of cats living in Victoria, Australia. J Clin Microbiol. 2008;46(2):618–26.
18. O'Brien CR, Malik R, Globan M, Reppas G, McCowan C, Fyfe JA. Feline leprosy due to Candidatus 'Mycobacterium tarwinense' further clinical and molecular characterisation of 15 previously reported cases and an additional 27 cases. J Feline Med Surg. 2017;19(5):498–512.
19. O'Brien CR, Malik R, Globan M, Reppas G, McCowan C, Fyfe JA. Feline leprosy due to Candidatus 'Mycobacterium lepraefelis': further clinical and molecular characterisation of eight previously reported cases and an additional 30 cases. J Feline Med Surg. 2017;19(9):919–32.
20. Appleyard GD, Clark EG. Histologic and genotypic characterization of a novel Mycobacterium species found in three cats. J Clin Microbiol. 2002;40(7):2425–30.
21. Miller MA, Fales WH, McCracken WS, O'Bryan MA, Jarnagin JJ, Payeur JB. Inflammatory pseudotumor in a cat with cutaneous mycobacteriosis. Vet Pathol. 1999;36(2):161–3.
22. Malik R, Hughes MS, James G, Martin P, Wigney DI, Canfield PJ, et al. Feline leprosy: two different clinical syndromes. J Feline Med Surg. 2002;4(1):43–59.

23. Reppas G, Fyfe J, Foster S, Smits B, Martin P, Jardine J, et al. Detection and identification of mycobacteria in fixed stained smears and formalin-fixed paraffin-embedded tissues using PCR. J Small Anim Pract. 2013;54(12):638–46.
24. Rhodes SG, Gruffydd-Jones T, Gunn-Moore D, Jahans K. Adaptation of IFN-gamma ELISA and ELISPOT tests for feline tuberculosis. Vet Immunol Immunopathol. 2008;124(3–4):379–84.
25. Rhodes SG, Gunn-Mooore D, Boschiroli ML, Schiller I, Esfandiari J, Greenwald R, et al. Comparative study of IFNgamma and antibody tests for feline tuberculosis. Vet Immunol Immunopathol. 2011;144(1–2):129–34.
26. O'Brien CR, Malik R, Globan M, Reppas G, Fyfe JA. Feline leprosy due to *Mycobacterium lepraemurium*: further clinical and molecular characterization of 23 previously reported cases and an additional 42 cases. J Feline Med Surg. 2017;19(7):737–46.
27. Roccabianca P, Caniatti M, Scanziani E, Penati V. Feline leprosy: spontaneous remission in a cat. J Am Anim Hosp Assoc. 1996;32(3):189–93.
28. England PH. Cases of TB in domestic cats and cat-to-human transmission: risk to public very low. 2014. Available from: https://www.gov.uk/government/news/cases-of-tb-in-domestic-cats-and-cat-to-human-transmission-risk-to-public-very-low.
29. Isaac J, Whitehead J, Adams JW, Barton MD, Coloe P. An outbreak of Mycobacterium bovis infection in cats in an animal house. Aust Vet J. 1983;60(8):243–5.
30. Phan TA, Relic J. Sporotrichoid Mycobacterium marinum infection of the face following a cat scratch. Australas J Dermatol. 2010;51(1):45–8.
31. Lloret A, Hartmann K, Pennisi MG, Gruffydd-Jones T, Addie D, Belak S, et al. Mycobacterioses in cats: ABCD guidelines on prevention and management. J Feline Med Surg. 2013;15(7):591–7.

Dermatophytosis

Karen A. Moriello

Abstract

Feline dermatophytosis is a superficial fungal skin disease of cats. The primary mode of transmission is via direct contact or traumatic fomite inoculation. *Microsporum canis* is the primary pathogen of cats although outdoor cats may contract *Trichophyton* spp. infections. Diagnosis is based upon use of complementary diagnostic tests. Evidence-based studies have concluded there is no one "gold standard diagnostic test." Contrary to popular belief, evidence-based studies found that Wood's lamp examinations are positive in >91% of untreated cats, making it a highly useful point-of-care diagnostic test when combined with direct examination of hair and scales. PCR analysis of infective material is also diagnostic. Fungal culture is needed for species identification. Topical antifungal therapy is necessary to disinfect hairs, minimize disease transmission, and prevent environmental contamination. Systemic antifungal therapy eradicates the disease within the hair follicle. Evidence-based studies have shown that environmental disinfection is easily done via continued removal of cat hair and debris. Spores do not multiply in the environment or invade homes; spores are easily removed from soft and hard surfaces via washing with a detergent. Over-the-counter home disinfectants (i.e., bathroom cleaners) labelled as efficacious against *Trichophyton* spp. are recommended over household bleach which can be a human and animal health hazard. This is a low-level zoonotic skin disease that may cause superficial skin lesions that are treatable and curable in people.

K. A. Moriello (✉)
School of Veterinary Medicine, University of Wisconsin-Madison, Madison, WI, USA
e-mail: Karen.moriello@wisc.edu

© Springer Nature Switzerland AG 2020
C. Noli, S. Colombo (eds.), *Feline Dermatology*,
https://doi.org/10.1007/978-3-030-29836-4_13

Introduction

Dermatophytosis is a contagious, superficial fungal skin disease of skin, hair, scales, and claws. It is non-life threatening, treatable, and curable and is a low-level zoonotic disease, i.e., it does not cause death and is easily treated. The disease will resolve without treatment in otherwise healthy animals. Treatment is recommended to shorten the course of the infection and limit the risk of transmission to other susceptible hosts. The two major goals of this chapter are to (1) summarize key aspects of this disease from recent evidence-based studies and (2) provide evidence to counter many "Internet" myths surrounding this disease that result in poor treatment, unwarranted client worries, and, in worst-case scenarios, euthanasia of cats and kittens.

Pathogens of Importance and New Classifications

Dermatophytes are aerobic fungi that invade and infect keratinized skin, hair, scales, and nails. These organisms are classified by host preference: anthropophilic (humans), zoophilic (animals), and geophilic (soil).

Dermatophytes are also classified by different names depending upon whether or not they are in an asexual state (anamorph) or a sexual state (teleomorph) [1, 2]. For example, *Microsporum canis* is an anamorphic species that belongs to the teleomorphic *Arthroderma otae* complex (*M. canis, M. ferrugineum, M audouinii*) [3]. The naming of anamorphs is based upon fungal culture macro- and micro-characteristics. Recently molecular testing has found many species to be one and the same. In 2011, the Amsterdam Declaration on Fungal Nomenclature (One Fungus = One Name) was adopted, and reclassification is currently underway [4]. *Trichophyton* and *Microsporum* are being reclassified into the genus *Arthroderma*. Clinicians need to be aware of this, as clinical manuscripts are increasingly using new nomenclature. This chapter will use the traditional names.

The most important pathogen of cats is *Microsporum canis*. Less commonly cats can be infected by *Trichophyton* spp. and *M. gypseum*. It has been well established that dermatophytes are not part of the normal fungal flora of cats by both traditional and molecular tests [5–7].

Prevalence

True disease prevalence is unknown as this is not a reportable disease. A recent review of 73 papers from 29 countries revealed prevalence data was highly biased depending upon the source of the cats, whether studies were prospective or retrospective, whether data was collected and interpreted before or after the recognition of fomite carriage, and other inclusion criteria [2].

The most helpful data on prevalence is from studies where true disease was confirmed. These studies consistently found an overall low prevalence (<3%) in clinical practice and shelters (Box 1). In one study from the United States ($n = 1407$ cats), the overall prevalence of confirmed disease was 2.4% [8], while in a Canadian study ($n = 111$ cats), it was 3.6% [9]. In a study from the United Kingdom ($n = 154$ cats), it was 1.3% [10]. More interestingly, in another study from the United Kingdom, the medical records of 142, 576 cats found that it was not even listed as a common skin disease, even though 10.4% of cats were presented for skin disease [11]. In a study looking at the chronic pruritic cats ($n = 502$), only 2.1% of cats were diagnosed with the disease [12]. Finally, in a retrospective study of cats admitted to an open admission animal shelter ($n = 5644$), disease prevalence was 1.6% over a consecutive 24-month period of time [13].

Box 1: Key Points on Disease Prevalence
- Dermatophytosis is an uncommon cause of skin lesions in cats (<3%).
- It is untrue that "it is ringworm until proven otherwise!"
- It is a common skin disease in kittens.

Risk Factors

Risk factors for dermatophytosis include warm, humid environments, young age, and group housing (e.g., animal shelters or catteries) [14–21]. Anecdotal reports of "old age" or "older cats" with underlying age-related diseases being predisposed to dermatophytosis were not supported by evidence [2]. Seropositive FeLV or FIV cats have not been shown to be at increased risk of infection [22]. The development of dermatophytosis in cats receiving immunosuppressive treatment for pemphigus foliaceus was not reported in two large studies [23, 24]. Given the widespread use of feline cyclosporine, there is only one cat reported to have developed disease while receiving this drug [25]. With respect to breed predispositions, Persian cats are often listed as being "predisposed"; however, this breed is over-represented in prevalence and treatment studies. Subcutaneous dermatophyte infections, although rare, are almost exclusively reported in long haired breeds.

Key Aspects of the Pathogenesis, Transmission, and Immune Response to Infection

Pathogenesis of Infection

The infective form of dermatophytes is an arthrospore which is formed by fragmentation of fungal hyphae into smaller infective units. There are three stages of a dermatophyte infection [2]. First, arthroconidia adhere to corneocytes which can occur within 2–6 hours of exposure [26–28]. Second, fungi start to germinate with germ

tubes emerging from the arthroconidia followed by penetration of the stratum corneum. Finally, there is invasion of keratinized structures; dermatophyte hyphae
invade and grow in multiple directions including the hair follicle unit. Hyphae can
start to form arthroconidia within 7 days. Obvious clinical lesions are usually seen
within seven to 21 days.

Transmission

Development of *M. canis* lesions has been studied in direct application models of
infection and in co-habitant natural exposure experiments and offers practical
insights on disease transmission [29–35]. In direct application models, it was
extremely difficult to establish infections unless there was a critical mass of infective spores ($>10^4$ spores per site). Successful infection required micro-trauma and
occlusion of the site. It was not possible to infect cats/kittens that were allowed to
remove the infective inoculum via grooming. Positive Wood's lamp fluorescence
was present in 100% of experimentally infected cats and was noted as early as 5 to
7 days post infection. In co-habitant models, a highly social infected cat was added
to a group of healthy cats, and lesion development followed a clear pattern. Lesions
developed over time in all cats, starting with the most social cats. All lesions began
on the face and ears and then progressed. In studies where healthy cats were housed
in contaminated environments but did not experience any skin micro-trauma, cats
became culture positive but did not develop lesions. Cats became culture negative
after washing or simply being moved to a clean room and allowed to groom.

It is now well established that the primary mode of disease transmission is via
direct contact with another infected animal. Grooming is an important innate protective mechanism against disease transmission. Micro-trauma is an important pre-
requisite for establishment of a successful infection. Increased micro-trauma to the
skin from pruritus or self-trauma, humidity, and ectoparasites all contribute to conditions optimal for disease development. Transmission from contaminated fomites
is a risk factor if it induces micro-trauma (e.g., grooming tools) or if the cat is in a
contaminated environment and self-traumatizes itself (e.g., is pruritic from ectoparasites). *Transmission from contaminated environments is not an efficient mode of
transmission in the absence of micro-trauma and moisture.*

Immunity and Recovery from Infection

Cats develop both a cell-mediated and humoral immune response to dermatophyte
infections [35–37]. Intradermal and in vitro studies show that recovery from infection depends upon the development of a strong cell-mediated immune response.
Cell-mediated immunity is important for protection against reinfection. Studies
have shown that reinfection of infected but cured cats was possible, but required a
greater number of spores, more occlusion, or both. The subsequent infections were
milder and resolved much sooner.

Clinical Findings

There are no "pathognomonic" clinical signs of feline dermatophytosis. The clinical signs of dermatophytosis reflect the pathogenesis of the disease: invasion of keratinized structures of the skin. Clinical signs are also impacted by the age and overall physiological health. For example, kittens with limited infections are at risk for developing more widespread lesions if they contract upper respiratory infections or gastrointestinal disease.

Practical Approach to "Clinical Signs"

Dermatophytosis severity reflects the overall global health of a cat. From that perspective, there are different clinical presentations of dermatophytosis: simple infections, complicated infections, and culture-positive lesion-free cats. Simple infections are any disease that occurs in an otherwise healthy cat. Lesion severity tends to be limited, and these cats respond well to treatment. It is likely that many kittens develop limited lesions of dermatophytosis that self-resolve and never get diagnosed. Complicated infections are more difficult to treat because lesions tend to be more severe and the cat/kitten has a concurrent medical disease, develops a concurrent disease shortly after diagnosis that explains the severity of the skin disease, and/or has some other complicating factor making treatment challenging (e.g., bandage requirements that make topical therapy options limited.) A "complicated infection" is any case where treatment is not straightforward. Culture-positive lesion-free cats are either cats that are fomite carriers or cats that have subtle lesions that were missed at the time of initial examination. When lesion-free, culture-positive cats are identified, re-examine the cat with a Wood's lamp and look for lesions. Fomite carriage is easily identified by simply washing the cat, moving it to a clean environment, and repeating a culture. True fomite carriage cats will be culture negative; a single bath will not remove infective spores from the hair coat with true infection.

Common Findings

Lesions tend to be asymmetrical. As mentioned above, observational studies on co-habitant infection models documented that lesions tend to start on the face, ears, and muzzle and then progress to the paws and tail (Figs. 1, 2, and 3) [34, 38]. Lesions can be focal or multifocal. Hair loss may be mild, and sometimes, the primary client concern is excessive hair loss. Some cats have a history of vomiting hair balls or constipation. Scaling is common and sometimes can be marked (Figs. 4 and 5). In severe cases, there can be exudative paronychia. The inflammatory reaction can vary from mild to marked, and diffuse erythema may be present. Follicular plugging and hyperpigmentation are somewhat uncommon in cats but most likely to be seen in cats with dermatophytosis. *Microsporum canis* can cause comedone-like lesions

Fig. 1 Facial lesions on a
young cat with
dermatophytosis.
(Courtesy of Dr. Rebecca
Rodgers)

Fig. 2 Alopecia and
scaling on the pinna of a
Persian cat with
dermatophytosis.
(Courtesy of Dr. Chiara
Noli)

in young cats. Pruritus is variable and can be intense and may mimic areas of eosin-ophilic pyotraumatic dermatitis.

Uncommon Presentations

Uncommon clinical presentations in cats include cases clinically identical to pem-phigus foliaceus, including symmetrical crusting over the face and ear and exuda-tive paronychia. Unilateral or bilateral pinnal pruritus is another unique clinical presentation. Infected hairs are on the ear margins or in the "bell" of the ear. Rarely,

Fig. 3 Alopecia on the back and tail in an advanced case of dermatophytosis in a Persian cat. (Courtesy of Dr. Chiara Noli)

Fig. 4 Same cat as in Figure 2: patch of alopecia with very mild scaling. (Courtesy of Dr. Chiara Noli)

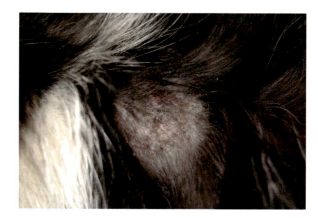

Fig. 5 Shorthaired cat with dermatophytosis showing a focal patch of alopecia and thick scales. (Courtesy of Dr. Chiara Noli)

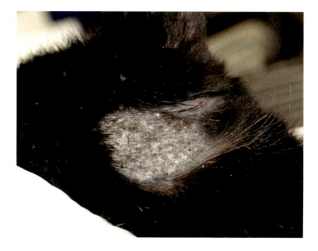

Fig. 6 Ulcerated nodule due to a dermatophytic mycetoma. (Courtesy of Dr. Andrea Peano)

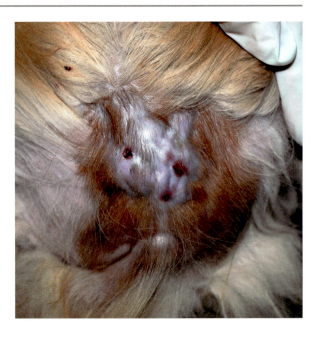

diffuse multifocal areas of waxy hyperpigmentation have been observed. Nodular dermatophytosis has been most commonly reported in Persian cats (Fig. 6). There is usually a history of prior dermatophytosis but not always. These may or may not ulcerate and drain.

Diagnosis

Dermatophytosis cannot be diagnosed based upon clinical signs. A recent evidence-based review concluded that no one test could be identified as the "gold standard." Current recommendations are to use multiple complementary diagnostics [2]. Diagnostics for dermatophytosis are divided into two major categories: point-of-care (POC) and reference laboratory (RL) testing. Complete blood counts, serum chemistry panels, urinalysis, and diagnostic imaging are not helpful for confirming the presence or absence of dermatophytosis. These tests are helpful when evaluating a cat with a complicated infection.

Point-of-Care Diagnostics

There are three key complementary POC diagnostic tests and tools: dermoscopy, Wood's lamp examination, and direct examination of hair/scales. Dermoscopy and Wood's lamp examinations are tools used to find suspect hairs for direct examination. If infection can be confirmed via direct examination of hair shafts and scale, treatment can be initiated at the time of presentation.

Dermoscopy

Dermoscopy (Figs. 7 and 8) is a non-invasive POC **tool** that allows for magnification and illumination of the skin. **The primary use of dermoscopy is to find hairs for direct examination.** This tool can be used with or without a Wood's lamp examination. Two studies found that *M. canis*-infected hairs have a unique appearance [39, 40]. Infected hairs are opaque, slightly curved, or broken with a homogenous thickness (Fig. 9). Hairs are easier to find in light colored cats than in darkly colored cats. The biggest obstacle to the use of this test is patient cooperation.

Wood's Lamp Examination

A Wood's lamp is a POC diagnostic **tool whose primary usage is to find hairs for direct examination or to lesion resolution in M. canis infected cats.** A Wood's lamp is a plug-in lamp with an ultraviolet light spectrum of 320–400 nm wavelength [41]. In

Fig. 7 Hand-held dermoscope. (Courtesy of Dr. Fabia Scarampella)

Fig. 8 Cat being examined with a dermoscope. (Courtesy of Dr. Fabia Scarampella)

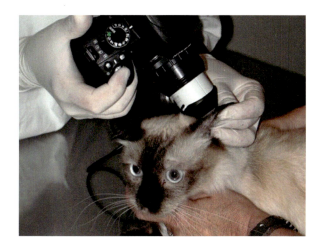

Fig. 9 Infected hairs visualized with a dermoscope. (Courtesy of Dr. Fabia Scarampella). Arrows required to indicate infected hairs

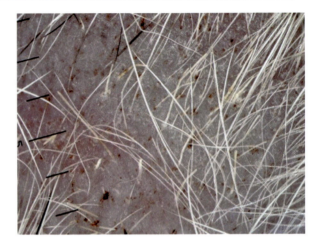

veterinary dermatology, the only fungal pathogen of importance that fluoresces is *M. canis*. The characteristic green fluorescence of *M. canis*-infected hairs is due to a water-soluble pigment located within the cortex or medulla of the hair [42–44]. The fluorescence is the result of a chemical interaction that occurs as a result of the infection and is not associated with spores or infective material.

An evidence-based review of literature has found that many commonly held beliefs are incorrect about the usefulness of Wood's lamp examination, prevalence of positive fluorescence, and overall usefulness as a point-of-care "test"; *this is tool and it cannot be emphaized enough*. Statements such as "less than 50% of strains fluoresce" are based upon retrospective studies of random source diagnostic specimens [15, 45–47]. When data from 30 experimental infection studies and spontaneous disease studies was examined, results were surprisingly different [2]. There was 100% fluorescence in cats with experimental infections, and in studies involving spontaneous disease in untreated animals, it was >91%. Not unexpectedly, positive fluorescence was less common in cats under treatment. Fluorescing "tips" are a common finding in cats that have been treated and cured. It is simply residual pigment left over from the infection within the hair follicle (Fig. 10).

In the author's experience, the use of a Wood's lamp is not unlike mastering sample acquisition for skin/ear cytology and using a microscope. See Box 2 for helpful hints on using a Wood's lamp. It cannot be stressed enough that this is a skill that can be learned. With the "right" Wood's lamp and practice, this is a helpful tool for finding suspect hairs (Figs. 11a, b). It is immensely helpful in finding lesions that are otherwise missed in room light (Fig. 12). Fluorescing hairs are commonly found in untreated infections; fluorescence may be more difficult to find in treated animals. False-positive and false-negative results are commonly due to inadequate equipment, lack of magnification, patient compliance, poor technique, or lack of

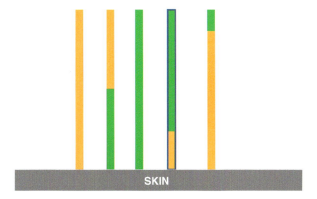

Fig. 10 Schematic of Wood's lamp positive hairs (from left to right). Positive fluorescence is found on the hair shafts only. Uninfected hairs show no fluorescence. Early infected hairs show fluorescence in the proximal part of the hair. As the infection progresses, the entire hair shaft will fluoresce. When the infection has been eradicated in the hair follicle, the proximal portion of the hair shaft will no longer fluoresce (see hair with blue outline). This is an indication of a good response to treatment. Cured cats will often have some residual "glowing tips" because the pigment is retained in the medulla or cortex (yellow = uninfected; green = fluorescent, thus infected). Glowing tip hairs may or may not be culture positive

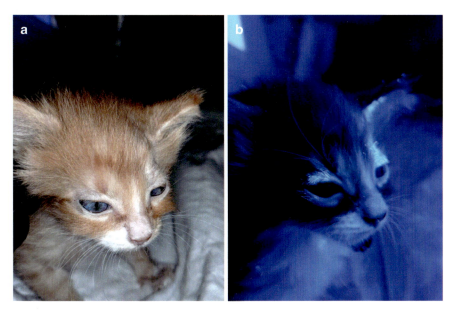

Fig. 11 Kitten with dermatophytosis: (**a**) mild lesions evident around the eyes; (**b**) evident fluorescence on the same sites with Wood light examination. (Courtesy of Dr. Laura Mullen)

Fig. 12 Wood's lamp positive hairs in the interdigital space. This lesion was not noticed during examination in room light

training. Wood's lamp can rapidly identify high risk cats. For example, in one shelter, 1226 cats were surrendered in a 7-month period [48]. Of these cats, 273 (22.3%) were culture positive, but only 60 of 273 were lesional, Wood's lamp positive, and direct examination positive. The 213 remaining culture-positive cats were nonlesional and Wood's lamp negative and were determined to be fomite carriers. The use of the Wood's lamp at intake allowed for rapid identification of infected cats (50 of 60 being kittens).

Box 2: Wood's Lamp Practice Tips
- Use a medical grade lamp with UV spectrum 320 to 400 nm wavelength
- Do not use hand-held battery-operated lamps
- Use a lamp with built-in magnification
- Lamps do not need to "warm up"
- Allow your eyes to light adapt to the dark
- *Use a positive control slide*
- *Hold lamp close to skin (2–4 cm); minimizes false fluorescence*
- Start at head and move slowly examining *hair shafts*
- Lift crusts and look for apple green fluorescing hair shafts
- Newly infected hairs are very short
- Worried about false fluorescence? Examine hair bulb

Direct Examination of Scales and Hairs

Direct examination is a POC diagnostic test that can confirm the presence of a dermatophyte infection at the time of initial presentation. Material can be collected with the aid of a dermoscope and Wood's lamp or via a combination of skin scraping and plucking of hairs. A recent study showed that the best way to collect

Fig. 13 Direct examination of infected hair (original magnification 4×). Note that infected hairs are pale and wider than normal hairs which appear as 'threads' in comparison.

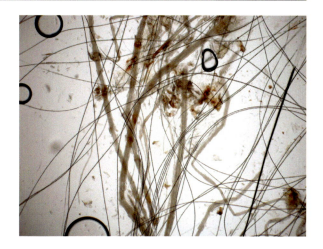

specimens is via both superficial skin scraping and plucking of hairs from lesions. In cats, combined hair plucking and skin scraping of lesion margins confirmed the diagnosis in 87.5% of cases [49]. In this study, a Wood's lamp was not used, and had it been, the results may have been higher. The authors used mineral oil for mounting of specimens and no clearing agent; clearing agents destroy fluorescence, cause artifacts, and damage microscope lens. The author routinely mounts specimens in mineral oil. This is a time and cost effective test to master as it allows for microscopic examination for mites with the same specimen. The most helpful aid in learning this technique is to have images comparing normal and abnormal hairs (Figs. 13, 14, and 15). Direct examination tips are summarized in Box 3.

Box 3: Direct Examination Practice Tips

Use a Wood's lamp or dermoscope to help identify suspect hairs

    Collect samples by both plucking and scraping of the lesion

    Use a skin scraping spatula to scrape margins of lesions

    Mount specimens in mineral oil; do not use a clearing agent

    Use glass coverslips

    Abnormal hairs are easily visible at 4× and 10×

    Infected shafts are wider, paler, and often retractile compared to normal hairs

    Tips to find hairs

- Use a picture guide that shows abnormal vs normal hairs
- Hold a Wood's lamp (2–3 cm) over the slide to help find the hairs
- Add lactophenol cotton blue or new methylene blue to the mineral oil; let sample set for 10 to 15 minutes before examination; infected hairs will be blue tinged

Fig. 14 Direct examination of infected (short thick arrow) and uninfected hairs (long thin arrow)

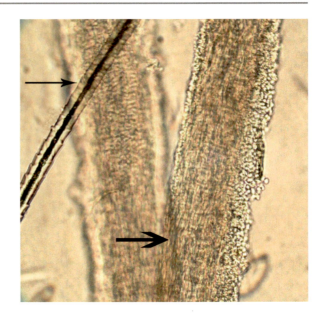

Fig. 15 Direct examination of infected hair in mineral oil and lactophenol cotton blue after 15 minutes

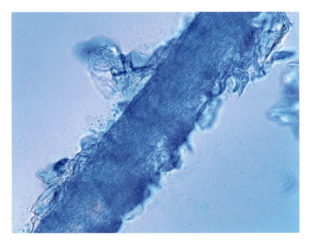

Skin Cytology

Macroconidia are never seen on cytological examination of skin cytology. However, *M. canis* arthrospores may be observed in cats with severe infections (Fig. 16).

Fungal Culture

Fungal cultures can be a POC or RL diagnostic test (see Box 5). If infection is confirmed via direct examination of hair and scale, fungal culture is used to confirm the dermatophyte species. Fungal culture can be used to confirm the diagnosis if POC diagnostics do not. A recent study revealed good correlation between point-of-care

Fig. 16 Cytologic examination of a skin cytology from a cat with dermatophytosis. Numerous arthroconidia are seen on the surface of the corneocyte

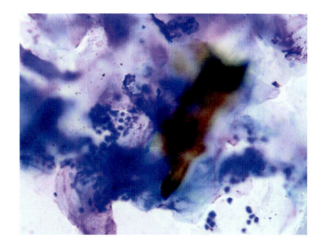

cultures and reference laboratories when both gross colony formation along with microscopic features were used to identify colonies [50]. However, there was an almost 20% error rate when color change alone was used.

The most commonly used POC fungal culture medium is Dermatophyte Test Medium (DTM) which consists of a nutrient medium plus inhibitors of bacterial and saprophytic growth and phenol red as a pH indicator. Several variants are available, some of which claim to speed growth of the culture, but a study found that all appeared to perform similarly [51]. From a practical perspective, use a POC plate with a large volume of medium; it is easy to inoculate with a toothbrush or hairs and is easily sampled for microscopic examination. The author discourages the use of DTM glass vials because they are hard to inoculate and sample. Vials that base their diagnostic value on "positive color change" are not recommended. Plates should be stored in individual plastic bags to prevent cross contamination and protect against desiccation and media mite infestation. The author stores samples in a plastic container and monitors temperature using an inexpensive digital thermometer for fish tanks. Dermatophyte colonies may appear as soon as 5 to 7 days after inoculation. Plates should be inspected daily for growth by holding the plate up to the light to look for colony growth (Fig. 17). To minimize contaminant growth, do not open plates until there is adequate growth to sample. It is common to see early colony growth using this backlighting technique several days before a red color change develops around the colony. The red color change is caused by a change in the pH of the medium and is not diagnostic: it merely identifies colonies for microscopic sampling (Fig. 17). The color change usually occurs at the time the colony is first visible, but may develop within 12 to 24 hours *after* visible fungal growth. All fungal growth, including non-pathogens, will eventually produce a red media color change after the colony has grown for several days to a week. Dermatophyte colonies are never green, gray, brown, or black. Pathogens are pale or buff in color and have a powdery to cottony mycelial growth. All suspect colonies should be examined microscopically (Figs. 18 and 19).

Fig. 17 DTM plate with initial growth of *M. canis*. Around the small white cottony colonies, the culture media color turned red. Note: Gloves should be worn when handling fungal culture plates. (Courtesy of Dr. Chiara Noli)

Fig. 18 Microscopic example of *M. canis* from a clear acetate tape preparation. The sample was allowed to stand for 15 min before examination, making macroconidia more visible. Note the tapered ends, rough surface, and thick walls

Fig. 19 High power (100×) of *M. canis* macroconidia. Note the thick walls as the most consistent feature

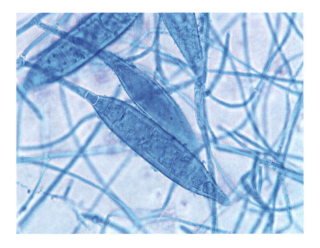

Unless there is a sampling error such as plucking only a few hairs rather than using a toothbrush for sampling lesions, large numbers of colonies of the dermatophyte will appear on the plate if the animal is truly infected. The number of colonies decreases as the infection resolves spontaneously or from treatment. One of the most common problems with in-house culturing is the lack of sporulation or growth, or both, of *M. canis* on DTM. A common cause of this is over-inoculation of the fungal culture plates. This is characterized by rapid swarming of the plate with fluffy colony growth, but only unsporulated hyphae are noted on microscopic examination. This can be avoided by limiting the number of toothbrush inoculations on the plate to six to eight stabs; individual impressions should be clearly visible.

A common question is how long to hold fungal culture plates. Recent research has shown that in cultures of human dermatophyte infections, 98.5% of fungal cultures were positive before day 17 [52]. A retrospective study of 2876 *M. canis* positive fungal cultures found that 98.2% were confirmed within 14 days of incubation [53]. The revised commendation is to consider a fungal culture negative if no pathogen or no growth has been isolated by day 14.

Reference Laboratory Diagnostics

PCR

There is increasing interest and use of PCR in animals to diagnose dermatophytosis. This is because PCR is commonly used to diagnose dermatophytosis of nail in people because of the difficulty in isolation of pathogens via fungal culture. Routine bathing and use of over-the-counter topical antifungal preparations make isolation of human *Trichophyton* infections via fungal culture challenging. If the reference laboratory has the protocol, PCR on tissue can be used to aid in the diagnosis of deep dermatophyte infections in cats [54, 55].

Commercial reference laboratories are increasingly offering PCR as a diagnostic test. The major advantage of this test over fungal culture is the rapid turnaround time. It is important to remember that PCR is very sensitive and will detect both viable and non-viable fungal DNA. In addition, like toothbrush fungal cultures, PCR cannot differentiate between fomite carriage and true disease. Field studies using a commercial PCR test in cats from animal shelters found that the test has high sensitivity and high specificity [56, 57]. Samples can be collected using the toothbrush technique, but it is important to sample only the target lesion ensuring an adequate amount of follicular hair is collected for analysis. Alternatively, avulsed crusts with hair shafts and hair bulbs can be collected and submitted for examination. In one field study, the qPCR assay for *Microsporum* spp. was more useful for initial disease confirmation, while the qPCR *M. canis* assay was more useful for determining mycological cure [57]. When using this test to monitor for mycological cure, it may be helpful to bathe and dry the cat's hair coat prior to sampling to minimize a positive PCR test from detection of non-viable DNA.

Histopathology

There are two clinical presentations when histological examination of tissue is helpful to diagnose dermatophytosis. The first is when cats present with unusual skin
lesions and routine point-of-care diagnostics do not identify a cause. Some cats with
dermatophytosis will develop lesions that look clinically similar to pemphigus foliaceus. The second is the investigation of non-healing wounds or nodules caused by
dermatophytosis (dermatophytic mycetoma or pseudomycetoma). It is important to
remember to submit as large a section of tissue as possible because processing can
result in marked shrinkage of the specimen [58]. The author routinely samples nodules via excisional biopsy or with a 6 to 8 mm skin biopsy punch. It is important to
tell the pathology laboratory that dermatophytosis is suspected because routine
stains (i.e., hematoxylin and eosin) are not as sensitive as periodic acid-Schiff (PAS)
or Gomori's methenamine silver (GMS) for detecting fungal elements in tissue. In
addition, tissue (4–6-mm punch or wedge) should be submitted to a reference laboratory for fungal culture.

Treatment

Dermatophytosis is a self-limiting disease in otherwise healthy animals. Treatment
is recommended to shorten the course of the infection because it is infectious and
contagious. Treatment is summarized in Table 1.

Confinement Considerations

Confinement needs to be reconsidered in the treatment of this disease. The recent
treatment consensus guidelines state that "Confinement needs to be used with care
and for the shortest time possible. Dermatophytosis is a curable disease, but behavior problems and socialization problems can be life-long if the young or newly
adopted animals are not socialized properly." [2] This disease is most common in
kittens at the same time that it is critical for socialization and bonding. Veterinarians
need to consider animal welfare and quality of life when making a recommendation
for confinement. The purpose of confinement is to limit the amount of work needed
to do routine cleaning. The living area should allow for 24/7 exercise, normal behavior (i.e., play and jumping), sleeping, eating, and socialization. It is important to
remember that disease transmission is limited via the use of concurrent systemic
antifungal therapy and, most importantly, topical therapy. Infection from contaminated environments is an inefficient and rare mode of transmission. The infection is
transmitted by direct contact with spores on the hair coat. Topical therapy and simple barrier protection (i.e., gloves and long shirt sleeves) and reasonable human
behavior (i.e., not "wearing the kitten/cat") will minimize the risk of transmission to
people.

Table 1 Quick summary of treatment recommendations

Confinement
Limit confinement to easily cleaned room. Kitten/cat must have 24/7 access to freely move and exercise
Human interaction and socialization must be possible
Topical therapy
Twice weekly whole body therapy with lime sulfur or enilconazole rinses or shampoo therapy with a miconazole-ketoconazole-climbazole/chlorhexidine shampoo
Daily focal miconazole 2% vaginal cream for lesions on the face
Daily otic medicaments that do not contain antibiotics for lesions in/on ears
Systemic therapy
Oral itraconazole 5 mg/kg orally once daily on a week on/week off basis: **Do not use compounded or reformulated itraconazole**
Terbinafine 30 to 40 mg/kg orally once daily is an option if itraconazole is not available
Do not use lufenuron, griseofulvin, or fluconazole
Cleaning
"Clean as if company is coming twice weekly"
Keep cat hair to a minimum and use disinfectant wipes between "cleanings"
Hard surfaces
Focus on removal of cat hair and debris, wash with a detergent until visibly clean, rinse and remove excess water, and use ready-to-use disinfectants labelled as effective against *Trichophyton* spp. (i.e., bathroom cleaners)
Bleach is no longer recommended
Soft surfaces
Wash laundry twice to disinfect; do not over-stuff the washer tub and use the longest cycle available
Vacuum carpets to remove cat hair; disinfect with steam cleaning or washing twice with a beater brush carpet shampooer
Monitoring
Start monitoring after a cat/kitten is lesion-free and there are no fluorescing hair shafts
One fungal culture is compatible with mycological cure in most cats, two may be needed in complicated situations, PCR can be used but may be associated with false positives

Clipping of the Hair Coat

There are no controlled studies comparing the number of days to cure in cats that have been clipped to those that have not been clipped. Based upon treatment outcomes in dedicated dermatophyte treatment programs in the United States, it is the author's experience that routine clipping of the hair coat is not necessary. Clipping of the hair coat requires sedation and can result in thermal burns which may not be clinically apparent until weeks later. Based upon experimental treatment studies, clipping of the hair coat can temporarily worsen lesions and/or result in the development of satellite lesions [30, 31]. If lesions or hair mats need to be removed, use children's round tipped metal scissors. Place the cat on newspaper to allow for easy disposal of infective material. If long haired cats are slow to cure and/or the owner cannot thoroughly soak the hair coat, scissor clipping to facilitate the penetration of topical antifungal treatment may be helpful. What is helpful is to brush the hair coat prior to the application of topical therapy to remove broken and easily shed hairs. Plastic flea combs are ideal for this purpose.

Topical Therapy

Topical therapy is as important as systemic antifungal therapy in the treatment of feline dermatophytosis. Systemic antifungal therapy eradicates infection within the hair follicle but does not kill infective spores on or in the hair shaft or on the hair coat. Topical therapy protects against disease transmission. Topical therapy minimizes shedding of infective spores into the environment which greatly decreases, if not prevents, the potential of positive fungal cultures due to fomite contamination [59, 60]. Concurrent topical therapy will decrease the overall length of treatment. It is important to remember that until the infection is eradicated within the hair follicle, the hair coat will be continually reseeded with infective spores.

Twice weekly whole body rinses or shampoos are recommended for the duration of treatment. Exposed uninfected cats and dogs should be treated with a whole body antifungal rinse or shampoo to minimize risk of transmission. The most consistently effective antifungal products *in vivo* studies are lime sulfur, enilconazole, or miconazole/chlorhexidine shampoos [60–66]. These products are fungicidal. The author has also successfully used combination miconazole/chlorhexidine and climbazole/chlorhexidine leave-on mousse formulations in cats that could not be wetted (i.e., cats with bandages, upper respiratory infections). In vitro studies have shown that antifungal shampoos containing miconazole, ketoconazole, climbazole, or accelerated hydrogen peroxide are antifungal when used with a minimum of a 3-minute contact time [67]. There is strong evidence from in vitro and in vivo studies on the antifungal activities of essential oil preparations as options [59, 68, 69].

Based upon treatment of cats in dermatophyte treatment centers, the most common reason for failure to cure is the presence of infective hairs in hard to treat areas. People are reluctant to apply antifungal rinses or shampoo to the face and in/near the ears of cats. Unfortunately, these are the sites of infective hairs when "failure to cure" cats are examined with a Wood's lamp. The author recommends the daily use of miconazole 2% vaginal cream on the face and periocular area if lesions are found. This particular product is used to treat fungal keratitis and is safe [70]. For infective hairs in the ears, antifungal otic products (chlorhexidine/miconazole or chlorhexidine/ketoconazole, or clotrimazole) can be used daily.

Systemic Antifungal Therapy

Systemic antifungal therapy eradicates the infection within the hair follicle and is used with concurrent topical therapy. Unless there is a contraindication, it is indicated in all cats with dermatophytosis.

The antifungal drug of choice for cats is itraconazole (Itrafungol, Elanco Animal Health). It is labelled for use at 5 mg/kg orally once daily on an alternating week on/ week off treatment schedule. An initial treatment schedule of 3 cycles is recommended, but additional cycles may be needed in some cats that have not reached mycological cure by 6 weeks [71]. Itraconazole accumulates in adipose tissue, sebaceous glands, and hair for weeks after administration, making it suitable for pulse therapy protocols [72].

Itraconazole has no age or weight limitations. Kittens as young as 10 days of age were treated with 5 mg/kg orally for four consecutive weeks, and no treatment-related side effects were reported [72]. The drug is well tolerated, and no treatment studies have reported death or adverse effects that required discontinuation of the drug when used at doses for treatment of feline dermatophytosis [2]. Side effects were rare and included salivation, mild anorexia, and vomiting [2, 71] Deaths associated with its use have not been reported. Target animal safety studies in which cats were given 5, 15, and 25 mg/kg itraconazole for 7 days on alternate days for 17 weeks with an 8-week recovery period noted dose-related hypersalivation, vomiting, and loose stool which were mild to moderate and self-resolving [73]. Elevations in hepatic enzymes above baseline were sporadic, dose-related, and rarely above laboratory normal ranges. In an extensive review of the literature, reports of severe adverse effects of itraconazole in cats were all traced to studies or case series treating cats with high doses for long periods of time [2].

The author is aware of many anecdotal reports of "itraconazole resistance," and, when investigated, compounded itraconazole was used. A recent paper compared the reference capsule, reference solution, compounded capsule, and compounded suspension in a randomized cross-over study [74]. The findings revealed that compounded formulations were poorly and inconsistently absorbed. *Compounded itraconazole should not be used.*

Terbinafine has been used successfully for the treatment of *M. canis* dermatophytosis [2]. Studies have shown that terbinafine is highly concentrated in the hair coat of cats after 14 days of continuous administration and is suitable for pulse therapy [75]. Doses in the literature range from 5 to 40 mg/kg per day; however, higher doses of 30–40 mg have been reported to be clinically more effective. The most common side effects are vomiting, diarrhea, and soft stools.

Griseofulvin was the first oral antifungal used to treat feline dermatophytosis, but it is no longer recommended given the superior efficacy of itraconazole and terbinafine. It is also a known teratogen and can cause dose-unrelated idiosyncratic bone marrow suppression. Ketoconazole is effective against dermatophytes but is poorly tolerated in cats and should not be used. Fluconazole has poor efficacy against dermatophytes and should not be used. Numerous well-controlled studies have shown that lufenuron has no efficacy and should not be used to treat dermatophytosis [32, 33, 76].

Fungal Vaccines

Antifungal vaccines for *M. canis* have shown no efficacy against challenge exposure but may be useful as adjuvant therapy. Commercial vaccines are limited in availability [2].

Disinfection of the Environment

The primary reason to disinfect the environment is to minimize fomite contamination of the hair coat which will make it difficult to determine mycological cure. Fomite contamination can lead to over-treatment of cats, over-confinement, expense, and in some cases euthanasia. Review of the literature found that contact with a contaminated environment alone *in the absence of concurrent micro-trauma* is a rare source of infection for people and cats [2]. *Severely* contaminated environments, e.g., hoarding situations, are a risk factor for cats under severe physiological stress or predisposed to skin micro-trauma (e.g., flea infestation).

Evidence-based studies on environmental decontamination have now shown that it is much easier to decontaminate an environment than past literature suggests and/ or what clients may find on Internet resources. Dermatophyte spores can only live and reproduce in keratin; they do not multiply or invade the environment as many clients believe. It is important to stress to clients that dermatophyte spores are not like mildew or mold that overgrows in homes after water damage. Clients will report reading that "ringworm lives" in the environment up to 24 months. This comment stems from a laboratory study where specimens ($n = 25$ total) were stored and sampled at various time points. In that study, three of six specimens stored between 13 and 24 months were viable on fungal culture medium [77]. This study did not document that stored specimens were able to cause disease. In a different study, stored specimens were only viable for 13 months, and it was not possible to induce infection in kittens [78]. In the author's experience with stored specimens for 25 years, isolates loose viability and become culture negative within months. In one experiment, 30% (45 of 150 specimens) were culture negative within 5 months, and all were culture negative by 9 months. Finally, spores in infective hairs and scales are very susceptible to moisture: 100 specimens were culture negative after being exposed to high humidity for 3 days.

Box 4 summarizes environmental cleaning recommendations. The environmental cleaning focus needs to be on mechanical cleaning and removal of debris coupled with washing of the target surface until visibly clean. The surface needs to be rinsed of detergent as this will inactivate many disinfectants. In addition, the surface needs to be free of any excess water as this will dilute disinfectants. A recent study showed that household bathroom cleaners labelled as efficacious against *Trichophyton* spp. are effective against the naturally infective form of *M. canis* and *Trichophyton* spp. [79] Clients should be strongly advised against the use of household bleach as a disinfectant due its lack of detergency, lack of penetration into organic material, and human/animal health hazard.

Box 4: Summary of Disinfection Recommendations

Key points to stress: Spores do not multiply in the environment, spores do not invade surfaces like mildew or black mold, spores are easily removed via cleaning, and spores are susceptible to moisture, i.e., they die quickly post exposure.

Key cleaning points: "If you can wash it, you can decontaminate it" and "clean as if company is coming."

Cleaning specifics:

- Laundry: Wash twice in the washer on hot or cold water; bleach is not necessary.
- Rugs: Keep pets off rugs and/or vacuum daily. Can be disinfected using "steam cleaning" or washing twice with a beater brush carpet scrubber.
- Keep pets in easily cleaned rooms, but do not over-confine. Close closets and drawers, and remove knick-knacks. Remove debris and pet hair daily using dusting cloths or 3 M Easy Trap dust cloth (these are sticky "swiffers") and then mop floors with a flat mop. Repeat two to three times weekly.
- Disinfectants do not take the place of mechanical cleaning and washing; spores are like dust and are easily removed via mechanical cleaning.
- Mechanical cleaning is most important, remove debris, wash with a detergent cleaner, rinse, and remove excess water. This alone can decontaminate surface.
- Disinfectants are needed for spores not removed by cleaning. For safety, only use ready-to-use commercial disinfectants labelled as efficacious against *Trichophyton* spp., thoroughly wet target non-porous surfaces, and let dry.
- Clean transport cages.
- Environmental sampling is NOT cost-effective and not recommended unless there is concern about fomite contamination.

Clients often ask what they can do to minimize environmental contamination in addition to cleaning. In addition to systemic antifungal therapy, the most important thing is to use topical antifungal therapy to disinfect the hair coat. In a recent study, proper cleaning combined with topical therapy resulted in homes being free of infective material within 1 week of starting treatment and remaining so throughout the study [69]. In a study of 70 homes contaminated by *M. canis* infected cats, only 3 of 69 homes needed more than one cleaning for complete decontamination. One home was never decontaminated due to admitted owner non-compliance [80].

Monitoring and Endpoint of Treatment

Mycological Cure

The term "mycological cure" was introduced into to the veterinary literature in 1959 and was defined as two negative fungal cultures at 2 weeks apart in a study using griseofulvin to treat long haired cats with *M. canis* feline dermatophytosis [81]. Because this is a contagious and infectious disease, *M. canis* is not part of the normal fungal flora of cats, and the disease is of zoonotic importance, it is reasonable to treat cats until the infectious agent is no longer detectable via fungal culture(preferred) or PCR. A recent study has found that the first negative fungal culture was predictive of mycological cure in >90% of cats that were otherwise healthy, where there was good compliance with cleaning, and topical and systemic treatment [82].

Treatment Length

A common question from owners is "how long until the cat is cured"? The response "as long as it takes" or "until mycological cure" is true, but irritating to owners. In a recent placebo controlled study using itraconazole at current label recommendations, lesion resolution, Wood's lamp examinations, and weekly fungal cultures were performed for 9 weeks [71]. In this study, cats did not receive any topical therapy. Mycological cure was documented as early as week four of treatment. At week nine, 39 of 40 (97.5%) cats had Wood's negative examinations. By the end of 9 weeks, 36 of 40 (90%) cats had at least 1 negative fungal culture, and 24 of 40 had two negative fungal cultures. These cats did not receive topical therapy. In a shelter study using 21 days of consecutive itraconazole and concurrent topical twice weekly lime sulfur, the mean number of days to mycological cure was 18 (range 10–49 days) in a group of 90 cats that were otherwise healthy [62]. In a later study involving random source cats with a multitude of concurrent illnesses, the mean number of days to cure was 37 days (range 10–93) [83]. Based upon these studies, it is reasonable to answer the question as follows: in otherwise healthy cats receiving itraconazole and topical therapy, mycological cure can be expected within 4 to 8 weeks. If the cat has concurrent illnesses, e.g., upper respiratory infection or poor nutrition, treatment will be longer.

Recommended Monitoring

Clinical Cure

It is well established that clinical cure precedes mycological cure. There should be resolution of clinical signs, and clients are usually capable of these observations. A lack of resolution and/or development of new lesions indicate a treatment problem or misdiagnosis. In the author's experience, cats receiving itraconazole show rapid resolution of clinical signs.

Wood's Lamp Examination

It is now known that Wood's lamp examinations are very useful for both detection of *M. canis*-infected hairs and for monitoring of infections. This tool is strongly recommended to monitor infections provided the user has a proper Wood's lamp, cats can be handled, and the room darkened. As the infection is eradicated in the hair follicle, fluorescence disappears from the proximal portion of the hair shaft (i.e., intra-follicular portion). As a new healthy hair grows, there is less and less fluorescence on the hair shaft. Residual pigment on the hair tips is common in cats that have recovered from dermatophytosis and reflects residual pigment deposited in the hair shaft at the time of initial infection.

PCR Testing

Commercial PCR testing for mycological cure can be used provided there is high confidence in the reference laboratory performing the test. It is important to remember that PCR will detect both viable and non-viable fungal DNA. Leave-on rinses or mousses will kill fungal spores, but because these are not "rinsed" off the hair coat, non-viable fungal DNA may be present. Fungal PCR testing should be considered only after there is clinical cure and lack of Wood's lamp hair shaft fluorescence, routine cleaning is in place, and only if topical therapy has been used concurrently with systemic antifungal therapy. If a leave-on antifungal has been used for topical therapy, wash the cat to remove any residual fungal DNA. qPCR *M. canis* assay was found to be more useful for detecting mycological cure than the qPCR *Microsporum* assay [57]. The use of cycle thresholds was found not to be helpful for determining mycological cure [84].

Fungal Culture

The most commonly used diagnostic test to determine mycological cure in cats is a toothbrush fungal culture. There is no established "best practice" for when to start monitoring response to treatment using toothbrush fungal cultures. What is important to note is that it is no longer acceptable to report fungal culture results as "positive" or "negative." The number of dermatophyte colony-forming units (cfu) per plate provides valuable information (Boxes 5 and 6). Clinical examination, fungal culture findings, and Wood's lamp examination are used to determine if a cat is infected or cured.

- Do weekly fungal cultures once the decision has been made to start evaluating for mycological cure.
- Fungal culture plates do not need to be held longer than 14 days; culture negative plates at day 14 should be considered negative.
- Do in-house cultures or use a reference laboratory that is familiar with the toothbrush inoculation technique and will provide weekly updates on cfu/plate.
- See Box 6 for use of cfu/plate and in practice.
- Continue topical therapy until the cat is mycologically cured (negative PCR or at least one negative toothbrush fungal culture).

Box 5: Fungal Culture Practice Tips and Use of Colony-Forming Units
Fungal culture plates

- Use large volume easy open plates.
- Do not over-inoculate plates; make sure bristles show a pattern on surface.
- Incubate in house in plastic bag to prevent cross contamination and minimize desiccation.
- Incubate at 25 °C to 30 °C.
- Examine daily for growth; use backlighting technique.
- No growth plates can be finalized at 14 days; not necessary to hold for 21 days.

Record growth twice weekly

- NG – no growth.
- C– contaminant growth bacterial or fungal.
- S – suspect growth (early growth of a pale colony or early growth of pale colony with a red color change.
- Pathogen – requires microscopic identification. In animals under treatment, the red color change may lag behind the growth of the pale colony especially as the animal approaches cure.

Count colony-forming units (only with toothbrush culture technique)

- The number of colony-forming units per plate can be used to monitor response to therapy. P or "pathogen score" is the nickname used for this system.
  - P3-≥ 10 cfu/plate (often too many to count!) – indicates high risk cat and active infection
  - P2 5–9/cfu plate – indicates need to continue treatment
  - P1 1–4 cfu/plate – most consistent with fomite exposure or exposure to another infected animal; continue topical therapy; improve cleaning of environment; consider if there is exposure to infected animal

Note: This system makes it easy to monitor culture results and provides a visual record of the pet's response to treatment. In most cases, animals with severe infections will have a starting culture score of P3. As treatment progresses, the P score becomes lower. Cured animals have cultures with no growth or just contaminant on culture. The scoring system is also very helpful in identifying pets undergoing treatment that are exposed to fomite contamination. These animals commonly will have cultures fluctuating from negative to P1. When this pattern is seen, the owner can be instructed to improve hygiene in the home. As fomite contamination is removed, the fungal cultures become negative. In addition to identification of fomite exposure, this system also rapidly alerts the clinician to animals that are failing therapy or are relapsing for one reason or another. Lack of response to therapy will be suggested by a persistently high P score. Relapses will be represented by a sudden increase in colony-forming units.

Box 6: Interpretation of P-score, Lesions, and Wood's Lamp Findings in Diagnosis and Treatment of *M. canis* Infections* [*]

P-score	Examination	Wood's lamp examination of hair shafts	Wood's examination of hair tips	Interpretation	Plan	Comments
P3 (>10 cfu/ plate)	Lesional/ nonlesional	Positive/ negative	Positive/ negative	High risk/not cured	Treat or continue treatment	A single infected hair can produce a P3 culture, examine carefully
P2 (5–9 cfu/ plate)	Lesional	Positive/ negative	Positive/ negative	High risk/not cured	Treat or continue treatment	
	Nonlesional	Positive	Positive/ negative	High risk/not cured	Treat or continue treatment	
	Nonlesional	Negative	Positive/ negative	Cured/low risk	Re-examine, apply whole body antifungal treatment, then repeat culture when dry	Likely represents a "dust mop" scenario
P1 (1–4 cfu/ plate)	Lesional	Positive/ negative	Positive/ negative	High risk/not cured	Treat or continue treatment	
	Nonlesional	Positive	Positive/ negative	High risk/not cured	Treat or continue treatment	
	Nonlesional	Negative	Positive/ negative (glowing tips are common in cured animals)	Cured/low risk	Re-examine, apply whole body antifungal treatment, then repeat culture when dry	If "dust mop" cat, repeat culture will be negative

Note

cfu colony forming unit; "dust mop" refers to a cat that is mechanically carrying spores from environmental contamination

*Adapted from the treatment and monitoring procedures used in the Felines In Treatment Program at the Dane County Humane Society, Madison, Wisconsin, USA

Reprinted with permission from [2]

Public Health Aspects

Dermatophytosis was a disease of major public health concern because until relatively recently, there was no effective and safe antifungal treatment. Animal-associated infections were common because of people's close association with agriculture and the lack of veterinary care for skin diseases of pet animals. The development of oral griseofulvin for use in people and small animals in the late 1950s was a major therapeutic advance for people and animals. The development of ketoconazole, itraconazole, terbinafine and a wide range of topical antifungals were further major advances.

Feline dermatophytosis is a pet-associated zoonosis, and it is a veterinarian's responsibility to inform clients of this risk and provide accurate information about the disease. The reader is referred to the references for a detailed discussion [2]. Key aspects to communicate to clients are the following:

- Dermatophytosis occurs in both animals and people. In people, it is commonly called "toe nail fungus" or athlete's foot fungus.
- It is the same disease, just a different pathogen. The primary pathogen of people is *Trichophyton*.
- The disease causes skin lesions and is treatable and curable.
- From cats, the disease is transmitted via direct contact with hair or skin lesions, and that is why topical therapy is so important. Topical therapy decreases the risk of disease transmission.
- Use reasonable barrier protection, e.g., as you would in handling an animal with infectious diarrhea.
- Risk of contracting the disease from the environment is low.
- Dermatophytosis is a common skin disease in immunocompromised people; however, literature review found that these infections are resurgences of pre-existing human dermatophyte infections [85]. Animal-associated dermatophytosis was rare.
- The most common complication of *M. canis* infection in immunocompromised people was a prolonged treatment time [86].

References

1. Weitzman I, Summerbell RC. The dermatophytes. Clin Microbiol Rev. 1995;8:240–59.
2. Moriello KA, Coyner K, Paterson S, et al. Diagnosis and treatment of dermatophytosis in dogs and cats.: Clinical Consensus Guidelines of the World Association for Veterinary Dermatology. Vet Dermatol. 2017;28:266–e68.
3. Graser Y, Kuijpers AF, El Fari M, et al. Molecular and conventional taxonomy of the Microsporum canis complex. Med Mycol. 2000;38:143–53.
4. Hawksworth DL, Crous PW, Redhead SA, et al. The Amsterdam declaration on fungal nomenclature. IMA Fungus. 2011;2:105–12.
5. Moriello KA, DeBoer DJ. Fungal flora of the coat of pet cats. Am J Vet Res. 1991;52:602–6.

6. Moriello KA, Deboer DJ. Fungal flora of the haircoat of cats with and without dermatophytosis. J Med Vet Mycol. 1991;29:285–92.
7. Meason-Smith C, Diesel A, Patterson AP, et al. Characterization of the cutaneous mycobiota in healthy and allergic cats using next generation sequencing. Vet Dermatol. 2017;28:71–e17.
8. Scott DW, Miller WH, Erb HN. Feline dermatology at Cornell University: 1407 cases (1988–2003). J Feline Med Surg. 2013;15:307–16.
9. Scott DW, Paradis M. A survey of canine and feline skin disorders seen in a university practice: Small Animal Clinic, University of Montreal, Saint-Hyacinthe, Quebec (1987–1988). Can Vet J. 1990;31:830.
10. Hill P, Lo A, Can Eden S, et al. Survey of the prevalence, diagnosis and treatment of dermatological conditions in small animal general practice. Vet Rec. 2006;158:533–9.
11. O'Neill D, Church D, McGreevy P, et al. Prevalence of disorders recorded in cats attending primary-care veterinary practices in England. Vet J. 2014;202:286–91.
12. Hobi S, Linek M, Marignac G, et al. Clinical characteristics and causes of pruritus in cats: a multicentre study on feline hypersensitivity-associated dermatoses. Vet Dermatol. 2011;22:406–13.
13. Moriello K. Feline dermatophytosis: aspects pertinent to disease management in single and multiple cat situations. J Feline Med Surg. 2014;16:419–31.
14. Lewis DT, Foil CS, Hosgood G. Epidemiology and clinical features of dermatophytosis in dogs and cats at Louisiana State University: 1981–1990. Vet Dermatol. 1991;2:53–8.
15. Cafarchia C, Romito D, Sasanelli M, et al. The epidemiology of canine and feline dermatophytoses in southern Italy. Mycoses. 2004;47:508–13.
16. Mancianti F, Nardoni S, Cecchi S, et al. Dermatophytes isolated from symptomatic dogs and cats in Tuscany, Italy during a 15-year-period. Mycopathologia. 2002;156:13–8.
17. Debnath C, Mitra T, Kumar A, et al. Detection of dermatophytes in healthy companion dogs and cats in eastern India. Iran J Vet Res. 2016;17:20.
18. Seker E, Dogan N. Isolation of dermatophytes from dogs and cats with suspected dermatophytosis in Western Turkey. Prev Vet Med. 2011;98:46–51.
19. Newbury S, Moriello K, Coyner K, et al. Management of endemic Microsporum canis dermatophytosis in an open admission shelter: a field study. J Feline Med Surg. 2015;17:342–7.
20. Polak K, Levy J, Crawford P, et al. Infectious diseases in large-scale cat hoarding investigations. Vet J. 2014;201:189–95.
21. Moriello KA, Kunkle G, DeBoer DJ. Isolation of dermatophytes from the haircoats of stray cats from selected animal shelters in two different geographic regions in the United States. Vet Dermatol. 1994;5:57–62.
22. Sierra P, Guillot J, Jacob H, et al. Fungal flora on cutaneous and mucosal surfaces of cats infected with feline immunodeficiency virus or feline leukemia virus. Am J Vet Res. 2000;61:158–61.
23. Irwin KE, Beale KM, Fadok VA. Use of modified ciclosporin in the management of feline pemphigus foliaceus: a retrospective analysis. Vet Dermatol. 2012;23:403–e76.
24. Preziosi DE, Goldschmidt MH, Greek JS, et al. Feline pemphigus foliaceus: a retrospective analysis of 57 cases. Vet Dermatol. 2003;14:313–21.
25. Olivry T, Power H, Woo J, et al. Anti-isthmus autoimmunity in a novel feline acquired alopecia resembling pseudopelade of humans. Vet Dermatol. 2000;11:261–70.
26. Zurita J, Hay RJ. Adherence of dermatophyte microconidia and arthroconidia to human keratinocytes in vitro. J Invest Dermatol. 1987;89:529–34.
27. Vermout S, Tabart J, Baldo A, et al. Pathogenesis of dermatophytosis. Mycopathologia. 2008;166:267–75.
28. Baldo A, Monod M, Mathy A, et al. Mechanisms of skin adherence and invasion by dermatophytes. Mycoses. 2012;55:218–23.
29. DeBoer DJ, Moriello KA. Development of an experimental model of Microsporum canis infection in cats. Vet Microbiol. 1994;42:289–95.
30. DeBoer D, Moriello K. Inability of two topical treatments to influence the course of experimentally induced dermatophytosis in cats. J Am Vet Med Assoc. 1995;207:52–7.

31. Moriello KA, DeBoer DJ. Efficacy of griseofulvin and itraconazole in the treatment of experi-
    mentally induced dermatophytosis in cats. J Am Vet Med Assoc. 1995;207:439–44.
32. Moriello KA, Deboer DJ, Schenker R, et al. Efficacy of pre-treatment with lufenuron for the
    prevention of *Microsporum canis* infection in a feline direct topical challenge model. Vet
    Dermatol. 2004;15:357–62.
33. DeBoer DJ, Moriello KA, Blum JL, et al. Effects of lufenuron treatment in cats on the estab-
    lishment and course of *Microsporum canis* infection following exposure to infected cats. J Am
    Vet Med Assoc. 2003;222:1216–20.
34. DeBoer DJ, Moriello KA. Investigations of a killed dermatophyte cell-wall vaccine against
    infection with *Microsporum canis* in cats. Res Vet Sci. 1995;59:110–3.
35. Sparkes AH, Gruffydd-Jones TJ, Stokes CR. Acquired immunity in experimental feline
    Microsporum canis infection. Res Vet Sci. 1996;61:165–8.
36. DeBoer DJ, Moriello KA. Humoral and cellular immune responses to *Microsporum canis* in
    naturally occurring feline dermatophytosis. J Med Vet Mycol. 1993;31:121–32.
37. Moriello KA, DeBoer DJ, Greek J, et al. The prevalence of immediate and delayed type hyper-
    sensitivity reactions to *Microsporum canis* antigens in cats. J Feline Med Surg. 2003;5:161–6.
38. Frymus T, Gruffydd-Jones T, Pennisi MG, et al. Dermatophytosis in cats: ABCD guidelines on
    prevention and management. J Feline Med Surg. 2013;15:598–604.
39. Scarampella F, Zanna G, Peano A, et al. Dermoscopic features in 12 cats with dermatophytosis
    and in 12 cats with self-induced alopecia due to other causes: an observational descriptive
    study. Vet Dermatol. 2015;26:282–e63.
40. Dong C, Angus J, Scarampella F, et al. Evaluation of dermoscopy in the diagnosis of naturally
    occurring dermatophytosis in cats. Vet Dermatol. 2016;27:275–e65.
41. Asawanonda P, Taylor CR. Wood's light in dermatology. Int J Dermatol. 1999;38:801–7.
42. Wolf FT. Chemical nature of the fluorescent pigment produced in *Microsporum*-infected hair.
    Nature. 1957;180:860–1.
43. Wolf FT, Jones EA, Nathan HA. Fluorescent pigment of Microsporum. Nature.
    1958;182:475–6.
44. Foresman A, Blank F. The location of the fluorescent matter in microsporon infected hair.
    Mycopathol Mycol Appl. 1967;31:314–8.
45. Sparkes A, Gruffydd-Jones T, Shaw S, et al. Epidemiological and diagnostic features of
    canine and feline dermatophytosis in the United Kingdom from 1956 to 1991. Vet Rec.
    1993;133:57–61.
46. Wright A. Ringworm in dogs and cats. J Small Anim Pract. 1989;30:242–9.
47. Kaplan W, Georg LK, Ajello L. Recent developments in animal ringworm and their public
    health implications. Ann N Y Acad Sci. 1958;70:636–49.
48. Newbury S, Moriello K, Coyner K, et al. Management of endemic *Microsporum canis* derma-
    tophytosis in an open admission shelter: a field study. J Feline Med Surg. 2015;17:342–7.
49. Colombo S, Cornegliani L, Beccati M, et al. Comparison of two sampling methods for
    microscopic examination of hair shafts in feline and canine dermatophytosis. Vet (Cremona).
    2010;24:27–33.
50. Kaufmann R, Blum SE, Elad D, et al. Comparison between point-of-care dermatophyte test
    medium and mycology laboratory culture for diagnosis of dermatophytosis in dogs and cats.
    Vet Dermatol. 2016;27:284–e68.
51. Moriello KA, Verbrugge MJ, Kesting RA. Effects of temperature variations and light expo-
    sure on the time to growth of dermatophytes using six different fungal culture media inocu-
    lated with laboratory strains and samples obtained from infected cats. J Feline Med Surg.
    2010;12:988–90.
52. Rezusta A, Gilaberte Y, Vidal-García M, et al. Evaluation of incubation time for dermatophytes
    cultures. Mycoses. 2016;59:416–8.
53. Stuntebeck R, Moriello KA, Verbrugge M. Evaluation of incubation time for Microsporum
    canis dermatophyte cultures. J Feline Med Surg. 2018;20:997–1000.
54. Bernhardt A, von Bomhard W, Antweiler E, et al. Molecular identification of fungal pathogens
    in nodular skin lesions of cats. Med Mycol. 2015;53:132–44.

55. Nardoni S, Franceschi A, Mancianti F. Identification of *Microsporum canis* from dermatophytic pseudomycetoma in paraffin-embedded veterinary specimens using a common PCR protocol. Mycoses. 2007;50:215–7.
56. Jacobson LS, McIntyre L, Mykusz J. Comparison of real-time PCR with fungal culture for the diagnosis of Microsporum canis dermatophytosis in shelter cats: a field study. J Feline Med Surg. 2018;20:103–7.
57. Moriello KA, Leutenegger CM. Use of a commercial qPCR assay in 52 high risk shelter cats for disease identification of dermatophytosis and mycological cure. Vet Dermatol. 2018;29:66.
58. Reimer SB, Séguin B, DeCock HE, et al. Evaluation of the effect of routine histologic processing on the size of skin samples obtained from dogs. Am J Vet Res. 2005;66:500–5.
59. Nardoni S, Giovanelli S, Pistelli L, et al. In vitro activity of twenty commercially available, plant-derived essential oils against selected dermatophyte species. Nat Prod Commun. 2015;10:1473–8.
60. Paterson S. Miconazole/chlorhexidine shampoo as an adjunct to systemic therapy in controlling dermatophytosis in cats. J Small Anim Pract. 1999;40:163–6.
61. Moriello K, Coyner K, Trimmer A, et al. Treatment of shelter cats with oral terbinafine and concurrent lime sulphur rinses. Vet Dermatol. 2013;24:618–e150.
62. Newbury S, Moriello K, Verbrugge M, et al. Use of lime sulphur and itraconazole to treat shelter cats naturally infected with *Microsporum canis* in an annex facility: an open field trial. Vet Dermatol. 2007;18:324–31.
63. Carlotti DN, Guinot P, Meissonnier E, et al. Eradication of feline dermatophytosis in a shelter: a field study. Vet Dermatol. 2010;21:259–66.
64. Jaham CD, Page N, Lambert A, et al. Enilconazole emulsion in the treatment of dermatophytosis in Persian cats: tolerance and suitability. In: Kwochka KW, Willemse T, Von Tscharner C, editors. *Advances in Veterinary Dermatology*. Oxford: Butterworth Heinemann; 1998. p. 299–307.
65. Hnilica KA, Medleau L. Evaluation of topically applied enilconazole for the treatment of dermatophytosis in a Persian cattery. Vet Dermatol. 2002;13:23–8.
66. Guillot J, Malandain E, Jankowski F, et al. Evaluation of the efficacy of oral lufenuron combined with topical enilconazole for the management of dermatophytosis in catteries. Vet Rec. 2002;150:714–8.
67. Moriello KA. In vitro efficacy of shampoos containing miconazole, ketoconazole, climbazole or accelerated hydrogen peroxide against *Microsporum canis* and *Trichophyton* species. J Feline Med Surg. 2017;19:370–4.
68. Mugnaini L, Nardoni S, Pinto L, et al. In vitro and in vivo antifungal activity of some essential oils against feline isolates of *Microsporum canis*. J Mycol Med. 2012;22:179–84.
69. Nardoni S, Costanzo AG, Mugnaini L, et al. Open-field study comparing an essential oil-based shampoo with miconazole/chlorhexidine for haircoat disinfection in cats with spontaneous microsporiasis. J Feline Med Surg. 2017;19:697–701.
70. Gyanfosu L, Koffuor GA, Kyei S, et al. Efficacy and safety of extemporaneously prepared miconazole eye drops in Candida albicans-induced keratomycosis. Int Ophthalmol. 2018;38:2089–210.
71. Puls C, Johnson A, Young K, et al. Efficacy of itraconazole oral solution using an alternating-week pulse therapy regimen for treatment of cats with experimental Microsporum canis infection. J Feline Med Surg. 2018;20:869–74.
72. Vlaminck K, Engelen M. An overview of pharmacokinetic and pharmacodynamic studies in the development of itraconazole for feline *Microsporum canis* dermatophytosis. Adv Vet Dermatol. 2005;5:130–6.
73. Elanco US I. Itrafungol itraconazole oral solution in cats. Freedom of Information Summary NADA 141–474, November 2016.
74. Mawby DI, Whittemore JC, Fowler LE, et al. Comparison of absorption characteristics of oral reference and compounded itraconazole formulations in healthy cats. J Am Vet Med Assoc. 2018;252:195–200.
75. Foust AL, Marsella R, Akucewich LH, et al. Evaluation of persistence of terbinafine in the hair of normal cats after 14 days of daily therapy. Vet Dermatol. 2007;18:246–51.

76. DeBoer D, Moriello K, Volk L, et al. Lufenuron does not augment effectiveness of terbinafine for treatment of *Microsporum canis* infections in a feline model. Adv Vet Dermatol. 2005;5:123–9.

77. Sparkes AH, Werrett G, Stokes CR, et al. *Microsporum canis*: Inapparent carriage by cats and the viability of arthrospores. J Small Anim Pract. 1994;35:397–401.

78. Keep JM. The viability of *Microsporum canis* on isolated cat hair. Aust Vet J. 1960;36:277–8.

79. Moriello KA, Kunder D, Hondzo H. Efficacy of eight commercial disinfectants against Microsporum canis and Trichophyton spp. infective spores on an experimentally contaminated textile surface. Vet Dermatol. 2013;24:621–e152.

80. Moriello KA. Decontamination of 70 foster family homes exposed to Microsporum canis infected cats: a retrospective study. Vet Dermatol. 2019;30:178–e55. https://doi.org/10.1111/vde.12722.

81. Kaplan W, Ajello L. Oral treatment of spontaneous ringworm in cats with griseofulvin. J Amer Vet Med Assoc. 1959;135:253–61.

82. Stuntebeck RL, Moriello KA. One vs two negative fungal cultures to confirm mycological cure in shelter cats treated for Microsporum canis dermatophytosis: a retrospective study. J Feline Med Surg. 2019. https://doi.org/10.1177/1098612X19858791.

83. Newbury S, Moriello KA, Kwochka KW, et al. Use of itraconazole and either lime sulphur or Malaseb Concentrate Rinse (R) to treat shelter cats naturally infected with *Microsporum canis*: an open field trial. Vet Dermatol. 2011; 22: 75–9.

84. Jacobson LS, McIntyre L, Mykusz J. Assessment of real-time PCR cycle threshold values in Microsporum canis culture-positive and culture-negative cats in an animal shelter: a field study. J Feline Med Surg. 2018;20:108–13.

85. Rouzaud C, Hay R, Chosidow O, et al. Severe dermatophytosis and acquired or innate immunodeficiency: a review. J Fungi. 2015;2:4.

86. Elad D. Immunocompromised patients and their pets: still best friends? Vet J. 2013;197:662–9.

Deep Fungal Diseases

Julie D. Lemetayer and Jane E. Sykes

Abstract

Deep mycotic infections are uncommon in cats. However, in endemic regions, cryptococcosis, sporotrichosis, and histoplasmosis occur regularly in immuno-competent cats. Cryptococcosis and sporotrichosis are more prevalent in cats than in dogs, and histoplasmosis is as prevalent or possibly slightly more preva-lent in cats than in dogs. Blastomycosis and coccidioidomycosis are rare in cats, even in highly endemic areas. Sino-nasal and sino-orbital aspergillosis are also infrequent worldwide but interestingly, brachycephalic cats appear to be predis-posed. Lastly, infections with saprophytic opportunistic fungi usually result from an accidental cutaneous inoculation in otherwise immune-competent cats and cause localized signs. Occasionally, however, disseminated infections can occur. Cats with systemic mycosis frequently have cutaneous manifestations. Reported cutaneous signs include multifocal ulcerated or nonulcerated cutaneous masses, subcutaneous masses, and draining abscesses, among others. The cutaneous signs are frequently associated with systemic signs of illness and/or other organ involvement, which should raise suspicion for fungal infection. The present chapter focus on the epidemiology, clinical signs, including cutaneous signs, diagnostic tests, and treatment of clinically important systemic fungal infections in cats. In addition, it reviews the antifungal drugs currently available for the treatment of these infections.

J. D. Lemetayer (✉) · J. E. Sykes
Veterinary Medical Teaching Hospital, University of California, Davis, CA, USA
e-mail: jesykes@ucdavis.edu

297

Introduction

Deep mycotic infections are uncommon to rare in cats worldwide. A 1996 study estimated a prevalence of seven deep mycotic infections per 10,000 cats in the USA [1]. Indeed, cats are relatively resistant to fungal infections and the prevalence of most fungal infections is lower in cats than in dogs except for cryptococcosis and sporotrichosis. Cats also may be slightly more susceptible to histoplasmosis than dogs [2]. This chapter will focus on the epidemiology, clinical signs, diagnostic tests for, and treatment of, clinically important systemic fungal infections in cats.

Box 1: Dimorphic Fungi
- Cryptococcosis, caused by *C. gattii* and *C. neoformans*, is the most common fungal disease in cats worldwide.
- Histoplasmosis is seen as frequently as or slightly more frequently in cats than in dogs in endemic regions
- Cats are also more susceptible to sporotrichosis.
- Blastomycosis and coccidioidomycosis are both infrequent in cats.
- Most cats are immunocompetent.
- Cutaneous signs are frequent in these dimorphic fungal diseases.
- Fluconazole is the first line treatment for most cases of cryptococcosis, and itraconazole is used for resistant cases of cryptococcosis (mostly *C. gattii* infections) and infections with other dimorphic fungal organisms.
- A combination with amphotericin B is recommended in severe cases.
- A short course of anti-inflammatory dose of glucocorticoids is recommended for CNS cases and animals with severe pulmonary disease.

Cryptococcosis

Epidemiology

The most common fungal infection in cats is cryptococcosis [1]. *Cryptococcus* spp. are dimorphic basidiomycetous fungi. Two main species cause cryptococcosis in cats: *Cryptococcus neoformans* and *Cryptococcus gattii*. Rarely, other species have been implicated. *Cryptococcus magnus* was isolated from a cat with otitis externa in Japan [3] and in a cat with a deep limb infection in Germany [4]. *Cryptococcus albidus* was isolated from a cat with disseminated cryptococcosis in Japan [5]. Two of these three cats were tested for feline immunodeficiency virus (FIV) and feline leukaemia virus (FeLV) and were negative [4, 5], and no other apparent underlying immunocompromise was identified in the three cats.

Cryptococcus neoformans is the most common species of *Cryptococcus* isolated worldwide and include two varieties: *C. neoformans* var. *neoformans* and

C. neoformans var. *grubii*. *Cryptococcus neoformans* var. *grubii* accounts for the majority of cases in Australia [6].

Cryptococcus gattii is mostly found in the west coast of the United States and in British Columbia, Canada; in South America, southeast Asia (New Guinea, Thailand), and in parts of Africa and Australia. While *C. neoformans* is more common than *C. gattii* in Australia, rural cats in Australia and cats from Western Australia seem to be more commonly infected by *C. gattii* than by *C. neoformans* [6, 7].

Cryptococcus neoformans can be found in avian guano, especially pigeon faeces; but is also found in other sources including milk, fermenting fruit juices, air, dust and decaying vegetation [6]. *C. gattii* is often found in the hollows of trees, especially *Eucalyptus* trees, in Australia, but has been associated with other hardwood tree species in other geographic locations.

Cryptococcus spp. have been divided into molecular types. *Cryptococcus neoformans* var. *grubii* isolates belong to molecular types VNI and VNII, whereas *C. neoformans* var. *neoformans* isolates belong to molecular type VNIV [7]. A hybrid variety of serotype AD has been classified as molecular type VNIII. *C. gattii* isolates are classified as VGI, VGII, VGIII, and VGIV. There is a proposal to rename *Cryptococcus* molecular types as separate species of *Cryptococcus*, which remains controversial.

Clinical Features in Cats

Siamese, Birman, Ragdoll, Abyssinian, and Himalayan breeds seem to be over-represented in studies evaluating cats with cryptococcosis [1, 6, 8–10], although this was not found in a study from California [11]. A male predisposition has been identified in a few studies [10, 12] but not in others [6, 9, 11]. Having access to the outdoors is also likely a risk factor but cats kept strictly indoors can also be affected [8]. Cats of all age are affected, and FIV or FeLV status does not seem to be a risk factor [6].

The incubation period is variable and can range from months to many years in animals which were initially able to control the disease [13]. After inhalation of the fungus, many cats have upper respiratory involvement with chronic sneezing, nasal discharge and nasal deformation and/or deformations of the structures adjacent to the nasal cavities such as the sinuses (Fig. 1). Involvement of the nasal cavity has been reported in 43–90% of cases [1, 6, 12]. Infections also involve the retina, draining lymph nodes and the central nervous system (CNS). Clinical signs include enlarged mandibular lymph nodes, blindness, dilated and fixed pupils, slow pupillary light reflexes, lethargy, ataxia, behavioural changes and disorientation. Single or multifocal ulcerated or nonulcerated cutaneous masses were seen in 31% and 41% of cases in two studies [1, 12]. The masses can be firm or fluctuant, raised, dome-shaped and erythematous. They frequently ulcerate and may ooze a greyish gelatinous exudate [1]. Other cutaneous lesions include plaques, miliary papules, firm, dome-shaped, alopecic and erythematous papules or nodules [14]. Cutaneous lesions are usually an extension of a sino-nasal pathology. Involvement of the skin

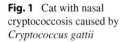
Fig. 1 Cat with nasal cryptococcosis caused by *Cryptococcus gattii*

and subcutis in multiple sites suggests dissemination of the infection. Other uncommon locations include the lungs (2–12% of cases), [1, 6, 15] gingiva [15], salivary glands [6], middle ear [16], kidneys, periarticular subcutaneous tissues, footpads and bones [6].

Diagnostic Tests

Changes on complete blood count (CBC), serum biochemistry and urinalysis are mild and non-specific. [17] A specific diagnosis of cryptococcosis can be obtained with antigen detection using latex agglutination assays on serum. The assay can also be used on pleural or peritoneal effusions, urine, and cerebrospinal fluid (CSF). The clinical sensitivity of the test on serum in cats ranges from 90% to 100% and specificity ranges from 97% to 100% [18]. The sensitivity appears to be lower in dogs. If the antigen test is negative, and cryptococcosis is still a possibility, tissue samples should be submitted for cytology, histology and culture [11]. When titres are <1:200, confirmatory tests are strongly recommended. Enzyme-linked immunosorbent assays (ELISAs) are also under study but data are not currently available.

Fig. 2 India ink negative stain highlighting the *Cryptococcus* polysaccharide capsule

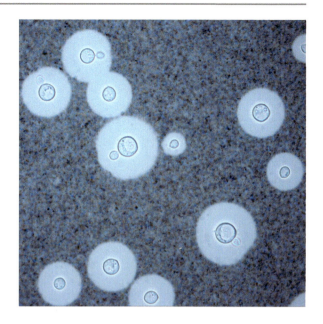

On cytology, cryptococcal yeasts are encapsulated, spherical to oval yeasts measuring 4–10 μm with narrow-based budding. The thick mucopolysaccharide capsule is a major virulence factor for the pathogen because it allows the organism to hide from the host immune system. It appears as a clear halo in stained smears and can be visualized using India ink negative stains (Fig. 2) [19]. However, the capsule can vary in size and in some patients the capsule can be thin [20] which can sometimes complicate the diagnosis. In these cases, there can be morphological overlaps between *Histoplasma* and *Cryptococcus* on cytology [20].

On histopathology, yeasts may be associated with well-ordered granulomas or pyogranulomas, sometimes with a background of a few eosinophils, lymphocytes and plasma cells. Lesions may also contain a large number of yeasts and only a mild degree of inflammation. This results in a "soap bubble" appearance on haematoxylin and eosin staining (H&E) [14] due to the organism's thick, non-staining, capsule. Macroscopically, these lesions are gelatinous masses (cryptococcomas).

In skin biopsies, numerous organisms are often present in the dermis, panniculus and subcutis [14] but occasionally, less typical lesions can complicate the diagnosis. For example, in a case series of four cats with cutaneous cryptococcosis, severe granulomatous to pyogranulomatous cutaneous lesions were reported with large numbers of eosinophils, but organisms could not be seen using H&E stain in three of the four cats, and the organisms were capsule deficient [14].

When yeasts are not seen using H&E stain, special stains such as Grocott's methenamine silver (GMS), periodic acid–Schiff (PAS), Fontana-Masson stain, Ziehl–Neelsen stain or Mayer's mucicarmine stains may help reveal organisms. Polymerase chain reaction (PCR) assays can also be applied to fresh biopsies or formalin-fixed paraffin-embedded tissues [14]. False-positive PCR results can occur from nasal tissues since subclinical nasal cavity colonization can occur, and therefore positive results should always be considered with the rest of the clinical picture [17].

Fresh tissue can also be submitted for fungal culture. *Cryptococcus* spp. grow on most laboratory media in 2 to 10 days. Because the organisms grow in culture as a yeast, rather than a mould, on routine fungal media they are less likely to represent a laboratory hazard than organisms that grow as moulds [17].

Treatment and Prognosis

The triazoles are the first line therapy in the treatment of cryptococcosis and can be used as a monotherapy for mild to moderate cryptococcal infections. Fluconazole is often preferred to itraconazole because of its good penetration in the brain, eye, and urinary tract; lower cost; and minimal adverse effects. However, the development of fluconazole resistance has been reported during treatment [21, 22]. Resistance to fluconazole can be due to overexpression or a change in the copy number of ERG11, the gene encoding for the target enzyme 14-α-demethylase [21, 22]. Efflux of triazole drugs via multidrug efflux transporters (AFR1 for *C. neoformans* and PDR11 for *C. gattii* VGIII) has also been identified. Isolates that are resistant to fluconazole remain susceptible to itraconazole but may demonstrate moderate voriconazole resistance [23].

In severe cases such as those with CNS involvement, addition of amphotericin B to either fluconazole or itraconazole is recommended [24]. While the penetration of amphotericin B into the CNS and vitreous is poor, the blood-brain or blood-eye barrier is compromised at the beginning of treatment so a clinical response can still occur. Flucytosine can also be used in combination with amphotericin B because of the synergy between the two anti-fungal drugs and because of flucytosine's good penetration in the CNS, however it may be cost prohibitive. See Tables 1 and 2 for more information regarding anti-fungal drugs in cats. A short course of prednisolone may improve outcome for cats with CNS infection because it decreases CNS inflammation at the beginning of anti-fungal treatment and may help limit neurological deterioration [25].

Typically, at least 6 to 8 months of treatment is necessary, and often treatment needs to be continued for years [17]. Serial monitoring of serum antigen titres should be used to evaluate treatment response as a decline in titre correlates with elimination of organisms [12]. Treatment should be continued until the titre is zero. Unfortunately, relapse can still occur after successful treatment and after titres have become negative, sometimes as long as 10 years after therapy is discontinued [24].

Prognosis is generally good, with the possible exception of cats with CNS infection [24, 25]. The prognosis may also depend on the *Cryptococcus* species and molecular type. For example, it is the authors' experience that infections with *C. gattii* VGIII tend to be less likely to be cured than those with VGII. While FeLV likely has a negative impact on the treatment response, the effect of FIV status on outcome is not clear. Good response to treatment is often seen despite a positive FIV status but these cats may have more severe disease and/or may respond more slowly to treatment [9, 10, 26].

Table 1 Azole antifungal drugs used in cats

	Ketoconazole	Fluconazole	Itraconazole	Voriconazole	Posaconazole
Mechanism of action	Inhibition of 14α-demethylase, a CYT P450-dependant fungal enzyme = accumulation of 14α-methylsterols and disruption of the fungal cell membrane				
Dose	50 mg/cat PO q 12–24 h	25–50 mg/cat PO or IV q 12–24 h	5 mg/kg PO q 12–24 h, 100 mg capsule q 48 h [129] or 3 mg/kg q 12–24 h with oral solution. Do not use compounded solutions [130]	Not available Possible dose: 12.5 mg/cat PO every 72 hours (with extreme caution as may be associated with severe toxicity) [131]	Oral suspension: 30 mg/kg once then 15 mg/kg q 48 h, or 15 mg/kg once then 5–7.5 mg/kg q 24 h [90, 120, 132]
Clinical use	*Malassezia* spp., Dimorphic fungi	*Candida* spp., *Malassezia* spp., *Coccidioides* spp., *Cryptococcus* spp., *Histoplasma* spp.	Dimorphic fungi, *Aspergillus* spp. and some other moulds	Yeasts, dimorphic fungi, most moulds especially *Aspergillus* spp.	Yeasts, dimorphic fungi, most moulds, including zygomycosis
Decreased antifungal activity	Poor activity against many moulds, including *Aspergillus* spp. *Histoplasma*: Development of resistance during treatment	*Aspergillus* spp. intrinsically resistant, poor activity against many moulds		*Sporothrix schenckii*, zygomycosis	

(continued)

Table 1 (continued)

	Ketoconazole	Fluconazole	Itraconazole	Voriconazole	Posaconazole
Tissue distribution	Good penetration into most tissues but not CNS	Widely distributed including eyes, CNS and kidneys/urine [133]	Good distribution into skin, bone & lungs. Limited penetration to CNS, eyes and kidneys/urine	Widely distributed including eyes, CNS and kidneys/urine	Widely distributed but probably not in urine
Adverse effects	Common: Gastrointestinal signs, hepatotoxicity	Well tolerated. Gastrointestinal signs, hepatotoxicity uncommon	Reported in 25% of cases [134]. Gastrointestinal signs, hepatotoxicity and lethargy	Visual changes (miosis), ataxia, paralysis, hypersalivation, hypokalaemia and arrhythmias [131]	Gastrointestinal signs and increased liver enzyme activities
Additional comments	Strong inducer CYT P450: many drug interactions	Very good oral absorption	For the capsules: give with food and avoid antacid medications. Therapeutic drug monitoring at steady state (14–21d) [135] recommended	Give without food. Therapeutic drug monitoring recommended (trough concentration). Strong inducer CYT P450: many drug interactions	Give with food and avoid antacid medications. Low oral absorption. Therapeutic drug monitoring recommended (trough concentration)

CNS central nervous system, *CYT P450* cytochrome P 450, *PO* per os, *h* hours, *d* days

Table 2 Other clinically important anti-fungal drugs in cats

	Amphotericin B	Terbinafine	Caspofungin	Flucytosine
Mechanism of action	Formation of pores in the fungal cell membrane by binding to sterols = leakage of ions	Inhibition of squalene epoxidase = reduction of ergosterol production in the fungal membrane	Inhibition of β-1,3-D-glucans = disturb the integrity of the fungal cell wall	Deamination of flucytosine to 5-fluorouracil = interfere with DNA replication and protein synthesis
Doses	Deoxycholate AmB: 0.25 mg/kg IV or 0.5 mg/kg SC AmB lipid complex and liposomal AmB: 1 mg/kg IV 3 times weekly (up to 12 treatments)	30–40 mg/kg PO q 24 h	1 mg/kg IV once then 0.75 mg/kg q 24 h [136]	25–50 mg/kg PO q 6–8 h
Clinical use	Yeasts, dimorphic fungi and most moulds	Dermatophytes, maybe useful in combination with other antifungal drugs for various mould infections	Invasive aspergillosis refractory to other antifungal therapy, invasive candidiasis. Some activity against *Histoplasma* spp. and *Coccidioides* spp. Variable activity against other filamentous fungi	*Cryptococcus* spp. and *Candida* spp.
Decreased antifungal activity	Some *Aspergillus* spp. Poor efficacy against *Pythium insidiosum*	Reported resistance for some dermatophytes [137] and *Aspergillus* spp. [138]	*Cryptococcus* spp., *Fusarium* spp., *Rhizopus* spp. and *Mucor* spp. are resistant [139]	Never used as sole agent because of rapid development of resistance
Tissue distribution	Poor penetration of the CNS & eyes. Liposomal and lipid complex formulations have better CNS penetration and less nephrotoxicity	Concentrate in skin nails and hairs	Widely distributed. Poor penetration of the CNS & eyes	Widely distributed including eyes, CNS

(continued)

Table 2 (continued)

	Amphotericin B	Terbinafine	Caspofungin	Flucytosine
Adverse effects	Cumulative nephrotoxicity (mostly AmB deoxycholate), rarely haemolytic anaemia [140]. Sterile injection site abscesses with SC injections	Well tolerated. Rarely, GI toxicity and facial pruritus	Possible anaphylactic reaction. Transient fever and diarrhoea reported [136]	Myelosuppression and GI signs
Additional comments	Liposomal and lipid complex formulations have better CNS penetration and less nephrotoxicity	Low oral absorption [141]		Avoid in animals with renal failure

AmB Amphotericin B, *CNS* central nervous system, *GI* gastro-intestinal, *IV* intravenous, *PO* per os, *SC* subcutaneous

Histoplasmosis

Epidemiology

Histoplasma capsulatum is a dimorphic, soil-borne fungus that is endemic in the USA (especially in the central and *eastern* states but has also been described in California and Colorado), Central and South America, Africa, India and Southeast Asia [27, 28]. It is found worldwide in various mammalian species but besides cases in these endemic areas, cases of histoplasmosis in cats have only been described in Ontario, Canada [29], Thailand [30] and Europe (Italy, Switzerland) [28, 31]. In a 1996 study, histoplasmosis was the second most common fungal disease in cats in the USA with an incidence of 0.01% of the total feline hospital population of the veterinary medical database [1].

Histoplasma capsulatum has been divided into eight to nine geographic clades by multi-locus sequence typing: North American-1, with possibly a related phylogenetically distinct strain isolated from non-endemic American areas; North American-2; Latin American group A; Latin American group B; Australian; Netherlands (of Indonesian origin); Eurasian; and African [32].

The primary reservoir of *H. capsulatum* is the intestinal tracts and guano of bats. It can also be found in decaying avian guano (especially around blackbird or starling roosts and chicken coops). After inhalation or ingestion, the fungus transforms into a yeast phase within the body of cats and is engulfed by phagocytic cells, primarily macrophages. Trafficking of these cells results in dissemination of the yeasts via the blood and lymphatics away from the lung and gastrointestinal tract to organs of the mononuclear phagocyte system mostly (lymph nodes, liver, spleen, and bone marrow) as well as other tissues. Yeast are 2–4 µm in diameter and are surrounded by a 4 µm thick wall and reside within mononuclear phagocytes [33].

Clinical Features in Cats

Cats of all age can be affected with a mean age, of 4 and 9 years in two studies [1, 34]. Persian cats may be slightly over-represented [1]. A sex predisposition has not been clearly identified, but females were over-represented in one case series [34]. Most cats are not concurrently infected with either FeLV or FIV. The disease seems to be diagnosed more often between the months of January to April [1] and can also affect cats that are housed exclusively indoors [35]. The reported duration of clinical signs before diagnosis of histoplasmosis ranged from 2 weeks to 3 months. [1]

When cats have clinical histoplasmosis, disseminated disease is the most commonly reported clinical presentation [36]. Clinical signs exhibited by cats with disseminated disease are mostly non-specific and include lethargy, weight loss, fever, anaemia, dehydration, weakness and anorexia [1, 34]. Respiratory signs such as dyspnoea and tachypnoea are common, but cough is rare. Other common clinical signs include hepatomegaly, icterus, lymphadenopathy and splenomegaly [36, 37], ocular signs (chorioretinitis, anterior uveitis, or retinal detachments) [1, 29,

38] and skeletal involvement (lameness or swelling of one or more limbs) [1, 38, 39]. Clinical signs of gastrointestinal tract involvement such as vomiting, diarrhoea, melaena or haematochezia are less common than in dogs [2]. Less common sites of infection include the skin [28, 38, 40, 41], CNS [42], oral mucosa [43] and urinary bladder [44].

Cutaneous signs consist usually of multiple papules and nodules which may be ulcerated and exude serosanguineous fluid. A case of cutaneous fragility secondary to disseminated histoplasmosis has also been described [41]. The cat had a large skin tear that developed over the dorsal cervical region with epidermal atrophy, dermal collagen separation, and infiltration in the dermis and subcutis with macrophages and intravascular monocytes containing *Histoplasma* yeasts on histology.

Diagnostic Tests

CBC findings include anaemia, which is often normocytic and normochromic and non-regenerative [10, 34]. Thrombocytopenia is also reported, as well as leucocytosis and leukopenia. Occasionally, *H. capsulatum* may be seen within phagocytic cells on peripheral blood smears from dogs and cats [1]. On serum biochemistry, hypoalbuminemia is a common finding. Cats with liver involvement can have increased liver enzyme activity and hyperbilirubinemia. Hyperglobulinemia and azotaemia are also reported in few cats [33], as well as hypercalcaemia [45].

Abnormalities on thoracic radiographs are common and may be subclinical [1, 44]. Radiographic patterns in cats with pulmonary histoplasmosis include fine, diffuse or linear interstitial patterns, bronchointerstitial patterns, diffuse miliary or nodular interstitial patterns, alveolar patterns and/or areas of pulmonary consolidation [33]. Sternal lymphadenopathy is also reported [46]. Bone lesions on radiographs are typically osteolytic, but there may be periosteal and endosteal proliferative lesions, which are mostly found in appendicular bones with a predilection for the elbow and stifle joints [39].

A definitive diagnosis of histoplasmosis is made by cytologic or histopathologic identification of *H. capsulatum* in tissues (Fig. 3). The organisms are usually identified intracellularly within macrophages but can sometimes be found free in necrotic exudates and may be confused with *Cryptococcus* spp. [20] As for *Cryptococcus* infections, a variety of stains can highlight the yeasts, such as Diff-Quik and Wright stains for cytology and GMS or PAS stains for histology.

The yeasts can be found on cytology of lymph nodes, lung, liver, spleen, skin or bone marrow. Serum antibody assays are available but their clinical utility has been limited by low sensitivity and specificity [47]. An antigen ELISA assay has been evaluated for the diagnosis and monitoring of histoplasmosis in cats when applied to serum and urine specimens [46, 47]. Sensitivities of 93–94% of the assay were reported in two studies when applied to urine whereas the sensitivity was only 73% when applied to serum [46, 47]. A specificity of 100% was found in one of these two studies, which included 20 cats diagnosed with other non-fungal diseases [47]. Based on the human literature, serologic cross-reactivity with other

Fig. 3 Cytology showing intracellular *Histoplasma* yeast organisms

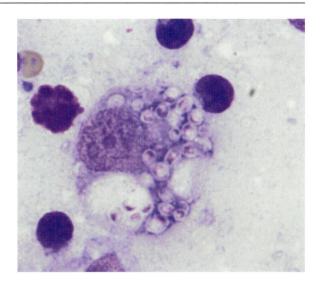

fungal pathogens such as *Blastomyces* spp. is however expected [47]. Antigen concentrations decrease with effective anti-fungal treatment and increase in cases that were not well controlled or following relapse [46]. However, antigen elimination sometimes preceded clinical remission and four cats still had measurable antigen concentrations at the time of remission.

Fungal culture and PCR can also be used to confirm a diagnosis of histoplasmosis. However, fungal culture is a hazard to laboratory workers and should therefore be performed only if necessary, and the laboratory should be warned of the possibility of a dimorphic fungal infection, so that appropriate precautions are taken. Although most cultures are positive within 2 or 3 weeks, growth may require up to 6 weeks of incubation. PCR is currently not used routinely for diagnosis but was used in a few cases to confirm the diagnoses in non-endemic areas [27, 28, 30, 48]. It can also be used when the identity of the fungus observed on histopathology is in doubt.

Treatment and Prognosis

Itraconazole is the treatment of choice for histoplasmosis [45]. Treatment is recommended for a minimum of 4 to 6 months and should be continued for at least 2 months after resolution of clinical signs and possibly until antigen assays are negative. The use of itraconazole may be cost-prohibitive for some clients and adverse effects are more common than with fluconazole, particularly hepatotoxicity [35]. A retrospective study comparing the outcome of 17 cats treated with fluconazole to 13 cats treated with itraconazole found no difference in mortality and relapse rate between the two groups suggesting that fluconazole may be a suitable alternative [35]. However, a lower efficacy of fluconazole compared to itraconazole, and development of fluconazole resistance during treatment has been described in people

[49] and in a cat [50]. The fluconazole-resistant isolates also had increased MICs to voriconazole but not to itraconazole or posaconazole.

Deoxycholate or lipid-complexed amphotericin B can be used initially to treat cats with severe acute pulmonary, acute disseminated, or CNS disease, after which treatment should be continued with either itraconazole or fluconazole. Other possible treatment options include posaconazole in cats that do not tolerate itraconazole or those that fail to respond to fluconazole. A short course of anti-inflammatory glucocorticoids may be useful for cats with severe pulmonary disease or CNS disease at the beginning of treatment.

The prognosis depends on the extent of disease, with reported survival rates varying from 66% to 100% [35, 45].

Blastomycosis

Epidemiology

Blastomyces dermatitidis is also a dimorphic fungus. It is found as a mycelium in the environment and as a thick-walled budding yeast in tissues [51]. Blastomycosis is a rare disease in cats and most feline infections are identified at necropsy. A 1996 study identified 41 cases over a 30–year time frame, and blastomycosis constituted 0.005% of all feline cases in the veterinary medical database [1]. In North America, cases of blastomycosis are mostly found in the eastern and southern parts of the USA, especially the Ohio and Mississippi River Valleys, and in the Great Lakes region, as well as in Canada, especially Quebec, Ontario, Manitoba, and Saskatchewan [1, 52–54]. Blastomycosis is also endemic in Africa and India [51]. Blastomycosis was also reported in a cat from Thailand [55].

In endemic regions, *B. dermatitidis* is found in localized regions where soils are moist and acidic with decaying vegetation or animal excreta [52]. Inhalation of conidia produced from the mycelial phase in soil or decaying matter is the primary route of infection [51]. Direct inoculation of the organism via skin puncture wounds occurs rarely.

From the lungs, the organism may disseminate via the vascular or lymphatic system, resulting in a granulomatous or pyogranulomatous inflammatory response in many organs, especially the lymph nodes, eyes, skin, bones and brain.

Clinical Features in Cats

A male predisposition to blastomycosis was found in one study and cats less than 4 years of age appear predisposed [1, 53, 56]. However in another case series of eight cats, most cases were female and over 7 years of age [52]. In addition, Siamese, Abyssinian, and Havana Brown cats may be predisposed [1]. Immunosuppression does not seem to play a role in predisposition to the disease [52], and cats housed strictly indoors can also be affected [52, 57, 58].

Fig. 4 Lateral thoracic radiograph of a cat with pulmonary blastomycosis. Courtesy University of California, Davis Veterinary Medical Teaching Hospital Diagnostic Imaging Service

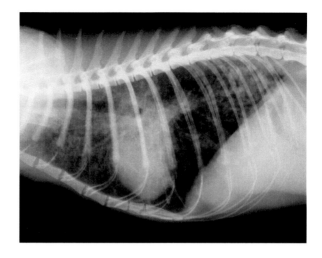

Duration of illness at diagnosis ranges from less than 1 week to 7 months and clinical signs include dyspnoea, cough, anorexia, lethargy, weight loss, peripheral lymphadenopathy, lameness, cellulitis of the limbs, CNS and cutaneous and ocular signs (Fig. 4) [1, 51]. Cutaneous involvement occurred in 23% of 22 affected cats [1], in 63% of eight cats in another case series [52] and in six additional cats [57, 59]. Cutaneous signs include non-ulcerated dermal masses, ulcerative skin lesions, draining abscesses or cellulitis [1, 52, 59].

Diagnostic Tests

Bloodwork findings in cats with blastomycosis are non-specific and indicative of an inflammatory process, such as mild non-regenerative anaemia [52]. Hypercalcaemia and increased calcitriol concentration have also been reported [59]. A specific diagnosis of blastomycosis is usually made using cytologic examination of impression smears, lavage specimens, or aspirates (skin, lymph nodes or lungs). Cytology was diagnostic in 4/6 cases and in 4/5 cases in two studies [53, 57]. *Blastomyces* yeasts are round to oval, 10–20 μm in diameter, have a basophilic cytoplasm, thick and double-contoured walls and display broad-based budding [51]. A pyogranulomatous inflammatory reaction is typical, but a suppurative response predominates on occasion. Special stains such as PAS, Gridley's fungal, and GMS can assist in the detection of the yeasts. Histology of tissue or bone biopsies, fungal culture or PCR can also be used to make a diagnosis of blastomycosis. However, culture is time-consuming and a hazard for laboratory staff. To date, PCR assays have primarily been used for research purposes [60, 61] but in one report, PCR was used to confirm a diagnosis of blastomycosis in a cat from a non-endemic country [55].

Serology has not been well evaluated in cats. In a study, only one of four cats with blastomycosis tested positive on agar gel immunodiffusion (AGID) using *Blastomyces* whole cell antigen [1].

Treatment and Prognosis

Cats with blastomycosis are usually treated with itraconazole. Fluconazole appears to be less effective than itraconazole, but it may be a more appropriate treatment choice for urinary tract, prostatic and CNS infections because of improved penetration of these organs. The addition of amphotericin B for cats with severe disease such as CNS or severe disseminated disease may also be valuable [56, 62]. However, two cats with severe disease treated with amphotericin B did not respond well to treatment [63]. Newer triazoles, including voriconazole, posaconazole and isavuconazole, have activity against *B. dermatitidis*. Voriconazole and posaconazole have both been successfully used to treat severe blastomycosis in humans, especially cases with CNS involvement [64], but in general voriconazole should be avoided in cats due to their susceptibility to voriconazole toxicity. Clinical signs and radiographic lesions can worsen in the first few days of treatment as a result of the inflammatory response to dying organisms. A short course of glucocorticoids at an anti-inflammatory dose should be considered for patients with CNS involvement and severe respiratory disease, although whether this ultimately improves outcome is not known.

In one study, 4 of 8 cats responded favourably to either itraconazole or fluconazole [52] and 4 of 7 responded favourably to surgical resection of cutaneous lesions and administration of ketoconazole and potassium iodide in another study [1]. However only 1 of 4 cats survived with itraconazole treatment in another study [57].

Coccidioidomycosis

Epidemiology

Coccidioidomycosis is also a rare disease worldwide in cats. The largest case series (48 cats) was reported from Arizona, including 41 cases that were diagnosed over 3 years [65]. Coccidioidomycosis is endemic to the semiarid desert regions of the southwestern USA, southern and central California, southern Arizona, southern New Mexico, western Texas, southern Nevada and Utah, as well as northern Mexico, and parts of Central and South America [66].

Coccidioidomycosis is caused by *Coccidioides immitis* and *Coccidioides posadasii*. *Coccidioides immitis* is mostly found in California and *C. posadasii* is found elsewhere [67]. No significant differences in morphology or disease course have been noted between the two organisms [68]. However, coccidioidomycosis is extremely rare in California, suggesting that cats may be less susceptible to *C. immitis* infections, or that feline exposure occurs to a greater degree in areas where *C. posadasii* resides [67].

Coccidioides spp. is present in soil as a mycelium which germinate to form arthroconidia that are released and dispersed when the soil is disturbed. Infection occurs following inhalation of these arthrospores but also rarely by direct inoculation of organisms into skin. Dissemination occurs when the immune system is incapable of containing replication of the organism to the lungs.

Clinical Features in Cats

The mean age of affected cats at diagnosis was 6.2 years in one study [1] and 9 years in another, [68] with age ranging from 3 to 17 years. Female cats were overrepresented in one study with 12 of 17 cases being female [68]. There is no breed predilection reported and immunosuppression does not seem to play a role in the development of the disease [68]. Although having access to outdoors is likely a risk factor, cats housed exclusively indoors can also develop coccidioidomycosis [68, 69]. The reported duration of clinical signs before diagnosis of coccidioidomycosis in cats ranged from less than 1 week to 1 year, with up to 86% of the animals having clinical signs for less than a month before diagnosis [1, 65, 68].

Cats are mostly diagnosed when the disease has disseminated, and dermatologic signs are the most common presenting complaints in cats with disseminated disease. Dermatological signs were seen in 56% of 48 cats [65]. Cutaneous lesions include plaque-like nodules, nodules with draining tracts, alopecia, scarring and induration with draining tracts, papules, pustules and lingual ulcerations [65, 68, 70]. Regional lymphadenopathy may also be present. More than half of the population of cats with dermatologic signs also had clinical signs of systemic illness such as fever, lethargy, weight loss, anorexia, lameness, or cough [68].

Respiratory signs such as cough or tachypnoea were noted in only 25% of cats with coccidioidomycosis [65]. However, lung involvement is probably more common since thoracic radiographs were not performed on many cats of this study and lung infection was found in nearly all cases at necropsy in another study [66]. Hilar lymphadenopathy, interstitial or mixed interstitial and bronchointerstitial pulmonary patterns and rarely pleural thickening or effusion are found on thoracic radiographs [65]. Other reported clinical signs include ocular signs such as chorioretinitis, anterior uveitis, retinal detachment, panophthalmitis; CNS signs (with intracranial or spinal cord lesions) such as seizures, hyperesthesia, behavioural changes, pelvic limb weakness and ataxia; and musculoskeletal signs such as lameness [65, 66, 69, 71, 72].

Restrictive pericarditis, pericardial effusion and right sided heart failure are reported in dogs with coccidioidomycosis but have not been described in cats [66]. However, pericardial involvement was found at necropsy in 26% of cats with coccidioidomycosis despite a lack of clinical signs referable to heart disease in all cats of the study [66].

Diagnostic Tests

In cats with coccidioidomycosis, laboratory abnormalities include nonregenerative anaemia, leucocytosis, leukopenia, hypoalbuminemia and hyperglobulinemia [65, 72]. In contrast to humans, eosinophilia has not been reported in cats with coccidioidomycosis [73]. The sensitivity and specificity of serology for the diagnosis of coccidioidomycosis in cats is unknown. Most commercial laboratories perform AGID assays for immunoglobulin G (IgG) and immunoglobulin M (IgM) antibodies. In 39 cats with coccidioidomycosis, all cats were seropositive at some point during their illness [65]. At the time of diagnosis, 29 cats were positive for IgM (tube precipitin) antibodies and six cats were negative for IgM but positive for IgG (complement fixating) antibodies. IgG antibody titres ranged from 1:2 to 1:128, with 31 cats having titres $\geq$1:16.

Cytologic confirmation may be made by evaluation of aspirates of affected lymph nodes, skin lesions, lungs, pleural effusion, as well as bronchoalveolar lavage but is relatively insensitive for the diagnosis of coccidioidomycosis when compared with other deep mycoses [67]. On cytology, a granulomatous or pyogranulomatous inflammation is often seen, with sometimes rare multinucleated giant cells, few eosinophils, and/or reactive lymphocytes. Occasionally a suppurative inflammatory response predominates. If seen, the organism appears as a large (10–80 μm) round, deeply basophilic, double-walled spherule that may contain endospores. The endospores are 2–5 μm in diameter, are surrounded by a thin, non-staining halo, and have small, round to oval, densely aggregated, eccentric nuclei. Diff-Quik and Wright stains can be used to facilitate visualization.

Similarly, multiple biopsy samples may need to be evaluated to identify the organism histologically. The use of special stains such as PAS or GMS stains may be required.

Coccidioides spp. structures were seen on cytology or histology of exudates or tissue specimens in only 56% of 48 cats [65]. Cultures of exudates or tissues grew *Coccidioides* in only 23% of these cats and therefore a negative culture does not rule out a diagnosis of coccidioidomycosis. *Coccidioides* spp. can be isolated on routine fungal media, but growth in culture represents a serious health hazard to laboratory personnel and should only be performed if necessary and by suitably equipped laboratories.

The use of PCR assays to aid in the diagnosis of coccidioidomycosis has not been reported in cats.

Treatment and Prognosis

Itraconazole or fluconazole are typically used for the treatment of feline coccidioidomycosis, but ketoconazole has been used historically. Of 53 cats diagnosed with coccidioidomycosis and treated mostly with ketoconazole, 67% survived with treatment [1]. The average treatment duration was 10 months.

Fluconazole may be used for patients with ocular and CNS involvement, because of its superior penetration of the eyes and CNS [71, 72]. Itraconazole is preferred for animals with bone involvement and is recommended for animals that fail to respond to treatment with fluconazole. In people, a trend toward slightly greater efficacy of itraconazole over fluconazole was found in non-meningeal cases but it was not statistically significant [74]. In addition, the use of amphotericin B is recommended for the treatment of human patients with very severe and/or rapidly progressing acute pulmonary or disseminated coccidioidomycosis, followed by fluconazole once patients have stabilized [75].

In human patients who fail standard therapy, treatment with posaconazole or voriconazole has been reported with approximately 70% of improvement and slightly better outcomes with posaconazole compared to voriconazole [72, 76]. The use of echinocandins in combination with voriconazole in refractory patients was also reported with success [77]. However, voriconazole should be avoided in cats due to their susceptibility to severe voriconazole toxicity.

Box 2: Mould Infections
- Mould infections are less prevalent in cats than in dogs.
- Cutaneous signs are rare with aspergillosis in cats. *Aspergillus* spp. cause sino-nasal aspergillosis and sino-orbital aspergillosis in cats with a predisposition for brachycephalic cats. They also rarely cause disseminated disease, which is usually seen in immunodeficient cats.
- Hyalohyphomycosis and phaeohyphomycosis are cutaneous and subcutaneous infections usually acquired from traumatic implantation of fungi from the environment. Most cats are immunocompetent.
- Zygomycosis is acquired by inhalation, ingestion or contamination of wounds. Concurrent immunosuppressive conditions are common.
- Pythiosis is caused by the penetration of acquired motile biflagellate zoospores in aquatic environments through damaged skin or GI mucosa. It is mostly manifested as cutaneous and subcutaneous lesions in cats.
- Complete surgical excision of the affected tissue with wide margins is the treatment of choice for hyalohyphomycosis, phaeohyphomycosis, zygomycosis and pythiosis as treatment with anti-fungal drugs is usually not curative.

Aspergillosis

Epidemiology

Aspergillus species are ubiquitous saprophytic moulds that are found worldwide in soil and decaying vegetation [78]. Species affecting cats are usually included in the *A. fumigatus* complex (*Aspergillus fumigatus*, *Neosartorya* spp., *Aspergillus lentulus* and *Aspergillus udagawae*) [79]. Members of the *A. fumigatus* complex cannot be identified reliably by phenotypic testing alone and require molecular techniques for identification [79]. *Aspergillus flavus*, *Aspergillus nidulans*, *Aspergillus niger*, *Aspergillus terreus*, *Aspergillus udagawae* and *Aspergillus felis* have also been detected in a few cases [78, 80, 81].

Clinical Features in Cats

The most common forms of aspergillosis in cats are sino-nasal aspergillosis (SNA) and sino-orbital aspergillosis (SOA). The development of localized sino-nasal or sino-orbital infection suggest defects of local defence mechanisms. In normal conditions, infections are prevented by physical barriers such as mucociliary clearance, and the local innate immune system (macrophages and neutrophils) [82]. Brachycephalic breeds, especially Persian or Himalayan cats, are predisposed [78]. It has been suggested that this may result from reduced mucociliary clearance [78]. Other possible risk factors include previous viral upper respiratory tract infections, inflammatory rhinitis and use of glucocorticoids or less likely previous antibiotic treatment [80, 82]. No association between aspergillosis and feline retrovirus infections has been reported [79]. Affected cats range in age from 1.5 to 13 years (median, 5 years) with no clear sex predisposition [79]. The duration of clinical signs before diagnosis ranged from less than 5 days to more than 6 weeks in one study [1].

Clinical signs of SNA include sneezing, uni- or bilateral serous to mucopurulent nasal discharge, and sometimes epistaxis and less commonly stertorous breathing, granuloma formation, soft tissue masses protruding from the nares, and bone lysis. SOA is a more invasive form of SNA, with involvement of the retrobulbar space. Clinical signs include unilateral exophthalmos, third eyelid prolapse, conjunctival hyperaemia and keratitis. With severe retrobulbar involvement, a mass may be observed in the caudal aspect of the oral cavity (Fig. 5). CNS involvement, regional lymphadenopathy and fever have also been described.

Systemic aspergillosis in cats is rare and is usually associated with immune-deficiency. *Aspergillus niger* pneumonia was reported in two cats with diabetes mellitus [83], and disseminated aspergillosis in 38 cats with more than half of these cats

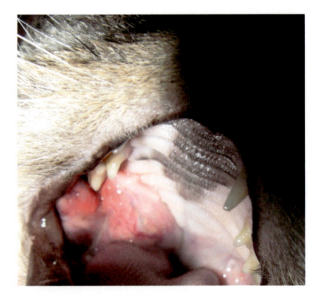

Fig. 5 Mass in the caudal aspect of the oral cavity of a cat with sino-orbital aspergillosis

having a concurrent immune-suppressive condition (mainly panleukopenia, FeLV and feline infectious peritonitis) [1, 84].

Cutaneous disease is also a very uncommon manifestation of aspergillosis in cats. Cutaneous involvement of the naso-ocular region of a cat with naso-sinusal aspergillosis was reported [85] and *Aspergillus vitricola* was cultured from an auricular lesion in another cat [86].

Diagnostic Tests

The diagnosis of SNA and SOA requires a combination of tests such as imaging studies, rhinoscopy, cytology and/or histology, and fungal cultures. Imaging modalities such as computerized tomography (CT) scan or magnetic resonance imaging (MRI) of the head can be used to evaluate for destruction of the nasal turbinates, nasal septum, cribriform plate, and involvement of the sinus and retrobulbar space. On rhinoscopy, destruction of the nasal turbinates and white-grey plaques may be seen [87].

Cytologic examination of blind or rhinoscopy-directed mucosal swabs, brush specimens of the nasal cavity, nasal biopsies from cats with SNA, or cytologic examination of ultrasound- or CT-guided aspirates of retrobulbar masses from cats with SOA often reveal mixed, predominantly pyogranulomatous inflammation. Sometimes, *Aspergillus* hyphae are seen but false-negative results are common.

Aspergillus fumigatus can usually grow in few days to few weeks on routine laboratory media and does not represent a significant hazard to laboratory personnel. In the absence of supportive rhinoscopic, cytologic or histopathologic findings, positive cultures from the nasal cavity require cautious interpretation because *Aspergillus* spp. are ubiquitous and therefore false positive results are not uncommon. Whenever possible, fungal culture should be submitted from samples collected by rhinoscopic guidance in order to increase sensitivity [88]. Growth of *Aspergillus* from aspirates or biopsy specimens from a normally sterile site such as a retrobulbar mass strongly suggests a diagnosis of SOA. In one study, fungal cultures were positive in 22/23 cats with SNA or SOA [79] but in another study the sensitivity of culture was lower. [89] The use of serologic tests (antibody and antigen tests) for diagnosis of aspergillosis in cats has been unreliable [78, 79, 81, 87, 89].

Treatment and Prognosis

Treatment of SNA in cats has been similar to treatment used in dogs. Intranasal infusion of clotrimazole for 1 hour was described in three cats with good outcomes [82, 87]. Treatment of SOA and disseminated aspergillosis requires systemic anti-fungal treatment (monotherapy or a combination of two anti-fungal treatments) but prognosis is guarded to poor. Anti-fungal drugs reported in the treatment of SOA and disseminated aspergillosis include itraconazole, amphotericin B, posaconazole, voriconazole, terbinafine, caspofungin and micafungin [79, 90–92]. Voriconazole is not recommended because of the potential for severe toxicity in cats. Fluconazole and flucytosine are not recommended because *Aspergillus* species are intrinsically resistant to these anti-fungal drugs [93]. In addition, high minimum inhibitory concentrations (MIC) for ketoconazole are common among *Aspergillus* species.

In a 2015 Australian study evaluating anti-fungal resistance in canine and feline isolates of *Aspergillus fumigatus*, the vast majority of isolates had low MICs for itraconazole, voriconazole, posaconazole, clotrimazole and enilconazole [93]. Interestingly, seven isolates had high MICs for amphotericin B.

Aspergillus felis exhibits high MICs to many anti-fungal drugs [94]. High MICs of *A. felis* isolates to at least one of the triazoles, and cross resistance among several triazoles were observed. In addition, a high MIC for caspofungin was described for one isolate.

Other Moulds

Hyalohyphomycosis

Hyalohyphomycosis is caused by non-dematiaceous (hyaline, nonpigmented) moulds. A retrospective study from the UK evaluating 77 cats with nodular granulomatous skin lesions caused by fungi found that the most frequent pathogens were hyalohyphomycetes [95]. Reported species associated with disease in cats include *Fusarium*, *Acremonium*, *Paecilomyces* spp. and *Metarhizium* spp., among others [95–101].

These are filamentous fungi found in soil and on plants and have a worldwide distribution.

Hyalohyphomycosis has been diagnosed in cats with cutaneous nodules, rhinosinusitis, pneumonia, pododermatitis and keratitis.

Diagnosis is made by cytology, histology and fungal culture. Cytological and histological examination usually reveals pyogranulomatous inflammation in association with nonpigmented, frequently septate, branching hyphae that are often pleomorphic. Culture and proper identification of the pathogen is recommended to guide in the choice of anti-fungal treatment because some species are predictably less susceptible to conventional anti-fungal drugs. However, because these fungi are common laboratory contaminants and can sometimes be isolated from the skin or hair of healthy animals, positive cultures from non-sterile sites should be considered in light of the clinical picture.

Complete surgical excision of the affected tissue with wide margins is the treatment of choice whenever possible, followed by anti-fungal therapy for 3–6 months.

Drugs used most often to treat hyalohyphomycosis in small animals include itraconazole and amphotericin B, but different fungal species vary in their susceptibility to anti-fungal drugs. Posaconazole and the echinocandins, such as caspofungin, may be more active against these fungi than itraconazole. *Fusarium* spp. are intrinsically resistant to glucan synthesis inhibitors such as caspofungin; however, in combination with amphotericin B, they can have a synergistic action [98].

Phaeohyphomycosis

Phaeohyphomycetes are dematiaceous filamentous fungi that contain a melanin-like pigment in the walls of the hyphae and occasionally cause opportunistic infections in cats. The pigment plays important role in virulence and pathogenicity of these pathogens as it aids fungal evasion of the host immune response by preventing hydrolytic enzymatic attack and scavenging of free radicals liberated by phagocytic cells during then oxidative burst [102]. Species that have caused disease in cats include *Exophiala* spp., *Alternaria* spp., *Cladosporium* spp., *Phialophora* spp., *Cladophialophora* spp., *Ulocladium* spp., *Microsphaeropsis* spp., *Fonsecae* spp., *Moniliella* spp. and *Aureobasidium* spp., among others [86, 98, 102–115]. *Cladosporium* spp. may be more likely to disseminate in immunocompetent cats. In addition, among the genus *Cladophialophora*, *Cladophialophora bantiana* shows a marked neurotropism in comparison with other fungi. [109, 115, 116]

Phaeohyphomycetes are found in soil, wood and decomposing plant debris worldwide. Infections usually result from cutaneous inoculation resulting in cutaneous and subcutaneous infections (Fig. 6). Most lesions in cats occur on the head or extremities and usually a single nodule is present. Systemic signs of illness are usually absent. Rarely, ingestion or inhalation of spores might also occur and cause deep infection [109].

Factors predisposing cats to phaeohyphomycoses may include treatment with immunosuppressive agents; concurrent disease; or age-related, non-specific loss of immunity. However, no obvious immunosuppression is found in most cases [103] although the majority of cats in reports were not tested for FIV and FeLV.

Diagnosis is made by cytology, histology and fungal culture. Cytology of exudates usually reveals pyogranulomatous inflammation that may contain pigmented fungal hyphae, pseudohyphae, and/or yeast-like cells. Fungal culture is recommended for proper diagnosis. An indirect ELISA has been developed for the

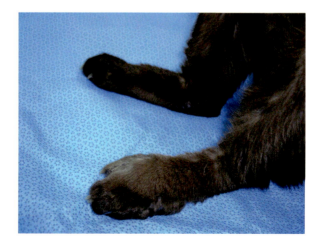

Fig. 6 Subcutaneous phaeohyphomycosis associated with swelling of the distal limb of a cat

detection of anti-*Alternaria* IgG antibodies in serum in domestic cats. However, cats with disease caused by *Alternaria* did not have significantly higher concentrations of antibody than healthy cats or cats with other diseases [117].

Complete surgical excision of the affected tissue with wide margins is the treatment of choice whenever possible, followed by anti-fungal therapy for 3 to 6 months. If complete surgical removal is not possible, the prognosis is guarded. Indeed, phaeohyphomycoses often have recrudescent clinical courses and are refractory to many anti-fungal drugs. In cases of disseminated or cerebral infection, treatment is rarely successful, and prognosis is poor. Ketoconazole, itraconazole, amphotericin B, flucytosine and terbinafine have been used with variable results to treat phaeohyphomycosis in cats [111, 113]. Combination therapy with terbinafine and an azole anti-fungal drug such as itraconazole or posaconazole has been suggested [114]. If improvement is seen, long-term treatment (6 to 12 months) is recommended to prevent recurrence of the lesions.

Zygomycosis

Zygomycetes are opportunistic organisms present in the soil, water, decaying matter and faeces. They include organisms that belong to the genera *Basidiobolus* and *Conidiobolus* in the order Entomophthorales, and the genera *Rhizopus*, *Absidia*, *Mucor*, *Saksenaea*, and others in the order Mucorales [118]. Infection is believed to be acquired by inhalation, ingestion or contamination of wounds. Rare reports exist of *Mucor* spp. infections in cats, including a case of cerebral mycosis, subcutaneous infection and duodenal perforation caused by *Rhizomucor* spp. [119–121] In addition, 12 cases of suspected mucormycosis diagnosed using histology were reported in a necropsy study [84]. Lesions in most of these cats involved the GI tract or lungs and 6 of the 12 cats had possible immune-suppressive conditions.

Conidiobolus infection was suspected in a 3-year-old cat with an ulcerative lesion of the hard palate [118].

A definitive diagnosis of zygomycosis is made based on cytological or histopathological examination in combination with fungal culture. Cytological and histological findings include pyogranulomatous, suppurative or eosinophilic inflammation. Broad (>8 µm), poorly septate hyphae with thick prominent eosinophilic sleeves are sometimes observed. Microscopic examination of macerated tissue that has been digested in 10% potassium hydroxide may be more likely to reveal the hyphal elements. Staining of histopathological specimens with GMS and PAS stains can also assist in visualization of hyphal elements.

Wide surgical excision (whenever possible) combined with long-term medical treatment is recommended for zygomycosis. Zygomycetes are variably susceptible to anti-fungal drugs. Posaconazole and amphotericin B are considered the most effective anti-fungal drugs for *Mucor* infections in humans [120]. Some cases of human zygomycosis also had good outcome with itraconazole treatment [122, 123].

Pythiosis

Pythiosis is caused by the aquatic oomycete *Pythium insidiosum*. Oomycetes are soil and aquatic organisms that are phylogenetically distant from the fungi, and more closely related to algae [118]. Chitin, an essential component of the fungal cell wall, is generally lacking in the oomycete cell wall, which instead contains predominately cellulose and β–glucan [118]. Ergosterol is also not a principal sterol in the oomycete cell membrane contrary to fungal organisms.

The infective form of *P. insidiosum* is a motile biflagellate zoospore, which is released into aquatic environments and causes infection by penetrating damaged skin or GI mucosa. Pythiosis is most commonly encountered in tropical and subtropical climates; however, infections in animals from temperate areas have also been reported [124]. It is endemic in the USA (primarily in the Gulf coast states), and cases also occur in Southeast Asia, eastern coastal Australia, New Zealand, and South America [118].

Pythiosis is extremely rare in cats and usually manifests as subcutaneous lesions (including in the inguinal, tail head, or periorbital regions), draining nodular lesions or ulcerated plaque-like lesions localized on the extremities [118]. One cat with nasal and retrobulbar mass [125], one cat with a sublingual mass [126] and two cats with gastro-intestinal pythiosis [124] were also reported. Specific breed and sex predilections have not been observed but young cats may be predisposed. In 10 cats with cutaneous lesions caused by *P. insidiosum*, five were younger than 10 months old, with an age range of 4 months to 9 years [118].

Cytological and histological examinations show eosinophilic and granulomatous inflammation with prominent fibrosis and necrosis [124]. To complicate the diagnosis, *P. insidiosum* typically does not stain with H&E stain and may be present in low numbers. The hyphae appear within necrotic areas and granulomas as clear round or oval to elongate structures delineated by a narrow rim of eosinophilic material. It also stains poorly with PAS but can be observed with GMS stain. The hyphae are infrequently septate, branching and measure from 2.5 to 8.9 mm in diameter with thick walls in comparison to the septa and with almost parallel sides [124]. Differentiation between pythiosis, lagenidiosis, and zygomycosis based on routine histologic examination is not usually possible because differences in their histologic characteristics are subtle, although *Mucor* spp. stain equally well using H&E, PAS, and GMS stains.

Culture of tissues or the use of immunohistochemistry, PCR and/or serology can aid in diagnosis [124]. Culture of exudates is usually unsuccessful and culture of tissues requires specific specimen handling (unrefrigerated tissues that are kept moist) and culture techniques. The identity of organisms isolated in culture can be confirmed using PCR sequencing [127].

In addition, immunoblot serology and ELISA techniques have been used successfully to support the diagnosis of pythiosis in a few cats [124, 125, 128].

The treatment of choice for pythiosis is aggressive surgical resection of infected tissues with 3–4-cm margins whenever possible. When used alone, medical therapy for pythiosis is typically unrewarding. This is likely because ergosterol, which is the target for most anti-fungal drugs, is generally lacking in the oomycete cell membrane. In dogs, a combination of itraconazole and terbinafine may be effective for resolution of incompletely resected or nonresectable lesions. Ketoconazole has also been used [125]. The short-term use of prednisone is also recommended in dogs with gastrointestinal pythiosis to improve clinical signs (vomiting, decreased appetite) [127].

References

1. Davies C, Troy GC. Deep mycotic infections in cats. J Am Anim Hosp Assoc. 1996;32:380–91.
2. Sykes JE, Taboada J. Histoplasmosis. In: Sykes JE, editor. Canine and feline infectious diseases. St Louis: Elsevier Saunders; 2014. p. 587–98.
3. Kano R, Hosaka S, Hasegawa A. First isolation of *Cryptococcus magnus* from a cat. Mycopathologia. 2004;157:263–4.
4. Poth T, Seibold M, Werckenthin C, Hermanns W. First report of a *Cryptococcus magnus* infection in a cat. Med Mycol. 2010;48:1000–4.
5. Kano R, Kitagawat M, Oota S, Oosumi T, Murakami Y, Tokuriki M, et al. First case of feline systemic *Cryptococcus albidus* infection. Med Mycol. 2008;46:75–7.
6. O'Brien CR, Krockenberger MB, Wigney DI, Martin P, Malik R. Retrospective study of feline and canine cryptococcosis in Australia from 1981 to 2001: 195 cases. Med Mycol. 2004;42:449–60.
7. Lester SJ, Malik R, Bartlett KH, Duncan CG. Cryptococcosis: update and emergence of Cryptococcus gattii. Vet Clin Pathol. 2011;40:4–17.
8. Pennisi MG, Hartmann K, Lloret A, Ferrer L, Addie D, Belak S, et al. Cryptococcosis in cats: ABCD guidelines on prevention and management. J Feline Med Surg. 2013;15:611–8.
9. McGill S, Malik R, Saul N, Beetson S, Secombe C, Robertson I, et al. Cryptococcosis in domestic animals in Western Australia: a retrospective study from 1995–2006. Med Mycol. 2009;47:625–39.
10. Malik R, Wigney DI, Muir DB, Gregory DJ, Love DN. Cryptococcosis in cats: clinical and mycological assessment of 29 cases and evaluation of treatment using orally administered fluconazole. J Med Vet Mycol. 1992;30:133–44.
11. Trivedi SR, Sykes JE, Cannon MS, Wisner ER, Meyer W, Sturges BK, et al. Clinical features and epidemiology of cryptococcosis in cats and dogs in California: 93 cases (1988–2010). J Am Vet Med Assoc. 2011;239:357–69.
12. Jacobs GJ, Medleau L, Calvert C, Brown J. Cryptococcal infection in cats: factors influencing treatment outcome, and results of sequential serum antigen titers in 35 cats. J Vet Intern Med. 1997;11:1–4.
13. Castrodale LJ, Gerlach RF, Preziosi DE, Frederickson P, Lockhart SR. Prolonged incubation period for *Cryptococcus gattii* infection in cat, Alaska, USA. Emerg Infect Dis. 2013;19:1034–5.
14. Myers A, Meason-Smith C, Mansell J, Krockenberger M, Peters-Kennedy J, Ross Payne H, et al. Atypical cutaneous cryptococcosis in four cats in the USA. Vet Dermatol. 2017;28:405–e97.
15. Odom T, Anderson JG. Proliferative gingival lesion in a cat with disseminated cryptococcosis. J Vet Dent. 2000;17:177–81.
16. Siak MK, Paul A, Drees R, Arthur I, Burrows AK, Tebb AJ, et al. Otogenic meningoencephalomyelitis due to *Cryptococcus gattii* (VGII) infection in a cat from Western Australia. JFMS Open Rep. 2015;1:2055116915585022.
17. Sykes JE, Malik R. Cryptococcosis. In: Sykes JE, editor. Canine and feline infectious diseases. St Louis: Elsevier Saunders; 2014. p. 599–612.

18. Trivedi SR, Malik R, Meyer W, Sykes JE. Feline cryptococcosis: impact of current research on clinical management. J Feline Med Surg. 2011;13:163–72.
19. Guess T, Lai H, Smith SE, Sircy L, Cunningham K, Nelson DE, et al. Size matters: measurement of capsule diameter in *Cryptococcus neoformans*. J Vis Exp. 2018;132:1–10.
20. Ranjan R, Jain D, Singh L, Iyer VK, Sharma MC, Mathur SR. Differentiation of histoplasma and cryptococcus in cytology smears: a diagnostic dilemma in severely necrotic cases. Cytopathology. 2015;26:244–9.
21. Sykes JE, Hodge G, Singapuri A, Yang ML, Gelli A, Thompson GR 3rd. In vivo development of fluconazole resistance in serial *Cryptococcus gattii isolates* from a cat. Med Mycol. 2017;55:396–401.
22. Kano R, Okubo M, Yanai T, Hasegawa A, Kamata H. First isolation of azole-resistant *Cryptococcus neoformans* from feline cryptococcosis. Mycopathologia. 2015;180:427–33.
23. Mondon P, Petter R, Amalfitano G, Luzzati R, Concia E, Polacheck I, et al. Heteroresistance to fluconazole and voriconazole in *Cryptococcus neoformans*. Antimicrob Agents Chemother. 1999;43:1856–61.
24. O'Brien CR, Krockenberger MB, Martin P, Wigney DI, Malik R. Long-term outcome of therapy for 59 cats and 11 dogs with cryptococcosis. Aust Vet J. 2006;84:384–92.
25. Sykes JE, Sturges BK, Cannon MS, Gericota B, Higgins RJ, Trivedi SR, et al. Clinical signs, imaging features, neuropathology, and outcome in cats and dogs with central nervous system cryptococcosis from California. J Vet Intern Med. 2010;24:1427–38.
26. Barrs VR, Martin P, Nicoll RG, Beatty JA, Malik R. Pulmonary cryptococcosis and *Capillaria aerophila* infection in an FIV-positive cat. Aust Vet J. 2000;78:154–8.
27. Balajee SA, Hurst SF, Chang LS, Miles M, Beeler E, Hale C, et al. Multilocus sequence typing of *Histoplasma capsulatum* in formalin-fixed paraffin-embedded tissues from cats living in non-endemic regions reveals a new phylogenetic clade. Med Mycol. 2013;51:345–51.
28. Fischer NM, Favrot C, Monod M, Grest P, Rech K, Wilhelm S. A case in Europe of feline histoplasmosis apparently limited to the skin. Vet Dermatol. 2013;24:635–8.
29. Percy DH. Feline histoplasmosis with ocular involvement. Vet Pathol. 1981;18:163–9.
30. Larsuprom L, Duangkaew L, Kasorndorkbua C, Chen C, Chindamporn A, Worasilchai N. Feline cutaneous histoplasmosis: the first case report from Thailand. Med Mycol Case Rep. 2017;18:28–30.
31. Mavropoulou A, Grandi G, Calvi L, Passeri B, Volta A, Kramer LH, et al. Disseminated histoplasmosis in a cat in Europe. J Small Anim Pract. 2010;51:176–80.
32. Kasuga T, White TJ, Koenig G, McEwen J, Restrepo A, Castaneda E, et al. Phylogeography of the fungal pathogen *Histoplasma capsulatum*. Mol Ecol. 2003;12:3383–401.
33. Bromel C, Sykes JE. Histoplasmosis in dogs and cats. Clin Tech Small Anim Pract. 2005;20:227–32.
34. Aulakh HK, Aulakh KS, Troy GC. Feline histoplasmosis: a retrospective study of 22 cases (1986–2009). J Am Anim Hosp Assoc. 2012;48:182–7.
35. Reinhart JM, KuKanich KS, Jackson T, Harkin KR. Feline histoplasmosis: fluconazole therapy and identification of potential sources of Histoplasma species exposure. J Feline Med Surg. 2012;14:841–8.
36. Atiee G, Kvitko-White H, Spaulding K, Johnson M. Ultrasonographic appearance of histoplasmosis identified in the spleen in 15 cats. Vet Radiol Ultrasoun. 2014;55:310–4.
37. Gingerich K, Guptill L. Canine and feline histoplasmosis: a review of a widespread fungus. Vet Med. 2008;103:248–64.
38. Clinkenbeard KD, Cowell RL, Tyler RD. Disseminated histoplasmosis in cats: 12 cases (1981–1986). J Am Vet Med Assoc. 1987;190:1445–8.
39. Wolf AM. *Histoplasma capsulatum* osteomyelitis in the cat. J Vet Intern Med. 1987;1:158–62.
40. Carneiro RA, Lavalle GE, Araujo RB. Cutaneous histoplasmosis in cat: a case report. Arq Bras Med Vet Zoo. 2005;57:158–61.
41. Tamulevicus AM, Harkin K, Janardhan K, Debey BM. Disseminated histoplasmosis accompanied by cutaneous fragility in a cat. J Am Anim Hosp Assoc. 2011;47:E36–41.

42. Vinayak A, Kerwin SC, Pool RR. Treatment of thoracolumbar spinal cord compression associated with *Histoplasma capsulatum* infection in a cat. J Am Vet Med Assoc. 2007;230:1018–23.
43. Lamm CG, Rizzi TE, Campbell GA, Brunker JD. Pathology in practice. *Histoplasma capsulatum* Infections. J Am Vet Med Assoc. 2009;235:155–7.
44. Taylor AR, Barr JW, Hokamp JA, Johnson MC, Young BD. Cytologic diagnosis of disseminated histoplasmosis in the wall of the urinary bladder of a cat. J Am Anim Hosp Assoc. 2012;48:203–8.
45. Hodges RD, Legendre AM, Adams LG, Willard MD, Pitts RP, Monce K, et al. Itraconazole for the treatment of histoplasmosis in cats. J Vet Intern Med. 1994;8:409–13.
46. Hanzlicek AS, Meinkoth JH, Renschler JS, Goad C, Wheat LJ. Antigen concentrations as an indicator of clinical remission and disease relapse in cats with histoplasmosis. J Vet Intern Med. 2016;30:1065–73.
47. Cook AK, Cunningham LY, Cowell AK, Wheat LJ. Clinical evaluation of urine *Histoplasma capsulatum* antigen measurement in cats with suspected disseminated histoplasmosis. J Feline Med Surg. 2012;14:512–5.
48. Klang A, Loncaric I, Spergser J, Eigelsreiter S, Weissenbock H. Disseminated histoplasmosis in a domestic cat imported from the USA to Austria. Med Mycol Case Rep. 2013;2:108–12.
49. Spec A, Connoly P, Montejano R, Wheat LJ. In vitro activity of isavuconazole against fluconazole-resistant isolates of *Histoplasma capsulatum*. Med Mycol. 2018;56:834–7.
50. Renschler JS, Norsworthy GD, Rakian RA, Rakian AI, Wheat LJ, Hanzlicek AS. Reduced susceptibility to fluconazole in a cat with histoplasmosis. JFMS Open Rep. 2017;3:2055116917743364.
51. Bromel C, Sykes JE. Epidemiology, diagnosis, and treatment of blastomycosis in dogs and cats. Clin Tech Small Anim Pract. 2005;20:233–9.
52. Gilor C, Graves TK, Barger AM, O'Dell-Anderson K. Clinical aspects of natural infection with *Blastomyces dermatitidis* in cats: 8 cases (1991–2005). J Am Vet Med Assoc. 2006;229:96–9.
53. Davies JL, Epp T, Burgess HJ. Prevalence and geographic distribution of canine and feline blastomycosis in the Canadian prairies. Can Vet J. 2013;54:753–60.
54. Easton KL. Cutaneous North American blastomycosis in a Siamese cat. Can Vet J. 1961;2:350–1.
55. Duangkaew L, Larsuprom L, Kasondorkbua C, Chen C, Chindamporn A. Cutaneous blastomycosis and dermatophytic pseudomycetoma in a Persian cat from Bangkok, Thailand. Med Mycol Case Rep. 2017;15:12–5.
56. Miller PE, Miller LM, Schoster JV. Feline blastomycosis – a report of 3 cases and literature-review (1961 to 1988). J Am Anim Hosp Assoc. 1990;26:417–24.
57. Blondin N, Baumgardner DJ, Moore GE, Glickman LT. Blastomycosis in indoor cats: suburban Chicago, Illinois, USA. Mycopathologia. 2007;163:59–66.
58. Houseright RA, Webb JL, Claus KN. Pathology in practice. Blastomycosis in an indoor-only cat. J Am Vet Med Assoc. 2015;247:357–9.
59. Stern JA, Chew DJ, Schissler JR, Green EM. Cutaneous and systemic blastomycosis, hypercalcemia, and excess synthesis of calcitriol in a domestic shorthair cat. J Am Anim Hosp Assoc. 2011;47:e116–20.
60. Meece JK, Anderson JL, Klein BS, Sullivan TD, Foley SL, Baumgardner DJ, et al. Genetic diversity in *Blastomyces dermatitidis*: implications for PCR detection in clinical and environmental samples. Med Mycol. 2010;48:285–90.
61. Sidamonidze K, Peck MK, Perez M, Baumgardner D, Smith G, Chaturvedi V, et al. Real-time PCR assay for identification of *Blastomyces dermatitidis* in culture and in tissue. J Clin Microbiol. 2012;50:1783–6.
62. Smith JR, Legendre AM, Thomas WB, LeBlanc CJ, Lamkin C, Avenell JS, et al. Cerebral *Blastomyces dermatitidis* infection in a cat. J Am Vet Med Assoc. 2007;231:1210–4.
63. Breider MA, Walker TL, Legendre AM, VanEe RT. Blastomycosis in cats: five cases (1979–1986). J Am Vet Med Assoc. 1988;193:570–2.
64. McBride JA, Gauthier GM, Klein BS. Clinical manifestations and treatment of blastomycosis. Clin Chest Med. 2017;38:435–49.

65. Greene RT, Troy GC. Coccidioidomycosis in 48 cats – a retrospective study (1984–1993). J Vet Intern Med. 1995;9:86–91.
66. Graupmann-Kuzma A, Valentine BA, Shubitz LF, Dial SM, Watrous B, Tornquist SJ. Coccidioidomycosis in dogs and cats: a review. J Am Anim Hosp Assoc. 2008;44:226–35.
67. Sykes JE. Coccidioidomycosis. In: Sykes JE, editor. Canine and feline infectious diseases. St Louis: Elsevier Saunders; 2014. p. 613–23.
68. Simoes DM, Dial SM, Coyner KS, Schick AE, Lewis TP. Retrospective analysis of cutaneous lesions in 23 canine and 17 feline cases of coccidiodomycosis seen in Arizona, USA (2009–2015). Vet Dermatol. 2016;27:346.
69. Foureman P, Longshore R, Plummer SB. Spinal cord granuloma due to *Coccidioides immitis* in a cat. J Vet Intern Med. 2005;19:373–6.
70. Amorim I, Colimao MJ, Cortez PP, Dias Pereira P. Coccidioidomycosis in a cat imported from the USA to Portugal. Vet Rec. 2011;169: 232a.
71. Bentley RT, Heng HG, Thompson C, Lee CS, Kroll RA, Roy ME, et al. Magnetic resonance imaging features and outcome for solitary central nervous system *Coccidioides* granulomas in 11 dogs and cats. Vet Radiol Ultrasoun. 2015;56:520–30.
72. Tofflemire K, Betbeze C. Three cases of feline ocular coccidioidomycosis: presentation, clinical features, diagnosis, and treatment. Vet Ophthalmol. 2010;13:166–72.
73. Alzoubaidi MSS, Knox KS, Wolk DM, Nesbit LA, Jahan K, Luraschi-Monjagatta C. Eosinophilia in coccidioidomycosis. Am J Resp Crit Care. 2013;187
74. Galgiani JN, Catanzaro A, Cloud GA, Johnson RH, Williams PL, Mirels LF, et al. Comparison of oral fluconazole and itraconazole for progressive, nonmeningeal coccidioidomycosis. A randomized, double-blind trial. Mycoses Study Group. Ann Intern Med. 2000;133:676–86.
75. Galgiani JN, Ampel NM, Blair JE, Catanzaro A, Geertsma F, Hoover SE, et al. 2016 Infectious Diseases Society of America (IDSA) Clinical Practice Guideline for the treatment of coccidioidomycosis. Clin Infect Dis. 2016;63:E112–E46.
76. Kim MM, Vikram HR, Kusne S, Seville MT, Blair JE. Treatment of refractory coccidioidomycosis with voriconazole or posaconazole. Clin Infect Dis. 2011;53:1060–6.
77. Levy ER, McCarty JM, Shane AL, Weintrub PS. Treatment of pediatric refractory coccidioidomycosis with combination voriconazole and caspofungin: a retrospective case series. Clin Infect Dis. 2013;56:1573–8.
78. Hartmann K, Lloret A, Pennisi MG, Ferrer L, Addie D, Belak S, et al. Aspergillosis in cats: ABCD guidelines on prevention and management. J Feline Med Surg. 2013;15:605–10.
79. Barrs VR, Halliday C, Martin P, Wilson B, Krockenberger M, Gunew M, et al. Sinonasal and sino-orbital aspergillosis in 23 cats: aetiology, clinicopathological features and treatment outcomes. Vet J. 2012;191:58–64.
80. Barachetti L, Mortellaro CM, Di Giancamillo M, Giudice C, Martino P, Travetti O, et al. Bilateral orbital and nasal aspergillosis in a cat. Vet Ophthalmol. 2009;12:176–82.
81. Whitney BL, Broussard J, Stefanacci JD. Four cats with fungal rhinitis. J Feline Med Surg. 2005;7:53–8.
82. Tomsa K, Glaus TA, Zimmer C, Greene CE. Fungal rhinitis and sinusitis in three cats. J Am Vet Med Assoc. 2003;222:1380–4.
83. Leite RV, Fredo G, Lupion CG, Spanamberg A, Carvalho G, Ferreiro L, et al. Chronic invasive pulmonary aspergillosis in two cats with diabetes mellitus. J Comp Pathol. 2016;155:141–4.
84. Ossent P. Systemic aspergillosis and mucormycosis in 23 cats. Vet Rec. 1987;120:330–3.
85. Malik R, Vogelnest L, O'Brien CR, White J, Hawke C, Wigney DI, et al. Infections and some other conditions affecting the skin and subcutis of the naso-ocular region of cats--clinical experience 1987–2003. J Feline Med Surg. 2004;6:383–90.
86. Bernhardt A, von Bomhard W, Antweiler E, Tintelnot K. Molecular identification of fungal pathogens in nodular skin lesions of cats. Med Mycol. 2015;53:132–44.
87. Furrow E, Groman RP. Intranasal infusion of clotrimazole for the treatment of nasal aspergillosis in two cats. J Am Vet Med Assoc. 2009;235:1188–93.
88. Sykes JE. Aspergillosis. In: Sykes JE, editor. Canine and feline infectious diseases. St Louis: Elsevier Saunders; 2014. p. 633–59.

89. Goodall SA, Lane JG, Warnock DW. The diagnosis and treatment of a case of nasal aspergillosis in a cat. J Small Anim Pract. 1984;25:627–33.
90. McLellan GJ, Aquino SM, Mason DR, Kinyon JM, Myers RK. Use of posaconazole in the management of invasive orbital aspergillosis in a cat. J Am Anim Hosp Assoc. 2006;42:302–7.
91. Smith LN, Hoffman SB. A case series of unilateral orbital aspergillosis in three cats and treatment with voriconazole. Vet Ophthalmol. 2010;13:190–203.
92. Kano R, Itamoto K, Okuda M, Inokuma H, Hasegawa A, Balajee SA. Isolation of *Aspergillus udagawae* from a fatal case of feline orbital aspergillosis. Mycoses. 2008;51:360–1.
93. Talbot JJ, Kidd SE, Martin P, Beatty JA, Barrs VR. Azole resistance in canine and feline isolates of *Aspergillus fumigatus*. Comp Immunol Microbiol Infect Dis. 2015;42:37–41.
94. Barrs VR, van Doorn TM, Houbraken J, Kidd SE, Martin P, Pinheiro MD, et al. *Aspergillus felis* sp nov., an emerging agent of invasive aspergillosis in humans, cats, and dogs. PLoS One. 2013;8:e64871.
95. Miller RI. Nodular granulomatous fungal skin diseases of cats in the United Kingdom: a retrospective review. Vet Dermatol. 2010;21:130–5.
96. Leperlier D, Vallefuoco R, Laloy E, Debeaupuits J, Thibaud PD, Crespeau FL, et al. Fungal rhinosinusitis caused by *Scedosporium apiospermum* in a cat. J Feline Med Surg. 2010;12:967–71.
97. Pawloski DR, Brunker JD, Singh K, Sutton DA. Pulmonary *Paecilomyces lilacinus* infection in a cat. J Am Anim Hosp Assoc. 2010;46:197–202.
98. Kluger EK, Della Torre PK, Martin P, Krockenberger MB, Malik R. Concurrent *Fusarium chlamydosporum* and *Microsphaeropsis arundini*s infections in a cat. J Feline Med Surg. 2004;6:271–7.
99. Sugahara G, Kiuchi A, Usui R, Usui R, Mineshige T, Kamiie J, et al. Granulomatous pododermatitis in the digits caused by *Fusarium proliferatum* in a cat. J Vet Med Sci. 2014;76: 435–8.
100. Binder DR, Sugrue JE, Herring IP. *Acremonium* keratomycosis in a cat. Vet Ophthalmol. 2011;14(Suppl 1):111–6.
101. Muir D, Martin P, Kendall K, Malik R. Invasive hyphomycotic rhinitis in a cat due to *Metarhizium anisopliae*. Med Mycol. 1998;36:51–4.
102. Overy DP, Martin C, Muckle A, Lund L, Wood J, Hanna P. Cutaneous Phaeohyphomycosis caused by *Exophiala attenuata* in a domestic cat. Mycopathologia. 2015;180:281–7.
103. Tennant K, Patterson-Kane J, Boag AK, Rycroft AN. Nasal mycosis in two cats caused by *Alternaria* species. Vet Rec. 2004;155:368–70.
104. Bostock DE, Coloe PJ, Castellani A. Phaeohyphomycosis caused by *Exophiala jeanselmei* in a domestic cat. J Comp Pathol. 1982;92:479.
105. Dion WM, Pukay BP, Bundza A. Feline Cutaneous phaeohyphomycosis caused by *Phialophora verrucosa*. Can Vet J. 1982;23:48–9.
106. Sisk DB, Chandler FW. Phaeohyphomycosis and cryptococcosis in a cat. Vet Pathol. 1982;19:554–6.
107. Kettlewell P, McGinnis MR, Wilkinson GT. Phaeohyphomycosis caused by *Exophiala spinifera* in two cats. J Med Vet Mycol. 1989;27:257–64.
108. Nuttal W, Woodgyer A, Butler S. Phaeohyphomycosis caused by *Exophiala jeanselmei* in a domestic cat. N Z Vet J. 1990;38:123.
109. Abramo F, Bastelli F, Nardoni S, Mancianti F. Feline cutaneous phaeohyphomycosis due to *Cladophyalophora bantiana*. J Feline Med Surg. 2002;4:157–63.
110. Beccati M, Vercelli A, Peano A, Gallo MG. Phaeohyphomycosis by *Phialophora verrucosa*: first European case in a cat. Vet Rec. 2005;157:93–4.
111. Knights CB, Lee K, Rycroft AN, Patterson-Kane JC, Baines SJ. Phaeohyphomycosis caused by *Ulocladium species* in a cat. Vet Rec. 2008;162:415–6.
112. McKenzie RA, Connole MD, McGinnis MR, Lepelaar R. Subcutaneous phaeohyphomycosis caused by *Moniliella suaveolens* in two cats. Vet Pathol. 1984;21:582–6.
113. Fondati A, Gallo MG, Romano E, Fondevila D. A case of feline phaeohyphomycosis due to *Fonsecaea pedrosoi*. Vet Dermatol. 2001;12:297–301.

114. Evans N, Gunew M, Marshall R, Martin P, Barrs V. Focal pulmonary granuloma caused by *Cladophialophora bantiana* in a domestic short haired cat. Med Mycol. 2011;49:194–7.
115. Bouljihad M, Lindeman CJ, Hayden DW. Pyogranulomatous meningoencephalitis associated with dematiaceous fungal (*Cladophialophora bantiana*) infection in a domestic cat. J Vet Diagn Investig. 2002;14:70–2.
116. Lavely J, Lipsitz D. Fungal infections of the central nervous system in the dog and cat. Clin Tech Small Anim Pract. 2005;20:212–9.
117. Dye C, Peters I, Tasker S, Caney SMA, Dye S, Gruffydd-Jones TJ, et al. Preliminary study using an indirect ELISA for the detection of serum antibodies to *Alternaria* in domestic cats. Vet Rec. 2005;156:633–5.
118. Grooters AM. Pythiosis, lagenidiosis, and zygomycosis in small animals. Vet Clin North Am Small Anim Pract. 2003;33:695–720. v
119. Ravisse P, Fromentin H, Destombes P, Mariat F. Cerebral mucormycosis in the cat caused by *Mucor pusillus*. Sabouraudia. 1978;16:291–8.
120. Wray JD, Sparkes AH, Johnson EM. Infection of the subcutis of the nose in a cat caused by *Mucor* species: successful treatment using posaconazole. J Feline Med Surg. 2008;10:523–7.
121. Cunha SC, Aguero C, Damico CB, Corgozinho KB, Souza HJ, Pimenta AL, et al. Duodenal perforation caused by *Rhizomucor* species in a cat. J Feline Med Surg. 2011;13:205–7.
122. Mahamaytakit N, Singalavanija S, Limpongsanurak W. Subcutaneous zygomycosis in children: 2 case reports. J Med Assoc Thail. 2014;97(Suppl 6):S248–53.
123. Eisen DP, Robson J. Complete resolution of pulmonary *Rhizopus oryzae* infection with itraconazole treatment: more evidence of the utility of azoles for zygomycosis. Mycoses. 2004;47:159–62.
124. Rakich PM, Grooters AM, Tang KN. Gastrointestinal pythiosis in two cats. J Vet Diagn Investig. 2005;17:262–9.
125. Bissonnette KW, Sharp NJ, Dykstra MH, Robertson IR, Davis B, Padhye AA, et al. Nasal and retrobulbar mass in a cat caused by *Pythium insidiosum*. J Med Vet Mycol. 1991;29:39–44.
126. Fortin JS, Calcutt MJ, Kim DY. Sublingual pythiosis in a cat. Acta Vet Scand. 2017;59
127. Grooters AM. Pythiosis, lagenidiosis, and zygomycosis. In: Sykes JE, editor. Canine and feline infectious diseases. St Louis: Elsevier Saunders; 2014. p. 668–78.
128. Thomas RC, Lewis DT. Pythiosis in dogs and cats. Comp Cont Educ Pract. 1998;20: 63.
129. Middleton SM, Kubier A, Dirikolu L, Papich MG, Mitchell MA, Rubin SI. Alternate-day dosing of itraconazole in healthy adult cats. J Vet Pharmacol Ther. 2016;39:27–31.
130. Mawby DI, Whittemore JC, Fowler LE, Papich MG. Comparison of absorption characteristics of oral reference and compounded itraconazole formulations in healthy cats. J Am Vet Med Assoc. 2018;252:195–200.
131. Vishkautsan P, Papich MG, Thompson GR 3rd, Sykes JE. Pharmacokinetics of voriconazole after intravenous and oral administration to healthy cats. Am J Vet Res. 2016;77:931–9.
132. Mawby DI, Whittemore JC, Fowler LE, Papich MG. Posaconazole pharmacokinetics in healthy cats after oral and intravenous administration. J Vet Intern Med. 2016;30:1703–7.
133. Vaden SL, Heit MC, Hawkins EC, Manaugh C, Riviere JE. Fluconazole in cats: pharmacokinetics following intravenous and oral administration and penetration into cerebrospinal fluid, aqueous humour and pulmonary epithelial lining fluid. J Vet Pharmacol Ther. 1997;20:181–6.
134. Medleau L, Jacobs GJ, Marks MA. Itraconazole for the treatment of cryptococcosis in cats. J Vet Intern Med. 1995;9:39–42.
135. Boothe DM, Herring I, Calvin J, Way N, Dvorak J. Itraconazole disposition after single oral and intravenous and multiple oral dosing in healthy cats. Am J Vet Res. 1997;58:872–7.
136. Leshinsky J, McLachlan A, Foster DJR, Norris R, Barrs VR. Pharmacokinetics of caspofungin acetate to guide optimal dosing in cats. PLoS One. 2017;12:e0178783.
137. Ghannoum MA. Antifungal resistance: monitoring for terbinafine resistance among clinical dermatophyte isolates. Mycoses. 2013;56:38.
138. Rocha EMF, Gardiner RE, Park S, Martinez-Rossi NM, Perlin DS. A Phe389Leu substitution in ErgA confers terbinafine resistance in *Aspergillus fumigatus*. Antimicrob Agents Chemother. 2006;50:2533–6.

139. Diekema DJ, Messer SA, Hollis RJ, Jones RN, Pfaller MA. Activities of caspofungin, itraconazole, posaconazole, ravuconazole, voriconazole, and amphotericin B against 448 recent clinical isolates of filamentous fungi. J Clin Microbiol. 2003;41:3623–6.
140. Ndiritu CG, Enos LR. Adverse reactions to drugs in a veterinary hospital. J Am Vet Med Assoc. 1977;171:335–9.
141. Wang A, Ding HZ, Liu YM, Gao Y, Zeng ZL. Single dose pharmacokinetics of terbinafine in cats. J Feline Med Surg. 2012;14:540–4.

Sporothrichosis

Hock Siew Han

Abstract

Sporothrix schenckii is currently recognized as a species complex consisting of *Sporothrix brasiliensis, Sporothrix schenckii* sensu stricto, *Sporothrix globosa,* and *Sporothrix luriei.* Due to divergent evolutionary process, each species possesses different virulence profiles, that allow it to thrive and persist in its niche. Currently the disease in cats is primarily caused by *S. brasiliensis, S. schenckii* sensu stricto and *S. globosa,* with cat fights and direct inoculation of the agent in the skin as the main mode of disease transmission. Expression of putative virulence factors, such as adhesins, ergosterol peroxide, melanin, proteases, extracellular vesicles and thermotolerance, determines the clinical manifestation in the feline patient, with thermotolerant *S. brasiliensis* exhibiting the highest pathogenicity, followed by *S. schenckii* sensu stricto, and *S. globosa.* Their ability to produce biofilm is documented, but their clinical significance remains to be elucidated. Despite comprehensive descriptions of the pathogenicity of the agent and of the disease, its prognosis remains guarded to poor, due to issues pertaining to cost, protracted treatment course, zoonotic potential and low susceptibility of some strains to antifungals.

Introduction

Sporothrix schenckii complex (also called *S. schenckii* sensu *lato*) causes a chronic, granulomatous, cutaneous or subcutaneous infection, mainly occurring in humans and cats. It has been recognised as an important cause of zoonotic subcutaneous mycosis since its description by Dr. Benjamin Schenk in 1896 [1].

H. S. Han (✉)
The Animal Clinic, Singapore

© Springer Nature Switzerland AG 2020
C. Noli, S. Colombo (eds.), *Feline Dermatology,*
https://doi.org/10.1007/978-3-030-29836-4_15

329

As a thermally dimorphic fungus, *Sporothrix schenckii* sensu *lato* exists as saprophyte in plant debris or decaying organic soil matter in its asexual filamentous form (25–30 °C). With favourable temperature and environment (35–37 °C), it phase transitions into its yeast form, and complete growth inhibition is achieved at 40 °C, with no sexual reproduction observed to date [2]. This characteristic underpins the epidemiology of clinical sporotrichosis where historically, the most common route of infection was reported to be the inoculation of conidia into broken skin via contaminated soil during horticultural activities. It is only in recent times that cats were perceived to be an important risk factor and disease propagators [3–7].

Etiologic Agent

Sporothrix schenckii is currently recognized as a species complex consisting of *Sporothrix brasiliensis*, *Sporothrix schenckii* sensu stricto, *Sporothrix globosa*, and *Sporothrix luriei* (Clinical clade) based on DNA sequencing, with each species having its own distinct virulence profiles and geographical distribution [8, 9]. *S. brasiliensis*, *S. s.* sensu stricto and S. *globosa*, in order of virulence, are the main species identified to cause pathology in cats [9]. *S. brasiliensis*, currently regionally restricted to Brazil, is characterised by its inherent thermotolerability which is responsible for causing systemic spread. This species was identified as the main cause of sporotrichosis epidemics in Rio de Janeiro and Sao Paolo, alongside *S. s.* sensu stricto *and S. globosa* [10–12]. *S. s.* sensu stricto is the second most pathogenic species with a worldwide distribution, especially in tropical or subtropical regions, with reports from the Americas, Africa, Australia and Asia. Zhou and colleagues demonstrated genetic diversity within this single species by subdividing *S. s.* sensu stricto into clinical clade C (most commonly isolated from Americas and Asia) and D (most commonly isolated from Americas and Africa), based on its internal transcribed spacer (ITS) [13]. The recent identification of a single clonal strain of *S. s.* sensu stricto clinical clade D from Malaysia (instead of the commonly isolated clinical clade C in Asia) suggests that this species is constantly evolving, with the ability to undergo a process of selection and subsequent population expansion, depending on local environmental or host selection pressure [14, 15]. *S. globosa* is commonly identified as the species responsible for sporotrichosis mainly in Asia and Europe, but is a rare cause in the Americas and Africa [11, 13, 16–20]. Exept *S. pallida*, Environmental clade associated sporothrix species such as *S.brunneoviolacea*, *S. lignivora*, *S. chilensis* and *S. mexicana* (*Sporothrix pallida* complex) have not been reported to cause disease in the feline patient at the time of writing [21]. These species are rare agents of sporotrichosis and normally causes low virulence, opportunistic infections from traumatic inoculation of fungus from soil into host tissue. This is in contrast to sporothrix species within the Clinical clade that is transmitted from animals.

Pathogenesis

Upon inoculation, the expression of putative virulence factors, such as adhesins, ergosterol peroxide, melanin, proteases, extracellular vesicles (EV) and thermotolerance, determines the pathogenicity and clinical presentation of sporotrichosis in the feline patient [22, 23]. The expression of adhesins and a 70 kDa glycoprotein (Gp70) on the cell wall mediates adhesion of the fungus to fibronectin, type II collagen and laminin in the host [24]. Upon invasion, the fungal cell wall composed of glucans, galactomannans, rhamnomannans, chitin, glycoprotein, glycolipids and melanin provides the ability to survive within host tissues and aids evasion from host innate immune response [25–27]. Melanin production in both mycelial and yeast form shields against a broad range of toxic insults. Melanin reduces susceptibility to antifungals and enzymatic degradation, and confers protection against oxygen nitrogen free radicals, macrophagic and neutrophilic phagocytosis [28]. The fungus readily produces ergosterol peroxide and proteinases (Proteinase 1 and 2), which allow it to evade phagocytosis and host immune response [29, 30]. EV (exosomes, microvesicles and apoptotic bodies) are membranous compartments composed of lipid bilayers, released by all living cells to the extracellular medium, that contain cargos of lipids (neutral glycolipids, sterols and phospholipids), polysaccharides (glucuronoxylomannan, alpha-galactosyl epitopes), proteins (lipases, proteases, urease, phosphatase) and nucleic acids (RNA) [31]. These cargos represent virulence factors that contribute to drug resistance, facilitate cell invasion and are eventually recognized by the innate immune system. EV contribution to fungal virulence was described in *Cryptococcus neoformans, Histoplasma capsulatum, Paracoccidioides brasiliensis, Malassezia sympodialis, Candida albicans* and, recently, also in *Sporothrix brasiliensis* [32–39]. Specifically, the EV cargos of *Sporothrix brasiliensis,* such as cell wall glucanase and heat shock proteins, were shown to increase phagocytosis but not pathogen elimination, stimulate cytokine production (IL-12p40 and TNFα) and favour the establishment of the fungus in the skin [38, 40, 41]. Current proteomic analyses revealed that 27% of EV proteins in *S. brasiliensis* and 35% in *S. schenckii* remain to be characterized, including the identification of their assigned biological process [38].

Thermotolerance, the ability of a fungus to grow or not at 37 °C, is another important virulence factor that has been identified in *Sporothrix* spp. Isolates that are able to grow at 35 °C but not at 37 °C in humans cause fixed cutaneous lesions, but those that grow at 37 °C (a close approximation to human and animal core body temperature) produce disseminated and extracutaneous lesions. Pathogenic thermotolerant species, such as *S. brasiliensis* have the ability to produce disseminated disease, compared to non-thermotolerant, less pathogenic species such as *S. globosa*. *S. s.* sensu stricto displays variable thermotolerability [14].

The ability of *Sporothrix schenckii* complex to produce biofilm has recently been documented, and an early report suggests that biofilm production alters the fungus sensitivity to antifungals, however, the full extent of its clinical significance has yet to be elucidated [42].

Both innate and adaptive immune responses play important roles in the prevention of disease progression. The first contact between fungal pathogen associated molecular pattern (PAMPs) and host pattern recognition receptors (PPRs) is mediated by toll-like-receptors (TLR)-4 and TLR-2 [43, 44]. During the initiation of infection, these receptors recognize lipid extracts from yeast cells that lead to an increased production of tumour necrosis factor alpha (TNF-alpha), interleukin (IL)-10 and nitric oxide (NO). While NO demonstrates antifungal activity in vitro, in vivo it is associated with immunosuppression during the initial and the terminal stages of the infection, due to its ability to increase apoptosis of immune cells [45]. The role of NO in the infection was also documented in histoplasmosis by *Histoplasma capsulatum* and paracoccidioidomycosis by *Paracoccidioides brasiliensis* [46, 47].

Yeast cells are also able to activate the antibody-dependent classical and alternative complement pathways [48, 49]. The main antigen recognized by antibodies is a 70 kDa cell wall glycoprotein, named Gp70 [50]. This protein plays a crucial role in fungal opsonisation, allowing macrophages to phagocytose and the production of pro-inflammatory cytokines [51]. Nevertheless, the cornerstone for an effective fungal eradication is based on an effectively coordinated innate and adaptive immune response (humoral and cell mediated) [52]. Recently, the nucleotide-binding oligomerization domain-like receptor pyrin domain-containing 3 (NLRP3) inflammasome was shown to be critical to link the innate immune response to the adaptive arm, contributing to effective protection against this infection by promoting the production of pro-IL1β [53]. Fungal interaction with dendritic cells drives a mixed Th1/Th17 immune response that activates macrophages, neutrophils and CD4+ T cells, that release IFN-gamma, IL-12 and TNF-alpha that ultimately culminates in the reduction of pathogen burden [54, 55].

Clinical Signs

Feline sporotrichosis occurs most commonly in young adult, free roaming intact male cats and is associated with fighting, with no known breed predisposition [4]. In the human patient, clinical signs of sporotrichosis may be classified into 3 forms: fixed cutaneous, lympho-cutaneous and disseminated forms, depending on the pathogenicity of the fungal species and the status of host immunity (Fig. 1). Such clear and distinct categorisation of clinical forms does not apply to cats and thus is seldom used.

In cats, chronic non-healing lesions such as nodules, ulcers and crusts are commonly found on the head, especially at the bridge of the nose (Fig. 2), on the distal limbs and tail base region (Fig. 3) and on the pinnae (Fig. 4). The majority of lesions occur in cooler regions of the host body such as at the nasal passages and ear tips.

Fig. 1 A human patient manifesting lymphocutaneous sporotrichosis after being bitten by a cat with sporotrichosis (nodule at base of thumb). Due to the lack of thermotolerability of the infectious agent, the lesion did not progress beyond the arm

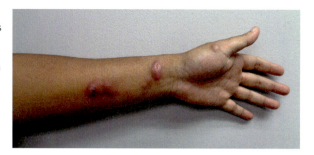

Fig. 2 Classical presentation of feline sporotrichosis: chronic non-healing wounds affecting the bridge of the nose

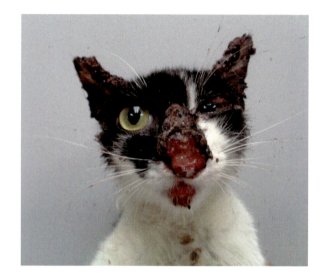

Fig. 3 Chronic non-healing wounds affecting the paws and the tail

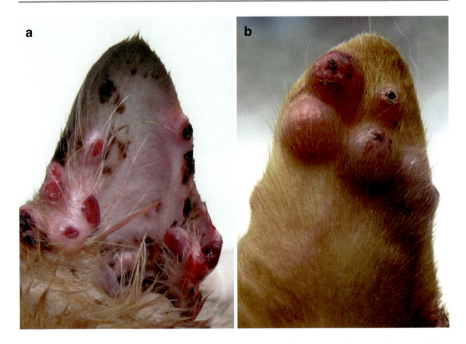

Fig. 4 (**a** and **b**) Concave and convex aspects, respectively, of the pinna of a cat with sporotrichosis presenting numerous ulcerated nodules

If nasal passages are affected, extracutaneous signs such as sneezing, dyspnoea and respiratory distress are commonly reported in tandem with cutaneous manifestations [5]. Cutaneous screwworm myiasis as secondary infestation was recently reported [56]. The fatal disseminated form of the disease is associated with *S. brasiliensis* infection. Co-infection with either feline immunodeficiency virus (FIV) or feline leukaemia virus (FeLV) has no significant effect on the clinical manifestations or on the prognosis of the disease [57].

Diagnosis

A definitive diagnosis of feline sporotrichosis requires the isolation and identification of the agent in culture. The species identification can be obtained by morphologic studies and physiologic phenotyping, as well as by polymerase chain reaction targeting the calmodulin gene [5]. At 25–30°, the fungus exists in its mycelial form and is seen as small and white or pale orange to orange-grey colonies with no cottony aerial hyphae. Later, the colony becomes black, moist, wrinkled, leathery or velvety with narrow white borders (Fig. 5). Some colonies are however black from the onset. At 35–37°, yeast colonies are cream or tan, smooth and yeast-like [2].

Cytologically, yeasts are found in abundance from cutaneous impression smears. They are located intra- and extracellularly, in pleomorphic shapes, ranging from

Fig. 5 In its mature mycelial form the fungi becomes black, moist, wrinkled, leathery or velvety with narrow white borders

Fig. 6 Cytologically, the yeasts are found in abundance intra- and extracellularly in pleomorphic shapes, ranging from the classical cigar-shaped to round or oval, measuring 3–5 μm in diameter with a thin, clear halo around a pale blue cytoplasm (Diff Quick, 1000×)

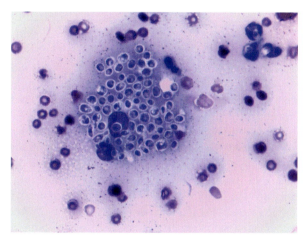

the classical cigar-shaped to round or oval bodies, measuring 3–5 μm in diameter with a thin, clear halo around a pale-blue cytoplasm (Fig. 6) [58]. The sensitivity of cytology to detect *Sporothrix* yeasts in the feline patient is estimated to range from 79% to 84.9% [59, 60].

On histology, a diffuse pyogranulomatous inflammation with large foci of necrosis is seen throughout the superficial and deep dermis, sometimes extending to the subcutis. There are abundant round to cigar-shaped organisms, 3–10 μm in length and 1–2 μm in diameter, seen both free and within macrophages. Commonly, organisms in cytoplasm of macrophages create large clear pockets full of yeast due to poorly visualized yeast cell wall (Fig. 7) [61]. Periodic acid of Schiff (PAS) stain may also be utilized to visualize yeasts as magenta stained organism on histological preparation. Other diagnostic techniques such as serology (enzyme-linked immunosorbent assay, ELISA) and polymerase chain reaction (PCR) may also be used for the diagnosis [62, 63].

Fig. 7 On histology there are abundant round to cigar-shaped organisms, 3–10 μm in length to 1–2 μm in diameter seen both free and within macrophages. Organisms in cytoplasm of macrophages create large clear pockets full of yeasts due to poorly visualized yeast cell wall

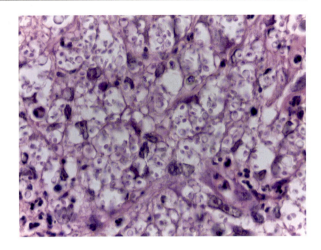

Treatment

Treatment of feline sporotrichosis requires several months and must be continued for at least 1 month beyond clinical cure. Luckily, despite a protracted treatment course, it is current understanding that the fungus does not develop resistance during treatment [14].

Due to the high cost of treatment, high risk of therapeutic side effects and of zoonosis and existence of low susceptibility strains, feline sporotrichosis carries a guarded to poor prognosis. Currently, potassium iodide, azolic antifungals (ketoconazole, itraconazole), amphotericin B, terbinafine, local heat therapy, cryosurgery and surgical resection have all been documented as treatment options in the feline patient. Potassium iodide has traditionally been the treatment of choice, either in its saturated form (saturated salt of potassium iodide, SSKI) or in its powder form re-packaged into capsules. Dosages range from 10 to 20 mg/kg every 24 hours [64, 65]. The powder form re-packaged into capsules is favoured over SSKI for the feline patient, due to the latter's tendency to cause hypersalivation. From a report of 48 cats receiving potassium iodide, 23 (47.9%) patients achieved clinical cure with treatment failure in 18 cats (37.5%), two reported deaths (4.2%) and treatment period averaging from 4 to 5 months. The most commonly observed side effects were hyporexia, lethargy, weight loss, vomiting, diarrhoea plus an increase in the liver enzyme alanine transaminase. No signs of iodism (lacrimation, salivation, coughing, facial swelling, tachycardia) nor thyroid hormone abnormalities were observed in this study [64]. Due to its low cost, potassium iodide is still often used either singularly or in conjunction with azole antifungals to treat feline sporotrichosis [65].

    Imidazoles such as ketoconazole and itraconazole currently represent the corner-
stone therapy for feline sporotrichosis. Itraconazole is favoured over ketoconazole
as the latter is commonly associated with a higher rate of side effects, such a vomiting, hepatic dysfunction and altered cortisol metabolism. Itraconazole at 5–10 mg/
kg has been used successfully to treat feline sporotrichosis, with a maximum plasma
concentration of 0.7 ± 0.14 mg/L achieved after a 5 mg/kg oral dosing [66]. Based
on the updated Clinical and Laboratory Standards Institute (CLSI) reference method
for broth dilution antifungal susceptibility testing of filamentous fungi (document
M38-A2), the minimum inhibitory concentration (MIC) of antifungals against *S.
brasiliensis*, *S. s* sensu stricto and *S. globosa* is presented in Table 1 [14, 19, 20, 67,
68]. Itraconazole may be the treatment of choice but there are isolates with MIC
above 4 mg/L, the putative breakpoint for this antifungal agent. This variability
in MIC values may reflect the extensive divergent evolutionary process within the
Sporotrix complex, where each species developed its own repertoire of virulence
factors allowing thriving and persisting in its niche. Clinically, this is reflected by
the fact that some cases of feline sporotrichosis are refractory to treatment and thus
protocols based on higher dosages of itraconazole and/or its combination with other
antifungals have been explored to treat these refractory cases [65, 69]. *Sporothrix
schenckii* sensu *lato* generally displays low susceptibility towards fluconazole and
exhibits species-dependent susceptibility towards terbinafine and amphotericin B
(Table 1). Despite reports of successful treatment of human sporotrichosis with
terbinafine, results are still inconclusive for the feline patient [70, 71]. The recent
description of the protective effects of pyomelanin and eumelanin, synthesized by
S. brasiliensis and *S. s.* sensu stricto, against the antifungal terbinafine may partially
explain why in vitro results do not always correlate with in vivo responses when
patients are treated with this drug [72]. The administration of amphotericin B is
associated with toxicity, high cost and side effects, such as localized sterile abscess
formation from intralesional injections [5]. It is interesting to note that *Sporothrix*
spp. displays variable susceptibility towards antifungals rarely used in veterinary
medicine such as micafungin, 5-flucytosine and even posaconazole, highlighting
the importance of susceptibility testing [14, 20, 68]. Resolving granulomas are
visually and tactile-wise indistinguishable from normal adjacent healthy skin under
normal room lighting, and may be better visualized when held against a bright light
source (Fig. 8). Treatment should be continued for 1 month beyond the resolution
of all granulomas. Localized heat therapy is based on the fact that the fungus does
not grow at temperatures above 40 °C. This treatment modality, however, is associated with issues of practicality and perhaps welfare concerns in its application on
animals and has not been pursued as a feasible treatment option in the feline patient.
Cryosurgery, used in conjunction with itraconazole has been used successfully to
treat and cure 11 of 13 cats with sporotrichosis, with treatment lasting 3–16 months
and a median of 8 months [73]. Surgical resection is possible for localized singular
lesions but unpractical for generalized, disseminated forms.

Table 1 All results are expressed in mg/L and based on the Clinical and Laboratory Standards Institute (CLSI) reference method for broth dilution antifungal susceptibility testing of filamentous fungi document M38-A2 (2008) in mycelial phase

	Origin	n	Itraconazole	Fluconazole	Amphotericin B	Terbinafine	References
S. globosa	Japan subgroup I	29	0.5–4	>128	1–4	Not tested	[20]
	Japan subgroup II	9	0.25–2	>128	2–4	Not tested	[20]
	Brazil	4	0.83 (0.06–16)	53.8 (16–128)	1 (0.2–4)	0.03 (0.01–0.06)	[67]
	Iran	4	8 (1–>16)	>64 32–>64	5.66 (4–8)	1.68 (1–2)	[19]
S. s. sensu stricto	Malaysia	40	1.3 (0.5–4)	>256	Not tested	2.85 (1–8)	[14]
	Japan	9	0.5–1	>128	2	Not tested	[20]
	Brazil	61	0.42 (0.03–16)	57.7 (8–128)	1.06 (0.03–2)	0.05 (0.01–0.50)	[67]
	Iran	5	0.76 (0.25–2)	>64	3.03 (1–8)	0.38 (0.13–1)	[19]
S. brasiliensis	Brazil	32	2	Not tested	1.2	0.1	[68]
	Brazil	23	0.36 (0.06–2)	56.7 (16–128)	1.03 (0.2–4)	0.06 (0.01–0.50)	[67]

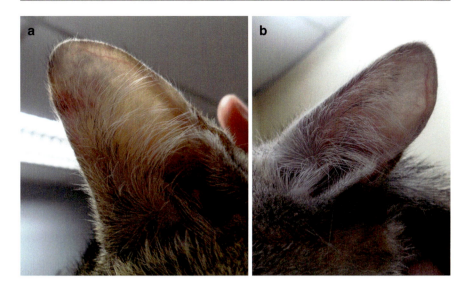

Fig. 8 The author utilizes a bright light source to evaluate and ascertain cure. (**a**) A resolving granulomatous reaction at the left ear tip, tactile and visually indistinguishable from adjacent normal tissue but is visualized with a bright light source. (**b**) Same patient after cure with complete resolution of granuloma

Conclusion

The prognosis of feline sporotrichosis remains guarded to poor due to cost, protracted treatment course, risk of zoonosis and low susceptibility of some strains. Despite the fact that antifungal susceptibility testing provides essential guidance for the treatment, its lack of commercial availability and validated breakpoints remains a stumbling block in the treatment of this disease. Unfortunately, the current repertoire of veterinary antifungals classes are inadequate to address the issue of fungal low fungal susceptibility.

References

1. Schenck BR. On refractory subcutaneous abscess caused by a fungus possibly related to the Sporotricha. Bull Johns Hopkins Hosp. 1898;9:286–90.
2. Larone DH. Identification of fungi in culture. In: Medically important fungi: a guide to identification. 5th ed. Washington, DC: ASM Press; 2011. p. 166–7.
3. Schubach A, Schubach TM, Barros MB, Wanke B. Cat-transmitted sporotrichosis, Rio de Janeiro, Brazil. Emerg Infect Dis. 2005;11(1):1952–4.
4. Rodrigues AM, de Hoog GS, de Camargo ZP. Sporothrix Species Causing Outbreaks in Animals and Humans Driven by Animal–Animal Transmission. PLoSPathog. 2016;12:e1005638. https://doi.org/10.1371/journal.ppat.100.

5. Gremião ID, Menezes RC, Schubach TM, Figueiredo AB, Cavalcanti MC, Pereira SA. Feline sporotrichosis: epidemiological and clinical aspects. Med Mycol. 2015;53(1):15–21.
6. Gremião IDF, Miranda LHM, Reis EG, Rodrigues AM, Pereira AS. Zoonotic epidemic of sporotrichosis: cat to human transmission. PLoS Pathog. 2017;13(1):1–7.
7. Tang MM, Tang JJ, Gill P, Chang CC, Baba R. Cutaneous sporotrichosis: a six-year review of 19 cases in a tertiary referral center in Malaysia. Int J Dermatol. 2012;51:702–8.
8. Marimon R, Cano J, Gene J, Sutton DA, Kawasaki M, Guarro J. *Sporothrix brasiliensis, S. globosa*, and *S. mexicana*, three new *Sporothrix* species of clinical interest. J Clin Microbiol. 2007;45:3198-206.
9. Arrillaga-Moncrieff CJ, Mayayo E, Marimon R, Marine M, Gene J, et al. Different virulence levels of the species of Sporothrix in a murine model. Clin Microbiol Infect. 2009;15:651–5.
10. Rodrigues AM, de Melo Teixeira M, de Hoog GS, TMP S, Pereira SA, Fernandes GF, et al. Phylogenetic analysis reveals a high prevalence of *Sporothrix brasiliensis* in feline sporotrichosis outbreaks. PLoS Negl Trop Dis. 2013;7(6):e2281.
11. Oliveira MME, Almeida-Paes R, Muniz MM, Barros MBL, Gutierrez-Galhardo MC, Zancope-Oliveira RM. Sporotrichosis caused by *Sporothrix globosa* in Rio de Janeiro, Brazil: case report. Mycopathologia. 2010;169:359–63.
12. Oliveira MME, Almeida-Paes R, Muniz MM, Gutierrez-Galhardo MC, Zancope-Oliveira RM. Phenotypic and molecular identification of Sporothrix isolates from an epidemic area of sporotrichosis in Brazil. Mycopathologia. 2011;172(4):257–67.
13. Zhou X, Rodrigues A, Feng P, Hoog GS. Global ITS diversity in the Sporothrix schenckii complex. Fungal Divers. 2013:1–13.
14. Han HS, Kano R, Chen C, Noli C. Comparisons of two in vitro antifungal sensitivity tests and monitoring during therapy of *Sporothrix schenckii sensu stricto* in Malaysian cats. Vet Dermatol. 2017;28:156–e32.
15. Kano R, Okubo M, Siew HH, Kamata H, Hasegawa A. Molecular typing of *Sporothrix schenckii* isolates from cats in Malaysia. Mycoses. 2015;58:220–4.
16. Watanabe M, Hayama K, Fujita H, Yagoshi M, Yarita K, Kamei K, et al. A case of Sporotrichosis caused by *Sporothrix globosa* in Japan. Ann Dermatol. 2016;28:251–2.
17. Yu X, Wan Z, Zhang Z, Li F, Li R, Liu X. Phenotypic and molecular identification of Sporothrix isolates of clinical origin in Northeast China. Mycopathologia. 2013;176:67–74.
18. Madrid H, Cano J, Gene J, Bonifaz A, Toriello C, Guarro J. *Sporothrix globosa*, a pathogenic fungus with widespread geographical distribution. Rev Iberoam Micol. 2009;26(3):218–22.
19. Mahmoudi S, Zaini F, Kordbacheh P, Safara M, Heidari M. Sporothrix schenckii complex in Iran: molecular identification and antifungal susceptibility. Med Mycol. 2016;54:593–9.
20. Suzuki R, Yikelamu A, Tanaka R, Igawa K, Yokodeki H, Yaguchi T. Studies in phylogeny, development of rapid identification methods, antifungal susceptibility and growth rates of clinical strains of Sporothrix schenckii Complex in Japan. Med Mycol J. 2016;57E:E47–57.
21. Thomson J, Trott DJ, Malik R, Galgut B, McAllister MM, Nimmo J et al. An atypical cause of sporotrichosis in a cat. Med Mycol Case Reports. 2019;23:72-6.
22. Barros MB, Paes RA, Schubach AO. *Sporothrix schenckii* and Sporotrichosis. Clin Microbiol Rev. 2011;24:633–54.
23. Rossato L, Moreno F, Jamalian A, Stielow B, Almeida R, de Hoog S, et al. Proteins potentially involved in immune evasion strategies in *Sporothrix brasiliensis* elucidated by high resolution mass spectrometry. mSphere. 2018;13:e00514–7.
24. Teixeira PA, de Castro RA, Nascimento RC, Tronchin G, Torres AP, Lazéra M, et al. Cell surface expression of adhesins for fibronectin correlates with virulence in *Sporothrix schenckii*. Microbiology. 2009;155:3730–8.
25. López-Esparza A, Álvarez-Vargas A, Mora-Montes HM, Hernández-Cervantes A, Del Carmen C-CM, Flores-Carreón A. Isolation of *Sporothrix schenckii* GDA1 and functional characterization of the encoded guanosine diphosphatase activity. Arch Microbiol. 2013;195:499–506.
26. Morris-Jones R, Youngchim S, Gomez BL, Aisen P, Hay RJ, Nosanchuk JD, et al. Synthesis of melanin-like pigments by *Sporothrix schenckii* in vitro and during mammalian infection. Infect Immun. 2003;71:4026–33.

27. Teixeira PA, De Castro RA, Ferreira FR, Cunha MM, Torres AP, Penha CV, et al. L-DOPA accessibility in culture medium increases melanin expression and virulence of *Sporothrix schenckii* yeast cells. Med Mycol. 2010;48:687–95.
28. Nosanchuk JD, Casadevall A. Impact of melanin on microbial virulence and clinical resistance to antimicrobial compounds. Antimicrob Agents Chemother. 2006;50:3519–28.
29. Sgarbi DB, da Silva AJ, Carlos IZ, Silva CL, Angluster J, Alviano CS. Isolation of ergosterol peroxide and its reversion to ergosterol in the pathogenic fungus *Sporothrix schenckii*. Mycopathologia. 1997;139:9–14.
30. Lei PC, Yoshiike T, Ogawa H. Effects of proteinase inhibitors on cutaneous lesion of *Sporothrix schenckii* inoculated hairless mice. Mycopathologia. 1993;123:81–5.
31. Joffe LS, Nimrichter L, Rodrigues ML, Del Poeta M. Potential roles of fungal extracellular vesicles during infection. mSphere. 2016;1:e00099–16.
32. Rodrigues ML, Nimrichter L, Oliveira DL, Frases S, Miranda K, Zaragoza O, et al. Vesicular polysaccharide export in *Cryptococcus neoformans* is a eukaryotic solution to the problem of fungal trans-cell wall transport. Eukaryot Cell. 2007;6:48–59.
33. Rodrigues ML, Nimrichter L, Oliveira DL, Nosanchuk JD, Casadevall A. Vesicular trans-cell wall transport in fungi: a mechanism for the delivery of virulence-associated macromolecules? Lipid Insights. 2008;2:27–40.
34. Albuquerque PC, Nakayasu ES, Rodrigues ML, Frases S, Casadevall A, Zancope-Oliveira RM, et al. Vesicular transport in *Histoplasma capsulatum*: an effective mechanism for trans-cell wall transfer of proteins and lipids in ascomycetes. Cell Microbiol. 2008;10:1695–710.
35. Vallejo MC, Matsuo AL, Ganiko L, Medeiros LC, Miranda K, Silva LS, et al. The pathogenic fungus *Paracoccidioides brasiliensis* exports extracellular vesicles containing highly immunogenic-galactosyl epitopes. Eukaryot Cell. 2011;10:343–51.
36. Vargas G, Rocha JD, Oliveira DL, Albuquerque PC, Frases S, Santos SS, et al. Compositional and immunobiological analyses of extracellular vesicles released by *Candida albicans*. Cell Microbiol. 2015;17:389–407.
37. Rayner S, Bruhn S, Vallhov H, Anderson A, Billmyre RB, Scheynius A. Identification of small RNAs in extracellular vesicles from the commensal yeast *Malassezia sympodialis*. Sci Rep. 2017;7:39742.
38. Ikeda MAK, de Almeida JRF, Jannuzzi GP, Cronemberger-Andrade A, Torrecilhas ACT, Moretti NS, et al. Extracellular vesicles from *Sporothrix brasiliensis* are an important virulence factor that induce an increase in fungal burden in experimental sporotrichosis. Front Microbiol. 2018;9:2286.
39. Huang SH, Wu CH, Chang YC, Kwon-Chung KJ, Brown RJ, Jong A. *Cryptococcus neoformans*-derived microvesicles enhance the pathogenesis of fungal brain infection. PLoS One. 2012;7:e48570.
40. Rossato L, Moreno F, Jamalian A, Stielow B, Almeida R, de Hoog S, et al. Proteins potentially involved in immune evasion strategies in *Sporothrix brasiliensis* elucidated by ultra-high-resolution mass spectrometry. mSphere. 2018;3:e00514–7.
41. Nimrichter L, de Souza MM, Del Poeta M, Nosanchuk JD, Joffe L, Tavares PM, Rodrigues ML. Extracellular vesicle-associated transitory cell wall components and their impact on the interaction of fungi with host cells. Front Microbiol. 2016;7:1034.
42. Brilhante RSN, de Aguiar FRM, da Silva MLQ, de Oliveira JS, de Camargo ZP, Rodrigues AM, et al. Antifungal susceptibility of *Sporothrix schenckii* complex biofilms. Med Mycol. 2018;56:297–306.
43. Carlos IZ, Sassá MF, da Graca Sgarbi DB, MCP P, DCG M. Current research on the immune response to experimental sporotrichosis. Mycopathologia. 2009;168:1–10.
44. Negrini Tde C, Ferreira LS, Alegranci P, Arthur RA, Sundfeld PP, Maia DC, et al. Role of TLR-2 and fungal surface antigen on innate immune response against *Sporothrix schenckii*. Immuno Invest. 2013;42:36–48.
45. Fernandes KS, Neto EH, Brito MM, Silva JS, Cunha FQ, Barja-Fidalgo C. Detrimental role of endogenous nitric oxide in host defense againsts *Sporothrix schenckii*. Immunology. 2008;123:469–79.

46. Brummer E, Division DA. Antifungal mechanism of activated murine bronchoalveolar or peritoneal macrophages for *Histoplasma capsulatum*. Clin Exp Immunol. 1995;102:65–70.
47. Bocca L, Hayashi EE, Pinheiro G, Furlanetto B, Campanelli P, Cunha FQ, et al. Treatment of *Paracoccidioides brasiliensis*-infected mice with a nitric oxide inhibitor prevents the failure of cell-mediated immune response. J Immunol. 1998;161:3056–63.
48. Torinuki W, Tagami H. Complement activation by Sporothrix schenckii. Arch Dermatol Res. 1985;277:332–3.
49. de Lima FD, Nascimento RC, Ferreira KS, Almeida SR. Antibodies against Sporothrix schenckii enhance TNF-alpha production and killing by macrophages. Scand J Immunol. 2012;75:142–6.
50. Ruiz-Baca E, Toreillo C, Perez-Torres A, Sabanero-López M, Villagómez-Castro JC, López-Romero E. Isolation and some properties of a glycoprotein of 70 kDa (Gp70) from the cell wall of Sporothrix shcenckii cell wall. Mem Inst Oswaldo Cruz. 2009;47:185–96.
51. Maia DC, Sassá MF, Placeres MC, Carlos IZ. Influence of Th1/Th2 cytokines and nitric oxide in murine systemic infection induced by Sporothrix schenckii. Mycopathologia. 2006;161:11–9.
52. Plouffe JF, Silva J, Fekety R, Reinhalter E, Browne R. Cell-mediated immune responses III sporotrichosis. J Infect Dis. 1979;139:152–7.
53. Goncalves AC, Ferreira LS, Manente FA, de Faria CMQG, Polesi MC, de Andrade CR, et al. The NLRP3 inflammasome contributes to host protection during *Sporothrix schenckii* infection. Immunology. 2017;151:154–66.
54. Tachibana T, Matsuyama T, Mitsuyama M. Involvement of CD4+ T cells and macrophages in acquired protection against infection with *Sporothrix schenckii* in mice. Med Mycol. 1999;37:397–404.
55. Flores-García A, Velarde-Félix JS, Garibaldi-Becerra V, Rangel-Villalobos H, Torres-Bugarín O, Zepeda-Carrillo EA, et al. Recombinant murine IL-12 promotes a protective TH1/cellular response in Mongolian gerbils infected with *Sporothrix schenckii*. J Chemother. 2015;27:87–93.
56. Han HS, Toh PY, Yoong HB, Loh HM, Tan LL, Ng YY. Canine and feline cutaneous screwworm myiasis in Malaysia: clinical aspects in 76 cases. Vet Dermatol. 2018;29:442–e148.
57. Schubach TM, Schubach A, Okamoto T, Barros MB, Figueiredo FB, Cuzzi T, et al. Evaluation of an epidemic of sporotrichosis in cats: 347 cases (1998–2001). J Am Vet Med Assoc. 2004;224(10):623–9.
58. Raskin RE, Meyer DJ. Skin and subcutaneous tissue. In: Canine and feline cytology. 2nd ed. St. Louis: Saunders Elsevier; 2010. p. 41–4.
59. Pereira SA, Menezes RC, Gremião ID, Silva JN, Honse Cde O, Figueiredo FB, et al. Sensitivity of cytopathological examination in the diagnosis of feline sporotrichosis. J Feline Med Surg. 2011;13:220–3.
60. Jessica N, Sonia RL, Rodrigo C, Isabella DF, Tânia MP, Jeferson C, et al. Diagnostic accuracy assessment of cytopathological examination of feline sporotrichosis. Med Mycol. 2015;53(8):880–4.
61. Gross TL, Ihrke PJ, Walder EJ, et al. Infectious nodular and diffuse granulomatous and pyogranulomatous diseases of the dermis. In: Skin disease of the dog and cat. 2nd ed. Oxford: Blackwell Science; 2005. p. 298–301.
62. Fernandes GF, Lopes-Bezerra LM, Bernardes-Engemann AR, Schubach TM, Dias MA, Pereira SA, et al. Serodiagnosis of sporotrichosis infection in cats by enzyme-linked immunosorbent assay using a specific antigen, SsCBF, and crude exoantigens. Vet Microbiol. 2011;147:445–9.
63. Kano R, Watanabe K, Murakami M, Yanai T, Hasegawa A. Molecular diagnosis of feline sporotrichosis. Vet Rec. 2005;156:484–5.
64. Reis EG, Gremião ID, Kitada AA, Rocha RF, Castro VP, Barros ML, et al. Potassium iodide capsule treatment of feline sporotrichosis. J Fel Med Surg. 2012;14:399–404.
65. Reis ÉG, Schubach TM, Pereira SA, Silva JN, Carvalho BW, Quintana MB, et al. Association of itraconazole and potassium iodide in the treatment of feline sporotrichosis: a prospective study. Med Mycol. 2016;54:684–90.
66. Liang C, Shan Q, Zhang J, Li W, Zhang X, Wang J, et al. Pharmacokinetics and bioavailability of itraconazole oral solution in cats. J Fel Med Surg. 2016;18:310–4.

67. Ottonelli Stopiglia CD, Magagnin CM, Castrillón MR, Mendes SD, Heidrich D, Valente P, et al. Antifungal susceptibility and identification of *Sporothrix schenckii* complex isolated in Brazil. Med Mycol. 2014;52:56–64.
68. Borba-Santos LP, Rodrigues AM, Gagini TB, Fernandes GF, Castro R, de Camargo ZP, et al. Susceptibility of *Sporothrix brasiliensis* isolates to amphotericin B, azoles, and terbinafine. Med Mycol. 2015;53:178–88.
69. Han HS. The current status of feline sporotrichosis in Malaysia. Med Mycol J. 2017;58E:E107–13.
70. Francesconi G, Valle AC, Passos S, Reis R, Galhardo MC. Terbinafine (250mg/day): an effective and safe treatment of cutaneous sporotrichosis. J Eur Acad Dermatol Venereol. 2009;23:1273–6.
71. Vettorato R, Heidrich D, Fraga F, Ribeiro AC, Pagani DM, Timotheo C, et al. Sporotrichosis by *Sporothrix schenckii sensu stricto* with itraconazole resistance and terbinafine sensitivity observed in vitro and in vivo: case report. Med Mycol Case Reports. 2018;19:18–20.
72. Almeida-Paes R, Figueiredo-Carvalho MHG, Brito-Santos F, Almeida-Silva F, Oliveira MME, Zancopé-Oliveira RM. Melanins protect *Sporothrix brasiliensis* and *Sporothrix schenckii* from the antifungal effects of terbinafine. PLoS One. 2016;11:e0152796. https://doi.org/10.1371/journal.pone.0152796.
73. De Souza CP, Lucas R, Ramadinha RH, Pires TB. Cryosurgery in association with itraconazole for the treatment of feline sporotrichosis. J Feline Med Surg. 2016;18:137–43.

Malassezia

Michelle L. Piccione and Karen A. Moriello

Abstract

Malassezia dermatitis/overgrowth is a superficial fungal (yeast) skin disease of cats. It has most often been reported in association with underlying hypersensitivity skin diseases, metabolic diseases, neoplasia and paraneoplastic syndromes. Common clinical signs include dark waxy debris associated with otitis externa, scaling, black waxy nail bed debris (paronychia), variable pruritus, erythema, and exudative dermatitis especially when complicated by bacterial pyoderma. The disease is most commonly diagnosed by cytological examination of the skin. *Malassezia pachydermatis* is the primary species isolated from cats; however, other lipid-dependent species can be isolated. Itraconazole is the treatment of choice along with topical antifungal shampoo therapy or leave-on antifungal products. Recurrent *Malassezia* dermatitis is a clinical sign of an underlying trigger, most of which are not life threatening. In cats with severe widespread disease, especially those with erythema, alopecia and/or marked scaling, *Malassezia* species overgrowth could be a clinical sign of systemic disease warranting a thorough systemic evaluation.

Introduction

Malassezia are yeast organisms and are part of the normal cutaneous microflora of humans and animals, including cats [1]. At the time of writing, at least 16 different human and animal species have been isolated. A review of the literature revealed a wide range of species isolated from cats but molecular diagnostics are reclassifying

M. L. Piccione (✉) · K. A. Moriello
School of Veterinary Medicine, University of Wisconsin-Madison, Madison, WI, USA
e-mail: mpiccione@wisc.edu; karen.moriello@wisc.edu

© Springer Nature Switzerland AG 2020   345
C. Noli, S. Colombo (eds.), *Feline Dermatology*,
https://doi.org/10.1007/978-3-030-29836-4_16

some of these species [2]. Several recent studies reconfirmed that the most commonly isolated species from cats are *M. pachydermatis, M. furfur, M. nana,* and *M. sympodialis* [3–6].

For the purposes of this chapter, the term "*Malassezia* dermatitis" and "*Malassezia* overgrowth" are synonymous and the former will be used for simplicity. *Malassezia* spp. dermatitis is increasingly being recognized as a complicating factor in many feline skin diseases, often in association with bacterial overgrowth. The goal of this chapter is to review the scientific literature available on *Malassezia* dermatitis in cats and summarize the key aspects of clinical signs, diagnosis, and treatment.

Etiology and Pathogenesis of Feline *Malassezia*

Biological Characteristics

The genus *Malassezia* are lipophilic yeast, are part of the cutaneous microflora of warm-blooded animals and tend to colonize skin rich in sebaceous glands. *Malassezia* belong to the basidiomycetous yeasts. They are characterized by a multilayered cell wall and reproduce by unilateral budding [7]. Bottle-shaped yeast may be globose, ovoid, or cylindrical. Buds form on a narrow or wide base [7]. Currently, the genus *Malassezia* includes 16 species, of which 15 are lipid-dependent and are most frequently recovered from humans, ruminants, and horses (*Malassezia furfur, M. globosa, M. obtusa, M. restricta, M. slooffiae, M. sympodialis, M. dermatis, M. nana, M. japonica, M. yamatoensis, M. equina, M. caprae* and *M. cuniculi, M. brasiliensis, M. psittaci*) [8]. The only non-lipid-dependent species, *M. pachydermatis,* is commonly recovered from cats and dogs [8]. With the exception of *M. pachydermatis,* the lipophilic yeast requires supplementation of long chain fatty acids in culture medium; utilizing the lipids is a source of carbon for survival [8].

Pathogenesis

M. pachydermatis is considered a nonpathogenic, commensal organism that can become an opportunistic pathogen when the environmental factors are appropriate and/or the host's defense mechanism fails. Factors involved in maintaining the balance of skin microflora include temperature, hydration, chemical constituents (sweat, sebum, and saliva) and pH [9]. When these factors are altered, *Malassezia* can overgrow and act as a pathogen on the skin of cats, inducing an inflammatory response. *M. pachydermatis* has been shown to adhere to human keratinocyte cells and keratinocytes respond by releasing pro-inflammatory mediators as a defense mechanism [10]. Humoral and cell-mediated responses to *Malassezia* have been documented and anything that interferes with or blunts these responses can result in overgrowth [11].

Prevalence

There are many studies investigating the carriage of *Malassezia* in healthy cats, cats with skin disease, in specific cat breeds, on various skin sites and in association with other diseases. It is important to keep in mind that reported studies used different methodologies (i.e., culture, cytology, combination of culture and cytology) making direct comparisons difficult.

In one study comparing 10 domestic short-haired (DSH) cats with no history of skin disease (controls) and 32 Sphynx cats, *Malassezia* was not isolated from the skin of control cats [12]. In Sphynx cats, it was isolated from 26 of 32 cats of which 5 were reported to have greasy skin (Fig. 1). There were 73 isolates of *Malassezia*, of which 69 were *M. pachydermatis*. Interestingly, *Malassezia* was not isolated from the ears of any of the 42 cats. In another study, carriage was compared between several groups of cats: 10 normal DSH, 33 Cornish Rex cats (5 normal, 28 with seborrheic skin disease), and 30 Devon Rex cats (21 normal and 9 with seborrheic skin disease) [13]. *Malassezia* was isolated in 5 of 10 normal cats, 5 of 15 Cornish Rex cats, and 27 of 30 Devon Rex cats. When normal and diseased cat data was pooled, *M. pachydermatis* was isolated from 70% of cats with seborrheic skin disease and only 17% of cats with normal skin. In this study, 121 of 141 isolates were *M. pachydermatis*.

The prevalence of *Malassezia* in the ear canal of cats is an area of interest, since otitis is a common problem in cats (Chapter, Otitis) (Fig. 2). In one study, *Malassezia* species were isolated from the ear canal of 63 of 99 (63.6%) cats with otitis and 12 of 52 (23%) normal cats [14]. In this study, *M. pachydermatis, M. globosa, and M. furfur* were the most common isolates. In another study, *Malassezia* was isolated from 9 of 17 cats with otitis externa and in 16 of 51 cats without otitis [15]. Again, *M. pachydermatis, M. globosa,* and *M. furfur* were the most common isolates. In

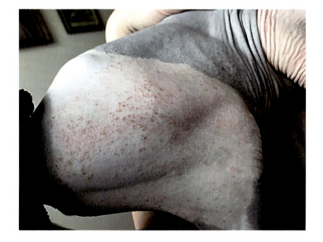

Fig. 1 Sphynx cat with seborrheic dermatitis and yeast overgrowth. This cat was very pruritic and yeast organisms were found on cytology from the papular eruption shown. The cat was diagnosed with environmental allergies

Fig. 2 Ear of cat with
yeast otitis. This is the
pinna of a cat with
hypersensitivity dermatitis.
The cat has extremely
pruritic ears and it
responded well to topical
steroid treatment

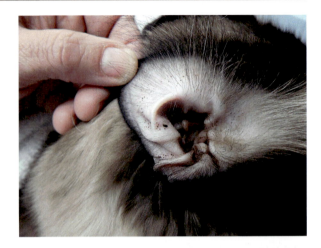

Fig. 3 Nail fold of a cat
with *Malassezia* dermatitis.
The owner reported that
this cat licked and chewed
at the paws and nail fold
area. The lesions resolved
with topical and systemic
antifungal treatment. The
cat had concurrent diabetes
mellitus

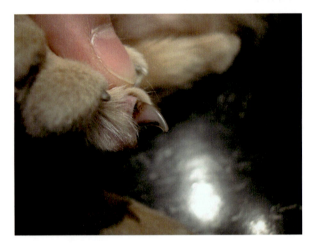

yet another study involving normal and affected cats, *Malassezia* was isolated in 7
of 25 and 15 of 20 cats [16]. Again, *M. pachydermatis* and *M. sympodialis* were the
most common isolates.

Another interesting site that has been investigated in cats is the nail fold (Fig. 3).
In one study, yeast were isolated from the claw fold of 26 of 29 Devon Rex cats [17].
In another study, *Malassezia* was found in 28 of 46 nail fold samplings from cats
[3]. Yeast were found in all 15 Devon Rex cats, 10 DSH cats, and 3 Persian cats.
Malassezia are also commonly found in the nail folds of Sphynx cats [4].

The facial fold of Persian cats has also been investigated for *Malassezia* dermatitis [18]. This breed is well known to have facial fold dermatitis which is often
idiopathic in origin. In one clinical case series, 13 Persian cats with idiopathic facial
fold dermatitis were investigated and *Malassezia* dermatitis was found in 6 of 13
cats. There was an incomplete response to treatment, suggesting *Malassezia* was
more of a complicating factor than a cause.

Prevalence studies show some common trends. First, *Malassezia* can be found on healthy cats but it is not common. Carriage is more common in cat breeds with genetically associated follicular dysplasia (Devon Rex, Cornish Rex, and Sphynx cat breeds). Interestingly, although Cornish Rex and Devon Rex cats share similar coat characteristics, the frequency and population of *Malassezia* isolated are different. The degree of colonization may be associated with the Devon Rex cat's predisposition to development of seborrheic dermatitis. *Malassezia* can be isolated from cats with and without otitis externa and from the nail folds of cats, particularly those with seborrheic or allergic skin disease. Finally, *M. pachydermatis* is the most common *Malassezia* isolate from cats.

Malassezia and Concurrent Diseases

Malassezia overgrowth/dermatitis is a common complication of skin diseases in other species and a similar picture is starting to emerge in cats.

Hypersensitivity skin disease is common in cats and the role of *Malassezia* dermatitis is increasingly being recognized (Fig. 4). In one study of 18 cats with hypersensitivity dermatitis, *Malassezia* dermatitis was found in all cats [19]. Sixteen cats showed marked reduction in pruritus after treatment. This suggests *Malassezia* overgrowth may be a contributing factor in some cats with hypersensitivity dermatitis. Not all cats with hypersensitivity skin disease have *Malassezia* dermatitis. A molecular study on fungal microbiota of allergic cats ($n = 8$) found *Malassezia* in only 21% of 54 samples from the 8 allergic cats [20].

A study of aural microflora in healthy cats ($n = 20$) compared with allergic cats ($n = 15$) and cats with systemic disease ($n = 15$) found that *Malassezia* colonization was more common in cats with hypersensitivity dermatitis and systemically ill cats compared to healthy cats [21]. In another study, *Malassezia* was more commonly isolated from retroviral positive cats than retroviral negative cats [22]. Although the cats were healthy, possibly retroviral infections interfered with the innate immune response. When the frequency of isolation of *Malassezia* on the skin of cats with

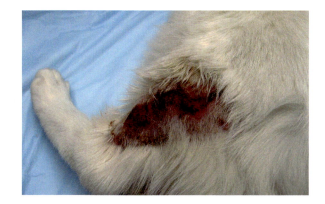

Fig. 4 This cat had hypersensitivity dermatitis and recurrent areas of eosinophilic dermatitis. Cytology revealed concurrent bacterial and *Malassezia* dermatitis. Lesions resolved with concurrent antibacterial and antifungal therapy

diabetes mellitus ($n = 16$), hyperthyroidism ($n = 20$, and neoplasia were compared ($n = 8$) to normal cats ($n = 10$), no difference was found [23].

There is increasing evidence that *Malassezia* dermatitis may be associated with paraneoplastic alopecia and/or be a cutaneous sign of systemic disease. In the above study, *Malassezia* was isolated from 9 sites in one cat with feline paraneoplastic syndrome and pancreatic adenocarcinoma [23]. In another case report, marked exfoliative dermatitis and yeast overgrowth was found in a cat with thymoma (Figs. 5 and 6) and, interestingly, there was complete resolution of clinical signs after complete surgical tumor resection [24]. One case report described a 13-year-old DSH cat with a history of progressively worsening paraneoplastic alopecia along with *Malassezia* overgrowth. Post-mortem results revealed a pancreatic adenocarcinoma with hepatic metastases [25]. In a retrospective study evaluating skin biopsy specimens from cats, 15 specimens contained large numbers of *Malassezia* organisms

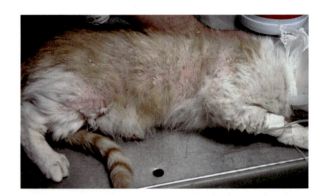

Fig. 5 This is a 13-year-old cat that presented with marked exfoliative dermatitis with severe *Malassezia* dermatitis found on cytology. The cat was systemically ill. Imaging revealed a thymoma

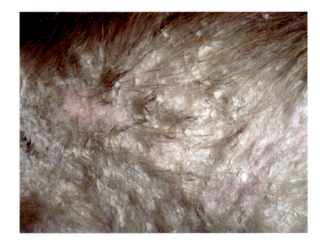

Fig. 6 This is a close-up view of the marked exfoliation on the skin of the cat in Fig. 5. Note marked erythema and large sheets of shed keratinocytes. This appearance of scales is highly suggestive of feline exfoliative dermatitis due to an underlying medical problem which may or may not be associated with a thymoma

in the epidermis or follicular infundibulum [26]. When clinical data was evaluated, 11 of 15 cats had acute onset of multifocal to generalized skin lesions. All 10 cats were euthanized and one died of metastatic carcinoma of the liver 2 months after the onset of clinical signs.

Clinical Signs

There are no pathognomonic clinical signs for *Malassezia* dermatitis in cats. Clinical signs reported and/or commonly noted by the authors are summarized in Box 1 and chapter images. Concurrent bacterial overgrowth is common. Scaling and an unkempt coat (Fig. 8) is a common finding and often *Malassezia* is found on cytology. Many cats with *Malassezia* dermatitis due to poor grooming will respond to coat hygiene and topical therapy alone. Nail bed involvement may vary in clinical appearance, it is usually brown black (Fig. 3) and may appear as marked seborrheic accumulations.

Box 1: Clinical Signs of *Malassezia* Dermatitis/Overgrowth
- Lesions can be generalized or localized
- Pruritus varies from none to marked
- Erythema
- Diffuse seborrhea that can be dry and/or oily
- Increased scaling
- Hairs pierced by scales and follicular casts
- Traumatic alopecia
- Hyperpigmentation characterized by brown waxy exudate
- Follicular plugging on ventral abdomen, especially around nipples
- Brown to reddish brown discoloration of nails
- Waxy debris under the nail folds
- Brown waxy debris adherent to lip folds
- Chin acne

Malassezia otitis

- Increased ceruminous debris
- Erythema of ear canal
- Swelling and narrowing of canal
- Pruritus of pinna and/or canal

Fig. 7 Cat with
Malassezia and bacterial
overgrowth. Note the
erythema, eruptions, and
scaling

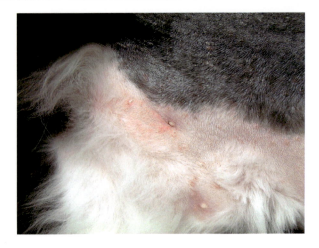

Fig. 8 *Malassezia*
dermatitis in a cat with an
unkempt coat

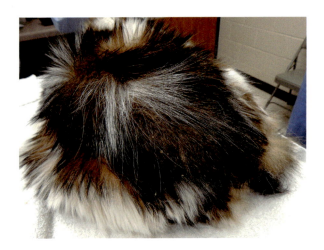

Diagnosis

The diagnosis of *Malassezia* dermatitis is based the identification of organism in light of a compatible/plausible history and clinical signs and a good response to antifungal treatment.

Cytology

Cytological examination of the skin is the single best technique for investigating whether or not *Malassezia* are present. There are no cytological criteria for determining the "normal number" of *Malassezia* organisms present on the skin of cats. The presence or absence of organisms can only be interpreted in light of the cat's clinical signs. The organisms are much larger than bacteria and can vary in size from 2–4 μm by 3 to 7 μm.

Skin cytology samples can be obtained using a clear acetate tape which allows for sampling in difficult areas, e.g., facial, interdigital, or nail folds. A clear piece of tape is pressed to the skin, the tape is then stained using in-house cytology stains (e.g., Diff Quik). It is important to avoid the fixative step and to stain the tape by holding with forceps, tweezers, or the authors' favorite tool, household clothes pins. Affixing unstained tape to a glass slide and then staining the slide results in poor staining and increased artifacts, and should be avoided. To make a proper preparation, put a drop of immersion oil on a glass slide, then mount a thoroughly dry stained piece of tape over the oil, and then examine it microscopically. For oil immersion, (recommended), a drop of immersion oil can be placed directly over the tape. Glass slide samples are the optimum tool to use when sampling the skin. To obtain the best possible sample, place the glass slide over the target area, gently lift the skin and squeeze the skin between two fingers. This will markedly increase the cellularity of the sample. Ears are best sampled with a cotton tip applicator. Nail beds are best sampled by gently scraping debris from under the claw fold using a skin scraping spatula, not a scalpel blade, and then smearing it onto a glass slide. In all cases, it is important to NOT heat fix slides as this will cause artifacts and/or damage other cells on the slide. It has been shown that increasing the number of dips in solution II (basophilic, blue) is all that is needed to improve visualization of yeast organisms [27, 28]. The authors routinely examine slides at 4× to find a cellular area, 10× and then 100×. *Malassezia* organisms are variable in size and may be seen easily on the slide or adherent to skin keratinocytes (Figs. 9 and 10).

Fig. 9 Note the large number of *Malassezia* organisms in this ear cytology. There are peanut- and ovoid-shaped organisms. Some of the organisms have not stained as deeply basophilic and this is common in samples from ear cytology

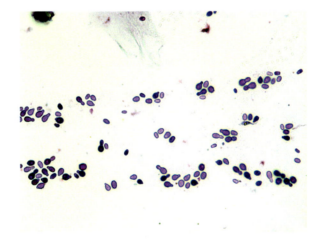

Fig. 10 Note the large number of *Malassezia* organisms adherent to skin cells. This sample was obtained from a cat with exfoliative dermatitis. Note the concurrent bacteria in this sample

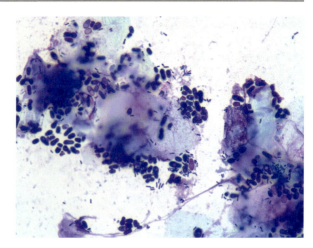

Fungal Culture

It is rare to need to culture this organism for diagnosis in clinical cases. If there is a need, e.g., suspected antifungal resistance, for research or if there is need to identify the species, there are two important things to remember. First, if using a culture swab to culture the skin, moisten the swab with the transport medium and aggressively rub the swab over a large area of the skin while rotating the head of the swab 360°. In the authors' experience, dry swab cultures of small areas are inadequate. If available, contact plate cultures are preferable. Secondly, it is common to isolate several different species from cats. It is important to inform the laboratory that both lipid-independent and lipid-dependent species are of clinical interest. *M. pachydermatis* is unique in that it grows well in Sabouraud dextrose agar at 32 °C to 37 °C without lipid supplementation; however, lipid-dependent *Malassezia* species will not grow on Sabouraud dextrose agar. Modified Dixon Agar and Leeming medium support growth of all *Malassezia* species. Due to the presence of lipid-dependent yeasts on the skin of cats, the use of lipid-supplemented media, especially the modified Dixon's medium or Leeming medium, is required [8, 29, 30].

PCR

PCR is not used in the routine diagnosis of *Malassezia* dermatitis in cats. Culture-based methods do not always allow species-specific identification and, if this is necessary, polymerase chain reaction (PCR) is a viable option. PCR uses laboratory methods that amplify DNA from a sample, even directly from skin or from culture with high accuracy and efficiency [31]. Recent findings show that the multiplex-real-time PCR was highly effective in identifying *Malassezia* species from animal and human samples [32].

Histopathology and Skin Biopsy

Skin biopsy is not routinely used to diagnose *Malassezia* dermatitis in cats. If the cat is otherwise healthy and yeast are noted on the skin biopsy, their presence is most likely due to the underlying skin disease, e.g., hypersensitivity disease or primary disorder of keratinization. However, if the cat is ill and has marked skin lesions, the presence of *Malassezia* organism should be interpreted as a sign of systemic illness and a thorough medical evaluation pursued. Histological sections reporting *Malassezia* yeasts often note their presence in the stratum corneum of the epidermis or follicular infundibulum [26]. In cases of severe exfoliation, they may be reported in areas of mild to severe orthokeratotic to parakeratotic hyperkeratosis [26].

Treatment

Treatment of *Malassezia* dermatitis in cats is individualized and depends on the severity of clinical signs and potential underlying cause. If the underlying cause is not identified and treated, *Malassezia* dermatitis will not resolve. If the underlying disease is chronic, e.g., hypersensitivity dermatitis, the owner should be warned that disease flares will cause relapses of *Malassezia* dermatitis.

Topical Therapy

The major obstacle to topical therapy for *Malassezia* dermatitis in cats is what the cat and owner can tolerate. Ideally, topical therapy is the treatment of choice. Attention to coat hygiene is important if there is matting or retained hairs. If the cat will tolerate bathing, the topical shampoos of choice are miconazole/chlorhexidine, ketoconazole/chlorhexidine, or climbazole/chlorhexidine combinations once or twice weekly. The authors have found it very helpful for owners to understand that *Malassezia* dermatitis is commonly associated with bacterial overgrowth, so combination products are the best choice. If the cat is otherwise healthy but has generalized lesions, whole body bathing is recommended. Given that this may not be possible, other options include the use of leave-on mousse products with the above ingredients. If lesions are focal, these combination products can be applied to just the affected areas. It is important to remember that grooming activities of cats put them at greater risk of adverse reactions to topical products.

Systemic Antifungal Therapy

Oral antifungal therapy is indicated if the topical therapy is impractical or ineffective. The oral antifungal of choice is itraconazole (Itrafungol, Elanco Animal Health). It is labelled for use in cats at 5 mg/kg orally once daily on an alternating week on/week off treatment schedule for dermatophytosis [33]. Itraconazole is generally well tolerated in cats and safe at this dose. The most common side effects reported were hypersalivation, decreased appetite, vomiting and diarrhea [33]. It is important to stress to clients that compounded itraconazole should not be used as there is strong evidence that it is not bioavailable [34].

The efficacy of oral itraconazole for treatment of *Malassezia* dermatitis was reported in two studies. In a retrospective study, 15 cats received 5 to 10 mg/kg itraconazole (Itrafungol/Janssen), administered orally once daily as the sole therapy [35]. Affected cats had either localized ($n = 8$) or generalized lesions ($n = 7$). Twelve of the cats had concurrent otitis externa. Itraconazole was effective in all cats with no reported side effects. In another study, pulse therapy (week on/week off) itraconazole was used to treat *Malassezia* dermatitis in 6 Devon Rex cats with concurrent seborrheic dermatitis [36]. There was a marked improvement in clinical signs with a reduction of inflammation and pruritus.

Yeast Otitis

Malassezia is a common cause of otitis externa in cats (Chapter, Otitis). It is most common in cats with hypersensitivity dermatitis (Fig. 2). Immediate treatment may include systemic itraconazole if there are large numbers of yeasts present and pruritus is severe. However, in most cases *Malassezia* otitis can be managed with weekly ear cleaning and topical application of an otic antifungal/glucocorticoid product. Long-term management of *Malassezia* otitis externa can be done successfully with once or twice weekly application of otic steroids. The authors frequently compound equal portions of injectable dexamethasone ear drops in saline or propylene glycol for the owner to apply. This avoids the unnecessary use of antimicrobials when the primary need is just an anti-inflammatory action.

Zoonotic Implications

Malassezia organisms are found on both people and animals. *Malassezia pachydermatis* is not a normal commensal organism. However, since cats can be colonized by both lipid-independent and lipid-dependent organisms, it is important for veterinary health care workers to practice good hand hygiene when handling cats and to remind owners to do the same.

Conclusion

Feline *Malassezia* dermatitis is a superficial fungal skin disease that can present with a wide range of clinical signs. Clinical signs are caused by overgrowth of normal body flora, and concurrent overgrowth of bacteria is common. *Malassezia* dermatitis is commonly associated with chronic skin diseases such as hypersensitivity disorders, seborrhea, and underlying metabolic diseases that can trigger changes in the skin immune system. Devon Rex cats appear particularly susceptible to both *Malassezia* colonization and *M. pachydermatis* associated seborrheic dermatitis, without evidence of systemic disease. Cytological examination is the most useful technique for assessment of the density of *Malassezia* yeasts on the skin surface. Additionally, contact-plate fungal cultures also provide a convenient technique for isolation and quantification of yeast colonies. PCR allows for rapid identification and speciation of samples analyzed as well. *M. pachydermatis* is the main species identified in cats, but lipid-dependent species, particularly in the claw folds, can also be found. Itraconazole is the systemic drug of choice along with concurrent topical therapy. Although rare, the finding of *Malassezia* dermatitis in cats with widespread skin lesions should prompt the clinician to consider whether or not this is an early marker of systemic disease.

References

1. Theelen B, Cafarchia C, Gaitanis G, et al. *Malassezia* ecology, pathophysiology, and treatment. Med Mycol. 2018;56:S10–25.
2. Cabañes FJ. *Malassezia* yeasts: how many species infect humans and animals? PLoS Pathog. 2014;10:e1003892.
3. Colombo S, Nardoni S, Cornegliani L, et al. Prevalence of *Malassezia spp.* yeasts in feline nail folds: a cytological and mycological study. Vet Dermatol. 2007;18:278–83.
4. Volk AV, Belyavin CE, Varjonen K, et al. *Malassezia pachydermatis* and *M nana* predominate amongst the cutaneous mycobiota of Sphynx cats. J Feline Med Surg. 2010;12:917–22.
5. Bond R, Howell S, Haywood P, et al. Isolation of *Malassezia sympodialis* and *Malassezia globosa* from healthy pet cats. Vet Rec. 1997;141:200–1.
6. Crespo M, Abarca M, Cabanes F. Otitis externa associated with *Malassezia sympodialis* in two cats. J Clin Microbiol. 2000;38:1263–6.
7. Guillot J, Gueho E, Lesord M, et al. Identification of *Malassezia* species: a practical approach. J Mycol Med. 1996;6:103–10.
8. Böhmová E, Čonková E, Sihelská Z, et al. Diagnostics of *Malassezia* Species: a review. Folia Vet. 2018;62:19–29.
9. Tai-An C, Hill PB. The biology of *Malassezia* organisms and their ability to induce immune responses and skin disease. Vet Dermatol. 2005;16:4–26.
10. Buommino E, De Filippis A, Parisi A, et al. Innate immune response in human keratinocytes infected by a feline isolate of *Malassezia pachydermatis*. Vet Microbiol. 2013;163:90–6.
11. Sparber F, LeibundGut-Landmann S. Host responses to *Malassezia spp.* in the mammalian skin. Front Immunol. 2017;8:1614.
12. Åhman SE, Bergström KE. Cutaneous carriage of *Malassezia* species in healthy and seborrhoeic Sphynx cats and a comparison to carriage in Devon Rex cats. J Feline Med Surg. 2009;11:970–6.
13. Bond R, Stevens K, Perrins N, et al. Carriage of *Malassezia spp.* yeasts in Cornish Rex, Devon Rex and Domestic short-haired cats: a cross-sectional survey. Vet Dermatol. 2008;19:299–304.
14. Nardoni S, Mancianti F, Rum A, et al. Isolation of *Malassezia* species from healthy cats and cats with otitis. J Feline Med Surg. 2005;7:141–5.
15. Crespo M, Abarca M, Cabanes F. Occurrence of *Malassezia spp.* in the external ear canals of dogs and cats with and without otitis externa. Med Mycol. 2002;40:115–21.
16. Dizotti C, Coutinho S. Isolation of *Malassezia pachydermatis* and *M. sympodialis* from the external ear canal of cats with and without otitis externa. Acta Vet Hung. 2007;55:471–7.

17. Åhman S, Perrins N, Bond R. Carriage of *Malassezia* spp. yeasts in healthy and seborrhoeic Devon Rex cats. Sabouraudia. 2007;45:449–55.
18. Bond R, Curtis C, Ferguson E, et al. An idiopathic facial dermatitis of Persian cats. Vet Dermatol. 2000;11:35–41.
19. Ordeix L, Galeotti F, Scarampella F, et al. *Malassezia spp.* overgrowth in allergic cats. Vet Dermatol. 2007;18:316–23.
20. Meason-Smith C, Diesel A, Patterson AP, et al. Characterization of the cutaneous mycobiota in healthy and allergic cats using next generation sequencing. Vet Dermatol. 2017;28:71–e17.
21. Pressanti C, Drouet C, Cadiergues M-C. Comparative study of aural microflora in healthy cats, allergic cats and cats with systemic disease. J Feline Med Surg. 2014;16:992–6.
22. Sierra P, Guillot J, Jacob H, et al. Fungal flora on cutaneous and mucosal surfaces of cats infected with feline immunodeficiency virus or feline leukemia virus. Am J Vet Res. 2000;61:158–61.
23. Perrins N, Gaudiano F, Bond R. Carriage of *Malassezia spp.* yeasts in cats with diabetes mellitus, hyperthyroidism and neoplasia. Med Mycol. 2007;45:541–6.
24. Hljfftee Ma M-V, Curtis C, White R. Resolution of exfoliative dermatitis and *Malassezia pachydermatis* overgrowth in a cat after surgical thymoma resection. J Small Anim Pract. 1997;38:451–4.
25. Godfrey D. A case of feline paraneoplastic alopecia with secondary *Malassezia* associated dermatitis. J Small Anim Pract. 1998;39:394–6.
26. Mauldin EA, Morris DO, Goldschmidt MH. Retrospective study: the presence of *Malassezia* in feline skin biopsies. A clinicopathological study. Vet Dermatol. 2002;13:7–14.
27. Toma S, Cornegliani L, Persico P, et al. Comparison of 4 fixation and staining methods for the cytologic evaluation of ear canals with clinical evidence of ceruminous otitis externa. Vet Clin Pathol. 2006;35:194–8.
28. Griffin JS, Scott D, Erb H. *Malassezia* otitis externa in the dog: the effect of heat-fixing otic exudate for cytological analysis. J Veterinary Med Ser A. 2007;54:424–7.
29. Guillot J, Bond R. *Malassezia pachydermatis*: a review. Med Mycol. 1999;37:295–306.
30. Peano A, Pasquetti M, Tizzani P, et al. Methodological issues in antifungal susceptibility testing of *Malassezia pachydermatis*. J Fungi. 2017;3:37.
31. Vuran E, Karaarslan A, Karasartova D, et al. Identification of *Malassezia* species from pityriasis versicolor lesions with a new multiplex PCR method. Mycopathologia. 2014;177:41–9.
32. Ilahi A, Hadrich I, Neji S, et al. Real-time PCR identification of six *Malassezia* species. Curr Microbiol. 2017;74:671–7.
33. Puls C, Johnson A, Young K, et al. Efficacy of itraconazole oral solution using an alternating-week pulse therapy regimen for treatment of cats with experimental *Microsporum canis* infection. J Feline Med Surg. 2018;20:869–74.
34. Mawby DI, Whittemore JC, Fowler LE, et al. Comparison of absorption characteristics of oral reference and compounded itraconazole formulations in healthy cats. J Am Vet Med Assoc. 2018;252:195–200.
35. Bensignor E. Treatment of *Malassezia* overgrowth with itraconazole in 15 cats. Vet Rec. 2010;167:1011–2.
36. Åhman S, Perrins N, Bond R. Treatment of *Malassezia pachydermatis*-associated seborrhoeic dermatitis in Devon Rex cats with itraconazole–a pilot study. Vet Dermatol. 2007;18:171–4.

Viral Diseases

John S. Munday and Sylvie Wilhelm

Abstract

Viruses are becoming increasingly recognized as an important cause of feline skin disease. Diseases associated with viruses in cats include hyperplastic and neoplastic skin disease caused by papillomaviruses, erosive and ulcerative skin disease caused by herpesviruses and poxviruses, and skin lesions that develop as a part of a more generalized viral infection as is seen due to calicivirus infection. Skin disease may also be seen in cats infected by feline leukemia virus and feline infectious peritonitis virus. In this chapter, the etiology and epidemiology of infection by each of the viruses that cause feline skin disease are reviewed along with the clinical disease presentation, the histological lesions, and other appropriate diagnostic tests. Additionally, the expected clinical course of the diseases and the currently recommended therapies are described.

Introduction

Viruses have traditionally been thought to rarely cause skin disease in cats. However, research in the last 30 years has expanded both the number of viruses that cause feline skin disease and the types of skin lesions caused by these viral infections. Feline viral infections can be broadly subdivided into those that cause hyperplastic or neoplastic skin disease (papillomaviruses), those that cause cell lysis and generally self-resolving inflammatory disease (herpesvirus, poxvirus), and those that infrequently cause skin lesions as part of a more generalized viral infection

J. S. Munday (✉)
Massey University, Palmerston North, New Zealand
e-mail: j.munday@massey.ac.nz

S. Wilhelm
Vet Dermatology GmbH, Richterswil, Switzerland

© Springer Nature Switzerland AG 2020
C. Noli, S. Colombo (eds.), *Feline Dermatology*,
https://doi.org/10.1007/978-3-030-29836-4_17

(calicivirus, feline leukemia virus, feline infectious peritonitis virus). While feline immunodeficiency virus is briefly discussed, it is currently uncertain whether or not this virus causes skin disease in cats.

Papillomaviruses

Papillomaviruses (PVs) are small, non-enveloped, circular double-stranded DNA viruses that typically infect stratified squamous epithelium. Their DNA contains seven open reading frames (ORFs), including five that code for the early (E) proteins and two that code for the late (L) proteins [1]. Their life cycle is dependent on the terminal differentiation, keratinization, and desquamation of epithelial cells, and feline PVs cause disease due to the ability of their E7 proteins to alter the normal growth and differentiation of these cells. Papillomaviruses are considered one of the oldest viral families and have co-evolved with their hosts over a long time. For this reason, the majority of PVs are species-specific and the overwhelming majority of PV infections are asymptomatic [2].

Papillomaviruses are classified by comparing the similarities of the L1 ORF [3]. Currently five PV types are recognized to infect cats, including *Felis catus* papillomavirus (FcaPV) type 1, which is classified in the *Lambdapapillomavirus* genus [4, 5]; FcaPV-2, which is classified in the *Dyothetapapillomavirus* genus [6]; and FcaPV-3, -4 and -5, which have not been fully classified, but will likely be grouped together in a novel feline PV genus [7–9].

Although most PV infections are asymptomatic, PVs were first proposed to be a cause of feline skin disease in 1990, when PV-induced cell changes were observed in a cutaneous plaque [10]. Since this time, the importance of PVs as a cause of skin disease has been increasingly recognized and PVs are currently thought to cause viral plaques/Bowenoid in situ carcinomas, a proportion of squamous cell carcinomas, feline sarcoids, a proportion of basal cell carcinomas and cutaneous viral papillomas [11].

Feline Viral Plaques/Bowenoid In Situ Carcinoma

These lesions have traditionally been thought of as two separate skin diseases of cats. However, as viral plaques and Bowenoid in situ carcinomas (BISCs) share many histological features and transitional lesions between the two lesions are often visible [12], it appears these two lesions are different severities of the same process.

Etiology and Epidemiology

FcaPV-2 is thought to be the predominant cause of these lesions [13, 14]. Current evidence suggests most kittens are infected from the dam within the first few weeks of life [15]. Infection by FcaPV-2 is probably lifelong and often does not stimulate an antibody response [16]. As most cats are infected by FcaPV-2, but few develop viral plaques/BISCs, it appears that host factors are important in determining whether or not a cat will develop clinical disease. While immunosuppressed cats may be at increased risk of viral plaque/BISC development, many cats have been reported to develop lesions without any detectable immunosuppression, and the factors that predispose a cat to lesion development are largely unknown [17]. The early development and severe manifestation of viral plaques/BISCs in Devon Rex and Sphinx cats suggests a genetic susceptibility, although the basis of this susceptibility is unknown [18]. Viral plaques/BISCs have also been associated with infection by FcaPV-3 and FcaPV-5. Currently little is known about the epidemiology of these viruses.

Clinical Presentation
Viral plaques/BISCs most often develop between the ages of eight and 14 years, although they have been reported in cats as young as 7 months of age [12, 19]. Cats with viral plaques tend to be younger than those with BISCs, supporting the hypothesis that some viral plaques progress to BISCs. Viral plaques most often develop on the trunk, head, or neck although in advanced cases lesions can develop anywhere on the body. They are often multiple and small, generally less than 1 cm in diameter, scaly papules or plaques that may be either pigmented or nonpigmented and can be covered by thin crusts (Fig. 1). While BISCs can appear clinically very similar to viral plaques, they tend to be larger, more markedly raised, and can be ulcerated or covered by a serocellular crust or a thick layer of keratin (Fig. 2). The head, neck, and limbs are most commonly affected. Viral plaques and BISCs can develop within pigmented or nonpigmented, haired or nonhaired skin and neither lesion is typically painful or pruritic [12].

Histopathology and Diagnosis
Histology of a viral plaque reveals a well-demarcated focus of mild epidermal hyperplasia. Cells retain their orderly maturation and no dysplasia is visible (Fig. 3). Histology of a BISC reveals a well-demarcated focus of marked epidermal

Fig. 1 Feline viral plaque. Plaques most frequently appear as focal, slightly raised lesions around the face of cats. Feline viral plaques and Bowenoid in situ carcinomas appear to be different severities of the same disease process with viral plaques the milder form of the disease. (Courtesy of Dr. Sharon Marshall, Veterinary Associates, Hastings, New Zealand)

Fig. 2 Feline Bowenoid in situ carcinoma. As with viral plaques, these often develop on the head of cats. Compared to a viral plaque, Bowenoid in situ carcinomas are larger, more markedly raised, and covered by increased quantities of keratin. However, as viral plaques and Bowenoid in situ carcinomas represent different severities of the same disease process and there is no clear distinction between the two lesions. (Courtesy of Dr. Richard Malik, Centre for Veterinary Education, University of Sydney, Australia)

Fig. 3 Feline viral plaque. Plaques appear as well-demarcated foci of mild to moderate epidermal hyperplasia. Little dysplasia is visible within the hyperplastic cells and orderly maturation of the cells is retained (HE, 200×)

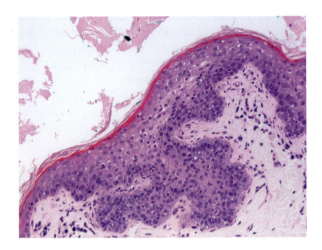

hyperplasia that can extend to involve follicular infundibula. The hyperplastic cells can form well-demarcated solid masses of basilar cells that bulge into the underlying dermis. Examination of deeper layers of the BISC reveals keratinocyte dysplasia with crowding of basal cells and cells with nuclei that are elongated vertically (windblown cells) [20]. Dyskeratosis is rarely visible within BISCs. Although significant atypia can be present, the cells remain confined by the basement membrane (Fig. 4). Viral replication can result in prominent PV-induced changes. However, keratinocyte dysplasia can prevent viral replication and PV-induced cell changes

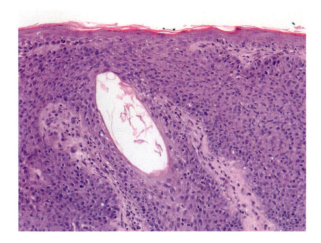

Fig. 4 Feline Bowenoid in situ carcinoma. Compared to the viral plaque, the hyperplasia is more marked with prominent involvement of follicular infundibula. There is moderate atypia within the cell population, but no penetration of the basement membrane. While papillomavirus-induced cell changes are prominent in this lesion, more advanced Bowenoid in situ carcinomas often contain little histological evidence of papillomavirus infection. (HE, 200×)

Fig. 5 Feline viral plaque. Papillomavirus-induced cell changes include the presence of keratinocytes that have dark nuclei surrounded by a clear halo (koilocytes; arrows) as well as the presence of cells that contain increased quantities of grey-blue smudged cytoplasm (arrowheads; HE, 400×)

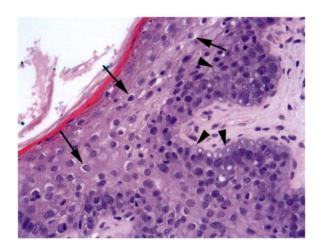

are rare in larger more developed BISCs. PV-induced changes include the presence of large keratinocytes with clear or blue-grey granular cytoplasm and/or shrunken nuclei that are surrounded by a clear halo (koilocytes; Fig. 5) [17]. Eosinophilic intranuclear inclusions can be visible, although care has to be taken to differentiate these from nucleoli. Hyperplasia of cells deeper within follicles or hyperplasia of sebaceous glands may be visible in viral plaques/BISCs that are caused by FcaPV-3 or -5. Additionally, these lesions contain prominent basophilic cytoplasmic inclusions that are often flattened against the nucleus [8, 21]. If no PV-induced changes are visible, then differentiation from actinic in situ carcinoma (actinic keratosis) is required. Features that support a BISC rather than an actinic lesion include the consistently altered nuclear polarity of the basal cells, the sharp demarcation between affected and normal epidermis, and the follicular involvement. In addition, actinic lesions will also often have solar elastosis visible in the underlying dermis.

Immunohistochemistry can be used in cases in which histological differentiation between a Bowenoid and an actinic in situ carcinoma is problematic. Antibodies to detect PV antigen can be used. However, antigens are only produced during viral replication and it is rare to have immunohistochemical evidence of PV infection in a lesion that does not contain PV-induced cell changes [17]. Therefore, p16^{CDK2NA} protein (p16) immunohistochemistry is recommended to investigate a PV etiology. The detection of a marked increase in p16 suggests a PV etiology because PVs cause cell dysregulation by mechanisms that consistently increase p16 (Fig. 6). In contrast, in actinic lesions, loss of cell regulation is caused by mechanisms that do not increase p16 [22]. When performing p16 immunohistochemistry it has to be remembered that only the G175-405 human p16 clone has been shown to cross-react with the feline p16 protein. As p53 immunostaining can be present in both actinic keratosis and BISCs, this will not be useful to differentiate between a Bowenoid and an actinic lesion [22]. Due to the frequency with which PVs asymptomatically infect skin, the detection of PV DNA in a lesion does not confirm a diagnosis of BISC or exclude a diagnosis of actinic in situ carcinoma.

Treatment

Viral plaques and BISCs can spontaneously resolve, persist without progressing, or slowly increase in size and number. In addition, all viral plaques/BISCs should be carefully monitored for progression to a SCC. Lesions in Devon Rex and Sphinx cats can rapidly progress to SCCs that have metastatic potential [18, 23].

Surgical excision of a viral plaque or BISC is expected to be curative, although additional lesions may subsequently develop at different locations. Imiquimod cream has been used to treat genital warts in people and has been suggested as a

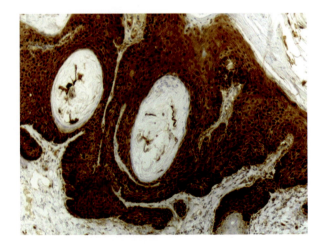

Fig. 6 Feline viral plaque. The use of antibodies against p16^{CDKN2A} protein reveals intense nuclear and cytoplasmic immunostaining throughout the hyperplastic epidermis (Hematoxylin counterstain 400×)

possible treatment. Imiquimod stimulates toll-like receptors and locally increases alpha interferon and tumor necrosis factor-α [24]. It is a topical therapy, usually applied three times per week for 8–16 weeks. In an uncontrolled study of 12 cats with BISCs, imiquimod resulted in partial resolution of at least one BISC in all 12 cats and complete remission of at least one BISC in 5 cats [25]. Side effects included local erythema and mild discomfort in five cats and potential signs of systemic toxicity, including neutropenia, elevated hepatic enzymes, anorexia, and weight loss were observed in two cats. While there is anecdotal evidence supporting the use of imiquimod cream, additional controlled studies are required to determine the efficacy and safety of this treatment. In humans, imiquimod has also been used to treat basal cell carcinomas and actinic lesions and this treatment does not appear to have a specific action against PV-induced lesions. Imiquimod is currently not recommended as a primary treatment for pre-neoplastic or neoplastic skin lesions in people, but may be effective if better established therapies are not available [26]. Likewise, in veterinary medicine, imiquimod has been used to treat viral and non-viral in situ carcinomas when other treatments were considered impractical, and investigation of a PV etiology may not be necessary prior to the use of imiquimod. Photodynamic therapy could be another treatment option as excellent response rates were recently reported, although there was no attempt in this study to differentiate between PV-induced and actinic in situ carcinomas [27].

Autologous vaccination has not been evaluated as a method of treating viral plaques/BISCs in cats. However, considering the immune response to a PV-induced lesion, this treatment modality is not expected to work. There is currently little evidence from any species that vaccination using autologous or virus-like particle vaccines has any significant efficacy in treating either PV-induced warts or pre-neoplastic lesions.

Cutaneous Squamous Cell Carcinomas

Squamous cell carcinomas (SCCs) are one of the most common skin neoplasms of cats and are a significant cause of morbidity and mortality (Chapter, Genetic Diseases for more information). While there can be no doubt that solar exposure is a significant cause of SCCs, there is evidence that PVs may also contribute to the development of some neoplasms. Evidence of a role of PVs includes the detection of FcaPV-2 DNA more frequently in cutaneous SCCs than in non-SCC skin samples [13]. Additionally, p16 immunostaining is visible within SCCs that contain PV DNA (Fig. 7) and SCCs that have p16 immunostaining demonstrate a different biological behavior, suggesting that they may have been caused by different carcinogenic pathways [28, 29]. Furthermore, FcaPV-2 RNA can be detected in a proportion of feline cutaneous SCCs and the proteins that are expressed by FcaPV-2 have been shown to have transforming properties in cell cultures [30, 31]. Overall current evidence suggests that PV infection causes most SCCs that develop in haired, pigmented skin and that PV infection, probably with UV light as a co-factor, could promote the development of between a third and a half of SCCs from nonhaired, nonpigmented

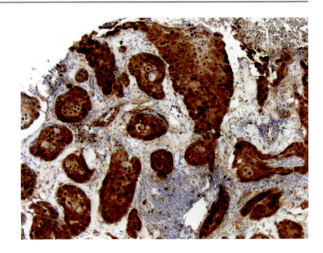

Fig. 7 Cutaneous squamous cell carcinoma. Immunostaining for p16^{CDKN2A} protein is diffusely present within the nucleus and cytoplasm of the neoplastic cells. Papillomavirus DNA was amplified from this neoplasm using PCR (Hematoxylin counterstain 200×)

skin [29]. However, as asymptomatic infection of the skin is extremely common in cats, it is currently impossible to definitively determine that role that FcaPV-2 plays in the development of cutaneous SCCs in cats.

Feline Sarcoids

Feline sarcoids are rare neoplasms in cats. They have also been called "fibropapillomas"; however, as fibropapillomas are considered hyperplastic rather than neoplastic lesions, the term "sarcoid" is preferred.

Etiology and Epidemiology

Bovine papillomavirus (BPV) type 14 has been consistently detected in feline sarcoids throughout the world [32–34], and infection by BPV-14 is thus considered to be the cause of this disease. BPV-14 is a *Deltapapillomavirus* that is most closely related to BPV-1 and -2, the causes of equine sarcoids [35]. The bovine deltapapillomaviruses have the unique ability to cause both self-resolving fibropapillomas in cattle and mesenchymal neoplasia in non-host species. Cows are commonly asymptomatically infected by BPV-14 [36], but BPV-14 was not detected in a large number of cutaneous and oral samples from cats [32]. This suggests that cats are probably dead-end hosts for the PV. It is currently unknown how BPV-14 is transmitted from cattle to cats. However, as this disease appears to be most common in cats that live in dairy barns, close contact with cattle appears to be necessary. It is also unknown whether any co-factors are required to allow BPV-14 to cause sarcoids. Evidence from horses suggests that mesenchymal cell proliferation may be important for equine sarcoid development and it is possible that cat fight wounds could be important in allowing introduction of the PV into the dermis and stimulating dermal mesenchymal proliferation.

Fig. 8 Feline sarcoid. The mass protrudesi from close to the nasal philtrum of this cat. (Courtesy of Dr. William Miller, Cornell University College of Veterinary Medicine, Ithaca, New York)

Clinical Presentation

Feline sarcoids have only been reported in outdoor cats from rural environments and are most common in younger male cats. They develop as solitary or multiple slow-growing exophytic nonulcerated nodules most frequently around the face, especially involving the nasal philtrum and upper lip, although sarcoids have also been reported in distal limbs and tail (Fig. 8) [33]. There is some evidence to suggest feline sarcoids may also rarely develop within the oral cavity.

Histopathology and Diagnosis

A feline sarcoid should be suspected if an exophytic mass is observed around the mouth or nose of a young cat that has contact with cattle. Unlike typical PV infections of the skin, infection by the PV is confined to the dermis [34]. Therefore, the predominant histological feature of a sarcoid is a proliferation of moderately well-differentiated mesenchymal cells within the dermis (Fig. 9). The proliferative dermal mass is covered by hyperplastic epidermis that extends into the mesenchymal cells by the formation of prominent rete pegs [33, 34]. As the sarcoid does not support viral replication, sarcoids do not contain any PV-induced cell changes and immunohistochemistry will not reveal the presence of PV L1 antigen [34]. The amplification of BPV-14 DNA from the lesion confirms a diagnosis of feline sarcoid.

Treatment

While there are few clinical reports of feline sarcoids, these neoplasms appear to be locally infiltrative, but do not metastasize. In the authors' experience, complete surgical excision is curative. However, as these lesions often develop around the nose and mouth, complete excision can be problematic and feline sarcoids often recur and show an increased growth rate after surgery. A cat with recurrent sarcoids was treated with topical imiquimod and intralesional cisplatin, but neither treatment appeared to alter the disease course and the cat was eventually euthanatized due to the local effects of the neoplasm [35].

Fig. 9 Feline sarcoid. The neoplasm consists of moderately well-differentiated fibroblasts that are covered by hyperplastic epidermis that forms prominent rete pegs (HE, 200×)

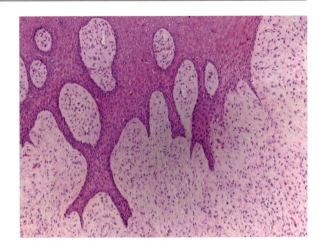

Basal Cell Carcinomas

These are rare neoplasms and only a limited number have been assessed for a PV etiology. However, a potential role of PVs in the development of feline basal cell carcinomas (BCCs) is supported by the observation that a proportion contain PV-induced changes [20, 37, 38]. Feline BCCs have not been reported to be caused by FcaPV-2. Instead feline BCCs have been associated with FcaPV-3 and a novel unclassified PV type [37, 38].

Cutaneous Papillomas

In cats, FcaPV-1 causes oral papillomas that typically develop on the ventral surface of the tongue [39]. There are also sporadic reports of cutaneous viral papillomas. While cross-species infection by a human PV was originally suspected [40], this appears unlikely and spread of FcaPV-1 from the mouth to the skin of cats appears to be a more likely cause of these rare lesions.

Herpesviruses

Feline herpesvirus 1 is a double-stranded DNA *Alphaherpesvirus* that is a common cause of upper respiratory tract disease and conjunctivitis in younger cats. In 1971, it was reported that herpesvirus infection may also cause dermatitis in cats [41], and feline herpesvirus dermatitis is now recognized as a distinct, albeit rare, manifestation of herpesvirus infection.

Etiology and Epidemiology

The rate of feline herpesvirus 1 infection is difficult to determine as many cats are vaccinated against this virus at an early age. Infection of an unvaccinated cat typically results in clinical signs of upper respiratory disease, such as rhinotracheitis, and conjunctivitis. While the clinical signs usually resolve within a few days or weeks, the herpesvirus infection can become latent, especially in the trigeminal ganglia. These latent infections can become recrudescent if the cat becomes immunosuppressed. It is hypothesized that recrudescence of previous herpesviral infections within the cutaneous nerves could cause feline herpesvirus dermatitis [42]. Due to the likely role of immunosuppression in the pathogenesis of disease, cats receiving glucocorticoids may be predisposed to disease development [42]. Cats that are in a household with numerous other cats also appear to be at increased risk, although it is uncertain whether this is because the cats are immunosuppressed due to stress or because the cats are more likely to be exposed to herpesvirus [42]. Feline herpesvirus dermatitis has not been associated with infection by feline immunodeficiency virus or feline leukemia virus. While previous infection by herpesvirus is thought to be key in the pathogenesis of this disease, herpesvirus dermatitis has been reported in cats that have a good vaccination history and in cats that do not have a history of previous respiratory disease [43].

Clinical Presentation

Herpesvirus dermatitis appears to be most common in cats around 5 years of age although this disease has been reported in cats 4 months to 17 years old [42, 44]. Most cats with herpesvirus dermatitis have lesions almost exclusively on the face with the dorsal muzzle to the bridge of the nose and periocular skin most frequently affected (Fig. 10). The lips can also be affected and, in rare cases, the lesions can become generalized over the body within a few days [42, 43, 45]. Lesions are typically erosions and ulcers that are covered by a thick serocellular crust and have been referred to as "ulcerative facial and nasal dermatitis and stomatitis syndrome." The lesions tend to be roughly spherical and are often asymmetrical, but the development of symmetrical lesions does not exclude a herpesviral etiology. Regional lymphadenopathy can be present [45]. Oral lesions are only rarely reported in cats

Fig. 10 Feline herpesviral dermatitis. This disease typically presents as multiple ulcerative lesions over the face, especially around the bridge of the nose. (Courtesy of Dr. Richard Malik, Centre for Veterinary Education, University of Sydney, Australia)

with herpesvirus dermatitis [45], and affected cats may or may not have active or historical evidence of respiratory disease. The skin lesions can be intensely pruritic, thus mimicking a wide range of possible differential diagnoses, especially in the absence of respiratory signs of disease. Depending on the localization of the lesions, differential diagnoses include allergic skin disease, calicivirus-associated dermatitis, autoimmune skin diseases, and erythema multiforme.

Exfoliative erythema multiforme is a rare disease that has been reported to develop following infection by herpesvirus. Clinical signs include widespread scaling (exfoliation) in combination with alopecia. Accompanying systemic symptoms are possible and the lesions spontaneously resolve after clearance of the herpesvirus infection [46].

Histopathology and Diagnosis

Histology of a lesion reveals full-thickness necrosis and loss of the epidermis. Underlying the areas of necrosis there are typically large numbers of inflammatory cells including a high proportion of eosinophils (Fig. 11). Necrosis of the epithelium can extend into the underlying follicular infundibula and adnexal glands. The epidermis adjacent to the areas of ulceration can be thickened and spongiotic. The lesions are covered by a marked serocellular crust that consists of degenerate inflammatory cells and fibrin. Careful examination of the intact epidermis adjacent to the necrosis, the follicles, and the adnexal glands may reveal the rare presence of intranuclear viral inclusions (Fig. 12). These are eosinophilic and surrounded by marginated nuclear material. Making a definitive diagnosis should not be problematic when inclusion bodies are present. However, in cases that do not have visible inclusions, additional techniques may be necessary. The most conclusive evidence supporting a diagnosis is the demonstration of herpesviral antigens within the lesions using immunohistochemistry [44]. The failure to amplify herpesviral DNA

Fig. 11 Feline herpesviral dermatitis. The dermis contains large numbers of inflammatory cells including a large proportion of eosinophils (HE, 200×)

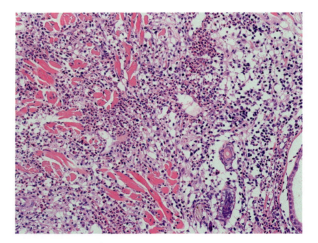

Fig. 12 Feline herpesviral dermatitis. The epidermis adjacent to areas of ulceration is thickened and spongiotic with some cells containing eosinophilic intranuclear viral inclusions (arrows; HE, 400×)

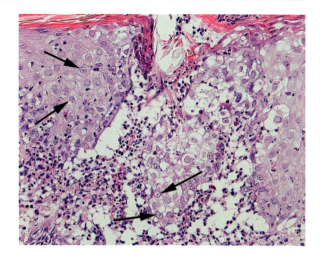

from a lesion using PCR excludes a diagnosis of herpesviral dermatitis. However, as DNA from latent or vaccinal herpesvirus infection or contamination from infected mucosa from grooming can be detected by PCR, this technique cannot be used to confirm a diagnosis of herpesviral dermatitis [47, 48]. Persico et al recently recommended that PCR can be used as a screening test for cases in which no viral inclusions are present, but immunohistochemistry was required to confirm a diagnosis of herpesviral dermatitis [48].

Treatment

Herpesvirus dermatitis may resolve spontaneously, although as few untreated cats are described in the literature, the frequency of self-cure is uncertain. In some cats, supportive care such as treatment of secondary bacterial infection may result in resolution of the clinical signs of disease [43]. As immunosuppression could contribute to disease development, any immunosuppressive treatments should be discontinued. Small lesions can be surgically excised, although whether or not the lesions would have spontaneously resolved if they had been left is unknown [42]. While herpesviral skin disease in humans typically spontaneously resolves, numerous treatments have been developed to accelerate disease resolution. Some of these antiviral drugs may also be beneficial in cats, but they generally have complicated pharmacokinetics that may render them ineffective or toxic in cats and none has consistently been found to be safe and effective [49, 50]. Currently, famciclovir has the greatest amount of evidence supporting efficacy, both in naturally infected cats and in a placebo-controlled study of experimentally infected cats. The used dosages vary from 40 to 90 mg/kg once or twice daily to 125 mg/kg every 8 hrs [51, 52]. Topical "cold sore" creams, especially those containing pencivovir may also be beneficial and can be used with systemic famciclovir therapy (R. Malik, pers. comm). Interferons (IFNs), including IFNα and recombinant IFNω [53], have also been suggested as potential treatments, although no controlled trials have been undertaken to assess their efficacy. Doses for IFNα again vary widely from $1MU/m^2$ subcutaneously three times a week to $0.01–1$ MU/kg once daily for up to 3 weeks. Most often 30 units/day have been used [54]. The effectiveness of lysine supplementation is highly controversial and no clinical benefit has been proven [50, 55].

Poxvirus

Poxviruses are large enveloped brick-shaped or oval linear double-stranded DNA viruses. Their DNA is 130–360 kb in length and encodes 130–320 proteins [56]. Most poxviruses are able to infect multiple species, and infection typically causes skin lesions due to the tropism of the viruses for epidermal keratinocytes. Skin disease due to poxviruses is rare in cats. Self-regressing proliferative skin disease due to infection with orf virus and pseudocowpox virus (both *Parapoxviruses*) has been sporadically reported in cats [57, 58]. However, the overwhelming majority of poxviral feline skin disease is caused by cowpox virus, an *Orthopoxvirus*, and the remainder of this section describes disease due to cowpox virus infection. In addition to the role of cowpox virus in diseases of cats, it also has to be noted that cats are important as a source of cowpox infection of humans.

Etiology and Epidemiology

Cowpox virus is a poor name for this virus as the reservoir hosts appear to be small mammals such as bank voles, short-tailed field voles, ground squirrels, and gerbils, with cattle, humans, and cats all only rarely infected [59]. The limited geographic distribution of the reservoir hosts explains why this disease is restricted to western Eurasia with the majority of clinical cases reported in the United Kingdom and Germany [59–61].

As cats are infected from a reservoir host during hunting, cowpox is limited to cats that have access to a rural environment, and increased numbers of cases are seen in autumn, due to increased hunting and greater numbers of reservoir host prey at this time [60, 62]. Skin disease is thought to develop after a reservoir host bites the cat. Systemic disease has also been reported, although it is uncertain if this results from inhalation of the virus or from systemic spread of the virus from an initial skin infection.

While disease due to cowpox virus is rare in cats, exposure to the virus appears to be much more common, with antibodies against orthopoxviruses detected in 2–17% of cats from Western Europe [60, 63, 64]. Unsurprisingly, rates of exposure were highest in populations of cats that had outside access and from areas in which clinical cases had been reported [60]. As behavioral factors that predispose to cowpox exposure also predispose to infection with feline immunodeficiency virus (FIV), it is possibly unsurprising that cats with antibodies against orthopoxvirus were more likely to also have antibodies against FIV [64].

Clinical Presentation

Lesions develop in younger to middle-aged cats that are able to come into contact with a reservoir host species [65]. As the lesions are initiated by a rodent bite, they typically start around the head or on the forelimbs. Lesions can subsequently spread to ears and paws, possibly by grooming and some cats can develop widespread lesions [62, 66].

The initial lesion is typically a single small raised ulcer covered by a serocellular crust at the site of inoculation by the rodent bite [62]. One to 3 weeks later, additional similar lesions may develop. These start as macules and small nodules that enlarge up to 1 cm in diameter (Fig. 13). They then become ulcerated, forming the typical crater-like skin lesion. These scab over and then gradually dry and exfoliate within 4–5 weeks. Pruritus is variable [67]. Up to 20% of cats may develop oral lesions, presumably due to the cat licking the skin lesions [66]. Cats that present with larger areas of dermal necrosis with extensive erythema, edema, abscessation, and cellulitis have also been reported (Fig. 14) [67, 68]. Whether this more severe presentation represents infection by a more virulent strain is uncertain. Cats may be transiently pyrexic and depressed during the viremic phase 1–3 weeks after infection, but most do not demonstrate signs of systemic disease on presentation [54, 66]. Rarely, cats may progress to develop signs of respiratory disease that can progress to a fatal pneumonia, although skin lesions are only variably present in cats that develop the respiratory form of cowpox disease.

Pruritic lesions require differentiation from allergic skin diseases. Other differentials for nodular skin diseases in cats include infection by fungi or higher-order bacteria as well as neoplastic skin disease.

Fig. 13 Feline cowpox dermatitis. This cat presented with numerous raised nodules on the forelimbs

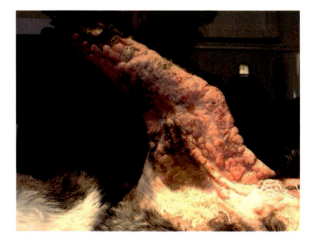

Fig. 14 Feline cowpox dermatitis. The lesions in this cat progressed to erythema, edema, and cellulitis involving the paw

Fig. 15 Feline cowpox dermatitis. Examination of the dermis reveals necrosis accompanied by large numbers of neutrophils. Epidermal cells are visible scattered within the inflammation. These cells show evidence of ballooning degeneration and some have prominent eosinophilic intracytoplasmic viral inclusions (arrows; HE, 200×)

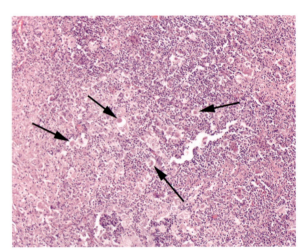

Histopathology and Diagnosis

Due to the nonspecific nature of the skin lesions observed, biopsy and histology are required for diagnosis. Examination of a lesion reveals necrosis of the epidermis with ulceration. Examination of adjacent epidermis and within the epithelium of the follicles will usually reveal marked ballooning degeneration (Fig. 15). The ballooning change within these cells often demonstrates the presence of prominent intracytoplasmic poxvirus inclusions. These inclusions are eosinophilic and round to oval. The lesions are covered by a serocellular crust and the underlying dermis often contains significant neutrophilic inflammation. Serology or electron microscopy can be used to confirm orthopoxvirus infection, but not the orthopoxvirus type. Immunohistochemistry can be performed using monoclonal antibodies specific for cowpox virus [69]. Alternatively, virus isolation from a fresh biopsy or scab material or amplification of viral DNA using PCR will enable a precise diagnosis to be made [70].

Treatment

Feline cowpox dermatitis has a good prognosis with most cases resolving spontaneously [66]. In cases with extensive or numerous lesions, supportive care may be required including the treatment of secondary bacterial infections. However, even cats with severe skin lesions most often make a full recovery, although scarring can occur [67]. The detection of serum antibodies against FIV does not influence the prognosis [66].

Why some cats develop respiratory disease is unknown. Treatment with immunosuppressive doses of corticosteroids is contraindicated as this may predispose to respiratory disease. While fatal pneumonia has been reported to develop in cats that were also infected by feline leukemia virus, feline immunodeficiency virus, or feline panleukopenia virus [66, 69], the role of these concurrent infections is uncertain and no underlying immunosuppression can be identified in most cases [61]. Cats with respiratory disease have a guarded prognosis and no specific treatments have been shown to be beneficial. Treatment with broad spectrum antibiotics to prevent secondary bacterial infection appears to be appropriate. Additionally, four cats with respiratory cowpox were treated with recombinant feline interferon omega. While two of the cats survived, it is currently impossible to determine whether interferon therapy is effective in the treatment of cowpox in cats [61].

The zoonotic potential of cowpox is an important consideration when formulating a treatment plan and infected cats are thought to cause around 50% of human cowpox infections [71]. Cowpox infection of people generally results in a transient, focal ulcerated lesion. However, life-threatening systemic cowpox virus infections can occur, particularly in immunosuppressed individuals [72], and in-clinic treatment may be most appropriate for cats owned by immunosuppressed clients.

Calicivirus

Feline calicivirus is a non-enveloped icosahedral single-stranded RNA virus that is classified in the *Vesivirus* genus. Feline calicivirus is well recognized to cause upper respiratory disease and oral ulcers in cats. Rarely, infection with caliciviruses can also cause skin lesions.

Etiology and Epidemiology

Although only one calicivirus infects cats, these viruses are generally quick to mutate and different viral strains can express different antigens and show marked differences in virulence [73]. Feline calicivirus commonly infects cats, it is transferred via direct contact from infected cats and is shed in ocular, nasal, and oral secretions. Calicivirus is one of the most common viral pathogens of cats worldwide [73]. Skin lesions can develop in association with the more typical upper respiratory calicivirus infections. However, generalized severe skin lesions are generally restricted to cats that develop virulent systemic disease. This rare presentation of caliciviral disease usually develops as an outbreak, presumably as multiple cats are exposed to a recently developed virulent calicivirus strain. Outbreaks of virulent systemic disease have been reported involving veterinary hospitals and animal shelters in North America and Europe, although considering the widespread distribution of this virus, outbreaks in other countries appear likely to occur [74]. Calicivirus vaccines may reduce the severity of clinical signs of disease, but do not appear to be protective against infection by highly virulent calicivirus strains [75, 76].

Clinical Presentation

Acute, nonvirulent, feline calicivirus infection is usually characterized by transient, self-limiting vesiculo-ulcerative lesions in the oral cavity (typically affecting the tongue), on the lips and nasal philtrum. Rarely ulceration can be detected on other body regions. Systemic signs, such as fever, depression, sneezing, conjunctivitis, oculonasal discharge, and arthropathy resulting in lameness that resolves in a few days (transient febrile limping syndrome) can also be observed [73].

Virulent systemic disease has been reported in cats from 8 weeks to 16 years of age, although adult cats may be more susceptible [74, 77, 78]. Cats that present with virulent systemic disease are unwell with fever, anorexia, lethargy, weakness, jaundice, or bloody diarrhea. Many cats will have oral ulcers [77]. Skin lesions include edema of the limbs and face with alopecia, ulcers and crusting lesions on the head (especially the lips, muzzle, and ears), and paw pads [73, 77]. Skin lesions have been reported less frequently on the abdomen and around the anus (Fig. 16) [76].

Two cats were reported to develop pustular skin disease due to calicivirus shortly after ovariohysterectomy. In these cats, lesions were restricted to the skin surrounding the wound. As both cats were also anorexic and depressed, it appears likely they both had systemic virulent disease. The presence of systemic disease was supported by the observation that one cat subsequently developed pleural effusion that necessitated euthanasia [79].

Fig. 16 Virulent systemic calicivirus infection. The cat has ulcers and crusting lesions on the ventral surface of the tail and surrounding the anus

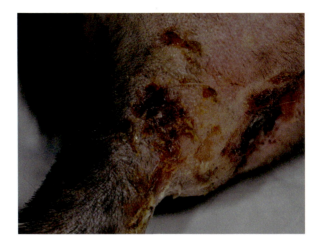

Fig. 17 Virulent systemic calicivirus infection. Histology reveals epidermal necrosis with ulceration. The ulcers are covered by a prominent serocellular crust (HE, 400×)

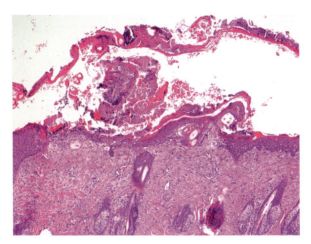

Histopathology and Diagnosis

Histology of the skin lesions reveals ballooning degeneration and necrosis of both surface and follicular epithelium with subsequent ulceration. Neutrophilic inflammation is often marked and ulcers are typically covered by a serocellular crust (Fig. 17) [77]. In cases in which the epidermis remains intact, intraepidermal and suprabasilar pustules may be visible [76]. Superficial dermal edema and vasculitis may also be detected.

Diagnosis of systemic virulent calicivirus infection is unlikely to be made solely on histological examination of skin samples. Instead, the skin histology will be interpreted along with the clinical evidence of severe systemic disease to make this diagnosis. A diagnosis of caliciviral dermatitis is supported by the demonstration of caliciviral antigens within the lesions using immunohistochemistry. The presence of viral nucleic acid can also be demonstrated using PCR, although since up to 30% of cats are carriers, the amplification of viral DNA should be interpreted with caution [54]. The detection of caliciviral nucleic acid from blood or lesions from a cat with clinical signs consistent with this disease is good evidence to support a diagnosis of caliciviral dermatitis [74]. Virus isolation and fluorescent antibody testing are also possible [73].

Treatment

Treatment is supportive and commercially available antivirals do not inhibit replication of caliciviruses [62]. The prognosis is good for the acute nonvirulent systemic disease cases. Controlling secondary infections, regular cleaning of discharges and mucolytic drugs (e.g., bromhexine), or nebulization with saline are helpful. Often cats do not eat due to the oral ulcers, and placement of a feeding tube and enteral nutrition may be necessary in severe cases [54].

Intensive supportive treatment is recommended for cats with virulent systemic calicivirus infection, but even with such treatments mortality rates of 30–60% have been reported [77]. A feline calicivirus-specific antiviral phosphorodiamidate morpholino oligomer (PMO) was developed and used to treat kittens in three outbreaks of severe calicivirus disease. In this trial, 47 of 59 kittens that were treated with the PMO survived, but only three of 31 untreated cats survived [80]. The success of this experimental targeted therapy suggests that newer methods of treating calicivirus infections in cats may become available in the future.

Feline Leukemia Virus

Feline leukemia virus (FeLV) is a retrovirus that is classified within the *Gammaretrovirus* genus. Infection is most commonly spread in saliva by mutual grooming, but can also be spread by bites and through the milk. The role of FeLV in the development of some lymphomas is well established. The role of this virus in the development of feline skin disease is less defined, but four cutaneous manifestations of FeLV have been proposed.

Immunosuppression

As FeLV can cause significant immunodeficiency [81], it is possible that infection could predispose to increased opportunistic skin infections. However, there is little direct evidence supporting an increase in skin disease in cats due to FeLV infection and cats with FeLV rarely, if ever, initially present at a veterinarian due to recurrent or difficult to treat opportunistic skin infections [81].

Giant Cell Dermatosis

This skin disease was first reported in six cats in 1994 and subsequently in an additional cat in 2005 [82, 83]. Evidence that the disease was caused by FeLV was the consistent detection of serum FeLV antigen and the presence of FeLV proteins in the lesions as demonstrated by immunohistochemistry. Interestingly, four of the cats had previously been vaccinated against FeLV and the authors speculated that some vaccines could contain infectious RNA that could result in the development of giant cell dermatosis [83]. To the authors' knowledge, no additional cases of feline giant cell dermatosis have been reported. This suggests that this is a rare manifestation of FeLV infection in cats. Additionally, due to the small numbers that have been reported, it is not possible to definitively confirm that FeLV causes giant cell dermatosis in cats.

Affected cats have been reported to present with variable clinical signs including multiple ulcers around the head, limbs, and paws [82], patchy alopecia and scaling starting on the dorsum, then progressing to become more generalized, or crusting skin lesions that were predominantly confined to the head and pinnae, but can also develop on the footpads and around the anus [83]. Pruritus was marked in many of the cats and concurrent gingivitis was frequently detected. Cats often presented with evidence of systemic disease, including pyrexia and anorexia.

This disease can only be diagnosed by histology. Histology of a lesion reveals epidermal hyperplasia with the presence of prominent multinucleate keratinocytes that can contain up to 30 nuclei. Giant cells can be present within the surface epidermis, sebaceous glands, or in the follicular infundibula [82, 83]. Disorganization and keratinocyte atypia may be present within the affected epidermis. Inflammation may be prominent within the underlying dermis, especially in cases in which secondary bacterial infections are present.

There are currently no treatments for this disease and all the cats that were reported to have this disease died shortly after diagnosis [82, 83].

Feline Paw Pad Cutaneous Horns

Affected cats develop multiple horn-like lesions that typically involve multiple pads on multiple digits (Fig. 18). While these were initially associated with FeLV [84], subsequent studies have identified cats with horns that were not infected by FeLV [85, 86]. Whether FeLV infection is strictly necessary for disease development is currently uncertain.

Cats present with multiple, elongated, conical or cylindrical masses involving the pads of multiple digits. The lesions consist almost entirely of keratin and so are typically grey, with a hard and dry texture. Histology reveals a well-demarcated column of dense pale orthokeratosis covering minimally to mildly hyperplastic epidermis (Fig. 19). Inflammation is typically minimal within the underlying dermis. Treatment is surgical excision although feline paw pad, and cutaneous horns often locally recur. As the lesions become bigger, they can develop fissures, which can result in secondary inflammation and pain.

Fig. 18 Feline paw pad cutaneous horn. Lesions are grey exophytic masses that often involve multiple paw pads

Fig. 19 Feline paw pad cutaneous horn. The horns consist of a column of dense orthokeratosis overlying comparatively normal epidermis (HE, 50×)

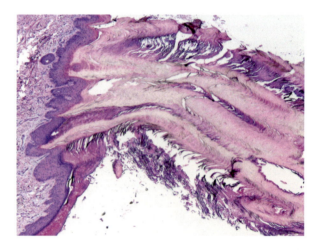

The location and appearance of these lesions typically allow clinical diagnosis to be made, although cutaneous horns that have developed secondary to Bowenoid in situ carcinomas or squamous cell carcinomas could require differentiation if they develop on the paw pad.

Cutaneous Lymphoma

Associations between FeLV and cutaneous lymphoma in cats have been inconsistently detected [87–89], and it currently appears that cats with FeLV infection are, at worst, at an only slightly increased risk of cutaneous lymphoma.

Feline Infectious Peritonitis Virus

Feline coronaviral-induced vasculitis (feline infectious peritonitis, FIP) rarely affects the skin. However, cats with FIP and skin lesions have sporadically been reported [90–92]. In all reported cases, the skin lesions developed late in the clinical course, in cats that also displayed more typical clinical signs such as pyrexia, lethargy, anorexia, or ocular lesions. The involvement of multiple cutaneous blood vessels results in the development of nonpruritic, nonpainful, raised papules over the neck and forelimbs or more generalized over the body. Histology reveals granulomatous vasculitis and immunohistochemistry can be used to confirm the presence of coronaviral antigens.

Feline Immunodeficiency Virus

Feline immunodeficiency virus (FIV) is a retrovirus within the *Lentivirus* genus. As the virus is typically spread by fighting, it is most common in free-roaming male cats. While experimental infection of cats can result in marked, fatal immunosuppression, natural infection of cats appears to be much less significant and the overall life-span of FIV-infected cats does not appear to be shorter than the life-span of uninfected cats [93].

Currently there is scant evidence that infection by FIV predisposes to feline skin disease [94]. While some initial cases of papillomaviral (PV) skin disease of cats were reported in cats with FIV, there have been no direct comparisons to determine if FIV-infected cats are disproportionately affected. Additionally, rates of PV infection were not higher in FIV-infected cats than in uninfected cats [95].

FIV has been associated with a mildly increased risk of lymphoma, although the precise role of the virus in neoplasm development is uncertain [96]. Cutaneous lymphoma has not been associated with FIV in cats. While an association between cutaneous SCCs and FIV infection was reported, this was suspected to be coincidental as higher rates of both SCCs and FIV infection are expected to be present in cats that spend significant amounts of time outside [97].

References

1. Munday JS, Pasavento P. Papillomaviridae and Polyomaviridae. In: NJ ML, Dubovi EJ, editors. Fenner's Veterinary Virology. 5th ed. London: Academic Press; 2017. p. 229–43.
2. Munday JS. Bovine and human papillomaviruses: a comparative review. Vet Pathol. 2014;51:1063–75.
3. de Villiers EM, Fauquet C, Broker TR, et al. Classification of papillomaviruses. Virology. 2004;324:17–27.
4. Tachezy R, Duson G, Rector A, et al. Cloning and genomic characterization of Felis domesticus papillomavirus type 1. Virology. 2002;301:313–21.

5. Terai M, Burk RD. Felis domesticus papillomavirus, isolated from a skin lesion, is related to canine oral papillomavirus and contains a 1.3 kb non-coding region between the E2 and L2 open reading frames. J Gen Virol. 2002;83:2303–7.
6. Lange CE, Tobler K, Markau T, et al. Sequence and classification of FdPV2, a papillomavirus isolated from feline Bowenoid in situ carcinomas. Vet Microbiol. 2009;137:60–5.
7. Dunowska M, Munday JS, Laurie RE, et al. Genomic characterisation of Felis catus papillomavirus 4, a novel papillomavirus detected in the oral cavity of a domestic cat. Virus Genes. 2014;48:111–9.
8. Munday JS, Dittmer KE, Thomson NA, et al. Genomic characterisation of Felis catus papillomavirus type 5 with proposed classification within a new papillomavirus genus. Vet Microbiol. 2017;207:50–5.
9. Munday JS, Dunowska M, Hills SF, et al. Genomic characterization of Felis catus papillomavirus-3: a novel papillomavirus detected in a feline Bowenoid in situ carcinoma. Vet Microbiol. 2013;165:319–25.
10. Carney HC, England JJ, Hodgin EC, et al. Papillomavirus infection of aged Persian cats. J Vet Diagn Investig. 1990;2:294–9.
11. Munday JS, Thomson NA, Luff JA. Papillomaviruses in dogs and cats. Vet J. 2017;225:23–31.
12. Wilhelm S, Degorce-Rubiales F, Godson D, et al. Clinical, histological and immunohistochemical study of feline viral plaques and bowenoid in situ carcinomas. Vet Dermatol. 2006;17:424–31.
13. Munday JS, Kiupel M, French AF, et al. Amplification of papillomaviral DNA sequences from a high proportion of feline cutaneous in situ and invasive squamous cell carcinomas using a nested polymerase chain reaction. Vet Dermatol. 2008;19:259–63.
14. Munday JS, Peters-Kennedy J. Consistent detection of Felis domesticus papillomavirus 2 DNA sequences within feline viral plaques. J Vet Diagn Investig. 2010;22:946–9.
15. Thomson NA, Dunowska M, Munday JS. The use of quantitative PCR to detect Felis catus papillomavirus type 2 DNA from a high proportion of queens and their kittens. Vet Microbiol. 2015;175:211–7.
16. Geisseler M, Lange CE, Favrot C, et al. Geno- and seroprevalence of Felis domesticus Papillomavirus type 2 (FdPV2) in dermatologically healthy cats. BMC Vet Res. 2016;12:147.
17. Munday JS. Papillomaviruses in felids. Vet J. 2014;199:340–7.
18. Ravens PA, Vogelnest LJ, Tong LJ, et al. Papillomavirus-associated multicentric squamous cell carcinoma in situ in a cat: an unusually extensive and progressive case with subsequent metastasis. Vet Dermatol. 2013;24:642–5, e161–642.
19. Sundberg JP, Van Ranst M, Montali R, et al. Feline papillomas and papillomaviruses. Vet Pathol. 2000;37:1–10.
20. Gross TL, Ihrke PJ, Walder EJ, et al. Skin diseases of the dog and cat: clinical and histopathologic diagnosis. 2nd ed. Oxford: Blackwell Science; 2005.
21. Munday JS, Fairley R, Atkinson K. The detection of Felis catus papillomavirus 3 DNA in a feline bowenoid in situ carcinoma with novel histologic features and benign clinical behavior. J Vet Diagn Investig. 2016;28:612–5.
22. Munday JS, Aberdein D. Loss of retinoblastoma protein, but not p53, is associated with the presence of papillomaviral DNA in feline viral plaques, Bowenoid in situ carcinomas, and squamous cell carcinomas. Vet Pathol. 2012;49:538–45.
23. Munday JS, Benfell MW, French A, et al. Bowenoid in situ carcinomas in two Devon Rex cats: evidence of unusually aggressive neoplasm behaviour in this breed and detection of papillomaviral gene expression in primary and metastatic lesions. Vet Dermatol. 2016;27:215–e255.
24. Miller RL, Gerster JF, Owens ML, et al. Imiquimod applied topically: a novel immune response modifier and new class of drug. Int J Immunopharmacol. 1999;21:1–14.
25. Gill VL, Bergman PJ, Baer KE, et al. Use of imiquimod 5% cream (Aldara) in cats with multicentric squamous cell carcinoma in situ: 12 cases (2002–2005). Vet Comp Oncol. 2008;6:55–64.
26. Love W, Bernhard JD, Bordeaux JS. Topical imiquimod or fluorouracil therapy for basal and squamous cell carcinoma: a systematic review. Arch Dermatol. 2009;145:1431–8.

27. Flickinger I, Gasymova E, Dietiker-Moretti S, et al. Evaluation of long-term outcome and prognostic factors of feline squamous cell carcinomas treated with photodynamic therapy using liposomal phosphorylated meta-tetra(hydroxylphenyl)chlorine. J Feline Med Surg 2018:1098612X17752196.

28. Munday JS, French AF, Gibson IR, et al. The presence of p16 CDKN2A protein immunostaining within feline nasal planum squamous cell carcinomas is associated with an increased survival time and the presence of papillomaviral DNA. Vet Pathol. 2013;50:269–73.

29. Munday JS, Gibson I, French AF. Papillomaviral DNA and increased p16CDKN2A protein are frequently present within feline cutaneous squamous cell carcinomas in ultraviolet-protected skin. Vet Dermatol. 2011;22:360–6.

30. Altamura G, Corteggio A, Pacini L, et al. Transforming properties of Felis catus papillomavirus type 2 E6 and E7 putative oncogenes in vitro and their transcriptional activity in feline squamous cell carcinoma in vivo. Virology. 2016;496:1–8.

31. Thomson NA, Munday JS, Dittmer KE. Frequent detection of transcriptionally active Felis catus papillomavirus 2 in feline cutaneous squamous cell carcinomas. J Gen Virol. 2016;97:1189–97.

32. Munday JS, Knight CG, Howe L. The same papillomavirus is present in feline sarcoids from North America and New Zealand but not in any non-sarcoid feline samples. J Vet Diagn Investig. 2010;22:97–100.

33. Schulman FY, Krafft AE, Janczewski T. Feline cutaneous fibropapillomas: clinicopathologic findings and association with papillomavirus infection. Vet Pathol. 2001;38:291–6.

34. Teifke JP, Kidney BA, Lohr CV, et al. Detection of papillomavirus-DNA in mesenchymal tumour cells and not in the hyperplastic epithelium of feline sarcoids. Vet Dermatol. 2003;14:47–56.

35. Munday JS, Thomson N, Dunowska M, et al. Genomic characterisation of the feline sarcoid-associated papillomavirus and proposed classification as Bos taurus papillomavirus type 14. Vet Microbiol. 2015;177:289–95.

36. Munday JS, Knight CG. Amplification of feline sarcoid-associated papillomavirus DNA sequences from bovine skin. Vet Dermatol. 2010;21:341–4.

37. Munday JS, Thomson NA, Henderson G, et al. Identification of Felis catus papillomavirus 3 in skin neoplasms from four cats. J Vet Diagn Investig. 2018;30:324–8.

38. Munday JS, French A, Thomson N. Detection of DNA sequences from a novel papillomavirus in a feline basal cell carcinoma. Vet Dermatol. 2017;28:236–e260.

39. Munday JS, Fairley RA, Mills H, et al. Oral papillomas associated with Felis catus papillomavirus type 1 in 2 domestic cats. Vet Pathol. 2015;52:1187–90.

40. Munday JS, Hanlon EM, Howe L, et al. Feline cutaneous viral papilloma associated with human papillomavirus type 9. Vet Pathol. 2007;44:924–7.

41. Johnson RP, Sabine M. The isolation of herpesviruses from skin ulcers in domestic cats. Vet Rec. 1971;89:360–2.

42. Hargis AM, Ginn PE. Feline herpesvirus 1-associated facial and nasal dermatitis and stomatitis in domestic cats. Vet Clin North Am Small Anim Pract. 1999;29:1281–90.

43. Sanchez MD, Goldschmidt MH, Mauldin EA. Herpesvirus dermatitis in two cats without facial lesions. Vet Dermatol. 2012;23:171–3, e135.

44. Lee M, Bosward KL, Norris JM. Immunohistological evaluation of feline herpesvirus-1 infection in feline eosinophilic dermatoses or stomatitis. J Feline Med Surg. 2010;12:72–9.

45. Suchy A, Bauder B, Gelbmann W, et al. Diagnosis of feline herpesvirus infection by immunohistochemistry, polymerase chain reaction, and in situ hybridization. J Vet Diagn Investig. 2000;12:186–91.

46. Prost C. P34 A case of exfoliative erythema multiforme associated with herpes virus 1 infection in a European cat. Vet Dermatol. 2004;15(Suppl. 1):41–69.

47. Holland JL, Outerbridge CA, Affolter VK, et al. Detection of feline herpesvirus 1 DNA in skin biopsy specimens from cats with or without dermatitis. J Am Vet Med Assoc. 2006;229:1442–6.

48. Persico P, Roccabianca P, Corona A, et al. Detection of feline herpes virus 1 via polymerase chain reaction and immunohistochemistry in cats with ulcerative facial dermatitis,

eosinophilic granuloma complex reaction patterns and mosquito bite hypersensitivity. Vet Dermatol. 2011;22:521–7.

49. Lamm CG, Dean SL, Estrada MM, et al. Pathology in practice. Herpesviral dermatitis. J Am Vet Med Assoc. 2015;247:159–61.

50. Maggs DJ. Antiviral therapy for feline herpesvirus infections. Vet Clin North Am Small Anim Pract. 2010;40:1055–62.

51. Malik R, Lessels NS, Webb S, et al. Treatment of feline herpesvirus-1 associated disease in cats with famciclovir and related drugs. J Feline Med Surg. 2009;11:40–8.

52. Thomasy SM, Lim CC, Reilly CM, et al. Evaluation of orally administered famciclovir in cats experimentally infected with feline herpesvirus type-1. Am J Vet Res. 2011;72:85–95.

53. Gutzwiller MER, Brachelente C, Taglinger K, et al. Feline herpes dermatitis treated with interferon omega. Vet Dermatol. 2007;18:50–4.

54. Nagata M, Rosenkrantz W. Cutaneous viral dermatoses in dogs and cats. Compendium. 2013;35:E1.

55. Bol S, Bunnik EM. Lysine supplementation is not effective for the prevention or treatment of feline herpesvirus 1 infection in cats: a systematic review. BMC Vet Res. 2015;11:284.

56. Delhon GA. Poxviridae. In: NJ ML, Dubovi EJ, editors. Fenner's Veterinary Virology. 5th ed. London: Academic Press; 2017. p. 157–74.

57. Fairley RA, Mercer AA, Copland CI, et al. Persistent pseudocowpox virus infection of the skin of a foot in a cat. NZ Vet J. 2013;61:242–3.

58. Fairley RA, Whelan EM, Pesavento PA, et al. Recurrent localised cutaneous parapoxvirus infection in three cats. NZ Vet J. 2008;56:196–201.

59. Chantrey J, Meyer H, Baxby D, et al. Cowpox: reservoir hosts and geographic range. Epidemiol Infect. 1999;122:455–60.

60. Appl C, von Bomhard W, Hanczaruk M, et al. Feline cowpoxvirus infections in Germany: clinical and epidemiological aspects. Berl Munch Tierarztl Wochenschr. 2013;126:55–61.

61. McInerney J, Papasouliotis K, Simpson K, et al. Pulmonary cowpox in cats: five cases. J Feline Med Surg. 2016;18:518–25.

62. Mostl K, Addie D, Belak S, et al. Cowpox virus infection in cats: ABCD guidelines on prevention and management. J Feline Med Surg. 2013;15:557–9.

63. Czerny CP, Wagner K, Gessler K, et al. A monoclonal blocking-ELISA for detection of orthopoxvirus antibodies in feline sera. Vet Microbiol. 1996;52:185–200.

64. Tryland M, Sandvik T, Holtet L, et al. Antibodies to orthopoxvirus in domestic cats in Norway. Vet Rec. 1998;143:105–9.

65. Breheny CR, Fox V, Tamborini A, et al. Novel characteristics identified in two cases of feline cowpox virus infection. JFMS Open Reports. 2017;3:2055116917717191.

66. Bennett M, Gaskell CJ, Baxbyt D, et al. Feline cowpox virus infection. J Small Anim Pract. 1990;31:167–73.

67. Godfrey DR, Blundell CJ, Essbauer S, et al. Unusual presentations of cowpox infection in cats. J Small Animal Pract. 2004;45:202–5.

68. O'Halloran C, Del-Pozo J, Breheny C, et al. Unusual presentations of feline cowpox. Vet Record. 2016;179:442–3.

69. Schaudien D, Meyer H, Grunwald D, et al. Concurrent infection of a cat with cowpox virus and feline parvovirus. J Comp Pathol. 2007;137:151–4.

70. Jungwirth N, Puff C, Köster K, et al. Atypical cowpox virus infection in a series of cats. J Comp Pathol. 2018;158:71–6.

71. Lawn R. Risk of cowpox to small animal practitioners. Vet Rec. 2010;166:631.

72. Czerny CP, Eis-Hubinger AM, Mayr A, et al. Animal poxviruses transmitted from cat to man: current event with lethal end. Zentralbl Veterinarmed B. 1991;38:421–31.

73. Radford AD, Addie D, Belák S, et al. Feline calicivirus infection: ABCD guidelines on prevention and management. J Feline Med Surg. 2009;11:556–64.

74. Deschamps J-Y, Topie E, Roux F. Nosocomial feline calicivirus-associated virulent systemic disease in a veterinary emergency and critical care unit in France. JFMS Open Reports. 2015;1:2055116915621581.

75. Pedersen NC, Elliott JB, Glasgow A, et al. An isolated epizootic of hemorrhagic-like fever in cats caused by a novel and highly virulent strain of feline calicivirus. Vet Microbiol. 2000;73:281–300.

76. Willi B, Spiri AM, Meli ML, et al. Molecular characterization and virus neutralization patterns of severe, non-epizootic forms of feline calicivirus infections resembling virulent systemic disease in cats in Switzerland and in Liechtenstein. Vet Microbiol. 2016;182:202–12.

77. Pesavento PA, Maclachlan NJ, Dillard-Telm L, et al. Pathologic, immunohistochemical, and electron microscopic findings in naturally occurring virulent systemic feline calicivirus infection in cats. Vet Pathol. 2004;41:257–63.

78. Hurley KE, Pesavento PA, Pedersen NC, et al. An outbreak of virulent systemic feline calicivirus disease. J Am Vet Med Assoc. 2004;224:241–9.

79. Declercq J. Pustular calicivirus dermatitis on the abdomen of two cats following routine ovariectomy. Vet Dermatol. 2005;16:395–400.

80. Smith AW, Iversen PL, O'Hanley PD, et al. Virus-specific antiviral treatment for controlling severe and fatal outbreaks of feline calicivirus infection. Am J Vet Res. 2008;69:23–32.

81. Hartmann K. Clinical aspects of feline retroviruses: a review. Viruses. 2012;4:2684.

82. Favrot C, Wilhelm S, Grest P, et al. Two cases of FeLV-associated dermatoses. Vet Dermatol. 2005;16:407–12.

83. Gross TL, Clark EG, Hargis AM, et al. Giant cell dermatosis in FeLV-positive cats. Vet Dermatol. 1993;4:117–22.

84. Center SA, Scott DW, Scott FW. Multiple cutaneous horns on the footpad of a cat. Feline Practice. 1982;12:26–30.

85. Komori S, Ishida T, Washizu M. Four cases of cutaneous horns in the foot pads of feline leukemia virus-negative cats. J Japan Vet Med Assoc. 1998;51:27–30.

86. Chaher E, Robertson E, Sparkes A, et al. Call for cases: cat paw hyperkeratosis. CVE Control and Therapy Series. 2016;282:51–4.

87. Burr HD, Keating JH, Clifford CA, et al. Cutaneous lymphoma of the tarsus in cats: 23 cases (2000–2012). J Am Vet Med Assoc. 2014;244:1429–34.

88. Fontaine J, Heimann M, Day MJ. Cutaneous epitheliotropic T-cell lymphoma in the cat: a review of the literature and five new cases. Vet Dermatol. 2011;22:454–61.

89. Roccabianca P, Avallone G, Rodriguez A, et al. Cutaneous lymphoma at injection sites: pathological, immunophenotypical, and molecular characterization in 17 cats. Vet Pathol. 2016;53:823–32.

90. Cannon MJ, Silkstone MA, Kipar AM. Cutaneous lesions associated with coronavirus-induced vasculitis in a cat with feline infectious peritonitis and concurrent feline immunodeficiency virus infection. J Feline Med Surg. 2005;7:233–6.

91. Martha JC, Malcolm AS, Anja MK. Cutaneous lesions associated with coronavirus-induced vasculitis in a cat with feline infectious peritonitis and concurrent feline immunodeficiency virus infection. J Feline Med Surg. 2005;7:233–6.

92. Bauer BS, Kerr ME, Sandmeyer LS, et al. Positive immunostaining for feline infectious peritonitis (FIP) in a Sphinx cat with cutaneous lesions and bilateral panuveitis. Vet Ophthalmol. 2013;16(Suppl 1):160–3.

93. Murphy B. Retroviridae. In: NJ ML, Dubovi EJ, editors. Fenner's Veterinary Virology. 5th ed. London: Academic Press; 2017. p. 269–97.

94. Backel K, Cain C. Skin as a marker of general feline health: cutaneous manifestations of infectious disease. J Feline Med Surg. 2017;19:1149–65.

95. Munday JS, Witham AI. Frequent detection of papillomavirus DNA in clinically normal skin of cats infected and noninfected with feline immunodeficiency virus. Vet Dermatol. 2010;21:307–10.

96. Magden E, Quackenbush SL, VandeWoude S. FIV associated neoplasms—a mini-review. Vet Immunol Immunopathol. 2011;143:227–34.

97. Hutson CA, Rideout BA, Pedersen NC. Neoplasia associated with feline immunodeficiency virus infection in cats of southern California. J Am Vet Med Assoc. 1991;199:1357–62.

Leishmaniosis

Maria Grazia Pennisi

Abstract

Leishmania spp. affecting cats include *L. infantum*, *L. mexicana*, *L. venezuelensis*, *L. amazonensis*, and *L. braziliensis*. *Leishmania infantum* is the species most frequently reported in cats and causes feline leishmaniosis (FeL). Cats exposed to *L. infantum* are able to mount a cell-mediated immune response that does not parallel antibody production. Cats with *L. infantum*-associated clinical disease have positive blood PCR and low to very high antibody levels. About half of the clinical cases of FeL are diagnosed in cats with impaired immunocompetence. Skin or mucocutaneous lesions are the most common clinical findings; however, FeL is a systemic disease. Skin or mucocutaneous lesions and lymph node enlargement are seen in at least half of cases, ocular or oral lesions and some aspecific signs (weight loss, anorexia, lethargy) in about 20–30% of cases, and many other clinical signs (e.g., respiratory, gastrointestinal) are sporadically observed. Ulcerative and nodular lesions due to diffuse granulomatous dermatitis are the most frequent skin manifestations, mainly distributed on the head or symmetrically on the distal limbs. Diagnosis can be obtained by cytology and histology, and immunohistochemistry is useful to confirm the causative role of *Leishmania* infection in the dermopathological manifestations; however, other skin diseases may coexist with FeL. Polymerase chain reaction is used in case of suggestive lesions with lack of parasites and for *Leishmania* speciation. Comorbidities, coinfections, and chronic renal disease influence the prognosis and should be investigated. Treatment is currently based on the same drugs used for canine leishmaniosis, and generally clinical cure is obtained; however recurrence is possible.

M. G. Pennisi (✉)
Dipartimento di Scienze Veterinarie, Università di Messina, Messina, Italy
e-mail: mariagrazia.pennisi@unime.it

© Springer Nature Switzerland AG 2020
C. Noli, S. Colombo (eds.), *Feline Dermatology*,
https://doi.org/10.1007/978-3-030-29836-4_18

Introduction

Leishmaniases are protozoan diseases caused by *Leishmania* spp. affecting humans and animals, but leishmaniosis is the term used for diseases in animals. Leishmaniosis caused by *Leishmania infantum* is a severe, zoonotic, vector-borne disease endemic in areas of the Old and New Worlds, with dogs as the main reservoir [1]. In fact, the majority of infected dogs do not develop clinical signs or clinico-pathological abnormalities, but they are chronically infected and infectious to sand fly vectors. Dogs may, however, develop a mild to severe systemic disease, with frequent skin lesions usually associated with other clinical and clinico-pathological abnormalities. Therefore, much research interest has been focused on canine leishmaniosis (CanL), in order to prevent the infection, understand the pathomechanisms driving infection to disease, make early and accurate diagnosis, and treat affected dogs. Conversely, until about 25 years ago, the cat was considered a resistant host species to *Leishmania* infections, based on very rare case reports, occasional post mortem finding of the parasite in cats from endemic areas, and results from an experimental infection study demonstrating limited infection rates [2]. Over the last decades, an increasing number of clinical cases have been reported, and investigations with more sensitive diagnostic techniques detected a variable, but not negligible, infection rate in cats living in endemic areas. Therefore, feline leishmaniosis (FeL) appears nowadays as an emergent disease, and the cat's role as reservoir host is revalued. We now know that the epidemiology of leishmaniosis is complex and the vectorial transmission in endemic areas involves multiple host species infectious to sand flies, including the cat. Tegumentary leishmaniosis caused by dermotropic *Leishmania* spp. is rarely reported in both dogs and cats. Dermotropic species infecting cats are *Leishmania tropica* and *Leishmania major* in the Old World and *Leishmania mexicana*, *Leishmania venezuelensis*, and *Leishmania braziliensis* in the Americas. Main reservoir hosts for dermotropic species are wild animals, such as rodents.

Etiology, Diffusion, and Transmission

Leishmania genus (Kinetoplastea: Trypanosomatidae) includes diphasic and dixenous protozoans replicating as promastigotes in the gut of phlebotomine sand flies, their natural vectors. When inoculated into vertebrate hosts by sand fly bites, promastigotes change to the non-flagellated amastigote form that multiplies by binary fission in macrophages. *Leishmania* spp. detected in cats are able to infect also other mammals (including dogs and humans) and belong to the subgenus *Leishmania* (*L. infantum*, *L. mexicana*, *L. venezuelensis*, *L. amazonensis*) or *Viannia* (*L. braziliensis*).

Leishmania infantum is the species most frequently reported in both dogs and cats in the Old World and in Central and South America. *Leishmania infantum* has been detected in cats in Mediterranean countries (Italy, Spain, Portugal, France, Greece, Turkey, Cyprus), Iran and Brazil [3–6]. Reported antibody and blood PCR prevalences are very variable (from nihil to >60%) and influenced by many factors such as the local level of endemicity, selection of tested cats and analytical differences [3]. However, *L. infantum* antibody and molecular prevalence is usually lower in cats compared to dogs and cases of FeL are rarer [3, 7]. Cases of both CanL and FeL are diagnosed in non-endemic areas in dogs or cats rehomed from or travelling to endemic areas [1, 8–13].

Sand fly transmission is the most important way of transmission of *Leishmania* to humans and animals, and several studies about the feeding habit of sand flies suggest that this is likely also in feline infection, but it has never been investigated [3, 14–16]. Non-vectorial transmission (vertical, by blood transfusion, mating, or bite wounds) of CanL is well known and responsible for autochthonous cases in non-endemic areas in dogs, but we have no evidence of these ways of transmission to and in cats [1, 10, 17, 18]. However, blood transfusion could be a source of infection in cats as proven in dogs and humans. In fact, healthy cats – similarly to healthy dogs and humans – are found blood PCR positive in endemic areas [4–7, 19–22].

Pathogenesis

Leishmania infantum

A great number of both experimentally and field controlled prospective studies performed on CanL provided information about immunopathogenesis of CanL, but we do not have similar studies performed on cats. In dogs, T helper 1 (Th1) immune response responsible for protective CD4+ T cell-mediated immunity is associated to resistance to the disease [1]. Conversely, progression of *L. infantum* infection and development of lesions and clinical signs in dogs and humans are associated with a predominant T helper 2 (Th2) immune response and the consequent non-protective antibody production and T cell exhaustion [23]. Depending on a variable balance between humoral and cell-mediated immunity in the infected dog, a wide and dynamic clinical spectrum is seen in CanL, including subclinical infection, self-limiting mild disease, or severe progressive illness [1, 24]. Sick dogs with severe clinical disease and high blood parasitemia show a high antibody level and lack in specific IFN-γ production [25]. Similarly to what occurs in mouse experimental models, a complex genetic background modulates the dog's susceptibility or resistance to CanL [1, 24]. In cats, the adaptive immune response elicited by *L. infantum* exposure in endemic areas was recently explored with measurement of specific antibody and IFN-γ production [26]. Some cats produced *L. infantum*-specific IFN-γ and were found blood PCR negative and antibody negative or in few cases borderline positive [26]. This means that, similarly to other mammals, cats exposed to *L. infantum* are able to mount a protective cell-mediated immune response that does not parallel antibody production. The relationship between immunological pattern and severity of disease is still unexplored in cats; however, we know that cats with *L. infantum*-associated clinical disease have a high blood parasitemia and low to very high antibody levels [3, 27–32]. Moreover, longitudinal studies found that progression of the infection toward disease is associated in cats with increasing antibody titers, and, on the other hand, clinical improvement obtained by anti-*L. infantum* therapy is associated with a significant reduction of antibody levels, similarly to CanL [33–36]. Coinfections with some vector-borne pathogens (e.g., *Dirofilaria immitis, Ehrlichia canis, Hepatozoon canis*) can influence parasite burden and progression of CanL [37–39]. In cats, the association between retroviral, coronavirus, *Toxoplasma*, or vector-borne coinfections and antibody and/or PCR positivity to *L. infantum* has been explored [5, 20, 40–50]. A significant association was found only between feline immunodeficiency virus (FIV) and *L. infantum* positivity in some cases [41, 46, 48]. Moreover, more than one third of cats with FeL and tested for retroviral coinfections were found positive to FIV [a few were also positive to Feline Leukemia Virus (FeLV)] [11, 12, 27–29, 31, 51–69]. Other FeL cases reported in FIV and FeLV negative cats were diagnosed in animals affected by immune-mediated diseases (and treated with immunosuppressive drugs), neoplasia, or diabetes mellitus, and we may assume that about half of the clinical cases of FeL were diagnosed in cats with impaired immunocompetence [12, 27–30, 34, 52, 59–61].

Despite the fact that skin or mucocutaneous lesions are the most common clinical findings, FeL is considered a systemic disease as CanL. Parasites can be detected in various other tissues, such as lymph nodes, spleen, bone marrow, eye, kidney, liver, and gastrointestinal and respiratory tract [8].

American Dermotropic *Leishmania* spp.

Some scanty information about adaptive immune response of cats toward American dermotropic *Leishmania* spp. can be inferred only from case reports of *L. mexicana* and from an experimental infection of cats with *L. braziliensis* [70–72].

Delayed-type hypersensitivity skin test with *L. donovani* antigen was repeatedly found negative in a cat affected by recurrent nodular dermatitis caused by *L. mexicana* infection, suggesting a lack of cell-mediated adaptive immune response in this cat [70]. Anti-*Leishmania* antibody production seems to be limited, as of five cats with *L. mexicana* tegumentary leishmaniosis only two were antibody positive at ELISA test, although Western blot test was positive in four [71]. Moreover, in cats intradermally infected with a human strain of *L. braziliensis*, a short-term antibody production was documented after the development of skin lesions, but, frequently, it appeared after the healing of lesions [72].

Clinical Picture

Leishmania infantum

Currently, in endemic areas FeL is far less frequently reported than CanL, but we are probably underestimating the disease, particularly the less frequent and less severe clinical presentations, as it occurred in the past with CanL. Furthermore, coinfections or comorbidities are frequently detected which can contribute to a clinical misrepresentation and misdiagnosis of FeL [3, 22, 27–32]. About a hundred of clinical cases were reported over the last 30 years – mostly in Southern Europe – and they are at present the only source of knowledge about FeL. We are therefore aware of the current low level of evidence (III–IV) for statements and recommendations concerning this disease.

Age range of affected cats is wide (2–21 years); however, they are mostly mature cats (median age 7 years) at diagnosis, with very few being 2–3 years old [3, 27, 28, 32, 51, 57, 73]. Both genders are similarly represented and almost all cases are reported in domestic short-hair cats.

Some clinical manifestations are very frequent at diagnosis – found in at least half of the cases – such as skin or mucocutaneous lesions and lymph node enlargement. Common presentations – found in one fourth to half of the cats – are represented by ocular or oral lesions and some aspecific signs (weight loss, anorexia, lethargy). Finally, there are many clinical signs seen in less than one fourth of the cases. Usually affected cats display more than one clinical sign and often develop different lesions with time.

Skin and Mucocutaneous Manifestations

Skin or mucocutaneous manifestations were found in about two thirds of reported cases, but they rarely were the only abnormality detected [3, 8, 27–30, 73]. In a study from a pathology laboratory from Spain, FeL was diagnosed in 0.57% of all skin and ocular biopsies (*n* = 2632) examined over a 4-year period [73].

Several dermatological entities have been described, and different presentations often coexisted or developed subsequently in the same cat. Most lesions were observed on the head. Pruritus was rarely reported, and in most cats manifesting pruritus, a concurrent dermatological disease was identified such as flea allergy, eosinophilic granuloma, pemphigus foliaceus, squamous cell carcinoma (SCC), or demodicosis [12, 67, 75, 76]. In one case, however, pruritus stopped after starting anti-*Leishmania* therapy [77].

Ulcerative dermatitis is the more commonly reported skin lesion and sometimes with a history of self-healing and recurrence of lesion. Crusty-ulcerative lesions with raised margins were seen on pressure points (hock, carpal, and ischiatic regions), often symmetrically, and were large up to 5 cm (Fig. 1) [27, 28, 54, 57, 64,

Fig. 1 Large ulcer with raised margins on right forelimb. A similar symmetrical lesion was present on the left forelimb

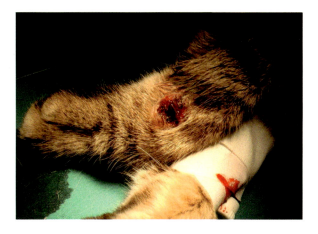

Fig. 2 Solitary focal ulceration on the face (white arrow) and conjunctival nodule (transparent arrow) observed in the same cat of Fig. 1

Fig. 3 Severe facial ulceration in a cat diagnosed with squamous cell carcinoma associated with *L. infantum* dermal infection

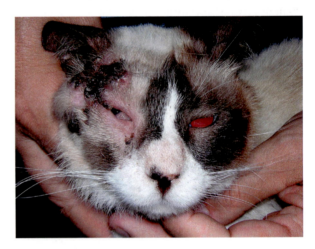

77]. Focal solitary or multiple smaller ulcers were reported on the face (Fig. 2), lips, ears, neck, or limbs [27, 28, 34, 64, 65, 73, 77–79]. In a few cases focal or diffuse ulcerative dermatitis affected face, trunk, or footpads [27, 63, 65, 79]. Ulceration of the *nasal planum* was also reported, and in one case it was associated with concurrent SCC [30, 54, 58, 67]. *Leishmania* infection and SCC were found associated in biopsied tissues obtained from a deep facial ulceration (Fig. 3) in other two cases [56, 76]. Unfortunately, the diagnosis of SCC was missed at first consultation in two cases when only *Leishmania* infection was detected by cytology or histology [30, 76]. Moreover, multifocal ulcerative dermatitis caused by *L. infantum* was diagnosed in a cat suffering from SCC at a different site [65]. Ulcerative dermatitis was found associated with eosinophilic granuloma complex, and in one other case *Leishmania* infection was confirmed (by serology and skin PCR) in a cat with pemphigus foliaceus [12, 73].

Nodular dermatitis is also a frequent dermatological manifestation, and single, multiple or diffuse, firm, alopecic, non-painful nodules were detected. They are usually small (<1 cm), mainly distributed on the head and, in descending order of frequency, on the eyelid, ear, chin, nose, lips, and tongue [11, 27, 28, 31, 55, 64, 66, 73, 80–83]. Nodules can be found also on limbs or rarely on the trunk or the anus [12, 55, 73]. In rare cases nodules were ulcerated [12, 66, 84].

Differently from CanL, facial or diffuse scaling and alopecia are less frequently reported in FeL, and in few of these cases, histopathological evaluation confirmed the presence of amastigotes in the affected skin [29, 63, 73]. Digital hyperkeratosis was found in one case only [27].

An atypical FeL presentation that is not reported in CanL is development of hemorrhagic bullae, observed in three cases, respectively, on the nasal planum, head, and margin of the pinna [34, 76]. However, the lesion developed on the nasal planum was histologically diagnosed as hemangioma [76]. The other two cases were cytologically evaluated and amastigotes were found [34].

Visceral Manifestations

Lymph node enlargement is the most frequent non-dermatological finding [3]. It is usually multicentric and can be symmetrical. Lymph nodes are firm and non-painful, and enlargement can be relevant mimicking neoplasia. Monolateral or bilateral ocular lesions were reported in about one third of cases, but a specialistic ophthalmic examination was not performed in all cats with FeL; therefore, some less severe ocular findings could have been missed. Conjunctivitis (including also conjunctival nodules) and uveitis are the most frequent ocular manifestations [11, 27, 31, 34, 60, 62, 64, 68, 73]. Keratitis, keratouveitis, and chorioretinitis were diagnosed in a few cats [27, 31, 34, 67, 78]. Panophthalmitis is the consequence of progressive extension of diffuse granulomatous inflammation in case of late diagnosis [60, 73].

Apart from single cases of gingival ulceration, nodular glossitis, or epulid-like lesions, chronic stomatitis and faucitis was found in about 20% of cats, and the parasite was detected in the inflamed oral tissue [27, 31–34, 52, 58, 60, 62, 66, 78, 83].

Non-specific manifestations as weight loss, anorexia, or lethargy were not very frequent [3], and occasionally gastrointestinal (vomiting, diarrhea) or respiratory (chronic nasal discharge, stertor, dyspnea, wheezing) signs were reported [3, 74]. Rare manifestations were icterus, fever, spleen or liver enlargement, and abortion [3]. Interestingly, chronic leishmanial rhinitis was confirmed in some cases [58, 64, 73–75].

American Tegumentary Leishmaniosis (ATL)

A limited number of cases of feline cutaneous leishmaniosis caused by dermotropic *Leishmania* species were reported in the Americas [70, 71, 85–91]. Not always *Leishmania* speciation was obtained from affected cats, and *L. mexicana* could be identified in nine cases [70, 71, 91], *L. braziliensis* in five [85–88], *L. venezuelensis* in four [90], and *L. amazonensis* in one [89]. They were all domestic short-hair cats and younger (age range: 8 months–11 years; median age 4 years) than cats with disease caused by *L. infantum*. The most common manifestation consisted in solitary or multiple firm nodules as large as 3 × 2 cm. They were alopecic, variably erythematous, or ulcerated and mainly distributed on the pinnae and the face (eyelids, *nasal planum*, muzzle) and rarely on the distal limbs or tail. A larger (6 cm) interdigital ovoid lesion was reported in a cat with *L. braziliensis* infection [88]. Nasal or ear ulcerations were seen in two cats with *L. mexicana* infection and in two others (*nasal planum* or medial canthus) with *L. braziliensis* infection [71, 86, 87]. Mucosal nodules may develop in the nasal cavity causing sneezing, stertor, and inspiratory dyspnea [71, 85]. No other manifestations were reported in cats with ATL; however, some followed up cases of *L. venezuelensis* or *L. mexicana* infections developed new nodular lesions at other sites [70, 90].

Diagnosis

Diagnostic testing of symptomatic cats aims to confirm *Leishmania* infection and to establish a causal relationship with the clinical picture. In case of dermatological or mucosal lesions, the cytological evaluation of impression smears from erosions and ulcers, of scrapings from margins of deep ulcers, and of fine needle punction of nodules can show a pyogranulomatous pattern and the presence of amastigotes (in the cytoplasm of macrophages or extracellularly) (Fig. 4) [3, 71]. Amastigotes have an elliptic shape with pointed ends, measure about 3–4 × 2 μm, and are characterized by the rod-shaped basophilic kinetoplast set perpendicular to the large nucleus. Morphology of amastigotes does not allow to differentiate between *Leishmania* species. In cats with leishmaniosis caused by *L. infantum*, amastigotes can be found also in cytological samples from enlarged lymph nodes, bone marrow, nasal exudates, liver, and spleen and rarely in circulating neutrophils [3].

Fig. 4 Cytology from the cutaneous lesion in Fig. 1. Macrophagic–neutrophilic inflammation with numerous intracellular (arrows) and extracellular amastigotes. In some extracellular amastigotes the basophilic rod-shaped kinetoplast is clearly visible (arrow heads) (May Grünwald–Giemsa stain 1000×)

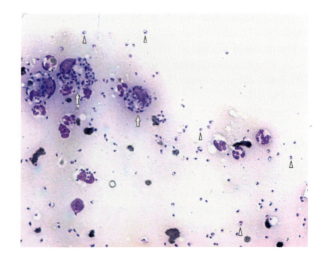

Biopsy of skin or mucosal lesions is required when cytology is inconclusive and in any case when the clinical presentation is compatible with neoplastic or immune-mediated diseases. Amastigotes are not easily detected by the conventional histological staining, and in suspected cases they should be investigated by immunohistochemistry (Fig. 5). However, immunohistochemistry does not allow the speciation of *Leishmania* amastigotes, which can be obtained by polymerase chain reaction (PCR) and sequencing of amplicons. PCR can be performed also from cytological slides, formalin-fixed and paraffin-embedded biopsies. Quantitative real-time PCR is very sensitive and can provide parasite load of samples.

Dermopathological evaluation (Fig. 6) shows dermal periadnexal to diffuse granulomatous inflammation with a diffuse infiltration of macrophages, a moderate number of amastigotes, and a variable number of lymphocytes and plasma cells [12, 73]. The overlying epidermis is affected by hyperkeratosis, acanthosis, and ulceration [73]. In nodular lesions giants cells may be seen [73]. A low number of parasites were found in nodular lesions, characterized by perifollicular granulomatous

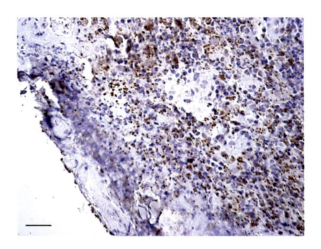

Fig. 5 Dark brown amastigotes evidenced by immunohistochemistry. Mayer's hematoxylin counterstain. Bar = 10 μm. (Courtesy of R. Puleio, IZS Sicilia, Italy)

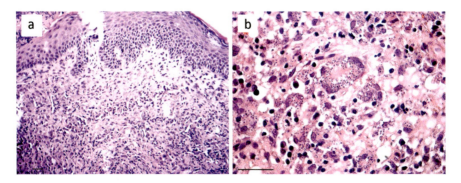

Fig. 6 Diffuse pyogranulomatous dermatitis (**a**) with numerous amastigotes within macrophages (**b**). HE. Bar = 10 μm. (Courtesy of R. Puleio, IZS Sicilia, Italy)

dermatitis and in a lichenoid interface dermatitis in a cat affected by scaly dermatitis [73]. Mucosal (and mucocutaneous) lesions harbor a higher parasite load and submucosal diffuse granulomatous inflammation is seen [62, 68, 73]. In some cases, a dermal, diffuse, granulomatous inflammation was found associated with lesions characteristic of feline eosinophilic granuloma complex [54, 73]. A transepidermal inflammatory infiltrate with parasitized macrophages was reported in the neoplastic tissue of a cat diagnosed with concurrent SCC [56]. In another case, a stromal infiltration of parasitized macrophages was observed adjacent to islands of SCC [30]. Nodular to diffuse granulomatous dermatitis with hyperkeratotic, hyperplastic, and often ulcerated epidermis is described in cases of ATL [71, 85, 91].

Anti-*L. infantum* antibody detection is performed by quantitative serology (IFAT, ELISA, or DAT) and Western blot (WB) techniques [3]. Cutoff setting for IFAT is established at 1:80 dilution, and almost all cats affected by clinical FeL caused by *L. infantum* have low to very high antibody levels [43, 92]. Conversely, sick cats with ATL may not have detectable circulating antibodies [71].

Culture of infected tissues provided feline strains that in most cases showed the same zymodemes and genotypes detected in dogs or humans [3, 30].

Clinico-pathological abnormalities more frequently reported at diagnosis in cats with FeL caused by *L. infantum* consisted in mild to moderate non-regenerative anemia, hyperglobulinemia, and proteinuria [3]. Chronic kidney disease (CKD), in most cases at an early stage (International Renal Interest Society [IRIS] stages 1 or 2), is often documented when a renal profile including urinalysis and the urine to protein concentrations ratio is performed [32, 75].

Clinico-pathological abnormalities of cats with ATL were rarely investigated and only eosinophilia and neutrophilia were found in one cat with *L. braziliensis* infection [70, 85].

Treatment and Prognosis

Treatment of cats with clinical FeL caused by *L. infantum* is empirical and based on off-label use of the most common drugs prescribed to dogs with CanL [3]. Long-term oral administration of allopurinol (10–20 mg/kg once or twice daily) as monotherapy or as maintenance treatment after a course of subcutaneous injections of meglumine antimoniate (50 mg/kg once daily for 30 days) are the most frequently used regimens. Clinical cure is usually obtained, but efficacy and safety of used protocols have never been evaluated in controlled studies; therefore cats should be monitored very carefully for adverse effects during treatment (particularly cats affected by renal disease) and for possible clinical recurrence after stopping the therapy [3, 27–32, 34, 74]. A cutaneous adverse drug reaction (head and neck erythema, alopecia, exfoliation, and crusting) was suspected few days after starting allopurinol in a cat [75]. The skin reaction rapidly solved after stopping allopurinol [75]. Increases in liver enzymes were observed in another cat, and they resolved after lowering dosage to 5 mg/kg twice a day [12]. In two further cases, acute kidney injury was diagnosed few weeks after starting allopurinol administration [32]. In another cat with concurrent IRIS stage 1-CKD at the time of FeL diagnosis, azotemia developed after meglumine antimoniate and afterward to miltefosine (2 mg/kg orally once daily for 30 days) administration. [75] This latter cat was hereafter maintained with a dietary supplementation of nucleotides and active hexose correlated compounds that was recently found effective in dogs as CanL maintenance treatment [75, 93].

Domperidone (0.5 mg/kg orally once daily) was recently used in two cats in association with allopurinol, and miltefosine was given in one other case [27, 29, 30].

Surgical removal of nodules was performed but generally they recurred [12, 27, 54, 81]. In one case an integrated approach between surgery and chemotherapy was needed for treating large ulcerations [28].

Clinical recurrence is associated with raised antibody titer and parasite load [34].

Cats with clinical FeL may live for several years after diagnosis, even those untreated and/or FIV positive, unless concurrent conditions (neoplasia) and complications (chronic kidney disease) occur or develop [32, 68].

Scant information is available about treatment and prognosis of ATL. Some cats with *L. mexicana* ATL were cured after surgical excision of nodules [91]. However, radical pinnectomy was not effective in a FIV- and FeLV-negative cat and lesions recurred at pinnectomy site in about 2 years [70]. Subsequently new lesions progressively involved the muzzle and finally the nasal mucosa, and the cat was euthanized over 6 years after ATL diagnosis due to a mediastinal lymphosarcoma [70].

Prevention of *L. infantum* Infection

Individual protection of exposed cats reduces their risk to be infected by sand fly bites and to develop the clinical disease [3, 22]. *Phlebotomus perniciosus* and *Lutzomyia longipalpis*, proven vectors of *L. infantum*, respectively, in the Old and New Worlds, were found infected after feeding on one single sick cat with FeL [33, 94]. This means that protection of cats at population level contributes to the regional control of *L. infantum* infection. In fact, the percentage of antibody and/or PCR-positive cats is often not negligible in endemic areas [3–6, 20, 21, 41, 42, 45, 47].

Pyrethroids are used in dogs for preventing the bites of sand flies, but most of them are toxic to cats [3, 95]. Collars containing a combination of 10% imidacloprid and 4.5% flumethrin are the only pyrethroid formulation licensed also for cats, and it was effective in reducing incidence of *L. infantum* infection in cats living in endemic areas [22].

According to current knowledge, testing of blood donors by antibody detection and blood PCR is the only advisable measure for preventing non-vectorial transmission in cats [96].

References

1. Solano-Gallego L, Koutinas A, Miró G, Cardoso L, Pennisi MG, Ferrer L, Bourdeau P, Oliva G, Baneth G. Directions for the diagnosis, clinical staging, treatment and prevention of canine leishmaniosis. Vet Parasitol. 2009;165:1–18.
2. Kirkpatrick CE, Farrell JP, Goldschimdt MH. *Leishmania chagasi* and *L. donovani*: experimental infections in domestic cats. Exp Parasitol. 1984;58:125–31.
3. Pennisi M-G, Cardoso L, Baneth G, Bourdeau P, Koutinas A, Miró G, Oliva G, Solano-Gallego L. LeishVet update and recommendations on feline leishmaniosis. Parasit Vectors. 2015;8:302.
4. Can H, Döşkaya M, Özdemir HG, Şahar EA, Karakavuk M, Pektaş B, Karakuş M, Töz S, Caner A, Döşkaya AD, İz SG, Özbel Y, Gürüz Y. Seroprevalence of *Leishmania* infection and molecular detection of *Leishmania tropica* and *Leishmania infantum* in stray cats of İzmir. Turkey Exp Parasitol. 2016;167:109–14.
5. Attipa C, Papasouliotis K, Solano-Gallego L, Baneth G, Nachum-Biala Y, Sarvani E, Knowles TG, Mengi S, Morris D, Helps C, Tasker S. Prevalence study and risk factor analysis of selected bacterial, protozoal and viral, including vector-borne, pathogens in cats from Cyprus. Parasit Vectors. 2017;10:130.
6. Metzdorf IP, da Costa Lima MS, de Fatima Cepa Matos M, de Souza Filho AF, de Souza Tsujisaki RA, Franco KG, Shapiro JT, de Almeida Borges F. Molecular characterization of *Leishmania infantum* in domestic cats in a region of Brazil endemic for human and canine visceral leishmaniasis. Acta Trop. 2017;166:121–5.
7. Otranto D, Napoli E, Latrofa MS, Annoscia G, Tarallo VD, Greco G, Lorusso E, Gulotta L, Falsone L, Basano FS, Pennisi MG, Deuster K, Capelli G, Dantas-Torres F, Brianti E. Feline and canine leishmaniosis and other vector-borne diseases in the Aeolian Islands: Pathogen and vector circulation in a confined environment. Vet Parasitol. 2017;236:144–15.
8. Pennisi MG. Leishmaniosis of companion animals in Europe: an update. Vet Parasitol. 2015;208:35–47.
9. Cleare E, Mason K, Mills J, Gabor M, Irwin PJ. Remaining vigilant for the exotic: cases of imported canine leishmaniosis in Australia 2000–2011. Aust Vet J. 2014;92:119–27.
10. Svobodova V, Svoboda M, Friedlaenderova L, Drahotsky P, Bohacova E, Baneth G. Canine leishmaniosis in three consecutive generations of dogs in Czech Republic. Vet Parasitol. 2017;237:122–4.
11. Richter M, Schaarschmidt-Kiener D, Krudewig C. Ocular signs, diagnosis and long-term treatment with allopurinol in a cat with leishmaniasis. Schweiz Arch Tierheilkd. 2014;156:289–94.
12. Rüfenacht S, Sager H, Müller N, Schaerer V, Heier A, Welle MM, Roosje PJ. Two cases of feline leishmaniosis in Switzerland. Vet Rec. 2005;156:542–5.
13. Best MP, Ash A, Bergfeld J, Barrett J. The diagnosis and management of a case of leishmaniosis in a dog imported to Australia. Vet Parasitol. 2014;202:292–5.
14. González E, Jiménez M, Hernández S, Martín-Martín I, Molina R. Phlebotomine sand fly survey in the focus of leishmaniasis in Madrid, Spain (2012–2014): seasonal dynamics, *Leishmania infantum* infection rates and blood meal preferences. Parasit Vectors. 2017;10:368.
15. Afonso MM, Duarte R, Miranda JC, Caranha L, Rangel EF. Studies on the feeding habits of *Lutzomyia* (*Lutzomyia*) *longipalpis* (Lutz & Neiva, 1912) (Diptera: Psychodidae: Phlebotominae) populations from endemic areas of American visceral leishmaniasis in Northeastern Brazil. J Trop Med. 2012;2012:1. https://doi.org/10.1155/2012/858657.
16. Baum M, Ribeiro MC, Lorosa ES, Damasio GA, Castro EA. Eclectic feeding behavior of *Lutzomyia* (*Nyssomyia*) *intermedia* (Diptera, Psychodidae, Phlebotominae) in the transmission area of American cutaneous leishmaniasis, State of Paranà. Brazil Rev Soc Bras Med Trop. 2013;46:547–54.
17. Karkamo V, Kaistinen A, Näreaho A, Dillard K, Vainio-Siukola K, Vidgrén G, Tuoresmäki N, Anttila M. The first report of autochthonous non-vector-borne transmission of canine leishmaniosis in the Nordic countries. Acta Vet Scand. 2014;56:84.
18. Naucke TJ, Amelung S, Lorentz S. First report of transmission of canine leishmaniosis through bite wounds from a naturally infected dog in Germany. Parasit Vectors. 2016;9:256.

19. Pennisi MG, Hartmann K, Addie DD, Lutz H, Gruffydd-Jones T, Boucraut-Baralon C, Egberink H, Frymus T, Horzinek MC, Hosie MJ, Lloret A, Marsilio F, Radford AD, Thiry E, Truyen U, Möstl K. European Advisory Board on Cat Diseases. Blood transfusion in cats: ABCD guidelines for minimising risks of infectious iatrogenic complications. J Feline Med Surg. 2015;17:588–93.
20. Persichetti M-F, Solano-Gallego L, Serrano L, Altet L, Reale S, Masucci M, Pennisi M-G. Detection of vector-borne pathogens in cats and their ectoparasites in southern Italy. Parasit Vectors. 2016;9:247.
21. Akhtardanesh B, Sharifi I, Mohammadi A, Mostafavi M, Hakimmipour M, Pourafshar NG. Feline visceral leishmaniasis in Kerman, southeast of Iran: Serological and molecular study. J Vector Borne Dis. 2017;54:96–102.
22. Brianti E, Falsone L, Napoli E, Gaglio G, Giannetto S, Pennisi MG, Priolo V, Latrofa MS, Tarallo VD, Solari Basano F, Nazzari R, Deuster K, Pollmeier M, Gulotta L, Colella V, Dantas-Torres F, Capelli G, Otranto D. Prevention of feline leishmaniosis with an imidacloprid 10%/flumethrin 4.5% polymer matrix collar. Parasit Vectors. 2017;10:334.
23. Esch KJ, Juelsgaard R, Martinez PA, Jones DE, Petersen CA. Programmed death 1-mediated T cell exhaustion during visceral leishmaniasis impairs phagocyte function. J Immunol. 2013;191:5542–50.
24. de Vasconcelos TCB, Furtado MC, Belo VS, Morgado FN, Figueiredo FB. Canine susceptibility to visceral leishmaniasis: A systematic review upon genetic aspects, considering breed factors and immunological concepts. Infect Genet Evol. 2019;74:103293. https://doi.org/10.1016/j.meegid.2017.10.005.
25. Solano-Gallego L, Montserrrat-Sangrà S, Ordeix L, Martínez-Orellana P. *Leishmania infantum*-specific production of IFN-γ and IL-10 in stimulated blood from dogs with clinical leishmaniosis. Parasit Vectors. 2016;9:317.
26. Priolo, V, Martínez Orellana, P, Pennisi, MG, Masucci, M, Foti, M, Solano-Gallego, L. *Leishmania infantum* specific production of IFNγ in stimulated blood from outdoor cats in endemic areas. In: Proceedings World Leish 6, Toledo-Spain (16th–20th May 2017), 2017: C1038.
27. Bardagi, M, Lloret, A, Dalmau, A, Esteban, D, Font, A, Leiva, M, Ortunez, A, Pena, T, Real, L, Salò, F, Tabar, MD. Feline Leishmaniosis: 15 cases. In: Proceedings 8th World Congress of Veterinary Dermatology, Toledo-Spain (31st May–4th June 2016), 2016: 112–113.
28. Basso MA, Marques C, Santos M, Duarte A, Pissarra H, Carreira LM, Gomes L, Valério-Bolas A, Tavares L, Santos-Gomes G, Pereira da Fonseca I. Successful treatment of feline leishmaniosis using a combination of allopurinol and N-methyl-glucamine antimoniate. JFMS Open Rep. 2016;2:205511691663000. https://doi.org/10.1177/2055116916630002.
29. Dedola, C, Ibba, F, Manca, T, Garia, C, Abramo, F. Dermatite esfoliativa associata a leishmaniosi in un gatto. Paper presented at 2° Congresso Nazionale SIDEV, Aci Castello-Catania, Italy, 17th–19th July 2015.
30. Maia C, Sousa C, Ramos C, Cristóvão JM, Faísca P, Campino L. First case of feline leishmaniosis caused by *Leishmania infantum* genotype E in a cat with a concurrent nasal squamous cell carcinoma. JFMS Open Rep. 2015;1:205511691559396. https://doi.org/10.1177/2055116915593969.
31. Pimenta P, Alves-Pimenta S, Barros J, Barbosa P, Rodrigues A, Pereira MJ, Maltez L, Gama A, Cristóvão JM, Campino L, Maia C, Cardoso L. Feline leishmaniosis in Portugal: 3 cases (year 2014). Vet Parasitol Reg Stud Reports. 2015;1–2:65–9. https://doi.org/10.1016/j.vprsr.2016.02.003.
32. Pennisi MG, Persichetti MF, Migliazzo A, De Majo M, Iannelli NM, Vitale F. Feline leishmaniosis: clinical signs and course in 14 followed up cases. Atti LXX Convegno SISVET. 2016;70:166–7.
33. Maroli M, Pennisi MG, Di Muccio T, Khoury C, Gradoni L, Gramiccia M. Infection of sandflies by a cat naturally infected with *Leishmania infantum*. Vet Parasitol. 2007;145:357–60.
34. Pennisi MG, Venza M, Reale S, Vitale F, Lo Giudice S. Case report of leishmaniasis in four cats. Vet Res Commun. 2004;28(Suppl 1):363–6.

35. Foglia Manzillo V, Di Muccio T, Cappiello S, Scalone A, Paparcone R, Fiorentino E, Gizzarelli M, Gramiccia M, Gradoni L, Oliva G. Prospective study on the incidence and progression of clinical signs in naïve dogs naturally infected by *Leishmania infantum*. PLoS Negl Trop Dis. 2013;7:e2225.
36. Solano-Gallego L, Di Filippo L, Ordeix L, Planellas M, Roura X, Altet L, Martínez-Orellana P, Montserrat S. Early reduction of *Leishmania infantum*-specific antibodies and blood parasitemia during treatment in dogs with moderate or severe disease. Parasit Vectors. 2016;9:235.
37. De Tommasi AS, Otranto D, Dantas-Torres F, Capelli G, Breitschwerdt EB, de Caprariis D. Are vector-borne pathogen co-infections complicating the clinical presentation in dogs? Parasit Vectors. 2013;6:97.
38. Morgado FN, Cavalcanti ADS, de Miranda LH, O'Dwyer LH, Silva MRL, da, Menezes, R.C, Andrade da Silva AV, Boité MC, Cupolillo E, Porrozzi R. Hepatozoon canis and Leishmania spp. coinfection in dogs diagnosed with visceral leishmaniasis. Braz J Vet Parasitol. 2016;25:450–8.
39. Tabar MD, Altet L, Martínez V, Roura X. *Wolbachia,* filariae and *Leishmania* coinfection in dogs from a Mediterranean area. J Small Anim Pract. 2013;54:174–8.
40. Ayllón T, Diniz PPVP, Breitschwerdt EB, Villaescusa A, Rodríguez-Franco F, Sainz A. Vector-borne diseases in client-owned and stray cats from Madrid. Spain Vector Borne Zoonotic Dis. 2012;12:143–50.
41. Pennisi MG, Masucci M, Catarsini O. Presenza di anticorpi anti-Leishmania in gatti FIV+ che vivono in zona endemica. Atti LII Convegno SISVET. 1998;52:265–6.
42. Pennisi MG, Maxia L, Vitale F, Masucci M, Borruto G, Caracappa S. Studio sull'infezione da *Leishmania* mediante PCR in gatti che vivono in zona endemica. Atti LIV Convegno SISVET. 2000;54:215–6.
43. Pennisi MG, Lupo T, Malara D, Masucci M, Migliazzo A, Lombardo G. Serological and molecular prevalence of *Leishmania infantum* infection in cats from Southern Italy. J Feline Med Surg. 2012;14:656–7.
44. Persichetti MF, Pennisi MG, Vullo A, Masucci M, Migliazzo A, Solano-Gallego L. Clinical evaluation of outdoor cats exposed to ectoparasites and associated risk for vector-borne infections in southern Italy. Parasit Vectors. 2018;11:136.
45. Sherry K, Miró G, Trotta M, Miranda C, Montoya A, Espinosa C, Ribas F, Furlanello T, Solano-Gallego L. A serological and molecular study of *Leishmania infantum* infection in cats from the Island of Ibiza (Spain). Vector Borne Zoonotic Dis. 2011;11:239–45.
46. Sobrinho LSV, Rossi CN, Vides JP, Braga ET, Gomes AAD, de Lima VMF, Perri SHV, Generoso D, Langoni H, Leutenegger C, Biondo AW, Laurenti MD, Marcondes M. Coinfection of *Leishmania chagasi* with *Toxoplasma gondii*, Feline Immunodeficiency Virus (FIV) and Feline Leukemia Virus (FeLV) in cats from an endemic area of zoonotic visceral leishmaniasis. Vet Parasitol. 2012;187:302–6.
47. Solano-Gallego L, Rodríguez-Cortés A, Iniesta L, Quintana J, Pastor J, Espada Y, Portús M, Alberola J. Cross-sectional serosurvey of feline leishmaniasis in ecoregions around the Northwestern Mediterranean. Am J Trop Med Hyg. 2007;76:676–80.
48. Spada E, Proverbio D, Migliazzo A, Della Pepa A, Perego R, Bagnagatti De Giorgi G. Serological and molecular evaluation of *Leishmania infantum* infection in stray cats in a nonendemic area in Northern Italy. ISRN Parasitol. 2013;2013:1. https://doi.org/10.5402/2013/916376.
49. Spada E, Canzi I, Baggiani L, Perego R, Vitale F, Migliazzo A, Proverbio D. Prevalence of *Leishmania infantum* and co-infections in stray cats in northern Italy. Comp Immunol Microbiol Infect Dis. 2016;45:53–8.
50. Vita S, Santori D, Aguzzi I, Petrotta E, Luciani A. Feline leishmaniasis and ehrlichiosis: serological investigation in Abruzzo region. Vet Res Commun. 2005;29(Suppl 2):319–21.
51. Britti D, Vita S, Aste A, Williams DA, Boari A. Sindrome da malassorbimento in un gatto con leishmaniosi. Atti LIX Convegno SISVET. 2005;59:281–2.
52. Caracappa S, Migliazzo A, Lupo T, Lo Dico M, Calderone S, Reale S, Currò V, Vitale M. Analisi biomolecolari, sierologiche ed isolamento in un gatto infetto da *Leishmania* spp. Atti. X Congresso Nazionale SIDiLV. 2008;10:134–5.

53. Coelho WMD, de Lima VMF, Amarante AFT, do, Langoni, H, Pereira VBR, Abdelnour A, Bresciani KDS. Occurrence of *Leishmania (Leishmania) chagasi* in a domestic cat (*Felis catus*) in Andradina, São Paulo, Brazil: case report. Rev Bras Parasitol Vet. 2010;19:256–8.
54. Dalmau A, Ossó M, Oliva A, Anglada L, Sarobé X, Vives E. Leishmaniosis felina a propósito de un caso clínico. Clínica Vet Pequeños Anim. 2008;28:233–8.
55. Fileccia, I. Qual'è la vostra diagnosi? Paper presented at 1st Congresso Nazionale SIDEV, Montesilvano (Pescara, Italia), 21–23, September 2012.
56. Grevot A, Jaussaud Hugues P, Marty P, Pratlong F, Ozon C, Haas P, Breton C, Bourdoiseau G. Leishmaniosis due to *Leishmania infantum* in a FIV and FELV positive cat with a squamous cell carcinoma diagnosed with histological, serological and isoenzymatic methods. Parasite Paris. 2005;12:271–5.
57. Hervás J, Chacón-M De Lara F, Sánchez-Isarria MA, Pellicer S, Carrasco L, Castillo JA, Gómez-Villamandos JC. Two cases of feline visceral and cutaneous leishmaniosis in Spain. J Feline Med Surg. 1999;1:101–5.
58. Ibba, F. Un caso di rinite cronica in corso di leishmaniosi felina. In: Proceedings 62nd Congresso Internazionale Multisala SCIVAC, Rimini-Italy (29th–31st May 2009), 2009:568.
59. Laruelle-Magalon C, Toga I. Un cas de leishmaniose féline. Prat Méd Chir Anim Comp. 1996;31:255–61.
60. Leiva M, Lloret A, Peña T, Roura X. Therapy of ocular and visceral leishmaniasis in a cat. Vet Ophthalmol. 2005;8:71–5.
61. Marcos R, Santos M, Malhão F, Pereira R, Fernandes AC, Montenegro L, Roccabianca P. Pancytopenia in a cat with visceral leishmaniasis. Vet Clin Pathol. 2009;38:201–5.
62. Migliazzo A, Vitale F, Calderone S, Puleio R, Binanti D, Abramo F. Feline leishmaniosis: a case with a high parasitic burden. Vet Dermatol. 2015;26:69–70.
63. Ozon C, Marty P, Pratlong F, Breton C, Blein M, Lelièvre A, Haas P. Disseminated feline leishmaniosis due to *Leishmania infantum* in Southern France. Vet Parasitol. 1998;75:273–7.
64. Pennisi, MG, Lupo, T, Migliazzo, A, Persichetti, M-F, Masucci, M, Vitale, F. Feline Leishmaniosis in Italy: Restrospective evaluation of 24 clinical cases. In: Proceedings World Leish 5, Porto de Galinhas, Pernambuco-Brazil (13th–17th May 2013), 2013:P837.
65. Pocholle E, Reyes-Gomez E, Giacomo A, Delaunay P, Hasseine L, Marty P. A case of feline leishmaniasis in the south of France. Parasite Paris Fr. 2012;19:77–80.
66. Poli A, Abramo F, Barsotti P, Leva S, Gramiccia M, Ludovisi A, Mancianti F. Feline leishmaniosis due to *Leishmania infantum* in Italy. Vet Parasitol. 2002;106:181–91.
67. Sanches A, Pereira AG, Carvalho JP. Um caso de leishmaniose felina. Vet Med. 2011;63:29–30.
68. Verneuil M. Ocular leishmaniasis in a cat: case report. J Fr Ophtalmol. 2013;36:e67–72.
69. Vides JP, Schwardt TF, Sobrinho LSV, Marinho M, Laurenti MD, Biondo AW, Leutenegger C, Marcondes M. *Leishmania chagasi* infection in cats with dermatologic lesions from an endemic area of visceral leishmaniosis in Brazil. Vet Parasitol. 2011;78:22–8.
70. Barnes JC, Stanley O, Craig TM. Diffuse cutaneous leishmaniasis in a cat. J Am Vet Med Assoc. 1993;202:416–8.
71. Rivas AK, Alcover M, Martínez-Orellana P, Montserrat-Sangrà S, Nachum-Biala Y, Bardagí M, Fisa R, Riera C, Baneth G, Solano-Gallego L. Clinical and diagnostic aspects of feline cutaneuous leishmaniosis in Venezuela. Parasit Vectors. 2018;11:141.
72. Simões-Mattos L, Mattos MRF, Teixeira MJ, Oliveira-Lima JW, Bevilacqua CML, Prata-Júnior RC, Holanda CM, Rondon FCM, Bastos KMS, Coêlho ICB, Barral A, Pompeu MML. The susceptibility of domestic cats (*Felis catus*) to experimental infection with *Leishmania braziliensis*. Vet Parasitol. 2005;127:199–208.
73. Navarro JA, Sánchez J, Peñafiel-Verdú C, Buendía AJ, Altimira J, Vilafranca M. Histopathological lesions in 15 cats with leishmaniosis. J Comp Pathol. 2010;143:297–302.
74. Altuzarra R, Movilla R, Roura X, Espada Y, Majo N, Novella R. Computed tomography features of destructive granulomatous rhinitis with intracranial extension secondary to leishmaniasis in a cat. Vet Radiol Ultrasound. 2018; https://doi.org/10.1111/vru.12666.
75. Leal RO, Pereira H, Cartaxeiro C, Delgado E, Peleteiro MDC. Pereira da Fonseca, I. Granulomatous rhinitis secondary to feline leishmaniosis: report of an unusual presentation and therapeutic complications. JFMS Open Rep. 2018;4(2):2055116918811374.

76. Laurelle-Magalon C, Toga I. Un cas de leishmaniose féline. Prat Med Chir Anim Comp. 1996;31:255–61.
77. Monteverde, V, Polizzi, D, Lupo, T, Fratello, A, Leone, C, Buffa, F, Vazzana, I, Pennisi, MG. Descrizione di un carcinoma a cellule squamose in corso di leishmaniosi in un gatto. Atti Congresso Nazionale ceedings of the 7th National Congress of the Italian Society of Veterinary Laboratory Diagnostics (SiDiLV). Perugia 9–10 November. 2006;7:329–330.
78. Ennas F, Calderone S, Caprì A, Pennisi MG. Un caso di leishmaniosi felina in Sardegna. Veterinaria. 2012;26:55–9.
79. Hervás J, Chacón-Manrique de Lara F, López J, Gómez-Villamandos JC, Guerrero MJ, Moreno A. Granulomatous (pseudotumoral) iridociclitis associated with leishmaniasis in a cat. Vet Rec. 2001;149:624–5.
80. Cohelo WM, Lima VM, Amarante AF, Langoni H, Pereira VB, Abdelnour A, Bresciani KD. Occurrence of *Leishmania (Leishmania) chagasi* in a domestic act (*Felis catus*) in Andradina, São Paulo, Brazil: case report. Rev Bras Parasitol Vet. 2010;19:256–8.
81. Costa-Durão JF, Rebelo E, Peleteiro MC, Correira JJ, Simões G. Primeiro caso de leish-maniose em gato doméstico (*Felis catus domesticus*) detectado em Portugal (Concelho de Sesimbra). Nota preliminar Rev Port Cienc Vet. 1994;89:140–4.
82. Savani ES, de Oliveira Camargo MC, de Carvalho MR, Zampieri RA, dos Santos MG, D'Auria SR, Shaw JJ, Floeter-Winter LM. The first record in the Americas of an autochtonous case of *Leishmania (Leishmania) infantum chagasi* in a domestic cat (*Felix catus*) from Cotia County, São Paulo State. Brazil Vet Parasitol. 2004;120:229–33.
83. Ortuñez, A, Gomez, P, Verde, MT, Mayans, L, Villa, D, Navarro, L. Lesiones granulomatosas en la mucosa oral y lengua y multiples nodulos cutaneos en un gato causado por *Leishmania infantum*. In: Proceedings Southern European Veterinary Conference (SEVC), Barcelona-Spain (30th September–3rd October 2011).
84. Attipa C, Neofytou K, Yiapanis C, Martínez-Orellana P, Baneth G, Nachum-Biala Y, Brooks-Brownlie H, Solano-Gallego L, Tasker S. Follow-up monitoring in a cat with leishmaniosis and coinfections with *Hepatozoon felis* and "*Candidatus* Mycoplasma haemominutum". JFMS Open Rep. 2017;3:205511691774045. https://doi.org/10.1177/2055116917740454.
85. Schubach TM, Figuereido FB, Pereira SA, Madeira MF, Santos IB, Andrade MV, Cuzzi T, Marzochi MC, Schubach A. American cutaneous leishmaniasis in two cats from Rio de Janeiro, Brazil: first report of natural infection with *Leishmania (Viannia) braziliensis*. Trans R Soc Trop Med Hyg. 2004;98:165–7.
86. Ruiz RM, Ramírez NN, Alegre AE, Bastiani CE, De Biasio MB. Detección de *Leishmania (Viannia) braziliensis* en gato doméstico de Corrientes, Argentina, por técnicas de biología molecular. Rev Vet. 2015;26:147–50.
87. Rougeron V, Catzeflis F, Hide M, De Meeûs T, Bañuls A-L. First clinical case of cutaneous leishmaniasis due to *Leishmania (Viannia) braziliensis* in a domestic cat from French Guiana. Vet Parasitol. 2011;181:325–8.
88. Passos VM, Lasmar EB, Gontijo CM, Fernandes O, Degrave W. Natural infection of a domes-tic cat (*Felis domesticus*) with *Leishmania (Viannia)* in the metropolitan region of Belo Horizonte, State of Minas Gerais. Brazil Mem Inst Oswaldo Cruz. 1996;91:19–20.
89. de Souza AI, Barros EM, Ishikawa E, Ilha IM, Marin GR, Nunes VL. Feline leishmaniasis due to *Leishmania (Leishmania) amazonensis* in Mato Grosso do Sul State. Brazil Vet Parasitol. 2005;128:41–5.
90. Bonfante-Garrido R, Valdivia O, Torrealba J, García MT, Garófalo MM, Urdaneta R, Alvarado J, Copulillo E, Momen H, Grimaldi G Jr. Cutaneous leishmaniasis in cats (*Felis domesti-cus*) caused by *Leishmania (Leishmania) venezuelensis*. Revista Científica, FCV-LUZ. 1996;6:187–90.
91. Trainor KE, Porter BF, Logan KS, Hoffman RJ, Snowden KF. Eight cases of feline cutaneous leishmaniasis in Texas. Vet Pathol. 2010;47:1076–81.

92. Persichetti MF, Solano-Gallego L, Vullo A, Masucci M, Marty P, Delaunay P, Vitale F, Pennisi MG. Diagnostic performance of ELISA, IFAT and Western blot for the detection of anti-*Leishmania infantum* antibodies in cats using a Bayesian analysis without a gold standard. Parasit Vectors. 2017;10:119.
93. Segarra S, Miró G, Montoya A, Pardo-Marín L, Boqué N, Ferrer L, Cerón J. Randomized, allopurinol-controlled trial of the effects of dietary nucleotides and active hexose correlated compound in the treatment of canine leishmaniosis. Vet Parasitol. 2017;239:50–6.
94. da Silva SM, Rabelo PFB, Gontijo, N. de F, Ribeiro RR, Melo MN, Ribeiro VM, Michalick MSM. First report of infection of *Lutzomyia longipalpis* by *Leishmania (Leishmania) infantum* from a naturally infected cat of Brazil. Vet Parasitol. 2010;174:150–4.
95. Brianti E, Gaglio G, Napoli E, Falsone L, Prudente C, Solari Basano F, Latrofa MS, Tarallo VD, Dantas-Torres F, Capelli G, Stanneck D, Giannetto S, Otranto D. Efficacy of a slow-release imidacloprid (10%)/flumethrin (4.5%) collar for the prevention of canine leishmaniosis. Parasit Vectors. 2014;7:327.
96. Pennisi MG, Hartmann K, Addie DD, Lutz H, Gruffydd-Jones T, Boucraut-Baralon C, Egberink H, Frymus T, Horzinek MC, Hosie MJ, Lloret A, Marsilio F, Radford AD, Thiry E, Truyen U, Möstl K. European Advisory Board on Cat Diseases. Blood transfusion in cats: ABCD guidelines for minimising risks of infectious iatrogenic complications. J Feline Med Surg. 2015;17:588–93.

Ectoparasitic Diseases

Federico Leone and Hock Siew Han

Abstract

Ectoparasitic skin diseases are extremely common in cats, and their correct identification is very important for both the cat's and the owner's welfare. In this chapter, the most important feline ectoparasitic diseases will be discussed, including the morphological features of the parasite, clinical signs, diagnostic techniques and therapeutic options. The majority of these diseases can be diagnosed with tests that can be easily performed during the clinical examination, such as the direct examination with a magnifying lens and microscopic examination of samples collected with clear cell tape, by superficial and deep skin scrapings and hair plucking and microscopic examination of ear cerumen. In some cases, the diagnostic techniques are not particularly sensitive, and a negative result doesn't allow ruling out the disease: a therapeutic trial is the only way to confirm or rule out the disease. The recent introduction on the market of new wide-spectrum parasiticidal drugs, effective to prevent flea and tick infestations and with acaricidal and insecticidal activity, will make ectoparasite control much easier. However, for many diseases there are no registered products and standardized protocols for the feline species.

Introduction

Ectoparasitic skin diseases caused by mites and insects are very important in feline dermatology as they are included in the differential diagnoses of many pruritic dermatological conditions. Their prevalence varies depending on the geographical area

F. Leone (✉)
Clinica Veterinaria Adriatica, Senigallia (Ancona), Italy

H. S. Han
The Animal Clinic, Singapore

© Springer Nature Switzerland AG 2020
C. Noli, S. Colombo (eds.), *Feline Dermatology*,
https://doi.org/10.1007/978-3-030-29836-4_19

considered and on the cat's lifestyle. Living in colonies of stray cats, breeding facilities or catteries or the potential contact with stray cats make the cat more susceptible to parasitic infestations. Some parasitic diseases may also involve the owner and, although these infestations are usually transient since the parasite is not adapted to man, these zoonoses should not be underestimated.

Notoedric Mange

Notoedric mange, also known as feline scabies, is a pruritic, contagious skin disease affecting the cat, caused by the mite *Notoedres cati*. The mite may affect other mammals, including man, and exceptionally the dog [1–3]. The disease prevalence is unknown; it is thought to be rare, however epidemics are still reported in some European countries [3, 4]. Kittens are more prone to the disease compared to adult cats.

Morphology

Notoedres cati has an oval body, ventrally flattened and dorsally convex; adult females are approximately 225 µm long and males 150 µm long. The head carries a short and squared rostrum. The limbs are short, with unjointed pretarsi ending with a sucker-like structure called pulvillus, present in females only on the two front limb pairs. The hind limbs are rudimental, do not extend beyond the mite's body, and carry long setae lacking suckers in both sexes (Fig. 1). The dorsal cuticle shows fingerprint-like concentrical rings, transversal rounded scales and no spines. The anal opening is dorsally located and the eggs are oval [4, 5].

Fig. 1 *Notoedres cati,* adult mite

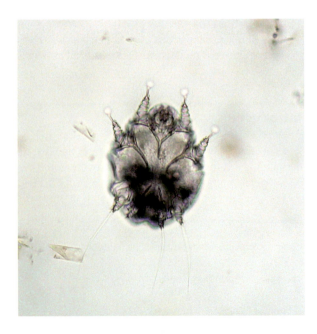

Life Cycle

The life cycle of *Notoedres cati* takes place entirely on the host (permanent parasitism). After mating on the skin surface, females burrow tunnels within the stratum corneum at a speed of 2–3 mm/day. Two to three eggs a day are laid in the tunnels for 2–4 weeks. The six-legged larva hatches from the egg, and after two moults as protonymph and tritonymph it becomes an adult mite. The life cycle spans 14–21 days, in favorable environmental conditions. The mite feeds on epidermal debris and interstitial fluid.

Epidemiology

Notoedric mange is extremely contagious and transmitted by direct contact. For this reason, cats leaving in breeding facilities, catteries or colonies are predisposed. Where feline colonies are maintained, the disease may persist and become established; this happens commonly in urban or extra-urban areas such as cemeteries and ruins, and in close proximity to hospitals and schools [1].

Notoedric mange is a zoonotic disease, and man can transiently be infested, showing pruritus, papules, vesicles and crusts especially on the limbs and trunk. In a study, 63% of people coming into contact with an infested cat showed clinical signs of notoedric mange. Mites were detected by skin scraping in 60% of patients examined [6]. Lesions resolve spontaneously within 3 weeks, once contact with the infected cat is stopped [6, 7].

Clinical Signs

Initial lesions are represented by papules or crusted papules and scales, which, with disease progression, evolve into gray-yellow thick crusts, extremely adherent to the skin surface (Fig. 2). Lesions initially appear on the ear pinnae margins and later involve the whole pinna, the face and the neck. With disease progression, the lesions may become generalized. Pruritus is usually severe and self-trauma is common, causing alopecia, erosions and ulcers and predisposing to secondary bacterial or yeast infections [1]. Grooming and the cat's curled up sleeping habit may cause diffusion to the limbs and perineum. If not treated, the cat may become lethargic and dehydrated and may die in rare cases [1, 8, 9].

Diagnosis

Diagnosis requires microscopic identification of the parasite and/or its eggs and/or its feces (round-shaped and brown) in samples collected by superficial skin scraping (Box 1). Mites are usually numerous and can be easily found, as opposite to

Fig. 2 Crusting of the pinnae margins in a cat with notoedric mange

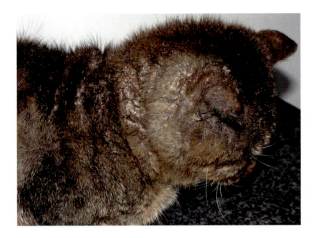

Fig. 3 Superficial skin scraping: adult mites, eggs, and mite feces are present

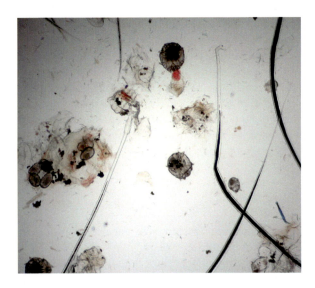

Sarcoptes scabiei (Fig. 3) [1, 8]. Recently, diagnosis of notoedric mange by microscopic examination of samples collected by using clear tape has been reported, with sensitivity comparable to skin scrapings. This technique is less traumatic and therefore indicated for difficult body sites such as lips and periocular regions [10].

Box 1: Superficial Skin Scraping: Practical Tips
- Select typical locations of the parasite (e.g., margins of ear pinnae in noto-edric mange)
- If clipping is necessary, use scissors instead of clippers and leave a few millimeters of hair, to avoid removal of material containing parasites (e.g., crusts)

- Apply a few drops of mineral oil on the skin
- Scrape a large area of skin superficially to avoid blood contamination
- Perform multiple skin scrapings
- If a large amount of material is obtained, divide it onto more slides
- Mix your sample on the glass slide adding a few drops of mineral oil, if necessary, and try to obtain a single layer
- Cover with a coverslip and observe the sample with the microscope, closing partially the diaphragm and reducing the light. This allows better visualization of the parasites

Treatment

Treatment of notoedric mange can be achieved with different acaricidal active ingredients. Registered products include a spot-on formulation containing eprinomectin, fipronil, (S)-methoprene and praziquantel [11] and a spot-on containing moxidectin and imidacloprid [12], which can be applied once or twice at 1-month interval. Other protocols involving active ingredients not registered for the disease involve the use of selamectin spot-on (6–12 mg/kg applied twice at 14 or 30 days' interval) [1, 13, 14], ivermectin (0.2–0.3 mg/kg subcutaneously at 14 days' interval) [1, 7, 15] and doramectin (0.2–0.3 mg/kg subcutaneously once) [16]. The new family of isoxazolines ectoparasiticidals has been shown to be effective in other diseases caused by mites. There are no specific studies on feline notoedric mange, but isoxazolines are likely to be effective. Notoedric mange is extremely contagious, and all in-contact cats must be treated to avoid re-infestations.

Otodectic Mange

Otodectic mange is a parasitic disease of the external ear canal caused by the mite *Otodectes cynotis*. The mite is not species-specific and may affect cats, dogs and other mammals. Fifty to eighty percent of cases of feline otitis externa is caused by *Otodectes cynotis*, which is present all over the world [5, 17].

Morphology

Otodectes cynotis has an oval body and a long, conical rostrum. Females are 345–451 μm long, while males are smaller (274–362 μm). The limbs are long, with short pedicles ending with a cup-shaped, sucker-like structure used by the parasite to move quickly within the ear cerumen. The adult mites show sexual dimorphism: males have four pairs of long limbs, ending beyond the body, and smaller abdominal lobes; in females, the fourth pair of legs is atrophic and does not extend beyond the body while abdominal lobes are bigger (Fig. 4). Eggs are oval, slightly flattened on one side and 166–206 μm long [4, 5].

Fig. 4 *Otodectes cynotis*, female mite

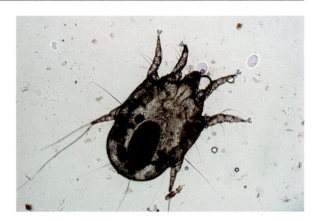

Life Cycle

The life cycle of *Otodectes cynotis* takes place entirely on the host (permanent parasitism). The mite lives on the surface of the external ear canals and does not burrow. After mating, female mites lay eggs, hatching in 4–6 days. Hexapod larvae actively feed for 3–10 days to moult in octopod protonymphae and then deutonymphae [4, 5]. Mating, often observed during microscopic examination, involves the male mite and the deutonympha: the male mite becomes attached to the deutonympha by using mating suckers and, if a female develops mating occurs, while when a male mite develops there is detachment [4, 8]. The life cycle requires 3 weeks to complete and adult mites survive on the host for approximately 2 months. Mites feed on skin debris and fluids stimulating production of large amounts of ear cerumen, occasionally mixed with blood [8]. Mites can survive for up to 12 days off the host, in ideal temperature conditions [18].

Epidemiology

Otodectic mange is extremely contagious and transmission occurs primarily by direct contact with infested cats. Common is also infestation of one ear from the other in the same cat [19]. The disease affects kittens and adults; however, juveniles are predisposed [19]. A temporary infestation of human beings may occur, with papules localized predominantly on the arms and trunk [20], while parasitic otitis is extremely rare [21].

Clinical Signs

Otodectic mange causes pruritic, erythematous and ceruminous otitis externa, almost always bilaterally. Otodectic otitis is characterized by large amounts of brown-black dry cerumen, resembling "coffee powder" (Fig. 5) [8]. In felines,

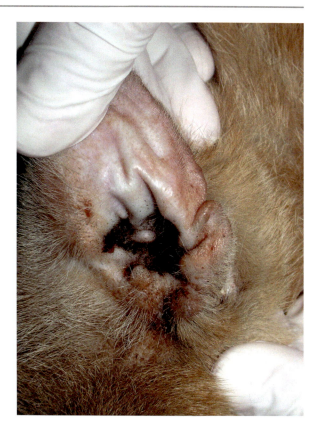

hypersensitivity to mites may occur and affected cats show severe pruritus, not proportionate to the number of mites in the ear canal [22]. On the other hand, some cats may have huge numbers of mites in the external ear canal without pruritus, and this may be explained by the absence of hypersensitivity phenomena [4]. Cats infested with *Otodectes cynotis* may be positive to intradermal testing for house dust mites such as *Dermatophagoides farinae*, *Dermatophagoides pteronyssinus* and *Acarus siro* [23]. Secondary bacterial or yeast infections are possible [24]. Pruritus severity is responsible for auto-traumatic lesions such as alopecia, erosions, ulcers and crusting affecting the preauricular regions, head, face and neck and for otohematomas [17].

Extra-auricular infestation may also occur, since the mite may leave the external ear canal and cause alopecia and miliary dermatitis in other body sites (ectopic mites) [4, 8].

Diagnosis

Diagnosis is made by microscopic observation of the mite or its eggs (Fig. 6). The preferred technique is microscopic examination of ear cerumen obtained with an

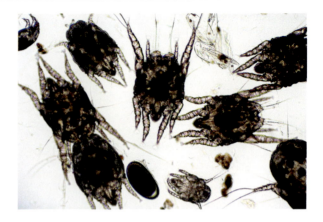

Fig. 6 Microscopic examination of ear cerumen: an egg and an adult mite mating with a deutonympha are visualized

ear swab (Box 2). Samples must be obtained before applying cerumenolytic products or cleaning the ear canal. To increase the sensitivity of the test, obtaining more samples collected from the horizontal canal by passing the swab through the otoscope cone is recommended. Mites can be visualized by otoscopic examination as white moving dots. Superficial skin scrapings allow to detect mites in cases with extra-auricular localization [4].

Box 2: Microscopic Examination of Ear Cerumen: Practical Tips
- Collect your sample from the ear canal with a swab
- Collect your sample before applying ceruminolytics or cleaning the ear canal
- To collect deeper samples, use the otoscope cone to guide the swab
- Dilute the sample in mineral oil, previously applied on a glass slide
- Cover with a coverslip and observe the sample on the microscope, closing partially the diaphragm and reducing the light. This allows better visualization of the parasites

---

Treatment

Many topical and systemic active ingredients are available to treat otodectic mange. Before treatment, cleaning the ear canals with a cerumenolytic product in order to mechanically remove the parasites and the excess of cerumen caused by the mites is recommended [8]. Topical therapy involves acaricidals such as permethrin or thiabendazole directly applied into the ear canals. These active ingredients have limited residual activity and require daily application for 3 weeks, to ensure that all eggs hatch and emerging larvae are exposed to the drug, despite the fact that they are usually registered to be used for 7–10 days [17, 25]. Otologic products not containing acaricidals are also effective, although their mechanism of action is unclear. It is hypothesized that mites die because they cannot move and/or breathe due to the product [26, 27]. Fipronil spot-on is not registered for otoacariasis; however, it proved effective when one drop is applied to each ear canal and the rest between the scapulae [28]. Systemic therapy has many advantages compared to topical therapy. Easiness of administration increases the owner's compliance to continuity of treatment. Systemic treatment is also effective in cases with ectopic mite localization [17]. Among non-registered active ingredients, ivermectin administered subcutaneously at 0.2–0.3 mg/kg twice at a 14-day interval or orally once weekly for 3 weeks has been shown to be effective [29]. Registered drugs include selamectin and moxidectin-imidacloprid spot-on, both administered twice at 1-month interval [30, 31]. An evidence-based review published in 2016 recommended the use of selamectin or moxidectin-imidacloprid spot-on once or twice at a 30-day interval for feline otodectic mange. There is not enough evidence to recommend other active ingredients [17].

Recently, new active ingredients belonging to the isoxazolines family have been marketed. Sarolaner and selamectin spot-on has been registered to treat otodectic mange and was effective with a single treatment [32]. Also as single treatments, fluralaner alone or in association with moxidectin as spot-on was shown to be effective [33, 34]. Afoxolaner, only registered for dogs, was successfully used in cats as single oral treatment [35]. A single application of a spot-on containing eprinomectin, fipronil, (S)-methoprene, and praziquantel was effective to prevent *Otodectes cynotis* infestation in cats [36]. Regardless of the treatment chosen, all in-contact animals must be treated due to the likelihood of contagion and presence of asymptomatic carriers [29].

Cheyletiellosis

Cheyletiellosis is a parasitic skin disease caused by mites belonging to the genus *Cheyletiella*. The majority of mites of the Cheyletiellidae family are predators feeding on other mites, while some species are only ectoparasitic. The three species of dermatological interest are *Cheyletiella blakei*, *Cheyletiella yasguri* and *Cheyletiella parasitivorax* [5]. *Cheyletiella* shows species preference, with *Cheyletiella blakei* adapted to the cat, *Cheyletiella yasguri* to the dog and *Cheyletiella parasitivorax* to the rabbit. However, there is no strict species specificity, and interspecies infestations are possible [4, 8].

Morphology

The adult mite is large (300–500 μm long); the body is hexagonal and, according to some authors, resembles a pepper or a shield [4]. The limbs are short and carry comb-like appendixes at the end, the rostrum is well-developed with palps ending with two prominent curved hooks, looking like Viking horns (Fig. 7). The three *Cheyletiella* species can be differentiated by observing the shape of the sensorial structure (solenidion) located on the third section of the first pair of legs. The solenidion is heart-shaped in *Cheyletiella yasguri*, conical in *Cheyletiella blakei*

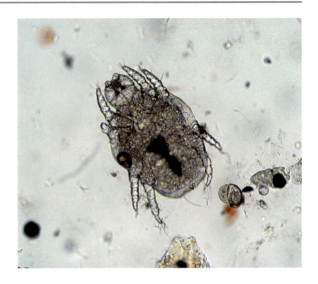

Fig. 7 *Cheyletiella blakei,* adult mite

and rounded in *Cheyletiella parasitivorax* [4, 5]. However, species differentiation is often difficult due to individual variation of the solenidion shape and artifacts due to fixation for microscopy [37]. The eggs are 235–245 μm long and 115–135 μm wide and elliptical and, unlike lice eggs, are non-operculated and loosely attached to hair shafts by thin filaments [4, 8].

Life Cycle

The life cycle of *Cheyletiella* spp. takes place entirely on the host (permanent parasitism). The mite lives in the stratum corneum at the base of hair shafts, moving quickly through the scales without burrowing and feeding on epidermal debris and fluids. The eggs are laid along the hair shafts at 2–3 mm distance from the skin surface. The hexapod larva develops within the egg; once hatched, it moults twice as nymphal stage and finally becomes an adult mite. The life cycle spans 14–21 days, when environmental conditions are favorable [4, 5, 8].

Epidemiology

Cheyletiellosis is extremely contagious and transmission is usually by direct contact [4, 8]. Less often, contagion occurs indirectly because adult female mites can survive for up to 10 days in the environment, while immature stages and males die quickly when off the host [4, 8]. *Cheyletiella* can also be carried by fleas, lice or flies [4]. The disease occurs more commonly in young animals coming from pet shops or colonies while in adult cats can be diagnosed in debilitated or systemically ill animals [5]. Cheyletiellosis is a zoonosis and man can be transiently infested, showing severely pruritic macules and papules on the limbs, trunk and buttocks [8, 38, 39]. When the affected animal is treated with an acaricidal, lesions in humans spontaneously regress within 3 weeks [8].

Clinical Signs

Clinical signs are of variable severity [8]. Most affected cats initially show exfoliative dermatitis affecting the dorso-lumbar area, with small and dry whitish scales easily detaching from the skin surface (Fig. 8) [4, 8]. The cat's grooming behavior may remove both the scales and the mites and initially the disease may be slowly progressive and remain undetected [8]. Later on, the exfoliation may become more severe and the hair coat may look "dusty." Many authors use the term "walking dandruff" to describe the mites moving on the skin surface. Mites are whitish in color and can be distinguished from scales because they move [4, 8]. Pruritus is of variable severity, from absent to severe, and not proportionate to the number of mites, increasing the suspicion of hypersensitivity phenomena in some cats [8, 37, 40]. Some animals present with self-traumatic lesions such as alopecia, excoriations, ulcers and crusts due to severe pruritus [8]. Miliary dermatitis or self-induced alopecia patterns can be observed [8, 29].

Fig. 8 Severe scaling on the dorsum of a cat with cheyletiellosis

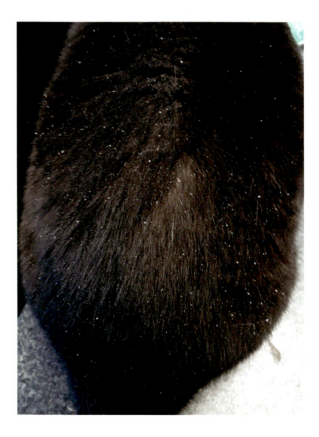

Fig. 9 *Cheyletiella* egg
attached to the hair shaft:
the egg is not operculated

Diagnosis

Diagnosis of cheyletiellosis can be made by microscopic observation of the parasite or its eggs, although the large size of the mite sometimes allows the direct observation on the cat's hair coat, with a magnifying lens [4, 8]. The preferred technique is microscopic examination of samples obtained with clear cell tape (Box 3). Collection of samples may be done directly on the cat's fur or after combing the coat with a flea comb. Another useful technique is superficial skin scraping, particularly if few mites are present. Microscopic examination of hair shafts allows observation of eggs attached to hair shafts (Fig. 9) [4, 8, 29, 40, 41]. Pruritic cats may ingest the mites or eggs due to overgrooming and a fecal flotation test may be diagnostic [4, 8, 40]. In feces, *Cheyletiella* eggs are similar to *Ancylostoma* eggs; however, they are three to four times bigger (230×100 μm) and often embryonated [4, 8]. Identification of mites can be difficult, and in some cases the diagnosis is confirmed by a therapeutic trial [8, 40].

Box 3: Microscopic Examination of Samples Collected with Clear Tape: Practical Tips
- Choose good-quality clear tape
- Collect your sample applying the tape to the skin multiple times (note: it is possible that parasites are not collected, especially in long-haired cats)
- Use the coat combing technique: brush the hair coat with a flea comb or with your hands so the sample falls on the table, which should be perfectly clean
- Collect the sample with clear tape directly from the table
- Apply a few drops of mineral oil on a glass slide and cover with the clear tape
- Cover with a coverslip and observe the sample with the microscope, closing partially the diaphragm and reducing the light. This allows better visualization of the parasites

Treatment

There is no registered active ingredient to treat cheyletiellosis in cats. Topical (fipronil spot-on as a single treatment) [42] or systemic (selamectin spot-on, three applications with 1-month interval [40, 43] or ivermectin, 0.2–0.3 mg/kg subcutaneously once every 2 weeks [41]) acaricidal products have been reported as effective.

Trombiculosis

Trombiculosis is a parasitic skin disease caused by larvae of mites belonging to the Trombiculidae family. The disease is also called "grass itch mites" or "chiggers" in North America, "scrub itch" in Australia and "harvest mites" in Europe [44]. Within the Trombiculoidea superfamily, the Trombiculidae family includes approximately 1500 species, of which only approximately 50 can infest birds, mammals and man. The most important species of veterinary interest belong to the genus *Trombicula*, which groups many subgenera such as *Neotrombicula* and *Eutrombicula*. In Europe, the most commonly involved species is *Neotrombicula autumnalis*, while in the Southeastern and Central USA, *Eutrombicula alfreddugesi* is most often diagnosed [5, 45]. The main feature of this family is that only the larval stage is parasitic (transient parasitism), while nymphae and adult mites are free living in the environment. The larvae are obliged parasites, are not host-specific, and can infest many species including man [4, 5].

Morphology

Exapod *Neotrombicula autumnalis* larvae are oval, 200–400 μm long and are characterized by a typical red-orange color (Fig. 10). The mouth parts include a well-developed rostrum and chelicerae with robust tweezer-shaped palps. The trunk carries a pentagonal dorsal scutum (rectangular in *Eutrombicula alfreddugesi*) and the body is covered by long feather-shaped setae. The limbs end with a trifurcated claw (bifurcated in *Eutrombicula alfreddugesi*) used to attach to the host [4, 5]. Adult mites are non-parasitic, approximately 1 mm long and are also red-orange in color [4].

Life Cycle

The female mite lays spherical eggs on the ground. Larvae emerge from the eggs in a week and move actively on the ground climbing on the grass and waiting for the host [5]. Larvae require 80% relative humidity, and for this reason they climb on plants less than 30 cm high [45]. Once on the host, the larvae attach with the chelicera and feed through a peculiar structure called stilosoma, which is made of solidified mite saliva. This structure allows the buccal apparatus to penetrate down to the

Fig. 10 *Neotrombicula autumnalis* hexapod larva; note the bright red-orange color

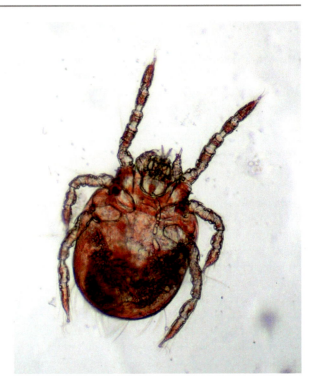

derma of the host and to feed on tissue fluids (extra-intestinal digestion) [4, 46]. During the time spent on the host, the larva grows from 0.25 mm to 0.75 mm, and its bright red-orange color becomes pale yellow [47]. After feeding for 3–15 days, the larvae fall on the ground to complete their life cycle in the environment. The nymphal and adult stages are free living and mobile, and feed on small arthropods or their eggs and fluids from plants. The life cycle spans over 50–70 days and is strongly influenced by the season [4, 5].

Epidemiology

In Europe female mites tend to lay eggs in spring and summer and larvae are very abundant at the end of summer and in autumn. However, depending on climate, more than one life cycle can be completed and larvae can also be found in different seasons [4, 48, 49].

Trombiculosis is not a zoonosis because humans get infested directly from the environment; however, direct transmission from animals to man cannot be excluded [47]. People working or spending time in the countryside or in forests during the larvae season are predisposed. Clinical signs are thought to be due to the irritant effect of the mite's saliva and to acquired hypersensitivity to salivary antigens. In non-sensitized individuals, pruritic macules and papules develop, while in sensitized patients the pruritus is severe and associated with urticaria, papules, vesicles, fever and enlarged lymph nodes. Lesions are mostly seen on the wrist, flexural surface of the arm, belt line, ankle, popliteal fossa and thigh [47, 49, 50]. In children, the "summer penile syndrome" is reported: an acute hypersensitivity reaction to mites with erythema, edema, and pruritus to the penis and dysuria due to partial phimosis with reduction of urinary output [51].

Neotrombicula autumnalis larvae are thought to be potential vectors of *Borrelia burgdorferi*, causing Lyme disease, and *Anaplasma phagocytophilum* (previously known as *Ehrlichia phagocytophila*), causing human granulocytic anaplasmosis, by trans-stage or trans-ovaric transmission [52–54].

Clinical Signs

Larvae climb on plants and wait for the host, to which they attach by direct contact. For this reason, parasites are preferentially found on body areas in contact with the ground, such as abdomen, interdigital spaces, claw folds, muzzle and pinnae, especially in the fold at the base of the pinnae margin (Henry's pocket). Mites can be visualized as red-orange aggregates (Fig. 11) [5, 48]. The facial location reflects the first contact site of the larvae with the host, directly linked to the feline exploratory behavior, while Henry's pocket location might be explained by the epidermal thinness which facilitates the stilosoma formation; moreover, the pocket protects the larvae [48].

In some cats, the infestation is completely asymptomatic, and mites can be incidentally noticed by the owner or observed during the clinical examination for annual vaccination [48].

Other cats show variable pruritus, from moderate to severe, possibly related to individual hypersensitivity which may persist after the larvae abandoned the host [5, 48]. Some cats show crusted papules and self-traumatic lesions such as alopecia, excoriations, ulcers and crusts, depending on pruritus severity. Miliary dermatitis or self-induced alopecia may be observed [48, 55, 56].

Fig. 11 Orange-colored collections of parasitic larvae can be seen to the naked eye on the head and pinna of a cat

Fig. 12 Superficial skin scraping: many *Neotrombicula autumnalis* larvae

Diagnosis

Diagnosis requires a compatible history and macroscopic and microscopic observation of the parasites. Hair coat examination with a magnifying lens allows to observe small aggregates of orange-colored larvae. Microscopic examination of samples collected with clear cell tape or superficial skin scraping allows parasite identification (Fig. 12) [48].

Treatment

There is currently no registered treatment for trombiculosis, and there are very few studies on the effectiveness of acaricidals to treat this disease in cats. It is a relatively easy disease to treat, as many ectoparasiticidal products are effective; however, re-infestation may be common in cats with free access to infested areas. Fipronil spray [48, 57], selamectin spot-on [48, 58], and imidacloprid-moxidectin spot-on [48] have been successfully used with a single application. These active ingredients seem to protect against environmental re-infestations.

Demodicosis

Feline demodicosis is an uncommon to rare parasitic skin disease caused by mites belonging to the genus *Demodex*. Currently, three species have been identified in cats, using molecular techniques: *Demodex cati*, *Demodex gatoi* and a third unnamed species [59, 60].

Morphology

Demodex cati is very similar to *Demodex canis*, with minimal taxonomic differences. The body is elongated and cigar-shaped. An adult male is 182 μm long and 20 μm wide, while an adult female is 220 μm long and 30 μm wide [4, 61, 62]. The gnathosoma, in the frontal part of the body, is trapezoidal and carries two chelicera and two palps. In the podosoma, the middle part of the body, there are four pairs of atrophied limbs, each one carrying one pair of tarsal claws, distally bifurcated with a large, caudally oriented dewclaw. The terminal part of the body is the opisthosoma, accounting for two thirds of the mite body, transversally striated and ending with a tapered point (Fig. 13) [61]. The female reproductive system is ventrally located, below the fourth pair of legs. In the male mite, it is in the dorsal half and corresponds to the second pair of legs. The eggs are oval and 70.5 μm long on average [4, 61].

Demodex gatoi is smaller and stubbier and morphologically similar to *Demodex criceti*, the hamster's parasite [8, 63]. Males are 90 μm long and females are 110 μm long [62–64]. The opisthosoma accounts for less than half of the total length of the body, is horizontally striated and caudally rounded (Fig. 14) [63, 65]. The eggs are oval and smaller than *Demodex cati* eggs [63].

Fig. 13 *Demodex cati*: the opisthosoma accounts for two thirds of the parasite body and the tip is tapered

Fig. 14 *Demodex gatoi*:
the opisthosoma length is
less than half of the entire
body and the tip is rounded

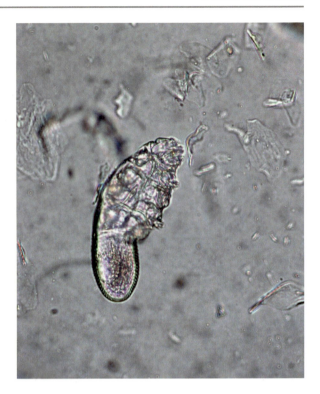

The third, still unnamed *Demodex* species is of intermediate size, with a body shorter and stubbier than *Demodex cati* but longer and more tapered than *Demodex gatoi* [62, 66, 67].

Life Cycle

Demodex cati lives in the hair follicle, often located close to the exit of the sebaceous gland duct, with its head directed downward [61]. Conversely, *Demodex gatoi* lives in the stratum corneum [62–64]. The environment of the third *Demodex* species is unknown, since it has never been described in histopathological samples [62, 66]. Information related to the life cycle are referred only to *Demodex cati* [61].

The life cycle takes place entirely on the host (permanent parasitism). Mating occurs on the skin surface; then the fertilized female moves into the hair follicle where it lays eggs. Six-legged larvae hatch and, after two nymphal stages, the second one moves back onto the skin surface and develops into adult, and more hair follicles are colonized [61].

Epidemiology

The way of transmission of *Demodex cati* is unknown. In the dog, transmission occurs from the mother to puppies within the first days of life, during lactation [8]. The morphologic and environmental similarities of *Demodex canis* and *Demodex cati* suggest that the way of transmission is identical. The disease is not contagious.

The disease caused by *Demodex gatoi* appears to be contagious among cats sharing the same environment, if there is enough parasitic pressure [64, 66, 68]. It is not known if the third *Demodex* species is contagious. *Demodex* spp. are host-specific mites and the disease is not zoonotic.

Clinical Signs

In *Demodex cati* demodicosis, a localized and a generalized form have been described [4, 8]. The localized form involves the head and neck, particularly the periorbital and perilabial regions and the chin [4, 8, 69]. The lesions are erythema, alopecia, scales and crusts. Pruritus is variable, generally mild to absent [8, 69–71]. When the disease involves the external ear canal, it causes a bilateral ceruminous otitis which is often reported in feline immunodeficiency virus (FIV)-positive cats [72, 73]. A localized form has also been reported in cats affected by asthma and chronically treated with glucocorticoids administered with aerosol [74].

The generalized form causes lesions similar to the ones observed in the localized form, but more severe and extensive, involving the muzzle, neck, trunk and limbs or the whole body (Fig. 15) [8, 65, 69–71]. The generalized disease is often associated with immunosuppressive therapies or concurrent systemic diseases

Fig. 15 Large alopecic area on the dorsum of a cat with generalized demodicosis due to *Demodex cati*

Fig. 16 Severe self-induced lesions in a cat with *Demodex gatoi*

such as diabetes mellitus, xanthomas, toxoplasmosis, systemic lupus erythematosus, hypercortisolism, retroviral infections and Bowenoid *in situ* carcinoma [69, 71, 75–79]. However, in some cases an underlying disease cannot be identified [80].

In *Demodex gatoi* infestation, the most common clinical sign is variable pruritus, from absent to severe, and in some cases mite hypersensitivity is suspected (Fig. 16) [8, 62, 64, 81]. Cats may show self-induced alopecia involving the trunk, abdomen flanks or limbs or self-traumatic lesions such as alopecia, excoriations, ulcers and crusts or papular and crusting dermatitis (miliary dermatitis) [64, 81]. This type of demodicosis is not associated with immunosuppression [81]. Infestation with different *Demodex* species in the same cat has been reported [62, 65].

Diagnosis

Diagnosis of feline demodicosis is confirmed by microscopic observation of the adult mite, its immature stages, or its eggs. Diagnostic techniques used are different depending on the involved mite species and its localization. *Demodex cati* lives in the hair follicle, and the preferred diagnostic method is the deep skin scraping, followed by microscopic examination of hair pluckings (Boxes 4 and 5) [81]. For *Demodex gatoi*, a superficially located species, the suggested methods are superficial skin scrapings or microscopic examination of samples obtained by clear cell tape [81]. These mites are small and transparent, and reducing the amount of light going through the microscope by partially closing the diaphragm to increase contrast is advised [64, 65]. In overgrooming cats, the observation of *Demodex gatoi* may be difficult, and some authors suggest a fecal examination with flotation [81, 82]. Moreover, some authors suggest to treat cats with an acaricidal whenever *Demodex gatoi* is suspected [64, 81].

> **Box 4: Microscopic Examination of Hair Pluckings: Practical Tips**
> - Carefully choose the hair shafts to examine
> - Use hemostats or your fingers to grab the hair base
> - Pluck the hair in the direction of growth
> - Align the hair shafts on a glass slide with a few drops of mineral oil
> - Cover with a coverslip and observe the sample on the microscope, closing partially the diaphragm and reducing the light. This allows better visualization of the parasites

> **Box 5: Deep Skin Scrapings: Practical Tips**
> - Choose your sample spot, avoiding ulcerated or fibrotic areas
> - Clip hair if required
> - Apply a few drops of mineral oil to the skin
> - Scrape the skin until capillary bleeding is observed
> - Perform multiple skin scrapings
> - If a large amount of material is obtained, divide it onto more slides
> - Mix your sample on the glass slide adding a few drops of mineral oil if necessary and try to obtain a single layer
> - Cover with a coverslip and observe the sample on the microscope, closing partially the diaphragm and reducing the light. This allows better visualization of the parasites

Treatment

There is no registered product for feline demodicosis and there are no standardized protocols. Various active ingredients have been used with variable results, depending on the mite species and the dosage administered. An evidence-based review recommended the use of weekly rinses with 2% calcium sulfur [83]; however, this product is not available in many countries.

Moderate evidence of effectiveness for both *Demodex* species was reported for once or twice weekly amitraz rinses (0.0125–0.025%), which may be toxic in felines, and for macrocyclic lactones [83]. Ivermectin may be administered both orally and subcutaneously and is effective for both species; however, failures have been reported in *Demodex gatoi* cases [62, 64, 81]. Doramectin (600 µg/kg subcutaneously once weekly for 2–3 weeks) is effective to treat *Demodex cati* [83, 84]. Both with ivermectin and doramectin, severe central nervous system toxicity has been described [62].

Milbemycin oxime has been shown to be effective against *Demodex cati* at 1–1.5 mg/kg orally once daily for 2–7 months [74, 76], and once weekly topical imidacloprid/moxidectin for eight applications is effective against *Demodex gatoi* [85].

Recently, single treatment with oral fluralaner has been reported to be effective for both *Demodex* species [86, 87]. *Demodex gatoi* is contagious and treatment of all in-contact cats is recommended [8].

Pediculosis

Pediculosis is a lice infestation. Lice are small, wingless insects, 0.5–8 mm long, dorso-ventrally flattened, with legs carrying strong claws to attach to the hair shafts [4, 88]. The majority of mammals, including man and birds and excluding mono-tremes and bats, are infested by at least one lice species [88]. As other insects, their body is segmented with head, thorax and abdomen; they have three pair of legs and one pair of antennae. They spend their whole life on the host and are highly host-specific, and many species have preferred body locations. The majority of lice belong to the suborder Anoplura, or sucking lice, infesting only placented mammals and to suborder Ischnocera, previously called Mallophaga, biting lice infesting mammals and birds. Sucking lice have a specialized buccal apparatus for sucking blood, while biting lice do not feed on blood but on epidermal debris and hair [4, 88]. *Felicola subrostratus* is the only lice infesting cats.

Morphology

Felicola subrostratus is a biting louse (suborder Ischnocera) 1–1.5 mm long, and its color is beige-yellowish with transverse dark bands. The head is wider than the chest and its shape is pentagonal and frontally pointed. On the ventral surface, the louse shows a longitudinal, median cleft adapted to the hair shaft. The antennae are similar in both sexes and comprised of three segments. The buccal apparatus is well-developed and helps the lice to remain attached to the hair shaft (Fig. 17). The legs are short, ending with a single claw [88].

Fig. 17 *Felicola subrostratus*, adult louse

Life Cycle

The life cycle takes place entirely on the host (permanent parasitism), where the female lays operculated eggs strongly attached to the hair shafts. A nymph emerges from the egg and three moults are required to become adult. The juvenile stages are similar to the adults, but smaller and sexually immature with undeveloped gonads (uncomplete metamorphosis). The whole cycle requires 2–3 weeks, and a female can lay up to 200–300 eggs in its life, which lasts for approximately 1 month [88].

Compared to other insects, lice do not have a high reproductive index; however, females during ovodeposition produce a sticky liquid which becomes solid, cementing the egg for all its length excluding the operculus (breathing opening) to the hair shaft. This reduces the loss of eggs and mortality of immature stages and increases the lice population on the host [88].

Epidemiology

Lice cannot survive for more than 1–2 days off the host and generally spend all their lives on the same host. Transmission occurs by direct contact between infested and susceptible cats, since lice leave their host only to move to another one [88]. Being highly host-specific, transmission occurs only among cats. In temperate climates, seasonal fluctuation with winter increase of infestations is reported, possibly due to the host hair coat characteristics. Long-haired cats are predisposed; however, the most severe cases are seen in malnourished cats or cats living in poor hygienic conditions [8, 88].

Clinical Signs

Lice can infest the whole body, with preferred localization on the head, neck and dorso-lumbar region [88]. Lesions observed on the cat vary depending on the number of parasites and severity of pruritus, which is absent to moderate [8, 88]. Some cats are asymptomatic: lice can be observed moving on the hair shafts and often only eggs can be seen, attached to the hair shafts and macroscopically similar to scales. Eggs can be correctly identified by their oval shape and whitish color on closer examination. The hair coat may appear dull, unkempt, and dirty (Fig. 18) [8]. In other cases, primary lesions such as papules and scaling may be seen, or self-traumatic secondary lesions (excoriations, crusts), self-induced alopecia or miliary dermatitis [8].

Diagnosis

Lice and their eggs are easily identified by close observation or using a magnifying lens. Microscopic examination of hair shafts and samples collected with clear

Fig. 18 Lice infestation in a cat: the hair coat is dull and unkempt and looks dirty

Fig. 19 Operculated louse egg firmly attached to the hair shaft

cell tape confirm the diagnosis [8]. Hair combing is also useful to collect samples from the examination table. When no adult lice are found, but just eggs, these must be differentiated from *Cheyletiella* spp. eggs, also attached to hair shafts. Lice eggs are much bigger than *Cheyletiella* eggs and the operculus is dorsal (Fig. 19). Moreover, lice eggs are attached for two thirds of their length to the hair shafts, while *Cheyletiella* eggs are loosely fixed by thin fibrils.

Treatment

Lice are susceptible to the majority of insecticides in the market [8]. Currently, registered active ingredients to treat feline pediculosis include fipronil (spot-on and spray) [89] and selamectin spot-on [90], recently made available also in association with sarolaner. A single treatment is recommended with all these products; however, eggs are resistant to the majority of insecticides. It is advisable repeating the treatment after 14 days to ensure that lice emerged from eggs after the first treatment are killed. Treatment must be extended to all in-contact cats [8, 88].

Lynxacariosis

Lynxacarus radovskyi (feline fur mites) are astigmatid mites of the Listrophoridae family which houses small long mites specialized for grasping of the hairs of mammals. Other notable fur mites include *Chirodiscoides caviae* and *Leporacarus gibbus* which infests the guinea pig and rabbit respectively. *Lynxacarus radovskyi* is characterized by a laterally compressed body, short anterior legs and has a characteristic ability to grasp the hair shaft using a modified specialized clasping structure comprised of the propodosomal flaps and palp coxae. Its legs terminate in ambulacral discs which are membraneous structures bearing remnants of claw that facilitates maximal contact for hair grasping. The male of the species possesses large anal suckers used to fasten onto the female during copulation. The female then lays eggs that hatch into six-legged larvae and then eight-legged nymphs, which finally moult into an adult mite (Fig. 20). *Lynxacarus radovskyi* feed on shed corneocytes, fungal spores, sebum and also pollen on the host. Exact life cycle has not been fully described and transmission is through direct contact. The mite has been reported in southern parts of the USA (Texas, Florida), Australia, New Zealand, New Caledonia, French Guyana, Caribbean, Fiji, Malaysia, Philippines, India, Singapore and also South America, but its incidence is thought to be under reported. *Lynxacarus radovskyi* is not zoonotic to humans or any other species except *Felis catus*.

Clinical Signs

Most infested cats are asymptomatic, but reports of a pathological response from susceptible host have been reported. In these cats, a self-induced, non-inflammatory, caudally directed alopecia has been described. Alopecia typically begins from

Fig. 20 A typical hair pluck demonstrating a female nymph of *Lynxacarus radovskyi* and an egg

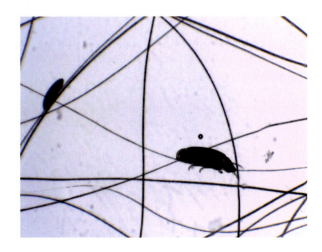

Fig. 21 A cat with lynxacariosis, presented with bilaterally symmetrical, non-inflammatory, self-induced alopecia

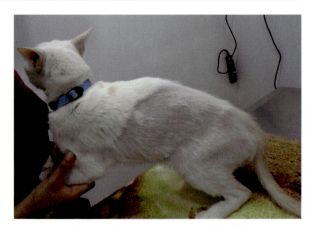

the perineum/tail base where the mites are thought to be most commonly isolated before spreading to the lateral thighs, abdomen, and flanks (Fig. 21) [91]. Infested cats are often presented with increased scale production and a dry, dull coat with easily epilated hairs. Other extracutaneous signs such as gingivitis, gastrointestinal disturbances (hairballs) and restlessness due to irritation may also be seen. As the mite can cause extracutaneous signs, it is important that attending veterinarians consider this parasite as a possible differential diagnosis, especially when these extra-cutaneous signs are present.

Diagnosis

The parasites can be detected via microscopic examination of hair plucks or adhesive tape technique obtained from the perineum, lateral hind limbs or cervical region where they are more readily observed [91]. Mites are easily demonstrated in overt heavy infestation but may be difficult to demonstrate in a patient that excessively self-grooms.

Treatment

The parasite is sensitive to all acaricidals. Published efficacy reports include fipronil, moxidectin plus imidacloprid and fluralaner [92, 93]. Topical selamectin, administered every fortnightly, is equally effective.

Feline Cutaneous Screwworm Myiasis

Myiasis is defined as the invasion of a living vertebrate animal by fly larvae, which may or may not be associated with feeding on the host tissue [94]. In cases of obligatory myiasis, fly species such as New-World screwworm (NWS) *Cochliomyia hominivorax* or the Old-World screwworm (OWS) *Chrysomya bezziana* lay their eggs on a living host regardless of species, and the disease they cause is referred to as cutaneous screwworm myiasis. Historically, the range of NWS extended from

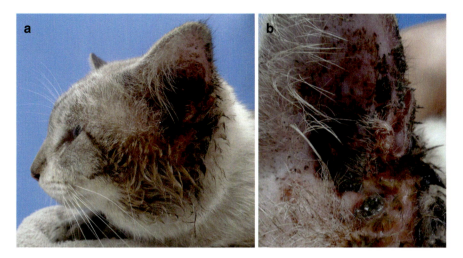

Fig. 22 (**a**) A cat presented with an exudative, ulcerative, swollen and erythematous wound at the base of the left ear, with characteristic putrid smell. (**b**) Upon closer inspection, burrowing larvae are clearly visualized within these lesions

the southern states of the USA, through Mexico, Central america, the Caribbean and northern countries of South America to Uruguay, northern Chile and northern Argentina. Its distribution contracts during the winter months and expanded during summer months, thus producing a seasonality at its edges and year round incidence in the central areas. With the successful implementation of sterile insect technique (SIT), NWS has been eradicted from USA, Mexico, Curacao, Puerto Rico, extending to Central American countries such as Guatemala, Belize, El Salvador, Honduras, Nicaragua to Panama. OWS, as the name suggest is confined to the Old World which includes much of Africa (from Ethiopia and sub-Saharan countries to northern South Africa), Middle East Gulf region, the Indian subcontinent and south east Asia (Malaysia, Singapore, Indonesia, Philipines to Papua New Guinea). OWS has recently been reported in Hong Kong and the southern autonomous region of Guangxi, in mainland China [95].

Clinical Signs

Female flies of *Cochliomyia hominivorax* and *Chrysomya bezziana* typically lay their eggs at wound edges. Eggs hatch within 12–24 hours. As the name screwworm suggests, the hatched larvae burrow or screw themselves head facing downward into host tissue and begin feeding. This results in exudative and ulcerative lesions with easily visualized maggots within the lesions, emanating a characteristic putrid odor (Figs. 22a, b). These foul putrid wounds then attract more oviposition, resulting in superinfestation that can lead to death due to sepsis in the untreated host. After approximately 7 days of feeding, the larvae drop onto the ground, burrow and pupate, and adult flies emerge from the puparium in approximately 7 days. Adult, intact male domestic short hair cats are predisposed to develop screwworm myiasis from inter-cat aggression, with some concurrently diagnosed with sporotrichosis in regions where these two diseases are reported. The most common sites for screwworm myiasis in the cats are the paws, followed by the tail and perineum [96]. Judging from the severity of tissue destruction, one would reasonably expect that the host-parasite relationship will be highlighted by the quick removal of these larvae by the fastidiously grooming host. However, the ability of the larvae to induce a state of immune supression renders the host extremely tolerant of the infestation and thus some patients are presented for medical attention in advanced stages of infestation.

Treatment

Nitenpyram (Capstar®, Elanco, IL, USA) at standard packaging dose, administered as recommended by manufacturer (with or without food), is the most common treatment modality in countries where the drug is available. Larvicidal efficacy is thought to range between 94.1% to 100% within 24 hours in dogs treated with nitenpyram with scarce data from cats [97–99]. Once the larvae have died, they are manually removed, the wound is debrided, and if no larvae are left inside the wound (foreign body), the wound typically heals quickly. In regions where nitenpyram is not available, extra-label use of systemic/topical ivermectin (0.3–0.6 mg/kg) and/or topical powder-based insecticides marketed for the treatment of cutaneous myiasis in farm animals, consisting of coumaphos, propoxur, and sulphanilamide (Negasunt™ Dusting Powder, Bayer Pharmaceuticals, Maharashtra, India), are used. There are very limited treatment options available for veterinarians to treat feline cutaneous myiasis other than nitenpyram. Due to this limitation, many cats are still treated with extra-label use of ivermectin and carbamates, originally meant for farm use.

References

1. Leone F, Albanese F, Fileccia I. Feline notoedric mange: a report of 22 cases. Prat Méd Chir Anim Comp. 2003;38:421–7.
2. Leone F. Canine notoedric mange. Vet Dermatol. 2007;18(2):127–9.
3. Foley J, Serieys LE, Stephenson N, et al. A synthetic review of notoedres species mites and mange. Parasitology. 2016;9:1–15.
4. Bowman DD, Hendrix CM, Lindsay, et al. The Arthropods. In: Bowman D, editor. Feline Clinical Parasitology. Ames: Iowa State University Press; 2002. p. 355–455.
5. Wall R, Shearer D. Mites (Acari). In: Veterinary Ectoparasites: biology, pathology & control. 2nd ed. Oxford: Blackwell Science; 2001. p. 23–54.
6. Chakrabarti A. Human notoedric scabies from contact with cats infested with Notoedres cati. Int J Dermatol. 1986;25(10):646–8.
7. Foley RH. A notoedric mange epizootic in an island's cat population. Feline Pract. 1991;19:8–10.
8. Miller WH, Griffin CE, Campbell KL. Parasitic skin disease. In: Muller and Kirk's small animal dermatology. 7th ed. St. Louis: Elsevier Mosby; 2013. p. 284–342.
9. Leone F, Albanese F, Fileccia I. Epidemiological and clinical finding of notoedric mange in 30 cats. Vet Dermatol. 2005;16(5):359.
10. Sampaio KO, de Oliveira LM, Burmann PM, et al. Acetate tape impression test for diagnosis of notoedric mange in cats. J Feline Med Surg. 2016;15:1–4.
11. Knaus M, Capári B, Visser M. Therapeutic efficacy of Broadline against notoedric mange in cats. Parasitol Res. 2014;113(11):4303–6.

12. Hellmann K, Petry G, Capari B, et al. Treatment of naturally Notoedres cati-infested cats with a combination of imidacloprid 10%/moxidectin 1% spot-on (advocate®/advantage® multi, Bayer). Parasitol Res. 2013;112(Suppl 1):57–66.

13. Itoh N, Muraoka N, Aoki M, et al. Treatment of Notoedric cati infestation in cats with selamectin. Vet Rec. 2004;154(13):409.

14. Fisher MA, Shanks DJ. A review of the off-label use of selamectin (stronghold/revolution) in dogs and cats. Acta Vet Scand. 2008;50:46.

15. Sivajothi S, Sudhakara Reddy B, et al. Notoedres cati in cats and its management. J Parasit Dis. 2015;39(2):303–5.

16. Delucchi L, Castro E. Use of doramectin for treatment of notoedric mange in five cats. J Am Vet Med Assoc. 2000;216(2):215–6.

17. Yang C, Huang HP. Evidence-based veterinary dermatology: a review of published studies of treatments for Otodectes cynotis (ear mite) infestation in cats. Vet Dermatol. 2016;27(4):221–e56.

18. Otranto D, Milillo P, Mesto P, et al. Otodectes cynotis (Acari: Psoroptidae): examination of survival off-the-host under natural and laboratory conditions. Exp Appl Acarol. 2004;32(3):171–9.

19. Sotiraki ST, Koutinas AF, Leontides LS, et al. Factors affecting the frequency of ear canal and face infestation by Otodectes cynotis in the cat. Vet Parasitol. 2001;96(4):309–15.

20. Herwick RP. Lesions caused by canine ear mites. Arch Dermatol. 1978;114(1):130.

21. Lopez RA. Of mites and man. J Am Vet Med Assoc. 1993;203(5):606–7.

22. Powell MB, et al. Reaginic hypersensitivity in Otodectes cynotis infestation of cats and mode of mite feeding. Am J Vet Res. 1980;41(6):877–82.

23. Saridomichelakis MN, Koutinas AF, Gioulekas D, et al. Sensitization to dust mites in cats with Otodectes cynotis infestation. Vet Dermatol. 1999;10(2):89–94.

24. Roy J, Bédard C, Moreau M, et al. Comparative short-term efficacy of Oridermyl(®) auricular ointment and revolution(®) selamectin spot-on against feline Otodectes cynotis and its associated secondary otitis externa. Can Vet J. 2012;53(7):762–6.

25. Ghubash R. Parasitic miticidal therapy. Clin Tech Small Anim Pract. 2006;21(3):135–44.

26. Scherk-Nixon M, Baker B, Pauling GE, et al. Treatment of feline otoacariasis with 2 otic preparations not containing miticidal active ingredients. Can Vet J. 1997;38(4):229–30.

27. Engelen MA, Anthonissens E. Efficacy of non-acaricidal containing otic preparations in the treatment of otoacariasis in dogs and cats. Vet Rec. 2000;147(20):567–9.

28. Coleman GT, Atwell RB. Use of fipronil to treat ear mites in cats. Aust Vet Pract. 1999;29(4):166–8.

29. Curtis CF. Current trends in the treatment of Sarcoptes, Cheyletiella and Otodectes mite infestations in dogs and cats. Vet Dermatol. 2004;15(2):108–14.

30. Shanks DJ, McTier TL, Rowan TG, et al. The efficacy of selamectin in the treatment of naturally acquired aural infestations of otodectes cynotis on dogs and cats. Vet Parasitol. 2000;91(3–4):283–90.

31. Fourie LJ, Kok DJ, Heine J. Evaluation of the efficacy of an imidacloprid 10%/moxidectin 1% spot-on against Otodectes cynotis in cats. Parasitol Res. 2003;90(Suppl 3):S112–3.

32. Becskei C, Reinemeyer C, King VL, et al. Efficacy of a new spot-on formulation of selamectin plus sarolaner in the treatment of Otodectes cynotis in cats. Vet Parasitol. 2017;238(Suppl 1):S27–30.

33. Taenzler J, de Vos C, Roepke RK, et al. Efficacy of fluralaner against Otodectes cynotis infestations in dogs and cats. Parasit Vectors. 2017;10(1):30.

34. Taenzler J, de Vos C, Roepke RKA, et al. Efficacy of fluralaner plus moxidectin (Bravecto® plus spot-on solution for cats) against Otodectes cynotis infestations in cats. Parasit Vectors. 2018;11(1):595.

35. Machado MA, Campos DR, Lopes NL, et al. Efficacy of afoxolaner in the treatment of otodectic mange in naturally infested cats. Vet Parasitol. 2018;256:29–31.

36. Beugnet F, Bouhsira E, Halos L, et al. Preventive efficacy of a topical combination of fipronil – (S)-methoprene – eprinomectin – praziquantel against ear mite (Otodectes cynotis) infestation of cats through a natural infestation model. Parasite. 2014;21:40.

37. Schmeitzel LP. Cheyletiellosis and scabies. Vet Clin North Am Small Anim Pract. 1988;18(5):1069–76.

38. Lee BW. Cheyletiella dermatitis: a report of fourteen cases. Cutis. 1991;47(2):111–4.
39. Wagner R, Stallmeister N. Cheyletiella dermatitis in humans, dogs and cats. Br J Dermatol. 2000;143(5):1110–2.
40. Chailleux N, Paradis M. Efficacy of selamectin in the treatment of naturally acquired cheyletiellosis in cats. Can Vet J. 2002;43(10):767–70.
41. Paradis M, Scott D, Villeneuve A. Efficacy of ivermectin against *Cheyletiella blakei* infestation in cats. J Am Anim Hosp Assoc. 1990;26(2):125–8.
42. Scarampella F, Pollmeier M, Visser M, et al. Efficacy of fipronil in the treatment of feline cheyletiellosis. Vet Parasitol. 2005;129(3–4):333–9.
43. Fisher MA, Shanks DJ. A review of the off-label use of selamectin (stronghold/revolution) in dogs and cats. Acta Vet Scand. 2008;50:46.
44. Takahashi M, Misumi H, Urakami H, et al. Trombidiosis in cats caused by the bite of the larval trombiculid mite Helenicula miyagawai (Acari: Trombiculidae). Vet Rec. 2004;154(15):471.
45. McClain D, Dana AN, Goldenberg G. Mite infestations. Dermatol Ther. 2009;22(4):327–46.
46. Shatrov AB. Stylostome formation in trombiculid mites (Acariformes: Trombiculidae). Exp Appl Acarol. 2009;49(4):261–80.
47. Caputo V, Santi F, Cascio A, et al. Trombiculiasis: an underreported ectoparasitosis in Sicily. Infez Med. 2018;26(1):77–80.
48. Leone F, Di Bella A, Vercelli A, et al. Feline trombiculosis: a retrospective study in 72 cats. Vet Dermatol. 2013;24(5):535–e126.
49. Guarneri C, Chokoeva AA, Wollina U, et al. Trombiculiasis: not only a matter of animals! Wien Med Wochenschr. 2017;167(3–4):70–3.
50. Guarneri F, Pugliese A, Giudice E, et al. Trombiculiasis: clinical contribution. Eur J Dermatol. 2005;15(6):495–6.
51. Smith GA, Sharma V, Knapp JF, et al. The summer penile syndrome: seasonal acute hypersensitivity reaction caused by chigger bites on the penis. Pediatr Emerg Care. 1998;14(2):116–8.
52. Fernández-Soto P, Pérez-Sánchez R, Encinas-Grandes A. Molecular detection of Ehrlichia phagocytophila genogroup organisms in larvae of Neotrombicula autumnalis (Acari: Trombiculidae) captured in Spain. J Parasitol. 2001;87(6):1482–3.
53. Kampen H, Schöler A, Metzen M, et al. Neotrombicula autumnalis (Acari, Trombiculidae) as a vector for Borrelia burgdorferi sensu lato? Exp Appl Acarol. 2004;33(1–2):93–102.
54. Literak I, Stekolnikov AA, Sychra O, et al. Larvae of chigger mites Neotrombicula spp. (Acari: Trombiculidae) exhibited Borrelia but no Anaplasma infections: a field study including birds from the Czech Carpathians as hosts of chiggers. Exp Appl Acarol. 2008;44(4):307–14.
55. Leone F, Cornegliani L, Vercelli A. Clinical findings of trombiculiasis in 50 cats. Vet Dermatol. 2010;21(5):538.
56. Fleming EJ, Chastain CB. Miliary dermatitis associated with Eutrombicula infestation in a cat. J Am Anim Hosp Assoc. 1991;27:529–31.
57. Nuttall TJ, French AT, Cheetham HC, et al. Treatment of Trombicula autumnalis infestation in dogs and cats with a 0.25 per cent fipronil pump spray. J Small Anim Pract. 1998;39(5):237–9.
58. Leone F, Albanese F. Efficacy of selamectin spot-on formulation against *Neotrombicula autumnalis* in eight cats. Vet Dermatol. 2004;15(Suppl.1):49.
59. Frank LA, Kania SA, Chung K, et al. A molecular technique for the detection and differentiation of Demodex mites on cats. Vet Dermatol. 2013;24(3):367–9. e82–e3
60. Ferreira D, Sastre N, Ravera I, et al. Identification of a third feline Demodex species through partial sequencing of the 16S rDNA and frequency of Demodex species in 74 cats using a PCR assay. Vet Dermatol. 2015;26(4):239–e53.
61. Desch C, Nutting WB. Demodex cati Hirst 1919: a redescription. Cornell Vet. 1979;69(3):280–5.
62. Löwenstein C, Beck W, Bessmann K, et al. Feline demodicosis caused by concurrent infestation with Demodex cati and an unnamed species of mite. Vet Rec. 2005;157(10):290–2.
63. Desch CE Jr, Stewart TB. Demodex gatoi: new species of hair follicle mite (Acari: Demodecidae) from the domestic cat (Carnivora: Felidae). J Med Entomol. 1999;36(2):167–70.
64. Saari SA, Juuti KH, Palojärvi JH, et al. Demodex gatoi-associated contagious pruritic dermatosis in cats – a report from six households in Finland. Acta Vet Scand. 2009;51:40.
65. Neel JA, Tarigo J, Tater KC, et al. Deep and superficial skin scrapings from a feline immunodeficiency virus-positive cat. Vet Clin Pathol. 2007;36(1):101–4.

66. Kano R, Hyuga A, Matsumoto J, et al. Feline demodicosis caused by an unnamed species. Res Vet Sci. 2012;92(2):257–8.
67. Moriello KA, Newbury S, Steinberg H. Five observations of a third morphologically distinct feline Demodex mite. Vet Dermatol. 2013;24(4):460–2.
68. Morris DO. Contagious demodicosis in three cats residing in a common household. J Am Anim Hosp Assoc. 1996;32(4):350–2.
69. Guaguere E, Muller A, Degorce-Rubiales F. Feline demodicosis: a retrospective study of 12 cases. Vet Dermatol. 2004;15(Suppl 1):34.
70. Stogdale L, Moore DJ. Feline demodicosis. J Am Anim Hosp Assoc. 1982;18:427–32.
71. Medleau L, Brown CA, Brown SA, et al. Demodicosis in cats. J Am Anim Hosp Assoc. 1988;24:85–91.
72. Kontos V, Sotiraki S, Himonas C. Two rare disorders in the cat: Demodectic otitis externa and Sarcoptic mange. Feline Pract. 1998;26(6):18–20.
73. Van Poucke S. Ceruminous otitis externa due to Demodex cati in a cat. Vet Rec. 2001;149(21):651–2.
74. Bizikova P. Localized demodicosis due to Demodex cati on the muzzle of two cats treated with inhalant glucocorticoids. Vet Dermatol. 2014;25(3):222–5.
75. White SD, Carpenter JL, Moore FM, et al. Generalized demodicosis associated with diabetes mellitus in two cats. J Am Vet Med Assoc. 1987;191(4):448–50.
76. Vogelnest LJ. Cutaneous xanthomas with concurrent demodicosis and dermatophytosis in a cat. Aust Vet. 2001;79(7):470–5.
77. Zerbe CA, Nachreiner RF, Dunstan RW, et al. Hyperadrenocorticism in a cat. J Am Vet Med Assoc. 1987;190(5):559–63.
78. Chalmers S, Schick RO, Jeffers J. Demodicosis in two cats seropositive for feline immunodeficiency virus. J Am Vet Med Assoc. 1989;194(2):256–7.
79. Guaguère E, Olivry T, Delverdier-Poujade A, et al. *Demodex cati* infestation in association with feline cutaneous squamous cell carcinoma *in situ*: a report of five cases. Vet Dermatol. 1999;10(1):61–7.
80. Bailey RG, Thompson RC, Nickels DG. Demodectic mange in a cat. Aust Vet J. 1981;57(1):49.
81. Beale K. Feline demodicosis: a consideration in the itchy or overgrooming cat. J Feline Med Surg. 2012;14(3):209–13.
82. Silbermayr K, Joachim A, Litschauer B, et al. The first case of Demodex gatoi in Austria, detected with fecal flotation. Parasitol Res. 2013;112(8):2805–10.
83. Mueller RS. Treatment protocols for demodicosis: an evidence-based review. Vet Dermatol. 2004;15(2):75–89.
84. Johnstone IP. Doramectin as a treatment for canine and feline demodicosis. Aust Vet Pract. 2002;32(3):98–103.
85. Short J, Gram D. Successful treatment of Demodex gatoi with 10% Imidacloprid/1% Moxidectin. J Am Anim Hosp Assoc. 2016;52(1):68–72.
86. Matricoti I, Maina E. The use of oral fluralaner for the treatment of feline generalised demodicosis: a case report. J Small Anim Pract. 2017;58(8):476–9.
87. Duangkaew L, Hoffman H. Efficacy of oral fluralaner for the treatment of Demodex gatoi in two shelter cats. Vet Dermatol. 2018;29(3):262.
88. Wall R, Shearer D. Lice. In: Veterinary Ectoparasites: biology, pathology & control. 2nd ed. Oxford: Blackwell Science; 2001. p. 162–78.
89. Pollmeier M, Pengo G, Longo M, et al. Effective treatment and control of biting lice, Felicola subrostratus (Nitzsch in Burmeister, 1838), on cats using fipronil formulations. Vet Parasitol. 2004;121(1–2):157–65.
90. Shanks DJ, Gautier P, McTier TL, et al. Efficacy of selamectin against biting lice on dogs and cats. Vet Rec. 2003;152(8):234–7.
91. Ketzis JK, Dundas J, Shell LG. *Lynxacarus radovskyi* mites in feral cats: a study of diagnostic methods, preferential body locations, co-infestations and prevalence. Vet Dermatol. 2016;27:425–e108.
92. Clare F, Mello RMLC. Use of fipronil for treatment of *Lynxacarus radovskyi* in outdoor cats in Rio de Janeiro (Brazil). Vet Dermatol. 2004;15(Suppl 1):50. (abstract)

93. Han HS, Noli C, Cena T. Efficacy and duration of action of oral fluralaner and spot-on moxidectin/imidacloprid in cats infested with *Lynxacarus radovskyi*. Vet Dermatol. 2016;27:474–e127.
94. Catts EP, Mullen G. Myiasis (*Muscoidea, Oestroidea*). In: Mullen G, Durden L, editors. Medical and veterinary entomology. Orlando: Academic Press; 2002. p. 317–43.
95. Fang Fang, Qinghua Chang, Zhaoan Sheng, Yu Zhang, Zhijuan Yin, Jacques Guillot, Chrysomya bezziana: a case report in a dog from Southern China and review of the Chinese literature. Parasitology Research.
96. Hock Siew Han, Peik Yean Toh, Hock Binn Yoong, Hooi Meng Loh, Lee Lee Tan, Yin Yin Ng, (2018) Canine and feline cutaneous screw-worm myiasis in Malaysia: clinical aspects in 76 cases. Vet Dermatol. 29(5):442–e148.
97. Clarissa P de Souza, Guilherme G. Verocai, Regina HR Ramadinha, (2010) Myiasis caused by the New World screwworm fly (Diptera: Calliphoridae) in cats from Brazil: report of five cases. J Feline Med Surg. 12(2):166–8.
98. Thaís R. Correia, Fabio B. Scott, Guilherme G. Verocai, Clarissa P. Souza, Julio I. Fernandes, Raquel M.P.S. Melo, Vanessa P.C. Vieira, Francisco A. Ribeiro, (2010) Larvicidal efficacy of nitenpyram on the treatment of myiasis caused by Cochliomyia hominivorax (Diptera: Calliphoridae) in dogs. Vet Parasitol. 173(1–2):169–72.
99. Hock Siew Han, Charles Chen, Carlo Schievano, Chiara Noli, (2018) The comparative efficacy of afoxolaner, spinosad, milbemycin, spinosad plus milbemycin, and nitenpyram for the treatment of canine cutaneous myiasis. Vet Dermatol. 29(4):312–e109.

Flea Biology, Allergy and Control

Chiara Noli

Abstract

Fleas are the most common ectoparasites and flea-bite allergy can develop in cats. The clinical signs are represented by pruritus, excoriations, self-induced alopecia, manifestations of the eosinophilic granuloma complex and miliary dermatitis, which often, but not exclusively, involves the caudal dorsal and ventral part of the body. The diagnosis is obtained with the clinical presentation and response to flea control. Flea control is based on adulticides, which kill adult fleas on the cat, and insect growth regulators (IGR), which inhibit the development of pre-adult stages in the environment.

Introduction

The flea species most frequently identified in cats is *Ctenocephalides felis felis* (Fig. 1). A comprehensive review on its biology and ecology has recently been published [1]. Fleas can be a cause and/or vector of a variety of diseases such as anemia in heavily infested kittens, tapeworm infestations, Lyme disease, pest, viruses, hemoparasites, cat-scratch disease and flea allergy [1, 2]. Recognition of some of these conditions, such as tapeworm in the cat or cat-scratch disease in the owner, is a sign of flea infestation, even if asymptomatic in the feline carrier host.

Flea-bite allergy is by far the most frequent disease caused by fleas in cats and its prevalence depends on the geographical region and local parasite prevention habits. In a recent multicenter European study, flea-bite allergy was found to account for about one third of all feline pruritic cases [3].

C. Noli (✉)
Servizi Dermatologici Veterinari, Peveragno, Italy

© Springer Nature Switzerland AG 2020
C. Noli, S. Colombo (eds.), *Feline Dermatology*,
https://doi.org/10.1007/978-3-030-29836-4_20

Fig. 1 Microscopic aspect of
the cat flea *Ctenocephalides
felis felis* (4×)

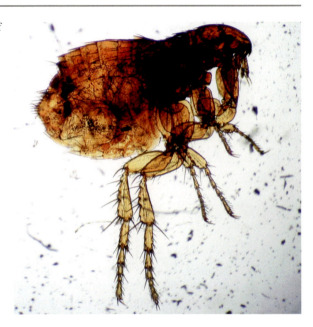

Pathogenesis of Flea Allergy

Cats are bitten by fleas several times a day [4]. Fleas insert their mouthparts through the epidermis in the dermis and suck blood from the capillaries. During this procedure, they deposit up to 15 salivary proteins within the epidermis and superficial dermis that soften tissues and prevent blood coagulation [5, 6]. Hypersensitivity to these proteins induces local edema and a cellular infiltrate, which constitutes the erythematous papule that may follow the bite. There are no specific studies yet that identify the precise allergenic components of flea saliva relevant for naturally sensitized cats. One study suggests that FSA1 (feline salivary antigen-1) may be a major flea saliva antigen in experimentally sensitized laboratory cats [7]. It is thought that non-allergic animals suffer little or no discomfort while being bitten and that only flea allergic subjects develop pruritus and skin disease.

Little is known about the pathogenesis of flea allergy in cats. Most flea allergic cats have immediate positive intradermal skin test reactions to flea allergens and delayed, type 4 reactions have been also described [8, 9]. As in dogs, allergen-specific IgE can be found in the serum of flea allergic cats by means of ELISA [8, 10]. Late-phase IgE-mediated cellular response and cutaneous basophil hypersensitivity have not yet been identified in cats.

Results of a study on early sensitization of 12-week-old kittens, which developed only mild clinical signs (10/18 kittens), suggest that cats exposed to fleas early in their life are less likely to develop flea allergy than cats exposed at a later age [11]. The authors suggested that early ingestion of fleas could induce tolerance, as cats experimentally exposed to fleas orally tended to have minimal clinical signs and lower in vivo and in vitro test scores, although this was not statistically different from the controls [11]. In the same study, cats continuously exposed to fleas from 16 to 43 weeks of age developed either immediate or late reactivity to live flea challenge. However, the same cats were not all positive on intradermal or serology testing. Immediate test reactivity was reported to persist for more than 90 days after experimental sensitization [7]. In a study specifically designed to clarify the role of intermittent exposure to flea bites, it was concluded that it had neither a protective nor a predisposing effect on the development of clinical signs of flea allergy [12].

Clinical Appearance

There is no age, breed, or sex predilection for the development of flea-bite hypersensitivity. In most cases flea control is either completely lacking or incomplete or wrongly performed. Clinical signs are usually worse in the warmer months, particularly at the end of the summer, when the flea population is at its highest point. In addition, many owners stop administering flea control in the same period, as they feel it is no longer needed.

Clinical signs of feline flea allergy are not different from those caused by other allergies in cats and include, alone (75% of cats) or in combination, pruritus, miliary dermatitis, self-induced alopecia, eosinophilic plaque and eosinophilic granuloma, lip ulcer and head and neck excoriations [3]. Please refer to Chapter, Feline Atopic Syndrome: Epidemiology and Clinical Presentation for a more extensive description of these clinical presentations. All of these signs could be reproduced in experimental sensitization studies [12]. Prevalence of lesions in flea-bite allergy is detailed in Table 1 [3]. A multicentric study on 502 pruritic cats reported a preferred localization of pruritus and lesions of miliary dermatitis on the caudal dorsum (Fig. 2) in cats with flea-bite hypersensitivity, if compared with other allergies [3]. In the same study, non-dermatological signs, such as conjunctivitis, rhinitis, vomiting, diarrhea, and soft feces, were observed in 30% of cats with flea-bite allergy, and otitis was observed in 3% [3].

Table 1 Prevalence of clinical sign of allergy in cats with flea bite hypersensitivity (reference Hobi)

Clinical sign	Prevalence	Most frequent distribution
Miliary dermatitis	35%	Caudal dorsum, caudal thighs or generalized
Symmetrical alopecia	39%	Caudal dorsum and flank Abdomen
Head and neck pruritus and excoriations	38%	Head and neck
Eosinophilic granuloma complex (including eosinophilic granuloma, eosinophilic plaque and lip ulceration)	14%	Granuloma: mouth, chin, caudal aspect of the hind legs Plaque: abdomen, groin Lip ulcer: upper lips

Fig. 2 Self-inflicted lesions on the back of a cat with flea-bite allergy

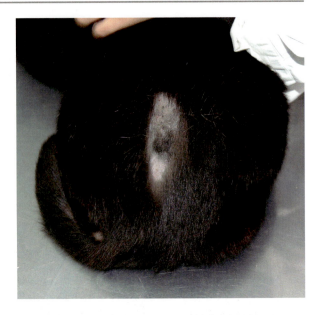

Fig. 3 Fleas and flea feces found in the coat of a non-allergic cat

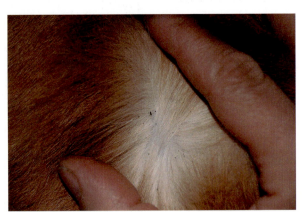

Differential Diagnoses and Diagnostic Approach

A dermatological examination of a cat should always include search for fleas and their feces, by means of a thorough fine-tooth combing of the whole patient (Figs. 3, 4 and 5). Flea feces are made of dry blood and can be easily recognized as they will leave a brown halo on a white moistened paper towel. Fleas or flea feces are not always found as cats are excellent groomers and can eliminate all fleas in a few hours [13]. Furthermore, the number of eggs that fall off flea allergic cats in the environment is lower, leading to a less obvious animal and environmental infestation [13]. For this reason, a lack of fleas or flea dirt in the coat does not exclude a diagnosis of flea allergy. The main differential diagnoses of flea allergy are other allergies, such as adverse reactions to food and environmental allergic dermatitis,

Fig. 4 Abundant flea feces and some adult fleas obtained by flea combing in a flea-infested cat

Fig. 5 Microscopic aspect of the same material of Fig. 4: flea feces appear as red, curled structures. They are made by over 90% of the cat's dry blood. This is an important parental investment by the female flea, as flea feces represent the main nourishment for the flea larvae

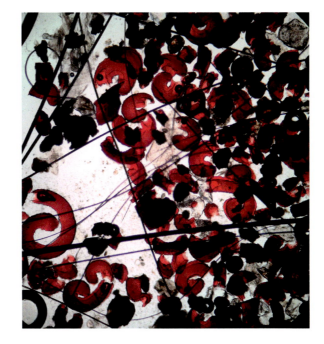

because they share all the abovementioned clinical manifestations. Other less frequent differentials are other parasitic diseases (Chapter, Ectoparasitic Diseases), psychogenic alopecia (Chapter, Psychogenic Diseases), dermatophytosis (for miliary dermatitis) and rare, pruritic, immune-mediated, autoimmune and neoplastic diseases.

A suspect diagnosis of flea allergy may be confirmed by performing an intradermal skin test. The flea allergen is injected (0.05 ml) intradermally together with a negative (saline) and a positive (histamine) control and reactions are read at 15 minutes and 48 hours. Current or recent administration of glucocorticoids or antihistamines (2 weeks for short-acting glucocorticoids and antihistamines, up to 8 weeks for depot glucocorticoids) may cause false-negative results. False-positive reactions in normal cats have been described: in one study, 36% of clinically normal cats that had been exposed to fleas had a positive immediate skin test reaction to flea antigens [14].

A positive predictive value of 85–100% was reported in earlier studies [9, 12, 15], while a more recent study performed with three different extracts obtained a sensitivity of 33% and a specificity of 78–100% [8]. In a study on experimental induction of flea hypersensitivity, the presence of positive immediate intradermal test reactions did not correlate with the development of clinical signs [11]. Allergens used in older studies were whole body flea extracts (1:1000 *w/v*), while more recently flea saliva or purified salivary antigens have been developed for a more sensitive in vivo test [5]. However, in experimentally induced feline flea-bite allergy, results of intradermal testing with purified allergens were not superior to crude extracts in the correlation with clinical signs [11, 12]. Furthermore, it is not known if the concentration used (1/1000 *w/v*) extrapolated from dogs is optimal for cats or if higher concentrations should be used [16].

In vitro serologic tests (ELISA) with whole body flea extracts or purified flea saliva or recombinant flea saliva antigens are available for determination of allergen-specific IgE in the feline serum. The readers should be warned that these tests may only identify animals with IgE-mediated disease and fail to diagnose those with a delayed reaction only. Furthermore, there are normal cats which may have allergen-specific IgE in the absence of clinical disease [8, 11, 12]. Sensitivity and specificity of serological tests performed with flea extracts were reported to be 88% and 77%, respectively, in one study [8] and 77% and 72% in another study [15], with a low positive predictive value of 0.58 in the latter one.

Flea saliva represents only 0.5% of whole flea extracts and in vitro tests performed in dogs with flea salivary antigens gave much better results than those performed with whole flea extracts [17]. In vitro test with salivary antigens and the use of high-affinity receptors FcεR1α gave an overall accuracy of 82% and may represent a more reliable tool for the diagnosis of flea allergy in cats [Mc Call].

In practice, the best approach to the correct diagnosis is to implement an effective ectoparasite control together with/followed by a good hypoallergenic diet. If an improvement is obtained, a dietary challenge can differentiate between flea-bite allergy and food hypersensitivity. If no improvement is obtained, then environmental allergy or other less frequent pruritic conditions can be taken into consideration (Chapter, Feline Atopic Syndrome: Diagnosis for a detailed description of the diagnostic approach to the pruritic cat).

Treatment

Flea control is pivotal for an effective treatment of flea-bite allergy. Adult fleas are obligate ectoparasites [4] and topical or systemic flea control on the cat is mandatory. However, the development of the life stages from egg through to pupa occurs in the immediate domestic environment of the infested pet rather than on the host and this requires adjunct environmental treatment [1, 4, 18]. Contact with other cats is another source of infection. Unfortunately, in a survey conducted on own-ers of flea-infested animals, only 71% of dogs and 50% of cats had been treated against fleas in the previous 12 months [19]. One of the most frequent challenges of treating fleas is that many owners, particularly in case of absence of parasites and feces on the cat's coat, will be skeptical and feel offended if faced with the assumption that there could be fleas on their pets and in their homes and will thus be unwilling to perform a thorough flea control. Explaining that there is no need to host high amounts of fleas to develop allergy and that only a well-conducted all-year-round flea control is able to prevent flea infestation can increase owner compliance.

The Flea Cycle and Ecology

Veterinarians should take time to thoroughly explain why and how to perform cor-rect flea control. This begins with telling the owner something about the flea cycle [1, 4]. Flea eggs are produced on the host and fall off within 8 hours of produc-tion. High egg counts have been found in places where the animal sleeps, eats, or spends most of its time. The eggs hatch after 1–10 days (Fig. 6). The larvae live freely in the environment and move actively under furniture and rugs, deep in carpet fibers, or under organic debris (grass, branches, leaves), in order to avoid light. After 5–11 days, the larvae produce a silk-like cocoon for their protection and camou-flage. Inside the cocoon the larvae develop into pupae and then in 5–9 days become young adults. The fleas in the cocoon are very well protected from insecticides and unfavorable environmental conditions and may survive in a quiescent state for up to 50 weeks. If a potential host is there, the fleas exit the cocoon and rapidly jump on it. If no host is available, the newly emerged fleas can survive several days (up to

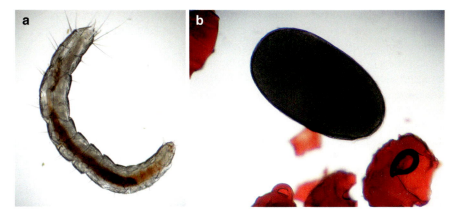

Fig. 6 Flea feces, eggs, and larvae obtained from the environment (couch) of a flea-infested cat. (4×)

2 weeks) in the environment. If the fleas do not find a domestic animal, they often bite humans before finding their preferred host. Adult fleas are permanent parasites of animals. As soon as they land on a host, they begin to feed. The first eggs are produced on the host after 36–48 hours. One single female is capable of producing up to 40–50 eggs a day and up to 2000 eggs in about 100 days of life. The minimum length of the whole cycle is 12–14 days, with an average of about 3–4 weeks in most household conditions, in winter as well. Adult fleas account for only 1–5% of all fleas in the cat's environment, 95–99% of the fleas being the egg, larval, or pupal stage. In fact, it is thought that in temperate climates, the house is the major source of re-infestation of small animals.

Flea Control

Factors important for successful flea control are efficacy and safety of the active ingredient, possibly with long residual activity. Molecules effective on fleas usually belong to one of two categories: either they kill adult fleas (adulticides) or they inhibit pre-adult stage development (insect growth regulators). Adulticides are needed on the animal, in order to kill adult fleas on the host, ideally before they bite and elicit the allergic reaction. Adulticides alone kill only 1–5% of the flea population and do not stop environmental (household) infestation, i.e., eggs, larvae and pupae, representing 95–99% of the entire flea population. Insect growth regulators are able to inhibit development of eggs and larvae and decrease environmental infestation, but cannot prevent the allergic animal from being bitten by an adult flea which comes from "outside" the house. Therefore both product types are needed *together* for effective flea control, especially for flea allergic animals, in order to break the flea life cycle in at least two stages. A list of antiparasitic products available on the market for cats, with their characteristics, is provided in Table 2.

Published trials on flea control measures have been recently extensively reviewed by Rust [1]. The best way to quickly and surely eliminate infestation in a cat is by administering an oral parasiticide. Nitenpyram is the most rapid one, as its effect is seen as soon as 15–30 minutes after administration [20]. Nitenpyram is thus an excellent means of diagnosing the presence of fleas, if given as soon as the cat enters the clinic, as fleas can be seen falling on the table during the consultation. However, being its duration of effect so short (48 hours in cats), it is not very practical as a flea prevention means (the drug should be administered every 48–72 hours). Other oral flea control products, with a slower onset of efficacy (8–12 hours) but with the advantage of a monthly duration, are spinosad and lotilaner [21, 22]. In one study, oral products were considered more effective than topical spot-ons applied by the owner in dogs [23], probably due to better reliability of the mode of administration. There is no data on cats.

Other common flea control measures are spot-on formulations containing an adulticide (imidacloprid, fipronil, selamectin, metaflumizone, dinotefuran, indoxacarb)

Table 2 Antiparasitic products available on the market for cats against fleas, at the time of writing

Name original product[a]	Active ingredient	Formulation[b]	Minimum age of use	Parasites[c]	IGR effect
Frontline	Fipronil	Spot-on	8 weeks	Fleas, ticks, lice, *Cheyletiella*	No
Frontline combo	Fipronil Methoprene	Spot-on	8 weeks	Fleas, ticks, lice, *Cheyletiella*	Yes
Effipro Duo	Fipronil Pyriproxyfen	Spot-on	10 weeks	Fleas, ticks, lice, *Cheyletiella*	Yes
Broadline	Fipronil Methoprene Eprinomectin Praziquantel	Spot-on	7 weeks	Fleas, lice, *Otodectes, Demodex*, heartworm, *Notoedres, Cheyletiella, Angiostrongylus*, GE nematodes, tapeworm	Yes
Advantage	Imidacloprid	Spot-on	8 weeks	Fleas	Yes
Advocate	Imidacloprid Moxidectin	Spot-on	9 weeks	Fleas, lice, *Otodectes, Demodex*, heartworm, *Notoedres, Cheyletiella*, GE nematodes	Yes
Stronghold/ Revolution	Selamectin	Spot-on	6 weeks	Fleas, lice, *Otodectes, Demodex*, heartworm, *Notoedres, Cheyletiella*, GE nematodes	Yes
Stronghold plus	Selamectin Sarolaner	Spot-on	8 weeks	Fleas, ticks, lice, *Otodectes, Demodex, Cheyletiella, Notoedres*, heartworm, myiasis, GE nematodes	Yes
Comfortis	Spinosad	Tablets (with food!)	14 weeks	Fleas, myiasis	No
Activyl	Indoxacarb	Spot-on	8 weeks	Fleas	No
Vectra felis	Dinotefuran Pyriproxyfen	Spot-on	7 weeks	Fleas	Yes
Bravecto	Fluralaner	Spot-on (12 weeks)	8 weeks	Fleas, lice, ticks, *Otodectes, Demodex, Notoedres, Cheyletiella*, myiasis	No
Bravecto plus	Fluralaner Moxidectin	Spot-on (12 weeks)	9 weeks	Fleas, ticks, lice, *Otodectes, Demodex, Cheyletiella, Notoedres*, heartworm (8 weeks), GE nematodes, myiasis	No
Credelio	Lotilaner	Tablet (with food!)	8 weeks	Fleas, ticks, lice, *Otodectes, Demodex, Cheyletiella, Notoedres*, myiasis	No
Seresto/ Foresto	Imidacloprid Flumethrin	Collar (6–8 months)	10 weeks	Fleas, ticks, sandflies, mosquitoes	Yes

(continued)

Table 2 (continued)

Name original product[a]	Active ingredient	Formulation[b]	Minimum age of use	Parasites[c]	IGR effect
Capstar	Nitenpyram	Tablet (activity 72 h)	4 weeks	Fleas, myiasis	No

[a]The original/first product marketed with this ingredient is reported in the table. Depending on the country, several other products are currently available containing fipronil, fipronil/methoprene, fipronil/pyriproxyfen, and imidacloprid
[b]Monthly administration unless stated otherwise
[c]Both label and "off-label" parasites are reported in this table

to be administered between the shoulder blades every 4 weeks. Pulicidal efficacy of each one of these drugs has been proven to be excellent (at least 90%) for up to 4 weeks in laboratory clinical trials [1]. Among these, indoxacarb is a pro-insecticide that must be bio-activated by insect enzymes to generate the active metabolite able to kill fleas and ticks. In mammals, indoxacarb is metabolized to inactive molecules by the liver and is not toxic, so that it is designated by the US Environmental Protection Agency as a "reduced risk" pesticide.

Recently, a new spot-on formulation for cats based on fluralaner, a member of a new class of antiparasitic agents, the isoxazolines, has been marketed with a residual activity against fleas for up to 3 months [24]. Fluralaner is absorbed transdermally and redistributed systemically, so that fleas will need to bite the cat to be killed. A 3-month duration time probably improves owner's compliance and can be preferred in allergic subjects.

There is one flea collar registered for use in the cat containing 10% imidacloprid and 4.5% flumethrin, with a 6–8-month-long pulicidal activity. This product has the advantages of being less expensive than spot-ons or tablets, with a higher compliance and repellent efficacy against fleas, ticks, mosquitos and sandflies vectors of leishmaniosis [25].

Some of the abovementioned insecticides also offer ovicidal and larvicidal activity (e.g., imidacloprid [26] or selamectin [27]), while others are formulated in association with an IGR, such as pyriproxyfen or methoprene. Insect growth regulators (IGR) interfere with the development of pre-adult flea stages, which account for the vast majority of the total flea population (up to 99%). They have a very low toxicity for mammals, because they act on very insect-specific metabolic pathways. The idea behind administering a product with IGR effect on the animal is that treated hairs shed in the environment are able to inhibit eggs' hatching and/or larval moulting. The use of an IGR is fundamental to reduce the environmental flea population, thus the flea burden on the cat and the consequent clinical symptomatology. IGR sprays, containing methoprene or pyriproxyfen, can also be used in the environment, especially in case of heavy or recurrent infestation. The principal strategic problem in trying to control a domestic flea population, however, is dealing with young adult fleas within the protective pupal case [23]. These can yield live, viable adults for periods of several months after all eggs, larvae and other adults have been killed, and repeated applications of environmental treatment may be necessary in some cases. Recently a 0.4% environmental dimeticone spray was able to prevent emergence of young adult fleas from cocoons and proved to be efficacious in immobilizing larvae and adults in the environment [28], with efficacy persisting for more than 3 weeks.

Certain physical measures can assist in flea control. Washable surfaces can be cleaned to remove organic matter and flea feces on which larvae feed. Vacuum cleaning will remove 20% of larvae and up to 60% of eggs as well as flea feces and organic matter. Vacuum cleaning assists spray penetration by raising the fibers in carpets. Bedding and other washable items should be laundered at the highest temperature possible. Carpets and soft furnishings should not be washed as increased humidity favors larval development.

How to Perform an Effective Flea Control and Causes of Failure

An *adulticide* has to be applied *to every animal* in the household *all year round* and an *IGR* has either to be applied in the environment or to all pets. Flea control must be thoroughly and constantly applied in order to be effective; thus, client compliance is the most important element for a successful flea control. Recurrence of signs usually depend on lack in flea control, which might be due to one or more of these factors [29]:

- Use of ineffective products
- Insufficient dosage or lack of application in the whole house or on all animals
- Use of adulticides without IGR or IGR without adulticides
- Too long period of time between administrations

Questioning the owner about how they perform flea control will nearly always identify the problem and it is our task to explain and convince them about the importance of a complete flea control.

Although flea control is mandatory, it may not be sufficient to result in complete control of the dermatosis in all cases, particularly where there is continued contact with untreated individuals. In such cases, anti-pruritic treatment will be necessary. Please refer to Chapter, Feline Atopic Syndrome: Therapy for a detailed discussion of anti-pruritic drugs in cats.

The potential for vaccination, either against the immunogenic salivary proteins of the flea or against concealed antigens within the flea gut, has been explored with variable results and could offer possibilities for the future management of flea allergy [7, 30–32].

Conclusion

Flea-bite hypersensitivity is one of the most important allergic skin conditions in cats, which can manifest with different clinical signs and has many possible differential diagnoses. Intradermal and in vitro allergy tests are not always reliable diagnostic tools and rigorous flea control, by means of adulticides and insect growth regulators, represents the best tool for diagnosing and treating this condition.

References

1. Rust MK. The biology and ecology of cat fleas and advancements in their Pest management: a review. Insects. 2017;8:118.
2. Shaw SE, Birtles RJ, Day MJ. Arthropod transmitted infectious diseases of cats. J Feline Med Surg. 2001;3:193–209.
3. Hobi S, Linek M, Marignac G, et al. Clinical characteristics and causes of pruritus in cats: a multicentre study on feline hypersensitivity-associated dermatoses. Vet Dermatol. 2011;22:406–13.
4. Dryden MW, Rust MK. The cat flea: biology, ecology and control. Vet Parasitol. 1994;52:1–19.
5. Frank GR, Hunter SW, Stiegler GL, et al. Salivary allergens of *Ctenocephalides felis*: collection, purification and evaluation by intradermal skin testing in dogs. In: Kwochka KW, Willemse T, von Tscharner C, editors. Advances in veterinary dermatology, volume 3. Oxford: Butterworth Heinemann; 1998. p. 201–12.
6. Lee SE, Johnstone IP, Lee RP, et al. Putative salivary allergens of the cat flea, Ctenocephalides felis felis. Vet Immunol Immunopathol. 1999;69:229–37.
7. Jin J, Ding Z, Meng F, et al. An immunotherapeutic treatment against flea allergy dermatitis in cats by co-immunization of DNA and protein vaccines. Vaccine. 2010;28:1997–2004.
8. Bond R, Hutchinson MJ, Loeffler A. Serological, intradermal and live flea challenge tests in the assessment of hypersensitivity to flea antigens in cats (*Felis domesticus*). Parasitol Res. 2006;99:392–7.
9. Lewis DT, Ginn PE, Kunkle GA. Clinical and histological evaluation of immediate and delayed flea antigen intradermal skin test and flea bite sites in normal and flea allergic cats. Vet Dermatol. 1999;10:29–38.
10. McCall CA, Stedman KE, Bevier DE, Kunkle GA, Foil CS, Foil LD. Correlation of feline IgE, determined by Fcε RIα-based ELISA technology, and IDST to *Ctenocephalides felis* salivary antigens in a feline model of flea bite allergic dermatitis. Compend Contin Educ Pract Vet. 1997;19(Suppl. 1):29–32.
11. Kunkle GA, McCall CA, Stedman KE, Pilny A, Nicklin C, Logas DB. Pilot study to assess the effects of early flea exposure on the development of flea hypersensitivity in cats. J Feline Med Surg. 2003;5:287–94.
12. Colombini S, Hodgin EC, Foil CS, Hosgood G, Foil LD. Induction of feline flea allergy dermatitis and the incidence and histopathological characteristics of concurrent indolent lip ulcers. Vet Dermatol. 2001;12:155–61.
13. McDonald BJ, Foil CS, Foil LD. An investigation on the influence of feline flea allergy on the fecundity of the cat flea. Vet Dermatol. 1998;9:75–9.
14. Moriello KA, McMurdy MA. The prevalence of positive intradermal skin test reactions to lea extracts in clinically normal cats. Comp Anim Pract. 1989;19:28–30.
15. Foster AP, O'Dair H. Allergy skin testing for skin disease in the cat *in vivo* vs *in vitro* tests. Vet Dermatol. 1993;4:111–5.
16. Austel M, Hensel P, Jackson D, et al. Evaluation of three different histamine concentrations in intradermal testing of normal cats and attempted determination of the irritant threshold concentrations of 48 allergens. Vet Dermatol. 2006;17:189–94.
17. Cook CA, Stedman KE, Frank GR, Wassom DL. The in vitro diagnosis of flea bite hypersensitivity: flea saliva vs. whole flea extracts. In: Proceedings of the 3rd veterinary dermatology world congress, 1996 Spet 11–14. Edinburgh; 1996. p. 170.
18. Osbrink WLA, Rust MK, Reierson DA. Distribution and control of cat fleas in homes in Southern California (Siphonaptera: Pulicidae). J Med Entomol. 1986;79:135–40.
19. Peribáñez MÁ, Calvete C, Gracia MJ. Preferences of pet owners in regard to the use of insecticides for flea control. J Med Entomol. 2018;55:1254–63.
20. Dobson P, Tinembart O, Fisch RD, Junquera P. Efficacy of nitenpyram as a systemic flea adulticide in dogs and cats. Vet Rec. 2000;147:709–13.

21. Cavalleri D, Murphy M, Seewald W, Nanchen S. A randomized, controlled field study to assess the efficacy and safety of lotilaner (Credelio™) in controlling fleas in client-owned cats in Europe. Parasit Vectors. 2018;11:410.

22. Paarlberg TE, Wiseman S, Trout CM, et al. Safety and efficacy of spinosad chewable tablets for treatment of flea infestations of cats. J Am Vet Med Assoc. 2013;242:1092–8.

23. Dryden MW, Ryan WG, Bell M, et al. Assessment of owner-administered monthly treatments with oral spinosad or topical spot-on fipronil/(S)-methoprene in controlling fleas and associated pruritus in dogs. Vet Parasitol. 2013;191:340–6.

24. Bosco A, Leone F, Vascone R, et al. Efficacy of fluralaner spot-on solution for the treatment of ctenocephalides felis and otodectes cynotis mixed infestation in naturally infested cats. BMC Vet Res. 2019;15:28.

25. Brianti E, Falsone L, Napoli E, et al. Prevention of feline leishmaniosis with an imidacloprid 10%/flumethrin 4.5% polymer matrix collar. Parasit Vectors. 2017;10:334.

26. Jacobs DE, Hutchinson MJ, Stanneck D, Mencke N. Accumulation and persistence of flea larvicidal activity in the immediate environment of cats treated with imidacloprid. Med Vet Entomol. 2001;15:342–5.

27. McTier TL, Shanks DJ, Jernigan AD, Rowan TG, Jones RL, Murphy MG, et al. Evaluation of the effects of selamectin against adult and immature stages of fleas (*Ctenocephalides felis felis*) on dogs and cats. Vet Parasitol. 2000;91:201–12.

28. Jones IM, Brunton ER, Burgess IF. 0.4% dimeticone spray, a novel physically acting household treatment for control of cat fleas. Vet Parasitol. 2014;199:99–106.

29. Halos L, Beugnet F, Cardoso L, et al. Flea control failure? Myths and realities. Trends Parasitol. 2014;30:228–33.

30. Heath AW, Arfsten A, Yamanaka M, et al. Vaccination against the cat flea *Ctenocephalides felis felis*. Parasite Immunol. 1994;16:187–91.

31. Halliwell REW. Clinical and immunological response to alum-precipitated flea antigen in immunotherapy of flea-allergic dogs: results of a double blind study. In: Ihrke PJ, Mason IS, White SD, editors. Advances in veterinary dermatology, vol. 2. Oxford: Pergamon Press; 1993. p. 41–50.

32. Kunkle GA, Milcarsky J. Double-blind flea hyposensitization trial in cats. J Am Vet Med Assoc. 1985;186:677–80.

Feline Atopic Syndrome: Epidemiology and Clinical Presentation

Alison Diesel

Abstract

Although very well defined and characterized in the dog, feline atopic syndrome remains less well understood with regard to disease pathogenesis and clinical presentations. While many similarities exist, questions remain whether atopic dermatitis is the same disease entity in dogs and cats. Atopic dermatitis in the cat is often referred to as "feline atopic syndrome" or "non-flea, non-food hypersensitivity dermatitis (NFNFHD)." Although the diagnostic process is similar for dogs and cats, with both being a diagnosis of exclusion, demonstration of immunoglobulin-E (IgE) involvement in feline atopic syndrome has been inconclusive. As with canine atopic dermatitis, pruritus remains a feature of the disease in cats; however, the distribution of pattern of pruritus and lesions is more variable in feline patients. Cats with feline atopic syndrome will typically present with at least one of four common cutaneous reaction patterns (head/neck/pinnal pruritus with excoriations, self-induced alopecia, miliary dermatitis, eosinophilic skin lesions). Additionally, non-cutaneous clinical signs may also be observed.

Introduction

Although very well defined and characterized in dogs and humans, feline atopic syndrome remains less well understood with regard to disease pathogenesis and clinical presentations. While many similarities exist, questions remain whether atopic dermatitis is the same disease entity in dogs and cats. In general, when allergic skin disease is compared across the two species, much less is known/documented

A. Diesel (✉)

College of Veterinary Medicine and Biomedical Sciences, Texas A&M University,
College Station, TX, USA
e-mail: ADiesel@cvm.tamu.edu

© Springer Nature Switzerland AG 2020
C. Noli, S. Colombo (eds.), *Feline Dermatology*,
https://doi.org/10.1007/978-3-030-29836-4_21

in cats, especially in regard to atopic dermatitis. While the term "feline atopy" has been a part of the veterinary literature since 1982 [1], this terminology has fallen out of favor when discussing the disease in cats. "Feline atopic dermatitis" was used initially to describe a clinical syndrome in feline patients with recurrent pruritic skin disease, positive reactions to several environmental allergens on intradermal testing and where other causes of pruritus (e.g., external parasites, infections) had been ruled out. Due to the lack of conclusive demonstration of immunoglobulin-E (IgE) involvement in the disease process, most veterinary dermatologists prefer either "feline atopic syndrome" (FAS) or "non-flea, non-food hypersensitivity dermatitis" (NFNFHD) when referring to what was historically referred to as feline atopic dermatitis (AD) [2].

While the condition remains a diagnosis of exclusion in both species, feline atopic syndrome presents a unique set of challenges for the veterinary practitioner. This includes not only quandaries in interpretation of diagnostic tests but also evaluation of the particular clinical syndromes unique to the feline patient and currently limited options for therapeutic intervention compared to the canine counterpart. This chapter aims to discuss what is presently known with regard to the pathogenesis of feline atopic syndrome, the epidemiology of disease, and observed clinical presentations. Subsequent chapters will present a discussion on diagnostic evaluations and current therapy.

Pathogenesis of Feline Atopic Syndrome

Compared to dogs and people where the pathogenesis of atopic dermatitis is relatively well characterized [3–5], there remains a paucity of information present in the literature with regard to the development of feline atopic syndrome. Although the body of information continues to grow in certain areas of disease pathogenesis for dogs and people (particularly in regard to influences in barrier function and more specific immunological factors), many of these foci have not yet been explored for the allergic feline patient. What has been documented, however, can be discussed with regard to the historical classic triad of factors involved in the development of atopic dermatitis (genetic influence, environmental factors, immunological abnormalities) and influences of barrier function on the course of disease.

Genetic Factors

In dogs and people, it is relatively well established that a genetic predisposition will often contribute to an allergic phenotype, specifically in relation to the development of atopic dermatitis. This has been shown in several human twin studies [6] and in the evaluation of the influence of filaggrin mutation as a contributory factor [7]. In the dog, specific phenotypes have been described for several commonly affected canine breeds [8]; however as in people, it is clear genetics is only part of the picture. The complex genotype of canine atopic dermatitis, with multiple genes involved in the genetic component of the disease development, indeed speaks of the multifaceted nature of the disease. That said, with certain documented genetic variations and improved understanding of the genetic influence for certain patients, targeted therapy aimed at specific molecules may be able to be developed and implemented in the future [9].

In the cat, however, genetic influence in the development of feline atopic syndrome has been only loosely documented [10]. While it seems plausible that indeed there is a genetic component to the disease in cats, to what degree this is apparent is far from known at this time.

Environmental Factors

As is seen with atopic dermatitis in dogs and people, exposure to environmental allergens exacerbates clinical signs in cats with atopic syndrome [11]. This is apparent in the naturally occurring disease presentation and has been supported with a clinical model. In a study utilizing a modified patch test with aeroallergens applied to the skin of healthy and allergic cats, only cats with atopic syndrome developed an inflammatory infiltrate similar to that seen in the lesional skin of cats with the spontaneous disease [12]. Whether application or exposure to aeroallergens in a laboratory setting would lead to more generalized lesions associated with atopic syndrome in the cat, as it does in dogs [13], has not been investigated.

Although a positive "allergy test" does not diagnose atopic dermatitis in any known species, the historical definition of atopic dermatitis in the cat [1] included the description of cats with several positive reactions to environmental allergens on intradermal allergen tests. Intradermal allergen testing (as well as serum allergen testing) for environmental allergens remains a cornerstone of support for the clinical diagnosis of feline atopic syndrome (see further discussion in Chapter, Feline Atopic Syndrome: Diagnosis). This combined with a favorable response to allergen immunotherapy in many cats with atopic syndrome further supports the influence of environmental factors in disease pathogenesis.

Immunological Findings in Cats with Atopic Syndrome

Pulling together all aspects of the disease, the current definition of canine atopic dermatitis describes "a genetically predisposed inflammatory and pruritic allergic skin disease with characteristic clinical features associated with IgE antibodies most commonly directed against environmental allergens" [14]. The influence of IgE has clearly been demonstrated in this species as well as in people; however, this association is less well defined for feline atopic syndrome. Indeed, the role of IgE remains an area of contention with regard to the immunological factors lending to disease development. Part of the argument stems from a lack of correlation with serum IgE levels in cats and clinical disease [15]; however, levels of allergen-specific IgE do not always correlate with clinical disease in canine atopic dermatitis either [16].

There is a reasonable body of evidence, however, that supports the influence of IgE in hypersensitivity dermatitis in the cat. Passive cutaneous anaphylaxis testing has been used in cats to demonstrate the transfer of allergen-specific cutaneous reactivity from a sensitized/allergic cat to a naïve feline via injection of serum from the allergic individual [17, 18]. This reactivity, however, does not occur if the serum is heated prior to injection. The heating process inactivates IgE but not other antibodies, thereby supporting IgE involvement [17, 19, 20]. When anti-IgE is injected into the skin of normal cats, immediate and delayed inflammatory responses occur [21], sharing many macroscopic and microscopic features of what has previously been reported in cats with spontaneously occurring allergic skin disease [10, 15]. A similar inflammatory response, however, was not observed with injection of IgG in this group [21], again supporting the involvement of IgE in feline hypersensitivity dermatitis. The role of IgE has been well established in other allergic diseases in the cat, most notably feline asthma [20, 22]. Given this condition occurs not infrequently in cats with (presumed) allergic skin disease [23], the suspected role of IgE in the phenotype of both conditions cannot be ignored.

Although there is still a bit of uncertainty in regard to the immunopathogenesis of feline atopic syndrome, there is a similar pattern of inflammatory infiltrate in the skin of allergic cats compared to that which is seen in humans and dogs with chronic atopic dermatitis [24]. Certain cell types involved in the innate and adaptive immune system can be seen in altered numbers in the skin of allergic cats compared to those without hypersensitivity dermatitis. Dendritic cells, including Langerhans cells, have been reported in higher numbers in allergic feline skin [24, 25]. These cells interface with the environment, lending to development of allergic inflammation, and have been implicated as well in the generation of atopic dermatitis in people [26]. Eosinophils, often seen with various allergic diseases across multiple species, are additionally increased in the skin of cats with allergy. Indeed, these cells are a conspicuous infiltrate in inflammatory lesions of feline allergic dermatitis, particularly in miliary dermatitis lesions, and are suspected to be the more specific indicator of a hypersensitivity response in cutaneous allergy in cats [27]. Tissue inflammation occurs secondary to the release of granule contents, including major basic protein, as well as inflammatory cytokine expression [28]. Although not specific to hypersensitivity dermatitis in the cat, mast cells are often increased in the skin of cats with allergies compared to healthy cat skin [27]. Additionally, as is seen in people with atopic dermatitis [29], mast cells in allergic cat skin undergo a change in granule content. In cats with allergic skin disease, a markedly lower number of mast cells have been observed staining for tryptase as opposed to chymase [27]. This is compared to healthy cat skin where all mast cells can be seen with tryptase staining and approximately 90% observed when staining for chymase [30].

It has been well documented that a skewed T-cell response in favor of T helper 2 (Th2) over Th1 is a part of the immunological development of atopic dermatitis in dogs and people. T cell involvement also appears to be involved in the immunopathogenesis of feline atopic syndrome. This has been seen with histopathological studies documenting increased populations of CD4+ T cells in allergic cat skin compared to that of CD8+; these cells are generally not observed in the skin from healthy cats [31]. Additionally, an increased number of IL-4 producing T cells have been found in the skin of allergic cats compared to that of healthy controls, supportive of a Th2 infiltrate [32]. This skewed population of T cells has not, however, been demonstrated in the peripheral blood of allergic cats compared to healthy controls [31]. The inflammatory cytokine profile has also not been well elucidated in the skin or peripheral blood of cats with feline atopic syndrome. Differences in the gene expression of various inflammatory interleukins and other cytokines could not be detected when comparing the skin of normal, lesional, and non-lesional allergic cat skin [33]. More recently, increased circulating levels of IL-31 have been demonstrated in sera from allergic cats compared to those without allergic skin disease [34] as has been shown in canine atopic dermatitis. This suggests involvement of this inflammatory cytokine in feline allergic dermatoses; however, a causative role has yet to be determined.

Skin Barrier and Other Factors

The role of barrier function in the skin of people and dogs with atopic dermatitis has become an increasingly important area of investigation. This factor, however, has not been well explored in cats with feline atopic syndrome. One study observed differences in transepidermal water loss (TEWL), skin hydration, and pH at various body sites in healthy cats [35]. Recently, a study examined the relationship between TEWL and severity of clinical symptoms in cats with feline atopic syndrome [36]. Using two scoring systems to assess skin lesions in allergic cats (Scoring Feline Allergic Dermatitis (SCORFAD) and Feline Extent and Severity Index (FeDESI)), a positive correlation was observed between TEWL and severity of clinical lesions at certain body sites, particularly when using the SCORFAD measurements. Less association was observed with FeDESI scoring. While there may indeed be differences in TEWL in allergic cats compared to healthy controls, the measurements may be less useful compared to what is seen in dogs and humans with atopic dermatitis.

In people and dogs with atopic dermatitis, bacterial infection and yeast overgrowth can exacerbate clinical signs of disease. The same appears to be true in some cats with feline atopic syndrome; secondary infections, however, with either bacteria or yeast tend to occur less frequently in allergic cats compared to allergic dogs or people. Although the exact implications have yet to be determined, there is a growing body of evidence documenting changes in the microbiome in atopic individuals. Indeed, this has been reported in both humans [37] and dogs [38], and more recently in allergic cats compared to healthy controls [39]. While there are some similarities across species (e.g., the *Staphylococcus* species is more abundant in allergic individuals compared to healthy controls), there are additionally species differences. Contrary to allergic dogs and humans, allergic cats seem to retain microbial diversity, in that the number of bacterial species was not significantly different in allergic compared to healthy individuals [39]. Furthermore, compared to dogs and people where differences in bacterial communities are seen at specific body locations in the face of an allergic "flare," in allergic cats, their entire body becomes colonized by an altered bacterial population independent of location sampled. This is postulated to be due to the fastidious grooming behavior of cats.

These differences may partially explain why secondary infections are less common in allergic cats compared to what is observed in dogs in people. What implication this dysbiosis has in disease development and/or response to therapeutic intervention remains yet to be discovered.

Epidemiology of Feline Atopic Syndrome

The exact prevalence of feline atopic syndrome in the general population has not been well described in the veterinary literature. A retrospective study on the population of cats seen at a teaching hospital in the United States identified "allergies" accounting for 32.7% of the feline skin diseases presented to the hospital during a 15-year period. "Atopic dermatitis" itself represented 10.3% of the feline dermatoses observed [40]. A similar study over a 1-year period at a university teaching hospital in Canada diagnosed "atopic dermatitis" in 7 of 111 (6.3%) presented for evaluation of dermatological disease [41]. In another study evaluating dermatological diseases seen in general practice in the United Kingdom, however, only 2 out of 154 (1.3%) cats were diagnosed with "atopic dermatitis." It is important to note, however, that other cutaneous reaction patterns (e.g., miliary dermatitis, eosinophilic granuloma complex) were observed in this population without a defined etiology [42]. This difference of prevalence may also partly be explained by differences in diagnoses obtained by a general practitioner compared to a specialist in dermatology.

Clinical Presentation of Feline Atopic Syndrome

As with canine atopic dermatitis, clinical signs of feline atopic syndrome revolve around the presence of pruritus in the cat. Comparatively, however, the distribution of pruritus and lesions is less well defined in the feline patient. With dogs, clinical signs of canine atopic dermatitis typically follow a very predictable pattern to include the face, concave pinnae, axillary and inguinal folds, ventrum, perineal skin, flexural surfaces, and paws [43, 44]. With cats, however, pruritus and lesions will generally include any one or more of the commonly recognized cutaneous reaction patterns reflecting a response to inflammation in feline skin [2]. While these patterns do not reflect a specific etiology, they often are indicative of underlying allergic skin disease.

Head/Neck/Pinnal Pruritus with Excoriations

Also referred to as cervicofacial pruritic dermatitis, lesions associated with this reaction pattern are restricted to the front part of the cat. From the neck directed caudally, the cat will generally appear normal. The face, ears, and neck, however, may be marked with excoriation, crusts, alopecia, and erythema (Fig. 1). In some cases, pruritus can be so severe that obvious self-trauma is apparent.

Fig. 1 Cat with cervicofacial pruritus secondary to feline atopic syndrome

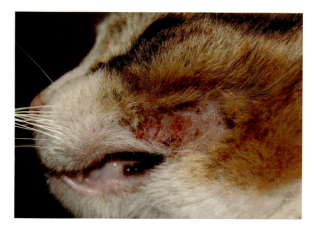

Fig. 2 Cat with self-induced alopecia secondary to feline atopic syndrome. Note barbered hair over site of forelimb amputation aligning with contralateral axillary alopecia

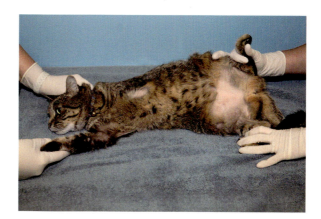

Self-Induced Alopecia

Historically, cats with self-induced alopecia (often referred to as "symmetrical alopecia," "fur mowing," or "barbering") have been overdiagnosed with behavioral abnormalities and psychogenic alopecia. With this reaction patterns, cats will remove hair by excessive licking, chewing, or pulling to the point of partial to near complete alopecia of the affected body region (Fig. 2). The hairs will frequently appear broken and rough where over-grooming has occurred. Concurrent erythematous skin and excoriation may or may not be present.

Miliary Dermatitis

Deriving its name from millet seeds (small grains), miliary dermatitis lesions in the cat will often better be palpated as opposed to visualized. Lesions most commonly present along the neck and dorsum; however, the sparsely haired region of

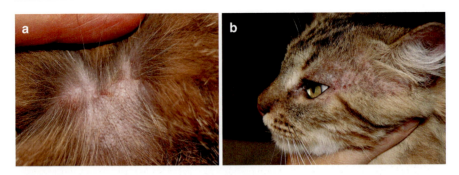

Fig. 3 (**a**) Miliary dermatitis lesions on dorsum of cat with flea allergy dermatitis. (**b**) Miliary dermatitis lesions on head of cat with feline atopic syndrome

preauricular skin can be the best location to visualize miliary dermatitis in the feline patient without having to clip the hair coat (Fig. 3a, b). When present, the lesions appear as small, pinpoint erythematous crusted papules. On palpation, the lesions will feel like small grits or grains under the skin, as if petting coarse sandpaper.

Eosinophilic Skin Lesions

Included in this group of lesions are eosinophilic granulomas, eosinophilic plaques, and lip ("indolent," "rodent") ulcers. This collection of lesions used to be referred to as the feline "eosinophilic granuloma complex"; however, this terminology has fallen out of favor with many dermatologists when describing these lesions in the cat due to their distinct clinical and histopathological appearance.

While they can appear on any given body surface, eosinophilic granulomas may appear most frequently on the caudal thigh (Fig. 4a) or the ventral surface of the chin (Fig. 4b). The previous may be referred to as "linear granulomas," while the latter may be termed "fat chin" or "pouty" cat lesions. Granulomas are typically semi-firm, rather well circumscribed, and may be seen in the presence or absence of pruritus. Additionally, granulomas may occur in the oral cavity secondary to feline atopic syndrome (Fig. 4c). Cats may initially present with clinical signs of dysphagia, drooling, decreased appetite, or even dyspnea depending on the size of the lesion present. Alternatively, they may be found on oral examination in the absence of any obvious clinical abnormalities.

Lip (indolent) ulcers may also present in the absence of clinical signs. These craterous, ulcerative lesions may be unilaterally or bilaterally present on the upper lips of affected cats (Fig. 5). Extension along the philtrum to the nasal planum is a fairly common finding.

Of the three lesions, eosinophilic plaques tend to be associated with severe pruritus and concurrent self-induced alopecia. The lesions may again be present on any part of the body and are most commonly visualized on the ventral abdomen.

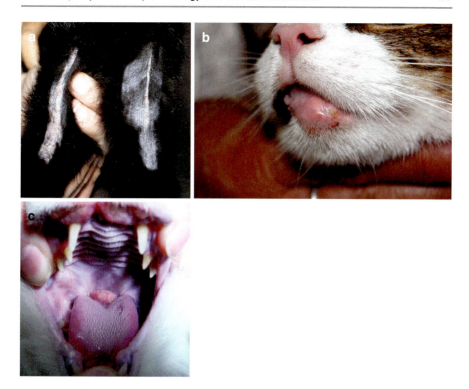

Fig. 4 (**a**) Eosinophilic granuloma lesions on caudal thighs of a cat with feline atopic syndrome. (**b**) Eosinophilic granuloma lesion on the chin of a cat with feline atopic syndrome. (**c**) Eosinophilic granuloma lesion in caudal oral cavity of cat with feline atopic syndrome. This cat had moderate dysphagia and respiratory stridor due to the size of the lesion

Fig. 5 Bilateral indolent ulcer on upper lip of a cat with feline atopic syndrome

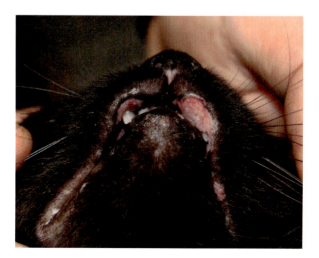

Eosinophilic plaques are generally well-circumscribed, erythematous plaque-like lesions with a glistening, moist surface. Lesions are often multifocal and may coalesce into a larger, single plaque (Fig. 6).

Extra-cutaneous Clinical Signs

While dermatological manifestation is the hallmark of feline atopic syndrome, other extra-cutaneous clinical signs may also present in the allergic cat. This may include allergic otitis, sinusitis, and conjunctivitis as well as feline small airway disease ("feline asthma") in certain patients. How frequently these diseases/clinical manifestations occur concurrently, however, is unknown.

As part of cervicofacial pruritic dermatitis, pinnal pruritus is a fairly common clinical finding in cats with atopic syndrome. On otoscopic examination, however, the external ear canals themselves are frequently normal in appearance. This is in contrast to dogs with canine atopic dermatitis as they frequently present with erythematous otitis externa secondary to allergic disease [2]. Commonly mistaken for ear mite infestation, many cats with atopic syndrome will present with recurrent ceruminous otitis externa, often in the absence of infectious organisms such as bacteria or yeast. This can contribute to the otic pruritus observed.

Commonly reported in cats with feline atopic syndrome, sneezing may be indicative of sinusitis in allergic cats. Although uncertain the exact prevalence, some sources site this concurrent clinical finding of upward of 50% in cats with feline atopic syndrome [45]. While there is only a single report of feline allergic rhinitis documented in the veterinary literature [46], this may be under-reported since a

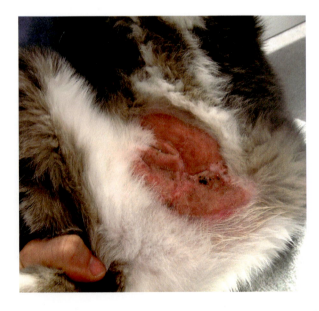

Fig. 6 Large eosinophilic plaque on the abdomen of a cat with feline atopic syndrome. Close inspection of the lesion shows where multiple smaller plaques coalesced to form the larger lesion seen on the patient

clinical suspicion may be addressed in the absence of a definitive (e.g., biopsy) diagnosis. However, the exact prevalence of these findings in cats with atopic syndrome is unknown since imaging studies have yet to investigate these concurrent disease presentations.

As with other extra-cutaneous manifestations of allergies, the prevalence of concurrent feline asthma in cats with atopic syndrome is uncertain. Feline small airway disease or feline asthma is a complex syndrome; however, many cats have an allergic pathogenesis [22]. In a pilot study evaluating the prevalence of positive reactions to inhaled allergens in cats with small airway disease [23], the presence of concurrent or pre-existing dermatologic abnormalities was quite high, making recruitment of patients for the study difficult. This finding may indicate a higher percentage of cats with both allergic airway disease and feline atopic syndrome. In some cases, the severity of respiratory signs may, however, overshadow the presence of concurrent skin disease, or treatment for feline asthma (i.e., glucocorticoids) may control the signs of cutaneous allergy and thereby mask the true clinical appearance. Further study is warranted to better elucidate the relationship between the two disease conditions.

Conclusions

Feline atopic syndrome draws several parallels to canine atopic dermatitis mostly in the involvement of pruritus in this clinical diagnosis. Much uncertainty remains, however, in regard to the similarities which can be identified between the two allergic conditions in regard to clinical manifestations of disease and the particular nature of disease pathogenesis. There remains quite a good amount of information yet to be discovered with regard to the patient with feline atopic dermatitis. Compared to their canine counterpart, studies are lacking in the veterinary literature for allergic cats.

References

1. Reedy LM. Results of allergy testing and hyposensitization in selected feline skin diseases. J Am Anim Hosp Assoc. 1982;18:618–23.
2. Hobi S, Linek M, Marignac G, Olivry T, Beco L, Nett C, et al. Clinical characteristics and causes of pruritus in cats: a multicentre study on feline hypersensitivity-associated dermatoses. Vet Dermatol. 2011;22:406–13.
3. Marsella R, De Benedetto A. Atopic dermatitis in animals and people: an update and comparative review. Vet Sci. 2017;4(3):37.
4. Peng W, Novak N. Pathogenesis of atopic dermatitis. Clin Exp Allergy. 2015;45(3):566–74.
5. Martel BC, Lovato P, Bäumer W, Olivry T. Translational animal models of atopic dermatitis for preclinical studies. Yale J Biol Med. 2017;90(3):389–402.
6. Elmose C, Thomsen SF. Twin studies of atopic dermatitis: interpretations and applications in the filaggrin era. J Allergy. 2015;2015:902359.
7. Amat F, Soria A, Tallon P, Bourgoin-Heck M, Lambert N, Deschildre A, Just J. New insights into the phenotypes of atopic dermatitis linked with allergies and asthma in children: an overview. Clin Exp Allergy. 2018;48(8):919–34.

8. Wilhem S, Kovalik M, Favrot C. Breed-associated phenotypes in canine atopic dermatitis. Vet Dermatol. 2010;22:143–9.
9. Nuttal T. The genomics revolution: will canine atopic dermatitis be predictable and preventable? Vet Dermatol. 2013;24(1):10-8.e.3-4.
10. Moriello KA. Feline atopy in three littermates. Vet Dermatol. 2001;12:177–81.
11. Prost C. Les dermatoses allergiques du chat. Prat Méd Chir Anim Comp. 1993;28:151–3.
12. Roosje PJ, Thepen T, Rutten VP, et al. Immunophenotyping of the cutaneous cellular infiltrate after atopy patch testing in cats with atopic dermatitis. Vet Immunol Immunopathol. 2004;101:143–51.
13. Marsella R, Girolomoni G. Canine models of atopic dermatitis: a useful tool with untapped potential. J Invest Dermatol. 2009;129(10):2351–7.
14. Halliwell R, the International Task Force on Canine Atopic Dermatitis. Revised nomenclature for veterinary allergy. Vet Immuno Immunopathol. 2006;114:207–8.
15. Taglinger K, Helps CR, Day MJ, Foster AP. Measurement of serum immunoglobulin E (IgE) specific for house dust mite antigens in normal cats and cats with allergic skin disease. Vet Immunol Immunopathol. 2005;105:85–93.
16. Lauber B, Molitor V, Meury S, Doherr MG, Favrot C, Tengval K, et al. Total IgE and allergen-specific IgE and IgG antibody levels in sera of atopic dermatitis affected and non-affected Labrador- and Golden retrievers. Vet Immunol Immunopathol. 2012;149:112–8.
17. Gilbert S, Halliwell RE. Feline immunoglobulin E: induction of antigen-specific antibody in normal cats and levels in spontaneously allergic cats. Vet Immunol Immunopathol. 1998;63:235–52.
18. Reinero CR. Feline immunoglobulin E: historical perspective, diagnostics and clinical relevance. Vet Immunol Immunopathol. 2009;132:13–20.
19. Gilbert S, Halliwell RE. Production and characterization of polyclonal antisera against feline IgE. Vet Immunol Immunopathol. 1998;63:223–33.
20. Lee-Fowler TM, Cohn LA, DeClue AE, Spinkna CM, Ellebracht RD, Reinero CR. Comparison of intradermal skin testing (IDST) and serum allergen-specific IgE determination in an experimental model of feline asthma. Vet Immunol Immunopathol. 2009;132:46–52.
21. Seals SL, Kearney M, Del Piero F, Hammerberg B, Pucheu-Haston CM. A study for characterization of IgE-mediated cutaneous immediate and late-phase reactions in non-allergic domestic cats. Vet Immunol Immunopathol. 2014;159:41–9.
22. Norris Reinero CR, Decile KC, Berghaus RD, Williams KJ, Leutenegger CM, Walby WF, et al. An experimental model of allergic asthma in cats sensitized to house dust mite or Bermuda grass allergen. Int Arch Allergy Immunol. 2004;135:117–31.
23. Moriello KA, Stepien RL, Henik RA, Wenholz LJ. Pilot study: prevalence of positive aeroallergen reactions in 10 cats with small airway disease without concurrent skin disease. Vet Dermatol. 2007;18:94–100.
24. Taglinger K, Day MJ, Foster AP. Characterization of inflammatory cell infiltration in feline allergic skin disease. J Comp Pathol. 2007;137:211–23.
25. Roosje PJ, Whitaker-Menezes D, Goldschmidt MH, et al. Feline atopic dermatitis. A model for Langerhans cell participation in disease pathogenesis. Am J Pathol. 1997;151:927–32.
26. Novak N. An update on the role of human dendritic cells in patients with atopic dermatitis. J Allergy Clin Immunol. 2012;129:879–86.
27. Roosje PJ, Koeman JP, Thepen T, et al. Mast cells and eosinophils in feline allergic dermatitis: a qualitative and quantitative analysis. J Comp Pathol. 2004;131:61–9.
28. Liu FT, Goodarzi H, Chen HY. IgE, mast cells, and eosinophils in atopic dermatitis. Clin Rev Allergy Immunol. 2011;41:298–310.
29. Jarvikallio A, Naukkarinen A, Harvima IT, et al. Quantitative analysis of tryptase- and chymase-containing mast cells in atopic dermatitis and nummular eczema. Br J Dermatol. 1997;136:871–7.
30. Beadleston DL, Roosje PJ, Goldschmidt MH. Chymase and tryptase staining of normal feline skin and of feline cutaneous mast cell tumors. Vet Allergy Clin Immunol. 1997;5:54–8.

31. Roosje PJ, van Kooten PJ, Thepen T, Bihari IC, Rutten VP, Koeman JP, et al. Increased numbers of CD4+ and CD8+ T cells in lesional skin of cats with allergic dermatitis. Vet Pathol. 1998;35:268–73.
32. Roosje PJ, Dean GA, Willemse T, et al. Interleukin 4-producing CD4+ T cells in the skin of cats with allergic dermatitis. Vet Pathol. 2002;39:228–33.
33. Taglinger K, Van Nguyen N, Helps CR, et al. Quantitative real-time RT-PCR measurement of cytokine mRNA expression in the skin of normal cats and cats with allergic skin disease. Vet Immunol Immunopathol. 2008;122:216–30.
34. Dunham S, Messamore J, Bessey L, Mahabir S, Gonzales AJ. Evaluation of circulating interleukin-31 levels in cats with a presumptive diagnosis of allergic dermatitis. Vet Dermatol. 2018;29:284. [abstract]
35. Szczepanik MP, Wilkołek PM, Adamek ŁR, et al. The examination of biophysical parameters of skin (transepidermal water loss, skin hydration and pH value) in different body regions of normal cats of both sexes. J Feline Med Surg. 2011;13:224–30.
36. Szczepanik MP, Wilkołek PM, Adamek ŁR, et al. Correlation between transepidermal water loss (TEWL) and severity of clinical symptoms in cats with atopic dermatitis. Can J Vet Res. 2018;82(4):306–11.
37. Sanford JA, Gallo RL. Functions of the skin microbiota in health and disease. Semin Immunol. 2013;25(5):370–7.
38. Rodrigues Hoffmann A, Patterson AP, Diesel A, Lawhon SD, Ly HJ, Elkins Stephenson C, et al. The skin microbiome in healthy and allergic dogs. PLoS One. 2014;9(1):e83197.
39. Older CE, Diesel A, Patterson AP, Meason-Smith C, Johnson TJ, Mansell J, et al. The feline skin microbiota: the bacteria inhabiting the skin of healthy and allergic cats. PLoS One. 2017;12(6):e0178555.
40. Scott DW, Miller WH, Erb HN. Feline dermatology at Cornell University: 1407 cases (1988–2003). J Fel Med Surg. 2013;15(4):307–16.
41. Scott DW, Paradis M. A survey of canine and feline skin disorders seen in a university practice: small animal clinic, University of Montréal, Saint-Hyacinthe, Québec (1987–1988). Can Vet J. 1990;31:830–5.
42. Hill PB, Lo A, Eden CAN, Huntley S, Morey V, Ramsey S, et al. Survey of the prevalence, diagnosis and treatment of dermatological conditions in small animal general practice. Vet Rec. 2006;158:533–9.
43. Griffin CE, DeBoer DJ. The ACVD task force on canine atopic dermatitis (XIV): clinical manifestations of canine atopic dermatitis. Vet Immunol Immunopathol. 2001;81(3–4):255–69.
44. Hensel P, Santoro D, Favrot C, Hill P, Griffin C. Canine atopic dermatitis: detailed guidelines for diagnosis and allergen identification. BMC Vet Res. 2015;11:196.
45. Foster AP, Roosje PJ. Update on feline immunoglobulin E (IgE) and diagnostic recommendations for atopy. In: August JR, editor. Consultations in feline internal medicine. 5th ed. St. Louis: Elsevier; 2006. p. 229–38.
46. Masuda K, Kurata K, Sakaguchi M, Yamashita K, Hasegawa A, Ohno K, Tsujimoto H. Seasonal rhinitis in a cat sensitized to Japanese cedar (Cryptomeria japonica) pollen. J Vet Med Sci. 2001;63:79–81.

Feline Atopic Syndrome: Diagnosis

Ralf S. Mueller

Abstract

Feline atopic syndrome is an aetiological diagnosis of a disease caused by environmental or dietary allergens. As such there is currently no single test reliably differentiating feline atopic syndrome from its differential diagnoses. This syndrome is associated with a number of clinical reaction patterns such as miliary dermatitis, eosinophilic granuloma, pruritus leading to non-inflammatory alopecia or ulcerative and crusty dermatitis. The diagnosis is confirmed by ruling out all differential diagnoses based on history and clinical examination. Hence, the diagnostic approach is different with the various reaction patterns. As adverse food reaction and flea bite hypersensitivity are differential diagnoses for all these reaction patterns, excellent ectoparasite control and an elimination diet are part of the recommended diagnostic work-up for all cats with suspected feline atopic syndrome. Depending on the clinical findings, other diagnostic tests such as cytology, Wood's lamp, trichogram, fungal culture or biopsy may be indicated.

Introduction

In contrast to canine atopic dermatitis, which has distinct clinical features, feline atopic syndrome is characterized by a number of cutaneous reaction patterns that look distinctively different [1, 2]. Miliary dermatitis, eosinophilic granuloma complex or pruritus without lesions, either leading to non-inflammatory alopecia or secondary excoriations with ulceration and crusting, can all be due to feline atopic syndrome. Similar to canine atopic dermatitis, feline atopic syndrome is a diagnosis

R. S. Mueller (✉)
Centre for Clinical Veterinary Medicine, München, Germany
e-mail: dermatologie@medizinische-kleintierklinik.de

© Springer Nature Switzerland AG 2020
C. Noli, S. Colombo (eds.), *Feline Dermatology*,
https://doi.org/10.1007/978-3-030-29836-4_22

based on history, clinical signs and exclusion of differential diagnoses [3]. However, each of the above-mentioned reaction patterns has a different list of differential diagnoses and consequently needs a slightly different approach. This chapter will discuss the differential diagnoses of the various cutaneous reaction patterns frequently associated with feline atopic syndrome as well as the diagnostic approach for each of those.

General Principles of Diagnosis

A thorough history and clinical examination are essential for the formulation of a list of differential diagnoses for each of the cutaneous reaction patterns seen regularly with feline atopic syndrome. Important questions to be asked depend on the individual reaction pattern and the differential diagnoses possibly responsible for it. Fleas, food or environmental allergens can cause all of the stated clinical signs and thus questions about current ectoparasite control, feeding habits, faecal consistency (when owners have access to the faeces) and attempted elimination diets are relevant for all those reaction patterns [1]. Other diagnoses are only associated with selected patterns. For example, an infestation with *Otodectes cynotis* has been reported as a cause of miliary dermatitis [4], but not with eosinophilic granuloma, and questions about previous ear disease and other affected animals in the household are important. Non-inflammatory alopecia may be caused by demodicosis [4] or rarely endocrine disease or alopecia areata, diseases not considered in a cat with miliary dermatitis. The age of the patient and careful questioning with regard to systemic signs may provide clinical clues for endocrine disease. Once the list of differential diagnoses and their order of priority based on history and clinical findings have been established, diagnostic tests to rule out or confirm those diagnoses are undertaken. The efforts spent to achieve a confirmed diagnosis as quickly as possible will of course also be determined by the owner and his or her willingness to invest time and money. In some patients, tests will be performed subsequently in the order of disease likelihood or necessity to rule out (e.g. a dermatophyte culture in a Persian cat), and other cat owners will choose to perform an array of tests at the same time to rule out a number of differential diagnoses fairly quickly. Once every other differential diagnosis has been ruled out, the diagnosis of feline atopic syndrome is confirmed.

Approach to the Cat with Miliary Dermatitis

Although miliary dermatitis (Fig. 1) is often assumed to be caused by allergic reactions against fleas and environmental or food allergens (and indeed those are the cause in the majority of cats), other causes are possible and may need to be considered in individual patients. Ectoparasites other than fleas, notably mites such

Fig. 1 Miliary dermatitis with small crusts in a domestic shorthaired cat

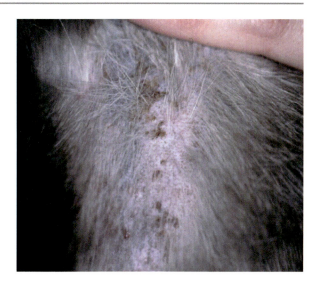

Table 1 Diseases causing or contributing to miliary dermatitis

Allergies	Flea bite hypersensitivity
	Environmental atopic syndrome
	Food-induced atopic syndrome
Infectious diseases	Dermatophytosis
	Bacterial infection
	Notoedres cati
	Otodectes cynotis
	Demodex cati
Immune-mediated diseases	Pemphigus foliaceus
Neoplastic diseases	Mast cell tumour
Nutritional deficiencies	Fatty acid deficiency

as *Cheyletiella blakei* or *Otodectes cynotis*, may also lead to miliary dermatitis (Table 1). Infections with dermatophytes or bacteria may cause or contribute to clinical signs of miliary dermatitis. On rare occasions, pemphigus foliaceus and mast cell tumours can present with similar small crusted papules. Finally, some nutritional deficiencies such as a lack of essential fatty acids in the diet may cause miliary dermatitis. In most circumstances, this is unlikely with current feeding practices. Depending on the clinical history and physical examination, skin scrapings, ectoparasite control, skin cytology or biopsy and elimination diets may be indicated to work up individual cats with miliary dermatitis. Skin biopsy is the diagnostic test of choice to differentiate allergic from immune-mediated or neoplastic skin disease. However, it is less likely to be able to differentiate between ectoparasitic and allergic causes and cannot reliably differentiate causes of allergy.

Approach to the Cat with Pruritus

Initial pruritus without lesions may lead either to non-inflammatory alopecia or to ulcerations and crusting due to self-trauma, usually in the area of the head and neck. Those reaction patterns have different possible aetiologies.

Non-inflammatory alopecia (Fig. 2) is most frequently caused by flea, food or environmental allergens or any combination thereof [1, 2]. A major differential diagnosis for a pruritic cat with non-inflammatory alopecia is psychogenic alopecia [5]. Major changes in the environment, such as a physical move from countryside to a town or city, a new cat moving into the neighbourhood, a new baby or animal in the household or changes in the work hours of the family, may cause excessive licking in some cats. Consulting a veterinary behaviourist may be helpful in such cases and is recommended in those cases which do not respond to an elimination diet or flea control. In early stages of dermatophytosis, associated mild scaling and a fine papular rash may not be present or may be overlooked [6], and fungal tests such as Wood's lamp, trichograms, cultures or PCR may be useful diagnostic options. However, keep in mind the possibility of false-positive PCR reactions due to transient environmental contamination. Very rarely, endocrine diseases such as hyperadrenocorticism may cause non-pruritic non-inflammatory alopecia in the cat, typically associated with other systemic signs [7–9]. Hormonal testing is needed in such cats. Unusual alopecias such as telogen effluvium (where a stressful event sends all hair follicles of a certain area into a synchronized telogen (the resting phase) and alopecia occurs 6–12 weeks later when the new hair is regrowing in the deep dermis) or anagen defluxion (where a severe metabolic disease or chemotherapy leads to the production of damaged hair shafts that break off within the follicular lumen leading to alopecia) can be ruled out or suspected by a thorough history. Ventral abdominal alopecia associated with demodicosis has been diagnosed in restricted geographical areas worldwide and associated with pruritic non-inflammatory alopecia (Table 2).

Head and neck pruritus leading sometimes to widespread crusting and ulceration (Fig. 3) may be due to environmental allergens, but food allergens, flea bite hypersensitivity and other ectoparasite infestations such as *Notoedres cati* or *Otodectes cynotis* may also be considered [1, 2]. Secondary infections with bacteria or yeast occur frequently. Infections with dermatophytosis may be associated with pruritus if the skin is inflamed. If there is a previous clinical history of an

Fig. 2 Non-inflammatory hypotrichosis and alopecia on the ventrum of a 7-year-old, male castrated domestic shorthaired cat

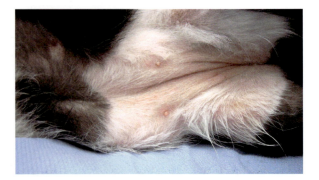

Table 2 Diseases causing alopecia in the cat

Allergies	Flea bite hypersensitivity
	Environmental atopic syndrome
	Food-induced atopic syndrome
Psychogenic diseases	Psychogenic alopecia
Infectious diseases	Dermatophytosis
Endocrine diseases	*Demodex cati*
	Hypothyroidism
	Hyperadrenocorticism
Drug reaction	Methimazole-induced alopecia
Miscellaneous diseases	Telogen effluvium
	Anagen defluxion

Fig. 3 A large crust and excoriations in a cat with severe pruritus on the head. (Courtesy of Dr. Chiara Noli)

upper respiratory tract infection and mucosal surfaces are affected as well, viral infections with feline herpes- or calicivirus should be considered. On rare occasions cowpox virus infections may also lead to variably pruritic, ulcerative and crusty skin disease with fever and anorexia [10]. Cowpox is inoculated through the bite wound of a rodent and initially a solitary lesion develops. Fever and multifocal cutaneous lesions result from a subsequent viremia. Such cats may be highly infectious and have been reported to die from viral pneumonia. PCR testing of crusts is the diagnostic tool of choice, it is fast, and the poxvirus inclusions are rich in the crust. On histopathology, cowpox virus is identified through intracytoplasmic eosinophilic inclusion bodies and herpes virus through an eosinophilic dermatitis, folliculitis and furunculosis and with careful searching for amphophilic intranuclear inclusion bodies.

Pruritus may be caused by flea, food or environmental allergens, but other diseases such as medication reactions or, in older cats, paraneoplastic pruritus may also be considered. If a hyperthyroid cat develops pruritus on methimazole, discontinuation of the medication typically leads to a quick resolution of the self-trauma, and an alternative treatment for hyperthyroidism should be elected. In older pruritic cats, particularly with *Malassezia* infections of the head or neck, paraneoplastic skin disease should be considered. Depending on history and physical examination, ultrasonography, radiography (and/or CT/MRI), lymph node aspirates, complete blood counts and serum biochemistry may all be indicated when paraneoplastic pruritus is considered.

Approach to the Cat with Lesions of the Eosinophilic Granuloma Complex

Lesions of the eosinophilic granuloma complex may be observed with feline atopic syndrome [1, 2]. *Indolent lip ulcers* (Fig. 4) (often asymmetrical ulcers frequently on the mucosal margin of the upper lip and often covered with a thick yellowish adherent exudate), *eosinophilic granulomas* (Fig. 5) (papular to linear lesions, often eroded or ulcerated, frequently found on the caudal thighs) and (typically highly pruritic) *eosinophilic plaques* (Fig. 6) found on the ventral abdomen and inner thighs can all be caused by flea, food or environmental allergens, and thus

Fig. 4 Indolent ulcer on the upper lip of a 6-year-old female domestic shorthaired cat. (Courtesy of Dr. Chiara Noli)

Fig. 5 A linear lesion of eosinophilic granuloma on the thigh. (Courtesy of Dr. Chiara Noli)

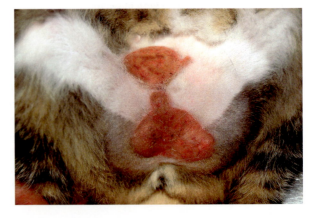

Fig. 6 Same cat as in Fig. 4: a large eosinophilic plaque on the abdomen. (Courtesy of Dr. Chiara Noli)

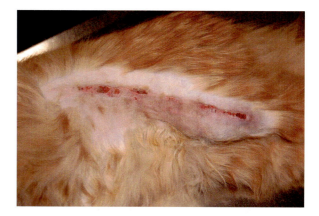

Fig. 7 Impression smear from a plaque: neutrophils with intra- and extracytoplasmatic cocci and rod-shaped bacteria. The latter probably origin from the oral flora (Diff Quick, 1000×). (Courtesy of Dr. Chiara Noli)

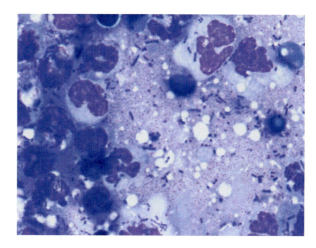

ectoparasite control and an elimination diet are part of the thorough diagnostic work-up of any cat with lesions belonging to the eosinophilic granuloma complex. On occasion, squamous cell carcinoma may be a differential diagnosis, particularly in older cats with lesions in non-pigmented or sparsely haired areas of the head. In those cats, a biopsy is also indicated.

Ruling Out Skin Infections

Although skin infections are rarer in cats than in dogs, they do occur in the feline species and may contribute significantly to both pruritus and clinical signs. They need to be recognized and treated to achieve optimal therapeutic outcome. A cytologic evaluation of an impression smear is the test of choice to identify bacterial or yeast infections [11]. If the skin and crusts are very dry, a better yield is often achieved by removing a few crusts and obtaining cytology from the underlying surface of the crusts. Neutrophils with intracellular bacteria (Fig. 7) confirm a bacterial skin infection without a doubt. The presence of bacteria or yeast has to be

Fig. 8 Numerous *Malassezia* yeasts from the skin of an allergic cat (Diff Quick 400×). (Courtesy of Dr. Chiara Noli)

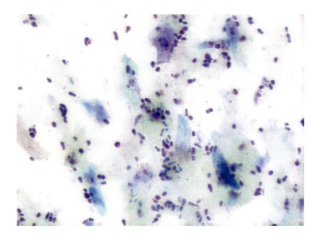

interpreted in light of the numbers of organisms, clinical signs and the sampling site. Large numbers of *Malassezia* spp. yeasts (Fig. 8) have been reported to be a possible clinical clue for internal malignancies in the cat; however, they can also be found with allergic skin disease. Examination of the skin with a Wood's lamp, trichograms, fungal cultures or PCR for fungal antigens may be useful in patients with possible dermatophytosis.

Ruling Out Ectoparasites

It is important to identify the type of ectoparasite control the owner conducts, which exact product is administered, how often and to which of the animals in the household. Some of the products on the market have a high efficacy for fleas and ticks; others can also be used to treat mite infestations. An effective and complete ectoparasite control should address not just fleas but also mites. Macrocyclic lactones or isoxazolines are examples of such ectoparasiticides. Whether additional environmental control is needed in addition to regular adulticides will depend on the individual patient, environment and climate. Flea proliferation is facilitated by warm and humid climates. With large numbers of immature stages in a conducive environment, spraying the house or apartment with an insect growth regulator such as methoprene or pyriproxyfen will hasten clinical improvement in affected cats. Similarly, such environmental control may be needed in households with multiple animals and consequently a large environmental load of immature flea stages such as eggs, larvae and pupae.

Performing an Elimination Diet

At this point, an elimination diet is the only reliable test to identify feline atopic syndrome caused by food antigens [12]. This involves – theoretically – the feeding of a protein source the animal has never received before. In cats, however, that seemingly simple condition is frequently difficult to achieve. First, many cats receive a far more varied diet than their canine counterparts, and it is not unusual that feline patients get a different protein source every day of the week. Second, cats are creatures of habit and more easily refuse to eat a new diet than dogs. In addition, refusal to eat for a few days increases the risk of development of hepatic lipidosis in the cat, and starvation until compliance is achieved is absolutely not recommended. Consequently, several different food sources may have to be trialled before a successful elimination diet can be conducted, and the author advises owners to have the choice of two protein sources available should the cat suddenly decide not to eat.

Owners can choose between a home-cooked elimination diet, commercial selected protein diets and hydrolysed diets. Many commercial selected protein diets have been shown to be contaminated with other, off-label protein sources, although the clinical relevance of those contaminations has not been evaluated. Consequently, the author prefers home-cooked or extensively hydrolysed diets for the diagnosis of food-induced feline atopic syndrome. Cats are obligate carnivores, so that when choosing a home-cooked diet, they can be fed pure protein. A carbohydrate source is not essential and may decrease palatability and compliance in cats. Ideally, the protein source is phylogenetically distant from the originally fed protein. If the cat received a predominantly chicken- and turkey-based cat food, then switching, for example, to duck may not be as suitable as rabbit or horse. Similarly, if the cat received a beef- or lamb-based diet, then deer or goat may not be the ideal alternative, as the chance of cross-reactivity between those allergens is probably much higher than if ostrich or crocodile is chosen for that particular cat. However, clinical cross-reactivities have not been established in cats with food allergies at this point in time.

The diet should be fed exclusively for approximately 8 weeks, during which time more than 90% of the cats with adverse food reactions will improve [13]. In those 8 weeks, no other protein sources should be permitted. A cat with access to outdoors technically needs to be confined indoors during the entire diet. If that is not possible and the cat does not respond, then an adverse food reaction cannot reliably be ruled out. However, this may indeed be too stressful for the cat. In the author's opinion, it may be still worthwhile conducting an elimination diet in some indoor-outdoor cats, because of the possibility that just a reduction of the amount of allergy-inducing protein may lower the pruritic threshold. In a multi-cat household, all cats should receive the elimination diet or the patient should be fed completely separately to avoid unintended intake of a different protein source.

If there is no clinical improvement after 8 weeks of an appropriate elimination diet, then an adverse food reaction is very unlikely. If however there was clinical improvement, then a re-challenge with the previous diet is essential as this improvement may be due to the diet, but may also be due to seasonal changes, different or more reliably administered concurrent treatments and other reasons not related to the diet. If the re-challenge with the previous diet leads to recurrence of clinical signs, which resolve again when the elimination diet is fed, the diagnosis of adverse food reaction is confirmed. Long-term, the offending allergen(s) can be identified by sequential re-challenges with individual proteins, the cat may be fed a commercial hydrolysed or selected protein diet, or the elimination diet may be continued. If the owner opts for the latter, it is recommended to consult a veterinary nutritionist in order to balance the home-prepared diet and avoid nutritional deficiencies.

Conclusion

Feline atopic syndrome is an aetiological diagnosis associated with a number of clinical reaction patterns such as miliary dermatitis, eosinophilic granuloma, pruritus leading to non-inflammatory alopecia or ulcerative and crusty dermatitis. The diagnosis is confirmed by ruling out all differential diagnoses based on history and clinical examination. As adverse food reaction and flea bite hypersensitivity are differential diagnoses for all these reaction patterns, excellent ectoparasite control and an elimination diet are part of the recommended diagnostic work-up for all cats with suspected feline atopic syndrome. Depending on the clinical findings, other diagnostic tests such as cytology, Wood's lamp, trichogram, fungal culture or biopsy may be indicated.

References

1. Hobi S, Linek M, Marignac G, Olivry T, Beco L, Nett C, et al. Clinical characteristics and causes of pruritus in cats: a multicentre study on feline hypersensitivity-associated dermatoses. Vet Dermatol. 2011;22:406–13.
2. Ravens PA, Xu BJ, Vogelnest LJ. Feline atopic dermatitis: a retrospective study of 45 cases (2001–2012). Vet Dermatol. 2014;25:95–102, e27-8
3. DeBoer DJ, Hillier A. The ACVD task force on canine atopic dermatitis (XV): fundamental concepts in clinical diagnosis. Vet Immunol Immunopathol. 2001;81:271–6.
4. Scheidt VJ. Common feline ectoparasites part 2: Notoedres cati, Demodex cati, Cheyletiella spp. and Otodectes cynotis. Feline Pract. 1987;17:13–23.
5. Waisglass SE, Landsberg GM, Yager JA, Hall JA. Underlying medical conditions in cats with presumptive psychogenic alopecia. J Am Vet Med Assoc. 2006;228:1705–9.
6. Scarampella F, Zanna G, Peano A, Fabbri E, Tosti A. Dermoscopic features in 12 cats with dermatophytosis and in 12 cats with self-induced alopecia due to other causes: an observational descriptive study. Vet Dermatol. 2015;26:282–e63.
7. Boord M, Griffin C. Progesterone secreting adrenal mass in a cat with clinical signs of hyperadrenocorticism. J Am Vet Med Assoc. 1999;214:666–9.
8. Rand JS, Levine J, Best SJ, Parker W. Spontaneous adult-onset hypothyroidism in a cat. J Vet Intern Med. 1993;7:272–6.
9. Zerbe CA, Nachreiner RF, Dunstan RW, Dalley JB. Hyperadrenocorticism in a cat. J Am Vet Med Assoc. 1987;190:559–63.
10. Appl C, von Bomhard W, Hanczaruk M, Meyer H, Bettenay S, Mueller R. Feline cowpoxvirus infections in Germany: clinical and epidemiological aspects. Berliner und Münchner Tierärztliche Wochenschrift. 2013;126:55–61.
11. Mueller RS, Bettenay SV. Skin scrapings and skin biopsies. In: Ettinger SJ, Feldman EC, Cote E, editors. Textbook of veterinary internal medicine. Philadelphia: W.B. Saunders; 2017. p. 342–5.
12. Mueller RS, Unterer S. Adverse food reactions: pathogenesis, clinical signs, diagnosis and alternatives to elimination diets. Vet J. 2018;236:89–95.
13. Olivry T, Mueller RS, Prelaud P. Critically appraised topic on adverse food reactions of companion animals (1): duration of elimination diets. BMC Vet Res. 2015;11:225.

Feline Atopic Syndrome: Therapy

Chiara Noli

Abstract

Feline allergic dermatitis is a chronic disease and allergen avoidance, when possible, is the best management option. If this is not possible, then a combination of aetiologic, symptomatic, topical, antimicrobial and nutritional therapy is implemented, depending on the individual case. Aetiologic therapy is based on allergy test and hyposensitization, which will be curative only in a minority of cases. All other cases will need some sort of symptomatic therapy, possibly avoiding long-term administration of glucocorticoids. Alternative systemic treatments include ciclosporin, antihistamines, oclacitinib, palmitoylethanolamide, maropitant and PUFAs: not all of these are effective in every case and some are not registered for the cat. Topical treatments are not easy to apply in cats and only a few studies confirm their efficacy. The pros and cons of allergy testing and hyposensitization, and of topical and/or systemic symptomatic treatment will be discussed in this chapter.

Introduction

Feline allergic dermatitis is a chronic disease. The clinician must make the client understand that unless the offending allergen(s) are identified and removed, a cure is rarely possible. The keys to a successful management of allergic dermatitis are client education, long-term commitment to the treatment protocol and a combination of aetiologic, symptomatic, topical, antimicrobial and nutritional therapy. The choice of the therapeutical plan will depend on the individual case, that is, on both cat (severity of the lesions and temper of the patient) and owner (economical possibilities, patience, time to devote to the cat, personal preferences). The pros and cons

C. Noli (✉)
Servizi Dermatologici Veterinari, Peveragno, Italy

© Springer Nature Switzerland AG 2020
C. Noli, S. Colombo (eds.), *Feline Dermatology*,
https://doi.org/10.1007/978-3-030-29836-4_23

of allergy testing and hyposensitization, and of topical and/or systemic symptomatic treatment should be clearly explained, including possible combinations – and costs – to help the owner make an informed choice. A practical guidance on how to therapeutically approach allergic cats is offered in Box 1.

Box 1: Practical Therapeutical Approach to the Pruritic Cat
1. Diagnostic period (from first presentation to end of elimination diet):

– Oral/topical flea control is advised in every case.
– For non-seasonal pruritus, the cat should undergo a 2-month-long elimination diet, better if with hydrolysed food.
– If pruritus is important and needs to be decreased, in the meanwhile the cat can be administered short-acting oral corticosteroids at tapering doses possibly every other day for the first 6 weeks. As an alternative oclacitinib or maropitant can be considered. At this stage ciclosporin should be avoided, as the lag time is very long and it takes also a long time for pruritus to come back after withdrawal. This makes it difficult to evaluate the diet.

2. First months of allergen specific immunotherapy (ASIT):

– Oral/topical flea control is advised in every case during the whole ASIT period.
– If pruritus is mild to moderate, consider antihistamines and/or ultramicronized PEA, EFAs, dermatological food, and topical hydrocortisone aceponate.
– If these do not work or if pruritus is moderate to severe, then consider ciclosporin or oclacitinib or maropitant during the first few months of ASIT (induction phase). While systemic corticosteroids can be administered for a few days at the beginning of the ASIT phase (particularly if ciclosporin is chosen as maintenance therapy), their long-term use should be avoided, as they could possibly interfere with the desensitization mechanism of ASIT. Every 2–3 months, antipruritic therapies could be withdrawn to better evaluate ASIT efficacy.

3. Long-term symptomatic management:

– Oral/topical flea control is advised in every case.
– If pruritus is mild to moderate, consider antihistamines and/or ultramicronized PEA, EFAs, dermatological food, and topical hydrocortisone aceponate, together or in combination.

- If these do not work or if pruritus is moderate to severe, then consider long-term ciclosporin administration. Corticosteroids can be associated in the first 2 weeks.
- If ciclosporin is not an option (e.g. for g.e. upset), then alternatives are low-dose corticosteroids given every other day (better if associated with steroid-sparing products, such as antihistamines, EFA or ultramicronized PEA) or oclacitinib or maropitant.

4. Management of the flare:

- Flares are best managed with short courses (5–15 days) of high-dose corticosteroids. Long-term managements (item no. 3) should then be instituted afterwards, if possible.

Therapy of Allergic Dermatitis and Quality of Life

Pruritus and self-induced skin lesions due to licking and scratching have a significant negative impact on the cat's and the owner's quality of life (QoL) [1], and therapy aiming at decreasing discomfort should be considered from the very first consultation. However, in two studies on the treatment of feline allergy, decrease of pruritus and lesions was always greater than improvement of QoL [2, 3]. This is due to the fact that administration of therapies and repeated visits to the veterinarian have a negative impact on the QoL of both cats and owners, as treating cats is certainly more difficult and a bigger source of psychological stress than treating dogs. This fact should be considered when designing a therapeutic plan for the allergic feline patient, and the plan should be sustainable by the cat and owner over a long period of time. Way of administration (oral, topical, injectable), formulation (tablets, oral liquid, lotion, spray) and frequency should be tailored to the individual patient and owner. Feeding a "dermatological" diet and/or essential fatty acids and/or palmitoylethanolamide (PEA) supplements mixed with food may be a non-traumatic way of decreasing inflammation and pruritus and the need (dose and frequency) of other antipruritic drugs.

Aetiologic Therapy

Identification of Allergens: Allergy Testing in Cats

Allergy testing is necessary for the identification of the allergens putatively responsible for pruritus and skin lesions observed in hypersensitive cats, but it cannot be used to diagnose allergy *per se*, as several healthy cats show positive results and some allergic ones have negative tests [4–7]. Intradermal testing as used in dogs is of limited value in the feline species, because wheals are small, soft and transient and tests are difficult to interpret. The use of fluorescein may improve readability and reliability of intradermal tests [8, 9]. One problem with feline allergy skin testing is that, until now, allergen solutions standardized for the dog's skin were used in cats, with limited knowledge regarding their suitability. Initial investigations have been published on allergen threshold concentrations in cats, albeit only for pollens and in healthy animals [10]. A further problem is that all cats need to be anaesthetized, as stress-related cortisol release interferes with wheal formation [11]. In order to overcome these problems, percutaneous (prick) testing is currently object of investigation in cats and is perceived to be a good alternative to intradermal testing [12, 13], but no kit is yet commercially available specific to the feline species. As in dogs, intradermal allergy tests should not be performed in cats being treated with corticosteroids.

Serum testing is easier to perform and is offered by several laboratories over the world. It has the advantage of being easy to carry out (just one blood sample), does not need anaesthesia and can be performed in cats treated with corticosteroids. While *in vitro* allergy tests are not able to differentiate allergic from normal cats [4–7], they can be useful for the choice of the allergens to be included in the ASIT solution. There are no studies determining that one methodology is preferable over another one. The frequently used and well-studied method based on the cloned alpha chain of the human high-affinity IgE receptor (FcE-RI) (Allercept®; Heska AG, Fribourg, Switzerland) can also be used in cats. A recent study found a strong agreement between results of a rapid screening immunoassay (Allercept® E-Screen 2nd Generation; Heska AG, Fribourg, Switzerland) and the complete Allercept panel; the screening assay may thus be beneficial for predicting the results of the complete-panel serum allergen-specific IgE assay [4].

Allergen Avoidance

Allergen avoidance is useful if the offending allergens have been correctly identified. Cats with indoor allergies (e.g. against house dust mites such as *Dermatophagoides* spp. and storage mites such as *Tyrophagus*, *Acarus* and *Lepidoglyphus* spp.) or danders can be allowed to spend more time outdoors. As house dust mite levels are much higher in bedrooms than in the rest of the house, limiting the cat's access to these rooms may be of help. In the case of indoor allergens, frequent vacuuming with a "high-efficiency particle air filter" (HEPA) vacuum cleaner may reduce the allergen load, or protective furniture covers designed for human asthmatics may be of value. Sprays or foggers (devices that produce a fine mist) containing acaricidal agents or insect growth regulators may be helpful in cases of house dust/storage mite allergy. Using benzoyl-benzoate sprays on a regular basis on beddings, carpets, rugs, furniture, etc. not only kills the mites but also degrades their metabolites (allergens). One study on house dust-/storage mite-sensitive dogs showed that the use of benzyl benzoate spray at home induces 48% resolution and 36% improvement of pruritus [14]. Unfortunately there are no studies yet on allergen avoidance in cats.

Immunotherapy

Allergen-specific immunotherapy (ASIT) is the aetiological treatment of choice in cases where the duration of pruritus is longer than 4 months a year. Allergens are administered at increasing concentration and dose and at decreasing frequency, generally by subcutaneous injection. The mechanism of action of immunotherapy has not been investigated in cats. In dogs and humans, it seems that a shift from a Th2- to a Th1-biased immune response and increase in T-regulatory lymphocytes is responsible for the development of tolerance [15]. Protocols vary depending on the producer and adjuvant. ASIT is considered safe and effective in cats, with good to very good responses (improvement by at least 50%) achieved in 50–80% of treated patients [16–20]. Adverse events such as increased pruritus or anaphylaxis are considered less common than in dogs [17].

As in dogs, clinical results measured by a decrease in pruritus and skin lesions are seen anywhere from 3 to 18 months after starting treatment. Thus, during the initial phase of immunotherapy, symptomatic treatment may be needed. If the treatment is effective, ASIT maintenance therapy is given for the rest of the patient's life. Only a proportion of the cases will be controlled by immunotherapy alone, while others will require adjunctive symptomatic therapy for at least part of the year. In a yet unpublished retrospective study conducted by the author, about 10% of the cats could achieve remission of the allergy after 4–5 years and could withdraw ASIT without relapses. A similar observation was also reported by Vidémont and Pin [21].

An alternative sublingual administration option is currently available, with anecdotal efficacy similar to the subcutaneous way; however there are no published reports yet in cats. Rush immunotherapy (administering the whole induction phase in a few hours, under medical control) was investigated in a small number of cats and was considered safe and effective [22].

Symptomatic Therapy

Doses, administration and adverse effects of the drugs mentioned hereunder are summarized in Table 1.

Glucocorticoids

Glucocorticoids are very effective in suppressing the signs of allergic dermatitis. Pharmacological data on glucocorticoids in felines are scarce: cats seem to need higher doses than dogs, as they have half the density of glucocorticoid receptors in the skin and liver [23] and metabolize the active prednisolone better than the prodrug prednisone [24].

Usual protocols suggest to administer oral prednisolone at 1–2 mg/kg or oral methylprednisolone 0.8–1.6 mg/kg daily until remission of pruritus (usually

Table 1 Main antipruritic and anti-inflammatory drugs used for feline allergic dermatitis. Antihistamines are reported in Table 2

Oral antipruritic drug	Dose	Indication	Contraindications	Side effects
Glucocorticoids: Prednisolone Methylprednisolone Triamcinolone Dexamethasone	Induction phase q24h: 1–2 mg/kg 0.8–1.6 mg/kg 0.1–0.2 mg/kg 0.1–0.2 mg/kg Maintenance phase: 1/2 to 1/4 of the induction dose q48-72 h	Quick decrease of pruritus and inflammation, resolution of lesions of the eosinophilic granuloma complex	Diabetes, kidney disease, liver disease, positive FIV and/or FeLV status	Skin fragility syndrome, diabetes mellitus, congestive heart failure, polyuria and polydipsia, increased susceptibility to bladder and skin infections, demodicosis and dermatophytosis
Ciclosporin	7 mg/kg q24h the first month, then q48h the second month, then twice weekly as maintenance dose, if signs are under control	Long-term use to keep pruritus and lesions in remission	Kidney disease, liver disease, positive FIV and/or FeLV status, malignancies, eating raw meat, hunting and eating the preys	Transitory vomit and/or diarrhoea (24%), weight loss, gingival hyperplasia (2%), hepatic lipidosis (2%), systemic toxoplasmosis
Oclacitinib (*off-label use*)	1 mg/kg q12h	Quick decrease of pruritus without the use of glucocorticoids	No information available. As a matter of caution the same as ciclosporin, suspect kidney disease	Limited information available. Increase kidney values in some cats in one study, not observed in another. Close monitoring is necessary.
Palmitoylethanolamide	10–15 mg/kg q24h Also in association with glucocorticoids as sparing agent	Mild pruritus and eosinophilic granuloma complex	None	None
Maropitant	2 mg/kg q24h	Pruritus	No information available for long-term use. Liver and heart disease	No information available for use longer than 2–4 weeks

3–15 days), and then the dose is reduced to every other day and then further reduced every week to the lowest dose that will control the clinical signs (generally 0.5–0.1 mg/kg every other day). If prednisolone or methylprednisolone do not seem to be effective, a good alternative in cats is oral dexamethasone or triamcinolone (both at 0.1–0.2 mg/kg), which should then be tapered to 0.02–0.05 mg/kg every second to third day for maintenance therapy. The use of methylprednisolone and triamcinolone at the doses mentioned above did not cause an increase of fructosamine above the reference range, while triamcinolone caused a higher increase of amylase compared to methylprednisolone [25].

The use of repositol methylprednisolone acetate (usually 15–20 mg/cat, SC), whose duration of action ranges from 3 to 6 weeks, should be considered only for refractory cats, when oral administration is not possible. Repeated repositol injections appear to become less and less effective with time, so that increased frequency and/or higher doses may become necessary, with increased risks of adverse effects development. In these cases, alternative therapies (such as oral or injectable ciclosporin) should be considered.

Cats are usually considered to tolerate glucocorticoids well; however adverse effects can occur and can be severe [26]. Among these there are cutaneous atrophy with skin fragility (Fig. 1), congestive heart failure, increased susceptibility to diabetes mellitus (particularly in obese cats), polydipsia and polyuria and increased susceptibility to bladder and skin infections, including development of dermatophytosis and demodicosis. A recent study, however, found no evidence of bacteriuria in cats treated with long-term oral or repositol glucocorticoids [27].

Hydrocortisone aceponate topical spray is useful to treat localized pruritus and reduce the need for systemic medication. This product has been proven to cause minimal thinning of the skin and local immunosuppression and has very low systemic absorption in dogs. An open pilot trial on ten cats determined that it is able to improve pruritus and skin lesions in allergic felines and keep them under control with daily or every other day maintenance administration [28].

Fig. 1 Large ulceration due to skin fragility in a cat being treated with 20 mg/cat injectable methylprogesterone acetate once monthly for 5 consecutive months. The cat has completely recovered after withdrawal of the drug

Ciclosporin

Ciclosporin is a polypeptide derived from the fungus *Tolypocladium inflatum*. Its mode of action is by inhibition of calcineurin. It has a variety of immunological effects on multiple components of the skin immune system and is active in the acute and chronic phase of allergic dermatitis. Ciclosporin has the same efficacy as prednisolone in the control of clinical signs of allergic dermatitis in cats [29]. Significant reduction in pruritus should be expected in 75–85% of cases within 1 month of treatment [30]. The initial oral dose in cats is 7 mg/kg/day [31]. This dose should be administered for at least 1 month before, if effective, tapering it to every other day. After another month of successful every other day administration, tapering to twice weekly can be tried. About 15% and 60% of cats with skin allergy can be kept under control, respectively, with every other day or twice weekly administration [32, 33]. A lag period of about 2–3 weeks, in which no response is seen, occurs after ciclosporin treatment is started, and owners should be warned about this. In dogs, association of 3 weeks of prednisone or oclacitinib with ciclosporin, to quickly decrease pruritus during the lag phase, has been described, [34, 35], but no such data are available for the feline species.

The proprietary feline product (Atopica® for Cats, Elanco) is a microemulsified ciclosporin liquid formulation (100 mg/ml), which cannot be mixed with water. To maximize absorption, ciclosporin should be administered 2 hours before a meal; however recent data suggested that giving ciclosporin with food does not alter clinical outcomes [36]. This formulation is not always palatable when mixed with food, and when administered directly in the mouth it can cause hypersalivation in some subjects. A syringe of fresh water may be dispensed after ciclosporin administration in order to overcome this problem. The successful use of injectable ciclosporin (50 mg/ml) at the dose of 2.5–5 mg/kg every 24–72 h was recently described [37] and could be considered for the treatment of refractory cats.

Ciclosporin is usually well tolerated by cats. Reported adverse effects are transitory vomiting and/or diarrhoea in up to one fourth of the cases, so that the owners should be warned about their possible occurrence [38]. The co-administration of maropitant (2 mg/kg) with ciclosporin during the first 2–3 weeks has been anecdotally suggested to decrease vomiting and provide a quick relief of pruritus (see later for anti-pruritic effects of maropitant). Other described adverse effects are weight loss (16%) and rarely gingival hyperplasia (Fig. 2), anorexia and hepatic lipidosis (each 2% of the cases) [38]. Cats should be FIV-FeLV negative and should not be allowed to hunt and eat raw meat, due to the risk of developing fatal toxoplasmosis [39]. Preventive or concurrent (during therapy) measurement of IgG and/or IgM anti-*Toxoplasma* serum titres does not seem to be useful to predict the development of toxoplasmosis. Clinicians should be alerted by the development of any neurologic and/or respiratory sign or significant weight loss (over 20%) in cats treated with ciclosporin.

Fig. 2 Gingival hyperplasia in a cat being treated with daily ciclosporin at 10 mg/kg for 3 months. The lesions greatly improved when the dose was lowered to 5 mg/kg every other day

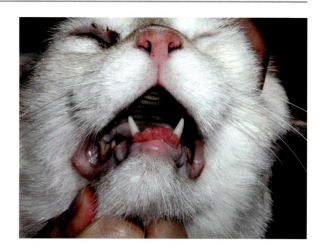

Table 2 Oral antihistamines reported to be used in allergic cats against pruritus

Antihistamine	Dose	Side effects	Reported efficacy (% of cats controlled)
Amitriptyline	5–10 mg/cat q12-24 h		
Cetirizine	1 mg/kg or 5 mg/cat q24h		Up to 41%
Chlorpheniramine	2–4 mg/cat q8-24h	Sleepiness	Up to 73%
Clemastine	0.25–0.68 mg/cat q12h	Sleepiness, soft stools	Up to 50%
Cyproheptadine	2 mg/cat q12h	Sleepiness, vomit, behavioural disturbances	Up to 40%
Diphenhydramine	1–2 mg/kg or 2–4 mg/cat q8-12h		
Fexofenadine	2 mg/kg up to 30–60 mg/cat q24h		
Hydroxyzine	5–10 mg/cat q8-12h	Behavioural disturbances	
Oxatomide	15–30 mg/cat q12h		Up to 50%
Promethazine	5 mg/cat q24h		

Antihistamines

Antihistamines inhibit the action of histamine by competitively blocking H1 receptors. As in dogs, the response to antihistamine therapy is variable, and it may be necessary to try several different agents for a period of 15 days each to determine which, if any, is more effective. The efficacy in terms of percentage of animals responding to antihistamines in cats is reported (in old, uncontrolled studies, summarized by Scott 1999 [40]) to be between 20% and 73% (Table 2).

In particular, cetirizine has been object of recent investigations in cats. Pharmacological studies determined that cetirizine is orally well absorbed in cats and is able to maintain high plasma concentrations for at least 24 h [41]. In an open study, 5 mg/cat q24h determined a reduction in pruritus in 41% (13/32) of allergic cats; however only a minority (1/13) of these improved more than 50%, while the majority (10/13) improved less than 25% [42]. A subsequent randomized, double-blinded, placebo-controlled, crossover trial on the use of cetirizine 1 mg/kg q24h in 21 allergic cats confirmed that only 10% of cats improved by more than 50% with cetirizine versus 20% of placebo-treated patients, with no statistical difference between groups for pruritus or lesions [43].

Oclacitinib

Oclacitinib (Apoquel®, Zoetis) is a JAK1 inhibitor registered for dogs, able to block intracellular metabolic pathways leading to the allergic activation of inflammatory cells and keratinocytes and to the elicitation of pruritus in neural fibres. Recently, the off-label use of oclacitinib in cats was investigated in a pilot [44] and in a methylprednisolone-controlled study [3]. Oclacitinib given at 1 mg/kg every 12 h has an efficacy similar to that of methylprednisolone given at the same dose, albeit with no obvious advantage. Given for 1 month, it was generally well tolerated; however mild increases in kidney values were observed in some cats [3]. In another study no clinical, hematological or biochemical alterations were observed in cats taking oclacitinib 1 or 2 mg/kg twice daily for 28 days [45]. Oclacitinib could be a useful alternative treatment when glucocorticoids are contraindicated and a rapid relief of pruritus is required. Readers should be warned that oclacitinib is not registered in cats and that its long-term safety is not known in this species. Regular haematology and biochemistry monitoring are advised for long-term maintenance therapies.

Palmitoylethanolamide (PEA)

Palmitoylethanolamide (PEA) is a natural-occurring bioactive lipid present in both animals and plants. PEA is produced by several different cell types in response to tissue damage and acts by controlling the functionality of mast cells (it inhibits degranulation) and other inflammatory cells such as macrophages and keratinocytes. Consequently, PEA decreases skin inflammation and nerve sensitization in animals with allergic dermatitis. An open pilot study on 17 cats with eosinophilic granuloma and eosinophilic plaque showed that PEA (10 mg/kg q24h for 30 days) improved pruritus, erythema and alopecia in 64.3% of cats and reduced the extent and severity of eosinophilic plaques and granulomas in 66.7% [46]. Recently a product containing ultramicronized PEA (PEA-um) with improved bioavailability and efficacy was released to the international veterinary market. A multicentre, placebo-controlled, randomized trial determined a glucocorticoid sparing effect of PEA-um (15 mg/kg q24h) in cats with non-seasonal allergic dermatitis [47]. In the same study, PEA-um was showed to be able to prolong the effects of a short course of oral glucocorticoids with virtually no significant adverse effects.

Maropitant

Maropitant is a neurokinin-1 receptor antagonist, able to block the interaction of substance P, a pruritogenic neurokinin, to its receptor. In an open pilot study at the dose of 2 mg/kg, it has been reported to be effective against pruritus and lesions in 11/12 allergic cats [48]. Maropitant was well tolerated if administered once daily for 2–4 weeks. There is no information about its safety for a long-term treatment.

Omega-3 and Omega-6 Fatty Acid Supplementation

There are only few old and uncontrolled studies investigating the efficacy of essential fatty acids (EFA) in miliary dermatitis and lesions of eosinophilic granuloma in cats [49–52]. These publications reported efficacy in 40–60% of treated animals. A lag period of 6–12 weeks occurs, before any benefits are seen. Probably, only a small minority of patients can be controlled with fatty acid therapy alone. EFA may have glucocorticoid- or ciclosporin-sparing effects, as determined in dogs, but no studies were conducted in cats to confirm this. Feeding a good dermatological diet may be an effective way of supplementing EFAs in allergic cats.

References

1. Noli C, Borio S, Varina A, et al. Development and validation of a questionnaire to evaluate the quality of life of cats with skin disease and their owners, and its use in 185 cats with skin disease. Vet Dermatol. 2016;27:247–e58.
2. Noli C, Ortalda C, Galzerano M. L'utilizzo della ciclosporina in formulazione liquida (Atoplus gatto®) nel trattamento delle malattie allergiche feline. Veterinaria (Cremona). 2014;28:15–22.
3. Noli C, Matricoti I, Schievano C. A double-blinded, randomized, methylprednisolone controlled study on the efficacy of oclacitinib in the management of pruritus in cats with nonflea nonfood induced hypersensitivity dermatitis. Vet Dermatol. 2019;30:110–e30.
4. Diesel A, DeBoer DJ. Serum allergen-specific immunoglobulin E in atopic and healthy cats: comparison of a rapid screening immunoassay and complete-panel analysis. Vet Dermatol. 2011;22:39–45.
5. Bexley J, Hogg JE, Hammerberg B, et al. Levels of house dust mite-specific serum immunoglobulin E (IgE) in different cat populations using a monoclonal based anti-IgE enzyme-linked immunosorbent assay. Vet Dermatol. 2009;20:562–8.
6. Gilbert S, Halliwell REW. Feline immunoglobulin E: induction of antigen-specific antibody in normal cats and levels in spontaneously allergic cats. Vet Immunol Immunopathol. 1998;63:235–52.
7. Taglinger K, Helps CR, Day MJ, et al. Measurement of serum immunoglobulin E (IgE) specific for house dust mite antigens in normal cats and cats with allergic skin disease. Vet Immunol Immunopathol. 2005;105:85–93.
8. Kadoya-Minegishi M, Park SJ, Sekiguchi M, et al. The use of fluorescein as a contrast medium to enhance intradermal skin tests in cats. Austr Vet J. 2002;80:702–3.
9. Schleifer SG, Willemse T. Evaluation of skin test reactivity to environmental allergens in healthy cats and cats with atopic dermatitis. Am J Vet Res. 2003;64:773–8.
10. Scholz FM, Burrows AK, Griffin CE, Muse R. Determination of threshold concentrations of plant pollens in intradermal testing using fluorescein in clinically healthy nonallergic cats. Vet Dermatol. 2017;28:351–e78.

11. Willemse T, Vroom MW, Mol JA, Rijnberk A. Changes in plasma cortisol, corticotropin, and alpha-melanocyte-stimulating hormone concentrations in cats before and after physical restraint and intradermal testing. Am J Vet Res. 1993;54:69–72.

12. Rossi MA, Messinger L, Olivry T, Hoontrakoon R. A pilot study of the validation of percutaneous testing in cats. Vet Dermatol. 2013 Oct;24:488–e115.

13. Gentry CM, Messinger L. Comparison of intradermal and percutaneous testing to histamine, saline and nine allergens in healthy adult cats. Vet Dermatol. 2016;27:370–e92.

14. Swinnen C, Vroom M. The clinical effect of environmental control of house dust mites in 60 house dust mite-sensitive dogs. Vet Dermatol. 2004;15:31–6.

15. Mueller RS, Jensen-Jarolim E, Roth Walter F, et al. Allergen immunotherapy in people, dogs, cats and horses – differences, similarities and research needs. Allergy. 2018;. early view online;73:1989. https://doi.org/10.1111/all.13464.

16. Carlotti D, Prost C. L'atopie féline. Le Point Vétérinaire. 1988;20:777–84.

17. Trimmer AM, Griffin CE, Rosenkrantz WS. Feline immunotherapy. Clin Techniques Small An Pract. 2006;21:157–61.

18. Ravens PA, Xu BJ, Vogelnest LJ. Feline atopic dermatitis: a retrospective study of 45 cases (2001-2012). Vet Dermatol. 2014;25:95–102.

19. Reedy LM. Results of allergy testing and hyposensitization in selected feline skin diseases. J Am Anim Hosp Assoc. 1982;18:618–23.

20. Löewenstein C, Mueller RS. A review of allergen-specific immunotherapy in human and veterinary medicine. Vet Dermatol. 2009;20:84–98.

21. Vidémont E, Pin D. How to treat atopy in cats? Eur J Comp An Pract. 2009;19:276–82.

22. Trimmer AM, Griffin CE, Boord MJ, et al. Rush allergen specific immunotherapy protocol in feline atopic dermatitis: a pilot study of four cats. Vet Dermatol. 2005;16:324–9.

23. Broek AHM, Stafford WL. Epidermal and hepatic glucocorticoid receptors in cats and dogs. Res Vet Sci. 1992;52:312–5.

24. Graham-Mize CA, Rosser EJ, Hauptman J. Absorption, bioavailability and activity of prednisone and prednisolone in cats. In: Hiller A, Foster AP, Kwochka KW, editors. Advances in veterinary dermatology, vol. 5. Oxford: Blackwell; 2005. p. 152–8.

25. Ganz EC, Griffin CE, Keys DA, et al. Evaluation of methylprednisolone and triamcinolone for the induction and maintenance treatment of pruritus in allergic cats: a double-blinded, randomized, prospective study. Vet Dermatol. 2012;23:387–e72.

26. Lowe AD, Campbell KL, Graves T. Glucocorticoids in the cat. Vet Dermatol. 2008;19:340–7.

27. Lockwood SL, Schick AE, Lewis TP, Newton H. Investigation of subclinical bacteriuria in cats with dermatological disease receiving long-term glucocorticoids and/or ciclosporin. Vet Dermatol. 2018;29:25–e12.

28. Schmidt V, Buckley LM, McEwan NA, Rème CA, Nuttall TJ. Efficacy of a 0.0584% hydrocortisone aceponate spray in presumed feline allergic dermatitis: an open label pilot study. Vet Dermatol 2012; 23: 11–6, e3–4.

29. Wisselink MA, Willemse T. The efficacy of cyclosporine a in cats with presumed atopic dermatitis: a double blind, randomized prednisolone-controlled study. Vet J. 2009;180:55–9.

30. King S, Favrot C, Messinger L, et al. A randomized double-blinded placebo-controlled study to evaluate an effective ciclosporin dose for the treatment of feline hypersensitivity dermatitis. Vet Dermatol. 2012;23:440–e84.

31. Roberts ES, Speranza C, Friberg C, et al. Confirmatory field study for the evaluation of ciclosporin at a target dose of 7.0 mg/kg (3.2 mg/lb) in the control of feline hypersensitivity dermatitis. J Feline Med Surg. 2016;18:889–97.

32. Steffan J, Roberts E, Cannon A, et al. Dose tapering for ciclosporin in cats with nonflea-induced hypersensitivity dermatitis. Vet Dermatol. 2013;24:315–22.

33. Roberts ES, Tapp T, Trimmer A, et al. Clinical efficacy and safety following dose tapering of ciclosporin in cats with hypersensitivity dermatitis. J Feline Med Surg. 2016;18:898–905.

34. Panteri A, Strehlau G, Helbig R, et al. Repeated oral dose tolerance in dogs treated concomitantly with ciclosporin and oclacitinib for three weeks. Vet Dermatol. 2016;27:22–e7.

35. Dip R, Carmichael J, Letellier I, et al. Concurrent short-term use of prednisolone with cyclosporine A accelerates pruritus reduction and improvement in clinical scoring in dogs with atopic dermatitis. BMC Vet Res. 2013;3(9):173.
36. Steffan J, King S, Seewald W. Ciclosporin efficacy in the treatment of feline hypersensitivity dermatitis is not influenced by the feeding status. Vet Dermatol. 2012;23(suppl. 1):64–5. (abstract)
37. Koch SN, Torres SMF, Diaz S, et al. Subcutaneous administration of ciclosporin in 11 allergic cats – a pilot open-label uncontrolled clinical trial. Vet Dermatol. 2018;29:107–e43.
38. Heinrich NA, McKeever PJ, Eisenschenk MC. Adverse events in 50 cats with allergic dermatitis receiving ciclosporin. Vet Dermatol. 2011;22:511–20.
39. Last RD, Suzuki Y, Manning T. A case of fatal systemic toxoplasmosis in a cat being treated with cyclosporin a for feline atopy. Vet Dermatol. 2004;15:194–8.
40. Scott DW, Miller WH Jr. Antihistamines in the management of allergic pruritus in dogs and cats. J Small Anim Pract. 1999;40:359–64.
41. Papich MG, Schooley EK, Reinero CR. Pharmacokinetics of cetirizine in healthy cats. Am J Vet Res. 2008;69:670–4.
42. Griffin JS, Scott DW, Miller WH Jr, et al. An open clinical trial on the efficacy of cetirizine hydrochloride in the management of allergic pruritus in cats. Can Vet J. 2012;53:47–50.
43. Wildermuth K, Zabel S, Rosychuk RA. The efficacy of cetirizine hydrochloride on the pruritus of cats with atopic dermatitis: a randomized, double-blind, placebo-controlled, crossover study. Vet Dermatol. 2013;24:576–681, e137-8.
44. Ortalda C, Noli C, Colombo S, Borio S. Oclacitinib in feline nonflea-, nonfood-induced hypersensitivity dermatitis: results of a small prospective pilot study of client-owned cats. Vet Dermatol. 2015;26:235–e52.
45. Lopes NL, Campos DR, Machado MA, Alves MSR, de Souza MSG, da Veiga CCP, Merlo A, Scott FB, Fernandes JI. A blinded, randomized, placebo-controlled trial of the safety of oclacitinib in cats. BMC Vet Res. 2019;15(1):137.
46. Scarampella F, Abramo F, Noli C. Clinical and histological evaluation of an analogue of palmitoylethanolamide, PLR 120 (comicronized Palmidrol INN) in cats with eosinophilic granuloma and eosinophilic plaque: a pilot study. Vet Dermatol. 2001 Feb;12(1):29–39.
47. Noli C, Della Valle MF, Miolo A, Medori C, Schievano C; Skinalia Clinical Research Group. Effect of dietary supplementation with ultramicronized palmitoylethanolamide in maintaining remission in cats with nonflea hypersensitivity dermatitis: a double-blind, multicentre, randomized, placebo-controlled study.Vet Dermatol. 2019;30:387–e117.
48. Maina E, Fontaine J. Use of maropitant for the control of pruritus in non-flea, non-food-induced feline hypersensitivity dermatitis: an open label uncontrolled pilot study. J Feline Med Surg. 2019;21:967–72.
49. Harvey RG. Management of feline miliary dermatitis by supplementing the diet with essential fatty acids. Vet Rec. 1991;128:326–9.
50. Harvey RG. The effect of varying proportions of evening primrose oil and fish oil on cats with crusting dermatosis (miliary dermatitis). Vet Rec. 1993a;133:208–11.
51. Harvey RG. A comparison of evening primrose oil and sunflower oil for the management of papulocrustous dermatitis in cats. Vet Rec. 1993b;133:571–3.
52. Miller WH, Scott DW, Wellington JR. Efficacy of DVM Derm caps liquid in the management of allergic and inflammatory dermatoses of the cat. JAAHA. 1993;29:37–40.

Mosquito-byte Hypersensitivity

Ken Mason

Abstract

Feline mosquito bite allergy has a worldwide distribution occurring where cats are seasonally exposed to mosquitoes. The distinctive skin lesions are punctate ulcers, crusts and pigmentary changes on the face, ears and nose. Associated pruritus causes face and nose pawing resulting in bleeding. Foot pad hyperkeratosis, crusts and pigmentary changes occur in some cats. Confining the cat inside a screened area and in late afternoon reduces severity of signs; intermittent corticosteroid with confinement also helps. Newer repellent pyrethroid/pyrethrins safe for cats are becoming available and prove useful to affected cats.

Introduction

Feline mosquito bite hypersensitivity is an uncommon, seasonal, visually distinctive pruritic dermatitis typically affecting the face, ears and footpads [1–4]. The disease was originally described in 1984 by Wilkinson and Bate as a seasonal variant of the eosinophilic granuloma complex that improved on hospitalization [5].

In 1991, Mason and Evans hypothesized that the cause was a mosquito bite hypersensitivity, when they realized that lesions were restricted to short-haired or non-haired areas, such as the nose and footpads [1]. The authors demonstrated that clipping the hair short on the forehead resulted in lesions, when the cat was

K. Mason (✉)
Specialist Veterinary Dermatologist, Animal Allergy & Dermatology Service,
Slacks Creek, QLD, Australia
e-mail: ken@dermcare.com.au

© Springer Nature Switzerland AG 2020   489
C. Noli, S. Colombo (eds.), *Feline Dermatology*,
https://doi.org/10.1007/978-3-030-29836-4_24

Fig. 1 At examination the cat presented with crusted ulceration on ear tips, erythema, crusts and depigmentation of the nasal bridge and paw and small punctate ulcers and depigmentation of the nasal planum

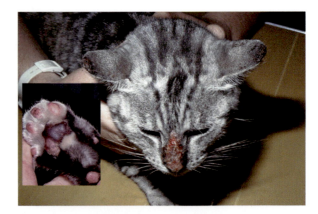

Fig. 2 After 1 week of hospitalization, the cat presents improvement of lesions. The forehead coat was clipped before returning to home environment to prove that areas with short coat are predisposed to lesions

exposed to the home environment. Only some cats in a multi-cat household developed lesions, further supporting an environmental hypersensitivity cause.

The final proof that mosquito bite caused the skin disease is demonstrated in the sequence of photos in Figs. 1, 2, 3 and 4.

Pathogenesis and Epidemiology

Similarly to flea allergy dermatitis (FAD), mosquito bite allergy is an IgE-mediated type I (immediate) hypersensitivity reaction [1, 2, 4]. The disease is seasonal, occurring intermittently in the spring and continuing through summer, waning in autumn and usually absent in winter. In subsequent years, the allergy and lesion severity may increase depending on weather pattern, favouring or not mosquito breeding. There is no age or sex predilection; usually affected cats are adult and have been

Fig. 3 Re-examination after 1 week at home environment: clipped areas present new lesions and previously improved lesions have flared up. The cat was then returned home and kept in a mosquito netting-covered cage outside

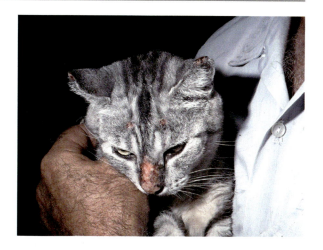

Fig. 4 When living at home in an outside environment in a mosquito proof netting-enclosed cage, lesions improved again. After a hole cut in the net, mosquitoes could enter and bite the cat again

exposed to more than one mosquito season. Occurrence is more commonly reported in cats that venture outside in geographic regions where mosquitoes are endemic. In multi-cat households, only one or a few cats are affected. The disease occurs anywhere cats are exposed to mosquitoes.

Clinical Signs

The striking and typical clinical manifestations of mosquito bite hypersensitivity are erythema, crusted and ulcerated ear margins, papules to small nodules with focal crusting on the haired ear surface, punctate ulcers to severely crusted lesions on the nasal bridge and erythema, ulceration and depigmentation of the nasal planum (Figs. 1 and 5). On the footpads, there may be hyperkeratosis often affecting the margins and variable pigmentation alteration. Pruritus can be intense during active mosquito challenge, leading to self-trauma and bleeding.

Fig. 5 After the mosquitoes bit the nose, the skin redeveloped inflammation, and skin biopsies were sampled again; sutures are visible

Fig. 6 Eosinophilic ulcer (indolent ulcer) eroding upper lip due to mosquito bite hypersensitivity

There are a variety of other less typical lesions, especially in cats living in severely mosquito-infested areas like swamps and irrigation areas, such as eosino-philic plaques, indolent lip ulcers (Fig. 6), hairless chin nodules and linear granu-lomas on the body. Eosinophilic keratoconjunctivitis is occasionally present and waxes and wanes with mosquito challenge.

Regional lymph nodes, particularly the submandibular ones, may be enlarged and the temperature may be slightly raised.

Differential Diagnoses and Diagnostic Tests

The clinical features are sufficiently characteristic to make the diagnosis in typi-cal cases; however there are potential alternative diagnoses such as squamous cell carcinoma and herpesvirus (FeHV-1) dermatitis, so that confirmation tests may be needed. Herpesvirus dermatitis can present with large crusts on the nasal bridge, and in squamous cell carcinoma, erosive, crusted lesions on the ear tip and nose may

Fig. 7 Photomicrograph histopathology section H&E stain showing follicular necrosis (arrow) and dermal inflammation of eosinophils and macrophages (star)

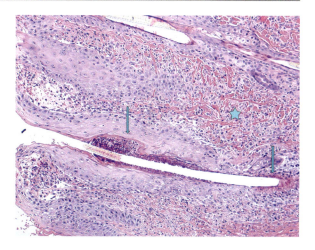

develop, particularly in white skin. Footpad hyperkeratosis can present a diagnostic dilemma, given the difficulty of making a diagnosis when confronted with discrete pad keratoses. In case of lesions of the eosinophilic granuloma complex, other allergic causes should be considered.

Intradermal skin test and blood immunoassay may be supportive if mosquito antigen is available. Blood haematology may show a raised eosinophil count. Similarly, cytology of lesions and lymph nodes can be supportive if dominated by eosinophils and may help to rule out alternative diseases, such as squamous cell carcinoma.

The diagnosis may be confirmed if isolation in a hospital or mosquito-free home or enclosure results in resolution of acute signs within days and when return to home outside causes a relapse of pruritus and lesions.

Histopathological examination of lesions is classically characterized by eosinophilic follicular necrosis. Common findings are eosinophilic folliculitis and furunculosis, surface serocellular crusts, hyperplastic spongiotic epidermis with eosinophil exocytosis and micro-pustules and a diffuse dermal eosinophilic inflammation, with a few lymphocytes and occasional flame figures (Fig. 7).

Treatment

Avoidance of mosquitoes as much as possible is the mainstay of treatment. Affected cats should be kept indoors behind insect screened enclosures, where mosquito exposure is prevented. Insect repellents designed for dogs or humans are toxic to cats [6]. However, natural pyrethrin from the chrysanthemum flowers and the newer synthetic flumethrin are safe for cats and could help to manage mosquito bite allergic patients. Very few products are approved for this disease in cats, but some, such as collars, have proven repellent activity against sandfly vectors of leishmaniosis [7–9]. Supplementary anti-pruritic glucocorticoids help in case of disease flares.

Yard environmental mosquito control may be helpful and is an important preventative human health consideration. Standing water should be eliminated to decrease mosquito breeding grounds.

Conclusion

The typical presentation of mosquito bite allergy is sufficiently distinctive to make a diagnosis without supportive tests. However, the less typical forms and footpad hyperkeratosis present a diagnostic challenge, and the aetiology can be overlooked, leading to chronic high doses of corticosteroids and subsequent major adverse effects. Keeping the cat inside in the late afternoon and overnight and flumethrin collars are likely to be beneficial.

References

1. Mason KV, Evans AG. Mosquito bite-caused eosinophilic dermatitis in cats. J Am Vet Med Assoc. 1991;198(12):2086–8.
2. Nagata M, Ishida T. Cutaneous reactivity to mosquito bites and its antigens in cats. Vet Dermatol. 1997;8(1):19–26.
3. Johnstone AC, Graham DG, Andersen HJ. A seasonal eosinophilic dermatitis in cats. N Z Vet J. 1992;40(4):168–72.
4. Ihrke PJ, Gross TL. Conference in dermatology—no. 2 mosquito-bite hypersensitivity in a cat. Vet Dermatol. 1994;5(1):33–6.
5. Wilkinson GT, Bate MJ. A possible further clinical manifestation of the feline eosinophilic granuloma complex. J Am Anim Hosp Assoc. 1984;20:325–31.
6. Dymond NL, Swift IM. Permethrin toxicity in cats: a retrospective study of 20 cases. Aust Vet J. 2008;86(6):219–23.
7. Stanneck D, Kruedewagen EM, Fourie JJ, Horak IG, Davis W, Krieger KJ. Efficacy of an imidacloprid/flumethrin collar against fleas and ticks on cats. Parasit Vectors. 2012;5:82.
8. Stanneck D, Rass J, Radeloff I, Kruedewagen E, Le Sueur C, Hellmann K, Krieger K. Evaluation of the long-term efficacy and safety of an imidacloprid 10% / flumethrin 4.5% polymer matrix collar (Seresto (R)) in dogs and cats naturally infested with fleas and/or ticks in multicentre clinical field studies in Europe. Parasit Vectors. 2012;5:66.
9. Brianti E, Falsone L, Napoli E, et al. Prevention of feline leishmaniosis with an imidacloprid 10%/flumethrin 4.5% polymer matrix collar. Parasit Vectors. 2017;10:334.

Autoimmune Diseases

Petra Bizikova

Abstract

Autoimmune skin diseases (AISDs) in cats are very rare and account for less than 2% of all skin diseases for which cats are seen by dermatologists. The most common AISD seen in this species is pemphigus foliaceus, for which numerous case reports and case series can be found in the literature. In contrast, other AISDs are very rare and limited to only few case reports published in the peer-reviewed literature over the last two decades. Many of these diseases are clinically and histologically homologous to diseases described in people and dogs, and, although the pathomechanism of these feline counterparts is unknown, similar mechanisms leading to the disruption of the epidermal cohesion or destruction of skin adnexae are hypothesized. Such mechanisms involve autoantibodies in diseases like pemphigus foliaceus, pemphigus vulgaris, paraneoplastic pemphigus, and autoimmune subepidermal blistering diseases, or autoreactive T cells in diseases like paraneoplastic pemphigus, cutaneous lupus, and vitiligo. This chapter will provide an overview of the current knowledge about feline AISDs available in the published literature.

Introduction

A healthy immune system protects the body from an onslaught of invading pathogens as well as from its own damaged or potentially neoplastic cells on a daily basis. Under specific circumstances (genetics, environment, infection, etc.), however, the same immune system may get awry and start targeting self-antigens. This break of

P. Bizikova (✉)

North Carolina State University, College of Veterinary Medicine, Raleigh, NC, USA

e-mail: pbiziko@ncsu.edu

© Springer Nature Switzerland AG 2020

C. Noli, S. Colombo (eds.), *Feline Dermatology*,

https://doi.org/10.1007/978-3-030-29836-4_25

495

a self-tolerance results in an injury to the body, which at the turn of the twentieth century was named a *horror autotoxicus* by Paul Ehrlich. Such autoimmune attack can be caused by autoantibodies (e.g., pemphigus) or by autoreactive T lymphocytes (e.g., cutaneous lupus). Autoimmune skin diseases (AISDs) are rare in cats and account for less than 2% of all skin diseases for which cats are seen by a dermatologist [1]. The most common AISD seen in this species is pemphigus foliaceus, for which numerous case reports and case series can be found in the literature. In contrast, other AISDs are very rare and limited to only few case reports published in the peer-reviewed literature over the last two decades. Due to the rarity of AISDs in cats, information about the identity of the autoantigen and the disease pathogenesis remains unknown.

Autoimmune Skin Diseases Affecting the Epidermal and Dermo-Epidermal Adhesion

An intact skin is a critically important organ that functions as a first-line defense mechanism against physical and chemical damage. Its integrity is dependent on complex structures maintaining cell-cell and cell-matrix adhesions [2, 3]. Several AISDs disrupting this cohesion have been recognized in cats. The mechanism by which this adhesion is disrupted varies depending on the type of disease.

(a) *Disruption of keratinocyte adhesion* – intra-epidermal blister formation due to desmosome dissociation (pemphigus foliaceus (PF), pemphigus vulgaris (PV), paraneoplastic pemphigus (PNP))
(b) *Disruption of basement membrane adhesion* – subepidermal blister formation due to dermo-epidermal separation (bullous pemphigoid (BP), mucous membrane pemphigoid (MMP))

Desmosome Autoimmunity

Pemphigus Foliaceus (PF)
Pemphigus foliaceus is the most common autoimmune skin disease in cats that accounts for about 1% of all skin diseases for which cats are seen by dermatologists [1]. Although the pathogenesis of feline PF has not been studied in such extent as it has been in dogs, it is believed that, like in dogs and people, antikeratinocyte IgG autoantibodies disrupt desmosomal adhesion between keratinocytes and induce subcorneal blisters in a form of pustules (Fig. 1a). Indeed, tissue-bound and circulating antikeratinocyte IgG have been detected in the majority of cats with PF (Fig. 2d) [4, 5]. The major target autoantigen that in people and dogs is desmoglein-1 and desmocollin-1, respectively, remains unknown for feline PF.

Signalment
Cats of different breeds have been reported to suffer with PF, but a breed predisposition has not been confirmed yet. The most commonly reported breeds

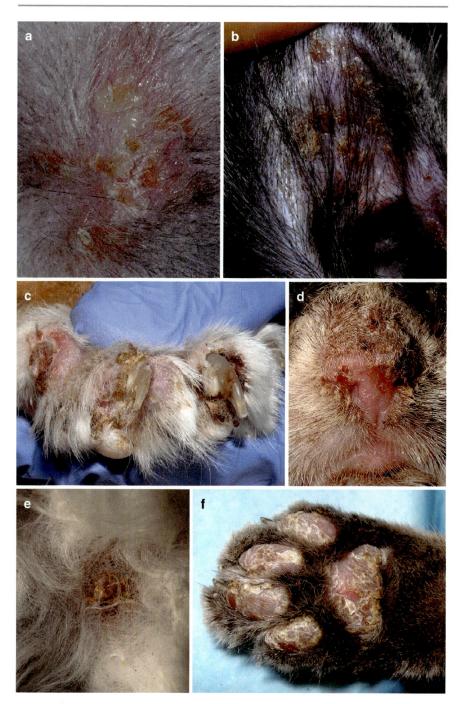

Fig. 1 Feline pemphigus foliaceus – clinical lesions: (**a**) Pustule and scale-crust; (**b**) well-demarcated scale-crust suggesting a pustular origin on the concave pinna; (**c**) claw skin fold with erythema, superficial erosions, scale-crust, and purulent exudate; (**d**) erosions and scale-crust on the nasal planum and dorsal muzzle; (**e**) scale-crust around the areola; (**f**) erosions and scale-crust on the footpad. (photo f – courtesy of Dr. Andrea Lamm)

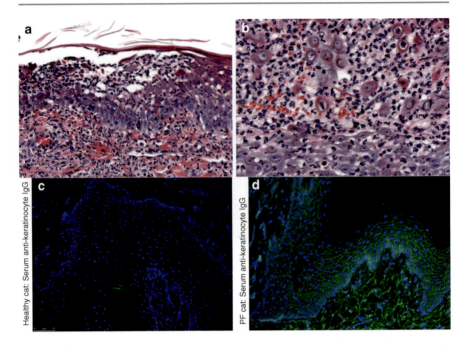

Fig. 2 Feline pemphigus foliaceus – histopathology and indirect immunofluorescence: (**a**) Subcorneal pustule with acantholytic keratinocytes; (**b**) close-up of individualized and clustered acantholytic keratinocytes (Courtesy of Dr. Keith Linder); (**c**) and (**d**) indirect immunofluorescence using healthy feline serum (**c**) and serum from a cat with pemphigus foliaceus (**d**). Note the intercellular, web-like immunofluorescence pattern in the PF-affected cat sample (**d**) caused by circulating anti-keratinocyte IgG antibodies

include domestic shorthaired, Siamese, Persian and Persian-crossbred, Burmese, Himalayan, and domestic medium-haired cats [6]. Pemphigus foliaceus affects usually adult cats (median age of onset about 6 years), though the range varies greatly (0.25–16 years) [4, 6–9]. A sex predilection has not been confirmed, but females appear to be marginally over-represented according to a recent review [6]. In most cats, a specific trigger precipitating the PF onset cannot be identified. Rare reports of a drug-triggered PF and PF associated with a thymoma can be found in the literature [8, 10–16].

Clinical Signs

The primary skin lesion of feline PF is a subcorneal pustule, which, because of its superficial nature, progresses rapidly into an erosion and crust. Indeed, the two latter skin lesions may represent the only clinical findings during the physical examination. Lesions are usually bilaterally symmetric with pinnae and claw skin folds being the most commonly affected body areas (Fig. 1b and c) [6]. Claw skin folds often exhibit accumulation of a thick purulent exudate, which is usually related to secondary bacterial infections seen in this body region more often than in others [17]. Other typically affected body areas include nasal planum, eyelids, pawpads, and periareolar areas (Fig. 1 d, e). Typical pawpads lesions are scaling, crusting, and hyperkeratosis, though they are usually not as prominent as in dogs (Fig. 1f). Pustules, if found, can be seen at the periphery of the pawpads not in contact with the ground. Most cats (81%) exhibit lesions on two or more body regions, while lesions localized to a single body area are less common (19%). More than a half of cats are pruritic and show systemic signs such as lethargy, fever, and/or anorexia.

Diagnostic Approach

The most critical and often challenging step in the diagnostic approach is the identification of a subcorneal pustular process. Indeed, while there are several erosive skin diseases in cats, the list of diseases presenting with primary subcorneal pustules with acantholysis is limited to PF and to anecdotal reports of pustular dermatophytosis; the latter has been reported to exhibit minimal to no acantholysis [18]. Bullous impetigo, a subcorneal pustular dermatitis with variable degree of acantholysis caused by *Staphylococcus aureus* and *pseudintermedius* in people and dogs, has not been described in cats yet [19, 20]. Lesions suggestive of a subcorneal pustular dermatitis include intact pustules, sharply demarcated, pinpoint to few millimeters large; superficial erosions; or scaling and crusting (Fig. 1a, b). The acantholytic nature of the disease can be confirmed by cytology taken from an intact pustule, from the underneath of a crust with an active erosion and exudation, or from the caseous pus around nails and/or by a biopsy of similar lesions. It is important to include crusts in the biopsy sample. Microscopic examination of biopsy samples reveals acantholytic keratinocytes, usually numerous, within a neutrophilic or mixed neutrophilic and eosinophilic, subcorneal or intragranular pustule (Fig. 2a, b). Ghost acantholytic cells can be found within the crusts, and, in many cases, may be the only histological evidence of the disease process. Aerobic bacterial culture should be considered in cases in which infection cannot be clinically ruled out, and fungal culture and special stains should be considered in cases in which dermatophytosis is suspected, particularly if a pustular folliculitis, lymphocytic mural folliculitis, and/or prominent hyperkeratosis is present in biopsy samples. An immunological testing for antikeratinocyte autoantibodies by direct or indirect immunofluorescence is not commercially available, nor the sensitivity and, particularly, specificity of such tests is known. Therefore, the current diagnosis of PF is based on the combination of (i) skin lesion character and distribution, (ii) exclusion of an infection, and (iii) supportive cytology and/or histopathology confirming acantholytic pustular dermatitis [21].

Treatment

Cats with PF have usually positive response to treatment, and the majority of them (93%) reach disease control (cessation of active lesions and healing of original lesions) within a few weeks (median time, 3 weeks) [6]. In most cats, the disease control can be achieved by a glucocorticoid monotherapy (e.g., prednisolone, 2–4 mg/kg/day; triamcinolone aceponate, 0.2–0.6 mg/kg/day; dexamethasone, 0.1–0.2 mg/kg/day). The dosage reduction is recommended only once the disease has been inactive for at least 2 weeks and most original skin lesions have healed (20–25% dosage reduction every 2–4 weeks, though faster reduction is possible). The use of non-steroidal agents is implemented in cats: (i) in which disease control is not achieved within 4 weeks using appropriate glucocorticoid dosages, (ii) exhibiting severe adverse effects related to glucocorticoids, or (iii) in which the dosage of glucocorticoids cannot be significantly reduced. Non-steroidal drugs reported to induce disease control in cats include ciclosporin (5–10 mg/kg/day), chlorambucil (0.1–0.3 mg/kg/day), azathioprine (1.1 mg/kg every other day) [7], and aurothioglucose (0.5 mg/kg/week). The two latter are not used commonly either because of the high risk of a bone marrow suppression (azathioprine) or its unavailability on the market (aurothioglucose). The risk of side effects in azathioprine-treated cats is dose-dependent, and, anecdotally, lower dosages (e.g., 0.3 mg/kg every other day) have been reported to be successful in managing other immune-mediated diseases [22].

According to the literature review, only a minority of cats (15%) appears to achieve a long-term disease remission off drugs [6]. Most cats require a long-term medical management with glucocorticoids or non-steroidal drugs such as ciclosporin or chlorambucil. The median maintenance dosages are usually lower than those required for the induction of the disease control (e.g., prednisolone, 0.5 mg/kg/day; dexamethasone, 0.03 mg/kg/day; or ciclosporin, 5 mg/kg/day). A combination of doxycycline and niacinamide has also been reported to be an effective maintenance therapy in cats with PF [17]. In refractory cases, other immunosuppressants (e.g., mycophenolate, leflunomide) could be considered, though the evidence of efficacy of these drugs in feline PF is currently lacking.

Box 1: Feline PF Treatment Outline Using Similar Principles than Those Used in Human Pemphigus [42]

(I) Induce rapid disease control (i.e., *time at which new lesions cease to form and old lesions start to heal*)

First line treatment: prednisolone or methylprednisolone 2–4 mg/kg/day (or its equivalent; e.g., triamcinolone aceponate, dexamethasone) until disease control is reached.

(II) Start gradual glucocorticoid dosage reduction (25% every 2 weeks) once the end of consolidation phase is reached (i.e., *time at which no new lesions have developed for at least 2 weeks and approximately 80% of original lesions have healed*). Consider also to taper to every-other-day administration before reducing the daily dose.

(III) Continue with the gradual glucocorticoid dosage reduction until the lowest effective dosage is identified or the cat is able to remain in remission off of drugs (a long-term remission off drugs has been reported in about 15% of cats).

III.a. Consider topical glucocorticoids (e.g., hydrocortisone aceponate) or tacrolimus to control minor, localized flares.

III.b. In case of a more severe flare-up, increase the dose of glucocorticoids back to the second to the last effective dosage. If the flare-up cannot be controlled within 2 weeks, increase the glucocorticoid dosage back to the initial immunosuppressive dosage.

(IV) Add non-steroidal immunosuppressive drug* if:

IV.a. The dosage of glucocorticoid cannot be reduced enough to limit the risk of side effects associated with a long-term glucocorticoid treatment.

IV.b. The patient suffers with intolerable side effects caused by glucocorticoids.

IV.c. Disease control cannot be achieved with glucocorticoid mono-therapy within 4–6 weeks.

Frequently selected non-steroidal immunosuppressive drugs in cats with PF are ciclosporin (5–10 mg/kg/day) or chlorambucil (0.2 mg/kg/every other day).

(V) V. Maintenance treatment:

V.a. Maintain the disease with the lowest possible dosage of drug(s); attempt to reduce the dosage of glucocorticoids as much as possible, or to complete replace them with a non-steroidal immunosuppressant.

V.b. Monitor for side effects (usually complete blood count, chemistry panel, urinalysis, and urine culture every 6–12 months, though the frequency and type of tests depends on the used drug(s), age of the cat, and its general health status).

V.c. Avoid potential flare-up triggers (e.g., UV light, etc.).

Pemphigus Vulgaris (PV)

In contrast to people, PV is considered to check if this should autoimmune dermatoses in animals, including the cat [4]. Because of the small number of described cases, breed, age, and sex predilections cannot be reliably estimated in cats. The clinical and histological homology between feline, canine, and human PV suggests similar pathomechanism; however, while desmoglein-3 has been confirmed to be the major target autoantigen in human and canine PV, the major target antigen in feline PV remains unknown.

Clinical Signs

Similarly to people, the primary lesion of animal PV is a flaccid vesicle rapidly progressing to a deep erosion (Fig. 3a–c). Erosions are seen more often due to the fragile nature of the vesicles, and further epithelial splitting beyond the preexisting erosion, even extending to a great distance (marginal Nikolsky's signs), can be elicited by pulling on the blister remnant. Crusts can develop over lesions at the mucocutaneous junctions or haired skin. The current knowledge about the lesion distribution in cats is extracted from less than a handful of described cases in the literature and from anecdotal reports [4, 23]. Like in people and dogs, lesions frequently involve the oral cavity, especially gums and hard palate, lips, nasal planum,

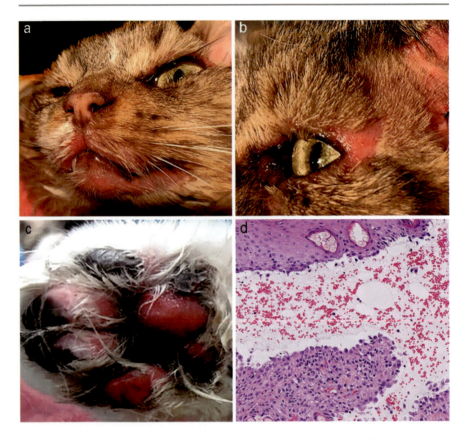

Fig. 3 Pemphigus vulgaris: Cats with pemphigus vulgaris with deep erosions affecting the (**a**) oral cavity, lips, and nasal philtrum; (**b**) periocular region; and (**c**) pawpads. A histopathology of feline pemphigus vulgaris showing the classic suprabasilar acantholysis (**d**). (Courtesy of Dr. Karen Trainor)

and philtrum (Fig. 3a, b). Haired skin involvement, as described in people, dogs, and horses [20, 24], and pawpad involvement (Fig. 3c) have been observed as well. Sialorrhea, halitosis, dysphagia, lethargy, and enlarged submandibular lymph nodes are common.

Diagnostic Approach
Because intact vesicles are rarely found, primary erosive diseases, especially those affecting oral cavity and mucocutaneous junctions, present the major differential diagnoses for feline PV. These include common diseases such as viral stomatitis caused by herpesvirus or calicivirus and chronic ulcerative stomatitis or rare diseases such as autoimmune subepidermal blistering skin diseases. The diagnosis of PV is confirmed by a biopsy, which shows suprabasilar acantholysis with basal cells remaining attached to the basement membrane (Fig. 3d). Samples for histopathology should be collected from the margin of the blister or erosion and should include

affected as well as intact tissue adjacent to the blister or erosion. The diagnostic value of direct and indirect immunofluorescence for antikeratinocyte autoantibodies has not been addressed in cats, and, so far, only tissue-bound, but not circulating, antikeratinocyte antibodies have been uncovered in cats with PV [4, 23].

Treatment
The information about the treatment and outcome of feline PV is very limited. Oral glucocorticoids (4–6 mg/kg/day of prednisolone) were reported to induce relatively rapid disease control. In cats in which prednisolone monotherapy is unable to control the disease, steroid-sparing drugs should be considered [23].

Paraneoplastic Pemphigus (PNP)
A single case of feline PNP associated with a thymoma can be found in the literature [25]. The cat exhibited progressive, deeply erosive dermatitis affecting concave pinnae, ventral abdomen and chest, perineum, and axillae [25]. In contrast to human and canine PNP, this cat did not exhibit mucosal involvement. Histopathology confirmed changes consistent with pemphigus vulgaris and erythema multiforme, which included suprabasilar acantholysis with lymphocytic interface dermatitis and keratinocyte apoptosis in multiple layers of the epidermis. Direct immunofluorescence revealed antikeratinocyte IgG antibodies in the patient's skin, and indirect immunofluorescence revealed circulating antikeratinocyte IgG autoantibodies binding oral mucosa as well as bladder epithelial cells. The latter observation suggested additional plakin autoreactivity, a feature seen in people and dogs with PNP [20, 26]. A removal of the neoplasia with or without a temporary immunosuppression should be curative in cases with PNP, but a poor response to treatment should be expected if the neoplasia cannot be successfully treated (Fig. 4).

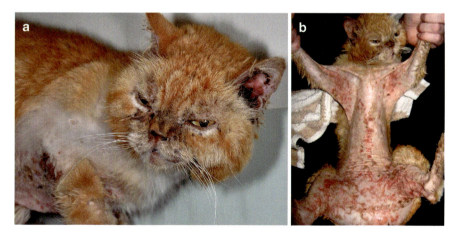

Fig. 4 Paraneoplastic pemphigus: Cat with paraneoplastic pemphigus due to thymoma. Deep erosions present on the concave pinnae and possibly philtrum (**a**) and on the caudal abdomen, medial thighs, perineum, ventral chest, and axillae (**b**). (Courtesy of Dr. Peter Hill)

Basement Membrane Autoimmunity

Mucous Membrane Pemphigoid (MMP)
Mucous membrane pemphigoid is a rare autoimmune subepidermal blistering disease involving preferentially mucosae and mucocutaneous junctions of people, dogs, and cats. Mucous membrane pemphigoid is immunologically heterogeneous with autoantibodies targeting components of the basement membrane zone such as collagen XVII, laminin-332, BP230, or integrins [27–29].

A naturally occurring MMP has been described in two adult cats, one of them being originally published under the diagnosis of bullous pemphigoid [29, 30]. Autoantibodies against collagen XVII (one cat) and laminin-332 (one cat) were identified in these cats. In both cats the disease had an adult-onset, though due to the small number of described cases, age, sex, or breed predisposition cannot be established.

Clinical Signs
Both cats exhibited vesicles and/or deep erosions and ulcers on mucosae and mucocutaneous junctions of the eyelids, lips and soft palate, and concave pinnae.

Diagnostic Approach
Because intact vesicles are rarely found, primary erosive diseases, especially those affecting oral cavity and mucocutaneous junctions, present the major differential diagnoses for feline MMP. These include common diseases such as viral stomatitis caused by herpesvirus or calicivirus and chronic ulcerative stomatitis or rare diseases such as pemphigus vulgaris. Histopathology is necessary to confirm the dermo-epidermal separation. In the two described cases, dermo-epidermal separation was accompanied by none to minimal dermal inflammation composed of dendritic/histiocytic cells and occasional neutrophils and eosinophils [29, 30]. Advanced immunological testing is not commercially available for veterinarians; however, because MMP is immunologically heterogeneous anyway, the diagnostic criteria include an adult-onset, blistering disease with a mucosae and mucocutaneous junction-dominant phenotype (Fig. 5a–c), histological confirmation of dermo-epidermal separation (Fig. 5d), and, ideally, demonstration of anti-basement membrane zone IgG.

Treatment
The information about treatment and outcome is available only for one of the two cats. In that cat, oral prednisone (4 mg/kg/day) resulted in a rapid disease control within a month, and the dosage of prednisone was gradually tapered within the following 6 months. The cat remained in a long-term remission off drugs [30]. In general, treatment principles outlined in pemphigus foliaceus section can be considered when managing this rare autoimmune skin disease.

Bullous Pemphigoid (BP)
In contrast to people in which BP is the most common autoimmune subepidermal blistering skin disease, BP is a rare diagnosis in animals. A single description of an

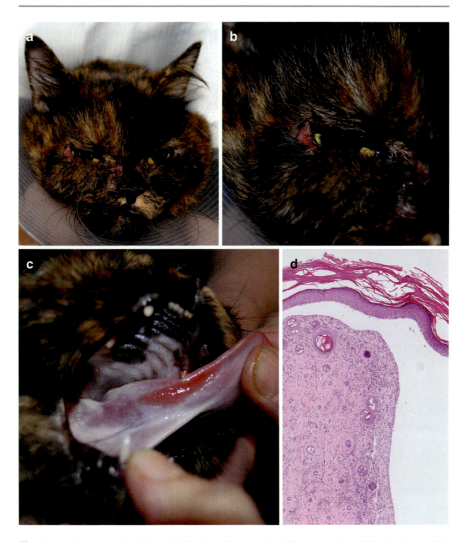

Fig. 5 Autoimmune subepidermal blistering disease: (**a–c**) Deep erosions affecting the eyelids and tongue (Courtesy of Dr. Chiara Noli); (**d**) histological confirmation of a dermo-epidermal separation. (Courtesy of Dr. Deborah Simpson/Judith Nimmo)

immunologically confirmed BP in a cat can be found in the literature (cat #2; cat #1 fulfills the diagnostic criteria of MMP (Olivry, personal communication)) [30]. Like in people and dogs, the BP-affected cat produced IgG against NC16A domain of collagen XVII [30].

Clinical Signs

The lesions of BP appear to be of minimal severity, with vesiculation and erosions occurring predominantly on the ears, trunk, and extremities. Mucosal involvement can be seen but appears to be mild.

Diagnostic Approach

Because intact vesicles are rarely found, other primary erosive diseases affecting the mucosae and haired skin present the major differential diagnoses for feline BP. These include herpesvirus stomatitis and dermatitis, other autoimmune subepidermal blistering skin diseases with haired skin involvement, pemphigus vulgaris, paraneoplastic pemphigus, and erythema multiforme. Histopathology is necessary to confirm the dermo-epidermal separation, which can be accompanied by a superficial dermal inflammation composed of mast cells, lymphocytes, and occasional eosinophils and neutrophils. Histopathology is unable to differentiate individual autoimmune subepidermal blistering diseases from each other, and antigen-specific immunological testing is not commercially available for veterinarians. The diagnostic criteria are therefore limited to an adult-onset, blistering disease with a haired skin-dominant phenotype, histological confirmation of dermo-epidermal separation, and, ideally, demonstration of anti-basement membrane zone IgG. The depth of the split could be assessed by a collagen IV staining of the patient's skin sample, which could assist in distinguishing BP from autoimmune subepidermal blistering diseases with collagen VII autoimmunity [31].

Treatment

The information about treatment and outcome is available only for one cat. In this cat, oral prednisone and doxycycline resulted in incomplete and temporary disease control [30]. In general, treatment principles outlined in pemphigus foliaceus section can be considered when managing this rare autoimmune skin disease.

Autoimmune Skin Diseases Causing Keratinocyte Injury

Lupus Erythematosus (LE)

Cutaneous and systemic lupus erythematosus are very rare and ill-defined entities in cats with a limited number of cases published during the last three decades [32, 33]. Affected cats are usually adult (>5 years), and Siamese cats appear to be more common among the systemic lupus erythematosus (SLE) cats, especially when compared with the hospital population. Cats suspected to suffer with cutaneous lupus erythematosus (CLE) were published under the diagnosis of discoid lupus erythematosus (DLE); however, their clinical signs were not always compatible with those described in human and canine DLE [32, 34, 35]. Among the published cats with systemic lupus erythematosus, only a single case that fulfilled the American Rheumatism Association criteria exhibited concurrent lupus erythematosus-specific skin lesions [36].

In cats, skin lesions compatible with CLE include erythema, alopecia, scaling, erosions, and crusts with or without dyspigmentation. In cases with compatible clinical lesions, an involvement of nasal planum, face, and trunk has been reported [33, 36]. Lupus erythematosus-specific skin lesions are histologically characterized by a lymphocytic interface dermatitis with hydropic degeneration of basal keratinocytes and their occasional single-cell necrosis. Lymphocytic interface folliculitis and hair follicle atrophy can be observed as well [36]. Thickened basement membrane zone can also be observed, and IgM deposition at the basement membrane zone has been detected in one cat [33]. Low antinuclear antibody (ANA) titers have been detected in the minority of these cats, though the diagnostic value of such findings is unknown because 30% of healthy, client-owned cats have low to high ANA titers as well [37]. Little is known about treatment of CLE in cats. The case with lesions localized to the nasal planum was successfully managed with topical steroids and sun avoidance [33], while the more generalized case with additional non-dermatological issues related was managed with immunosuppressive dosage of prednisone (4 mg/kg/day initially, 2 mg/kg every other day eventually) [36].

Autoimmune Skin Diseases Targeting Skin Melanocytes

Vitiligo

Vitiligo is a rare condition that has been described in cats. This condition is characterized by a progressive loss of melanocytes from the skin. The etiology is not fully understood even in people, and multiple theories had been proposed historically. Recently, however, the convergence theory that intertwines existing theories into a single one has become accepted [38]. Much attention has been paid to T cells and the inflammatory milieu in the vitiligo skin, which have been shown to possess cytotoxic properties against melanocytes and to be involved in the disease progression. Not surprisingly, treatments targeting lymphocytes (glucocorticoids, calcineurin inhibitors, JAK inhibitors) and promoting tolerance have shown promising results in managing this disease in people. Our knowledge about feline vitiligo is limited [39, 40]. Siamese cats have been reported more often [39, 41]. The characteristic clinical feature is leukoderma, frequently affecting the nasal planum, lips, and eyelids, but footpad depigmentation as well as patchy to more generalized leukoderma and leukotrichia can be seen (Fig. 6). A visible skin inflammation is often missing.

Fig. 6 Vitiligo characterized by a depigmentation of the nose and of the pawpads. (Courtesy of Dr. Silvia Colombo)

Histology demonstrates reduction or complete loss of melanocytes in the epidermis and hair follicles. Lymphocytes migrating into the lower epidermis may be seen. Vitiligo is considered a cosmetic disease in cats, and a successful treatment has not been published yet.

References

1. Scott DW, Miller WH, Erb HN. Feline Dermatology at Cornell University: 1407 Cases (1988–2003). J Feline Med Surg. 2013;15:307–16.
2. LeBleu VS, Macdonald B, Kalluri R. Structure and function of basement membranes. Exp Biol Med (Maywood). 2007;232:1121–9.
3. Delva E, Tucker DK, Kowalczyk AP. The desmosome. Cold Spring Harb Perspect Biol. 2009;1:a002543.
4. Scott DW, Walton DL, Slater MR. Immune-mediated dermatoses in domestic animals: ten years after – Part I. Comp Cont Educ Pract. 1987;9:424–35.
5. Levy B, Mamo LB, Bizikova P. Circulating antikeratinocyte autoantibodies in cats with pemphigus foliaceus (Abstract). Austin: North America Veterinary Dermatology Forum; 2019.
6. Bizikova P, Burrows M. Feline pemphigus foliaceus: original case series and a comprehensive literature review. BMC Vet Res. 2019;15:1–15.
7. Caciolo PL, Nesbitt GH, Hurvitz AI. Pemphigus foliaceus in eight cats and results of induction therapy using azathioprine. J Amer An Hosp Assoc. 1984;20:571–7.
8. Preziosi DE, Goldschmidt MH, Greek JS, et al. Feline pemphigus foliaceus: a retrospective analysis of 57 cases. Vet Dermatol. 2003;14:313–21.
9. Irwin KE, Beale KM, Fadok VA. Use of modified ciclosporin in the management of feline pemphigus foliaceus: a retrospective analysis. Vet Dermatol. 2012;23:403–9.
10. McEwan NA, McNeil PE, Kirkham D, Sullivan M. Drug eruption in a cat resembling pemphigus foliaceus. J Small Anim Pract. 1987;28:713–20.
11. Prelaud P, Mialot M, Kupfer B. Accident Cutane Medicamenteux Evoquant Un Pemphigus Foliace Chez Un Chat. Point Vet. 1991;23:313–8.
12. AFFOLTER VK, TSCHARNER CV. Cutaneous drug reactions: a retrospective study of histopathological changes and their correlation with the clinical disease. Vet Dermatol. 1993;4:79–86.
13. Barrs VR, Beatty JA, Kipar A. What is your diagnosis? J Small Anim Pract. 2003;44(251):286–7.
14. Salzo P, Daniel A, Silva P. Probable pemphigus foliaceus-like rug reaction in a cat (Abstract). Vet Dermatol. 2014;25:392.
15. Biaggi AF, Erika U, Biaggi CP, Taboada P, Santos R. Pemphigus foliaceus in cat: two cases report. 34th World Small Animal Veterinary Association Congress. Brazil: São Paulo, 21–24 July 2009.
16. Coyner KS. Dermatology how would You handle this case? Vet Med. 2011;106:280–3.
17. Simpson DL, Burton GG. Use of prednisolone as monotherapy in the treatment of feline pemphigus foliaceus: a retrospective study of 37 cats. Vet Dermatol. 2013;24:598–601.
18. Gross TL, Ihrke PJ, Walder EJ, Affolter VK. Pustular diseases of the epidermis (Superficial pustular dermatophytosis). In: Skin diseases of the dog and cat. 2nd ed. Oxford, UK: Blackwell Science Ltd; 2005. p. 11–3.
19. Gross TL, Ihrke PJ, Walder EJ, Affolter VK. Pustular diseases of the epidermis (Impetigo). In: Skin diseases of the dog and cat. 2nd ed. Oxford, UK: Blackwell Science Ltd; 2005. p. 4–6.
20. Olivry T, Linder KE. Dermatoses affecting desmosomes in animals: a mechanistic review of acantholytic blistering skin diseases. Vet Dermatol. 2009;20:313–26.
21. Olivry T. A review of autoimmune skin diseases in domestic animals: I – superficial pemphigus. Vet Dermatol. 2006;17:291–305.

22. Willard MD. Feline inflammatory bowel disease: a review. J Feline Med Surg. 1999;1:155–64.
23. Manning TO, Scott DW, Smith CA, Lewis RM. Pemphigus diseases in the feline: seven case reports and discussion. JAAHA. 1982;18:433–43.
24. Winfield LD, White SD, Affolter VK, et al. Pemphigus vulgaris in a welsh pony stallion: case report and demonstration of antidesmoglein autoantibodies. Vet Dermatol. 2013;24:269–e60.
25. Hill PB, Brain P, Collins D, Fearnside S, Olivry T. Putative paraneoplastic pemphigus and myasthenia gravis in a cat with a lymphocytic thymoma. Vet Dermatol. 2013;24:646–9, e163-4.
26. Kartan S, Shi VY, Clark AK, Chan LS. Paraneoplastic pemphigus and autoimmune blistering diseases associated with neoplasm: characteristics, diagnosis, associated neoplasms, proposed pathogenesis, treatment. Am J Clin Dermatol. 2017;18:105–26.
27. Xu HH, Werth VP, Parisi E, Sollecito TP. Mucous membrane pemphigoid. Dent Clin N Am. 2013;57:611–30.
28. Olivry T, Chan LS. Spontaneous canine model of mucous membrane pemphigoid. In: Chan LS, editor. Animal models of human inflammatory skin diseases. 1st ed. Boca Raton: CRC Press; 2004. p. 241–9.
29. Olivry T, Dunston SM, Zhang G, Ghohestani RF. Laminin-5 is targeted by autoantibodies in feline mucous membrane (cicatricial) pemphigoid. Vet Immunol Immunopathol. 2002;88:123–9.
30. Olivry T, Chan LS, Xu L, et al. Novel feline autoimmune blistering disease resembling bullous pemphigoid in humans: igg autoantibodies target the NC16A ectodomain of type XVII collagen (BP180/BPAG2). Vet Pathol. 1999;36:328–35.
31. Olivry T, Dunston SM. Usefulness of collagen IV immunostaining for diagnosis of canine epidermolysis bullosa acquisita. Vet Pathol. 2010;47:565–8.
32. Willemse T, Koeman JP. Discoid lupus erythematosus in cats. Vet Dermatol. 1990;1:19–24.
33. Kalaher K, Scott D. Discoid lupus erythematosus in a cat. Feline Pract. 1991;17:7–11.
34. Kuhn A, Landmann A. The classification and diagnosis of cutaneous lupus erythematosus. J Autoimmun. 2014;48–49:14–9.
35. Olivry T, Linder KE, Banovic F. Cutaneous lupus erythematosus in dogs: a comprehensive review. BMC Vet Res. 2018;14:132–018-1446-8.
36. Vitale C, Ihrke P, Gross TL, Werner L. Systemic lupus erythematosus in a cat: fulfillment of the American Rheumatism Association Criteria with Supportive Skin Histopathology. Vet Dermatol. 1997;8:133–8.
37. Abrams-Ogg ACG, Lim S, Kocmarek H, et al. Prevalence of antinuclear and anti-erythrocyte antibodies in healthy cats. Vet Clin Pathol. 2018;47:51–5.
38. Kundu RV, Mhlaba JM, Rangel SM, Le Poole IC. The convergence theory for vitiligo: a reappraisal. Exp Dermatol. 2019;28:647–55.
39. López R, Ginel PJ, Molleda JM, et al. A clinical, pathological and immunopathological study of vitiligo in a Siamese cat. Vet Dermatol. 1994;5:27–32.
40. Alhaidari Z. Cat Vitiligo. Ann Dermatol Venereol. 2000;127:413.
41. Alhaidari, Olivry, Ortonne, et al. Vet Dermatol. 1999;10:3–16.
42. Murrell DF, Pena S, Joly P, et al. Diagnosis and management of pemphigus: Recommendations by an international panel of experts. J Am Acad Dermatol. 2018 Feb 10. pii: S0190-9622(18)30207-X. doi: 10.1016/j.jaad.2018.02.021. [Epub ahead of print]

Immune Mediated Diseases

Frane Banovic

Abstract

As the spectrum of feline immune-mediated cutaneous diseases has expanded markedly in the recent two decades, veterinarians are encouraged to become familiar with the characteristic clinical features of various immune-driven skin disorders to permit early diagnosis and appropriate treatment. This article describes the signalment, clinical signs, laboratory and histopathology findings, as well as treatment outcomes in different immune-mediated skin diseases in cats.

Erythema Multiforme, Stevens-Johnson Syndrome, and Toxic Epidermal Necrolysis

First described by von Hebra in 1860, erythema multiforme (EM) had long been considered as part of a spectrum of diseases that included Stevens-Johnson syndrome (SJS) and toxic epidermal necrolysis (TEN) [1]. The currently accepted clinical classification in humans [2], dogs [3], and cats defines SJS and TEN as variants of the same disease spectrum that are different from subsets of EM in characteristic clinical appearance and causality. Drugs are thought to induce most cases of SJS/TEN, but infectious triggers dominate in EM, which implicates drug withdrawal as a crucial requirement for the treatment and prognosis of SJS/TEN [2, 3].

F. Banovic (✉)
University of Georgia, College of Veterinary Medicine, Department of Small Animal Medicine and Surgery, Athens, GA, USA
e-mail: fbanovic@uga.edu

© Springer Nature Switzerland AG 2020
C. Noli, S. Colombo (eds.), *Feline Dermatology*,
https://doi.org/10.1007/978-3-030-29836-4_26

511

Erythema Multiforme

Erythema multiforme is an acute, immune-mediated disorder that affects the skin and/or mucous membranes, including the oral cavity [4]. In humans, typical EM represents a blistering and ulcerative skin disease characterized by target or iris lesions (i.e., three different zones of color) distributed symmetrically on the extremities [2, 4, 5]. Atypical EM features widespread, large, round, bullous lesions (i.e., atypical targets with two different zones of color) that affect the trunk. Erythema multiforme is divided into minor (EMm) and major (EMM) forms based on mucosal involvement and systemic signs of illness present in the latter [2, 4, 5]. The lesions in typical and atypical EM can be confluent, but they do not lead to large areas of epidermal sloughing as in SJS/TEN. In cases of EM, the skin detachment is limited to 1–3% body surface in EMm (i.e., with acral distribution), but it may be more extensive in atypical EMM with a widespread distribution up to 10% body surface area [2, 4, 5].

Only a small number of feline EM cases can be found in the veterinary literature [6–12]. Localized multifocal maculopapular and target lesions affecting the ventral body were described in three out of the eight reported cases, and widespread crusts and/or ulcerations with or without mucocutaneous or buccal mucosa involvement were described in three other cats (Fig. 1a, b). Interestingly, severe ulcerations with crusts affecting >50% of the body area, including footpads and nail beds, were described in one case [10]. Similarly to canine EM [3], the feline EM lesions described as erythematous macules with widespread sloughing/ulcerations and secondary crusting may represent SJS being published as cases of EMM.

The proposed pathomechanism of EM includes autoreactive T cell generation and activation by antigen (viral, bacterial, drug)-loaded epithelial cells resulting in epidermal damage from lysis of surrounding keratinocytes [13]. A drug causality was implicated in seven feline EM patients; however there was preceding laryngotracheitis of unknown origin or a vaccination with feline rhinotracheitis-calicivirus-panleukopenia virus vaccine reported in three cases. Similarly to human and canine EM patients, there are doubts regarding the accuracy of feline EM diagnosis in several reports of drug-induced EM [10], as some of these cases may have exhibited SJS or SJS/TEN overlap.

Interestingly, a putative feline herpes-associated EM was proposed in a single case where feline herpes virus 1 (FHV1) DNA was isolated from skin biopsies of a cat presenting with widespread exfoliative dermatitis, scaling, and a history of an upper respiratory tract infection 2 weeks before presentation [12]. The lesions in human herpes simplex virus (HSV)-induced EM are virus-free (i.e., viral cytopathic changes are not present) but contain HSV DNA fragments, most often comprising sequences that express polymerase gene *Pol* [13]. Viral protein expression in the skin (notably Pol, rarely thymidine kinase) initiates lesion development through recruitment of a Vβ-restricted population of virus-specific CD4 helper T cells, type 1, that produce interferon (IFN)-γ. This early virus-specific response is followed by an amplified inflammatory cascade, characterized by enhanced cytokine production

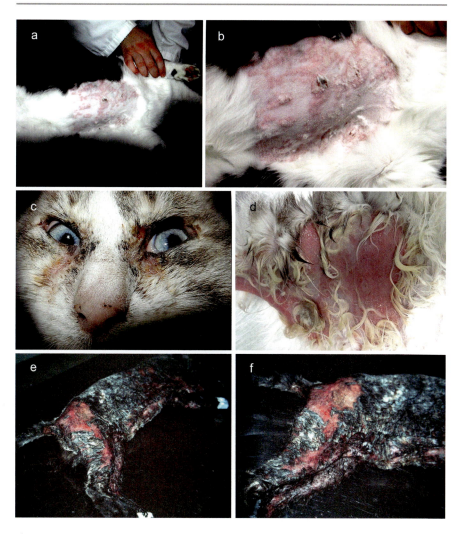

Fig. 1 Feline erythema multiforme, Stevens-Johnson syndrome, and toxic epidermal necrolysis. (**a, b**) Multifocal maculopapular and target lesions affecting the ventral body in a cat with erythema multiforme. (**c**) Severe bilateral periocular erythematous erosions with sloughing of epidermis at the medial and lateral canthus. (**d, e, f**) Erythematous erosions with detached or easily detachable epidermis are present on ventral abdomen and inguinal area. (Courtesy of Drs. Chiara Noli and Silvia Colombo)

and the accumulation of T cells that respond to auto-antigens, which are likely released by lysed or apoptotic virus-infected cells [13]. Although proposed as an analogue of the human herpes-associated EM, there is a need for proposed cases of feline herpes-associated EM to be better characterized and described for viral replication and the presence of inclusion bodies that would differentiate between infection and true EM-like pattern.

Histologically, EM is the prototypical cytotoxic interface dermatitis showing transepidermal keratinocyte apoptosis with hydropic changes and dyskeratosis of basal keratinocytes [13]. Importantly, the diagnosis of EM is mainly based on the history and clinical presentation, as histopathologic features are not pathognomonic for the disease [13]. Depending on the biopsy site and the stage of the clinical disease, full-thickness necrosis may be the dominant lesion in a biopsy from the center of a target lesion, whereas interface dermatitis with vacuolar change may be seen at the margin ("zonal changes") [13]. Similarly to humans and dogs [2, 3], the diagnosis of EM in cats is clinicopathological. Erythema multiforme may be difficult to differentiate histologically from other cytotoxic interface dermatoses in cats, particularly SJS/TEN, thymoma-associated paraneoplastic exfoliative dermatosis, non-thymoma-associated exfoliative dermatitis, and potentially some variants of cutaneous lupus erythematosus (CLE).

The treatment of EM varies according to disease severity and causality; the clinical course of proposed feline herpes-associated EM is usually self-limiting, resolving within weeks without significant sequelae [12]. Any drug suspected to have precipitated EM should be promptly discontinued. In the severe form of EM, systemic glucocorticoids in conjunction with antiviral therapy (e.g., famciclovir) are advised depending on the etiology. Some cats with the persistent form of EM may respond to oral immunosuppressants ciclosporin (5–7 mg/kg every 24 hours) or mycophenolate mofetil (10–12 mg/kg every 12 hours).

Stevens-Johnson Syndrome/Toxic Epidermal Necrolysis Spectrum

Stevens-Johnson syndrome and TEN are rare, predominantly drug-induced, severe cutaneous T cell-mediated immune reactions characterized by widespread sloughing of the epidermis and mucosal epithelium [2, 3]. The two terms describe variants of the same disease spectrum, in which SJS is the less extensive (with less than 10% of the body surface affected) and TEN the more widespread form (more than 30% of the body surface is involved) [2, 3].

Clinical signs of the SJS/TEN spectrum of disease in humans, dogs, and cats are homologous [2, 3, 14, 15]: patients exhibit painful, irregular, and flat erythematous/purpuric macules and patches that blister into confluent and larger areas of epidermal sloughing (Fig. 1c, d, e, f). However, only very few in-depth characterized cases of feline SJS/TEN cases can be found in the veterinary literature [14, 16–18]. Lesions can affect the skin diffusely over the body; mucocutaneous junctions, mucosae (e.g., oral, rectal, conjunctival), and footpads are frequently involved. The denuded dermis exudes serum, and the lesions can become secondarily infected with crust development and potential sepsis. In severe cases, necrolysis of the respiratory and gastrointestinal epithelium is accompanied by bronchial obstruction, profuse diarrhea, and a variety of systemic complications, including multi-organ failure.

The main clinical sign distinguishing TEN from SJS is the amount of body surface area with epidermal detachment, which is defined as any necrotic skin that is already detached (e.g., blisters, erosions) or which is detachable (i.e., areas with a positive pseudo-Nikolsky sign) at the worst stage of the disease [2, 3]. Some cats with TEN initially exhibit epidermal detachment over less than 10% of their body surface, an extent typical of SJS, but the severity progresses to 30% of the body surface area in a few days, which is more typical of TEN [14]. Despite the striking clinical presentation of SJS/TEN, a number of disorders can present with a macular rash and blistering of the skin and mucous membranes in cats. Some of the clinical differential diagnoses include erythema multiforme major, thermal burns, vasculitis, exfoliative dermatitis secondary to thymoma, and non-thymoma exfoliative dermatitis.

Drugs are reported as the leading cause of SJS/TEN, with the risk of a hypersensitivity reaction developing in the first few weeks after drug ingestion [19].

A new disease-specific algorithm, the Assessment of Drug Causality in Epidermal Necrolysis (ALDEN), has been recently validated for human patients with SJS/TEN, and it shows superiority to previous algorithms [19]; a recent feline SJS/TEN case report also utilized ALDEN for the drug causality evaluation [14]. Strong associations between SJS/TEN and several drugs exist in dogs such as beta-lactam and trimethoprim-potentiated sulfonamide antibiotics, phenobarbital, and carprofen [3, 15]. In cats, a strong causal association has been reported for beta-lactam antibiotics, organophosphate insecticide, and d-limonene [14, 16–18]. At the time, the precise molecular and pathogenic cellular mechanisms leading to the development of SJS/TEN are partially understood. The lesions of SJS/TEN are characterized by widespread epithelial keratinocyte apoptosis and necrosis, a process initiated by drug- or drug/peptide-specific cytotoxic T-lymphocytes (CTL) and/or natural killer (NK) cells [20]. Drugs can stimulate the immune system by directly binding to the class I major histocompatibility complex, resulting in the clonal expansion of a specific population of CTLs, which infiltrate the skin and secrete soluble proapoptotic factors like granulysin, Fas ligand, perforin, and granzymes [20].

Although a diagnosis of SJS/TEN is suggested by the history and clinical signs, a skin biopsy is necessary to support the clinical assessment and exclude other blistering dermatoses. Histopathological examination reveals lymphocytic interface dermatitis with apoptosis at multiple epidermal levels, progression to epidermal coagulation necrosis, and epidermal detachment with ulcerations (Fig. 2) [14]. Similarly to human and canine SJS/TEN, there is a potential histologic overlap between feline EM and SJS/TEN [14]. Therefore, a pathologist's microscopic interpretation should be restricted to an umbrella diagnosis of an EM-TEN epidermal necrotizing disease, and the further subclassification of the different entities should depend upon patient history, clinical signs, and skin lesion extent [15]. In suspect SJS/TEN feline cases, clinicians should be encouraged to take multiple biopsies since some skin biopsies in these patients may lack epithelium and may not be useful for diagnosis (Fig. 2c, d). Dermal necrosis is not present in feline TEN skin biopsies despite large ulcer development and bacterial colonization in some cases [15]. This is important because the exact depth of skin necrosis, when shallow, can be difficult to determine clinically, and histologic examination aids disease classification.

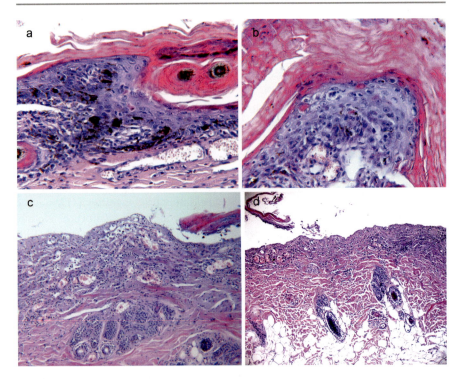

Fig. 2 Histopathology of toxic epidermal necrolysis. (**a, b**) Lymphocytic interface dermatitis with extravasation of lymphocytes into the epidermis, lymphocytic satellitosis of apoptotic keratinocytes, and superficial epidermal hydropic degeneration is the predominant inflammatory pattern in TEN. (**c, d**) If ulcerated areas are biopsied in TEN, complete ulceration and loss of epithelium are commonly observed, which is non-diagnostic. (Courtesy of Drs. Chiara Noli and Silvia Colombo)

Although rare, SJS/TEN is a devastating disease; the mortality for SJS is <10%, with the figure rising to 40% for TEN in humans [19]. Significant mortality is associated with TEN in cats, which confirms TEN as one of the few dermatological diseases that constitute an actual medical emergency [14, 16–18]. Early recognition and prompt and appropriate management are necessary and can be lifesaving. Immediate withdrawal of suspected medication (most commonly beta-lactams, sulfa drugs, NSAIDs) and referral to an emergency center are crucial requirements for improving SJS/TEN prognosis. An extensive epidermal loss results in a massive fluid, electrolyte, and plasma protein losses. Supportive care similar to that for burn patients (aggressive fluid replacement, antimicrobial therapy, wound care, analgesia, and nutritional support) is required. The use of immunosuppressive agents (e.g., glucocorticoids, ciclosporin, mycophenolate) has been controversial, but recent evidence shows a possible beneficial role for ciclosporin in humans during early disease development [21].

Pseudopelade

Pseudopelade is a rare hair disorder in cats, presumably immune-mediated and characterized by non-pruritic permanent alopecia that grossly is non-inflammatory [22–24].

Pathogenesis

The pathogenesis of pseudopelade in humans is not completely understood. Some of the suspected factors include acquired autoimmunity, *Borrelia* infection, and senescence of follicular stem cell reservoir; however, case series of familial pseudopelade indicate that genetic heredofamilial factors may play a role [22].

A key feature is perifollicular lymphocytic infiltrate that targets the mid-isthmus of the hair follicle and damages the follicular bulge stem cells leading to permanent scarring hair loss [22–24]. Immunologic studies conducted on skin biopsies of a single cat affected by pseudopelade revealed a predominance of cytotoxic CD8+ lymphocytes within the epithelium of the hair follicle isthmus, whereas the perifollicular dermis was rich in CD4+ and CD8+ lymphocytes as well as CD1+ dendritic antigen-presenting cells [24]. Circulating autoantibodies (IgG class) against multiple hair follicle proteins (including hair keratins) and trichohyalin were demonstrated in the cat. However, it is hypothesized that the humoral immune response occurs after exposure of cryptic follicular epitopes secondary to follicular destruction [24].

Clinical Signs

Lesions have been characterized by a more or less symmetrical, non-inflammatory, patchy to diffuse alopecia that may begin on the face and then spread to the ventrum, legs, and paws (Fig. 3a) [24]. The pruritus is absent and broken-off hair shafts are not observed. Interestingly, onychorrhexis may be present on some claws.

Diagnosis

The possible differential diagnoses include almost any skin disease characterized by relatively non-inflammatory, asymptomatic (non-pruritic) alopecia such as alopecia areata, follicular dysplasia, psychogenic alopecia, endocrinopathies, and dermatophytosis [23, 24]. Definitive diagnosis is based on skin biopsy, and multiple specimens should be obtained from regions of maximal alopecia, margin to healthy haired areas, as well as healthy haired skin. The characteristic early histopathologic

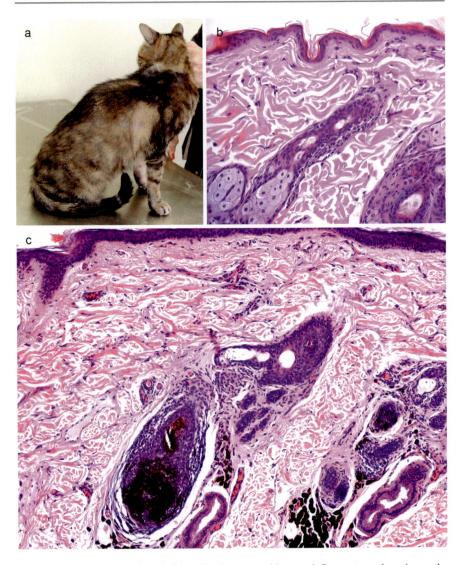

Fig. 3 Feline pseudopelade. (**a**) Generalized, non-pruritic, non-inflammatory alopecia on the flanks, neck, and abdomen. (**b, c**) Histopathology reveals moderate follicular inflammation in and around the isthmus; hair bulb and adnexae remain unaffected. (Courtesy of Dr. Chiara Noli)

findings include a variably severe accumulation of inflammatory cells including lymphocytes, histiocytes, and fewer plasma cells predominantly at the follicular isthmus (Fig. 3b, c). In late lesions, inflammation is mild, and hair follicles undergo atrophy and are replaced by fibrosing tracts [23, 24].

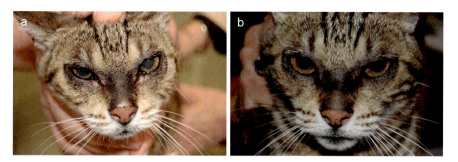

Fig. 4 Pseudopelade in a cat before (**a**) and after treatment (**b**) with oral ciclosporin at 5 mg/kg twice daily for 30 days. (Courtesy of Dr. Chiara Noli)

Clinical Management

There is no definite management to halt the pseudopelade progression in humans, presumably because of follicular bulge stem cell destruction. This condition has failed to respond to topical and systemic glucocorticoids in humans [22]. Transient hair regrowth following oral ciclosporin administration (5 mg/kg every 12 h) was observed in a single cat (Fig. 4) [24].

Auricular Chondritis

Auricular chondritis is a rare disease of cats characterized by inflammation and destruction of the auricular cartilage [25–32].

Pathogenesis

In humans, relapsing polychondritis (RPC) is an inflammatory connective tissue immune-mediated disease characterized by recurrent episodes of inflammation and destruction involving both articular and non-articular cartilaginous structures, resulting in progressive anatomical deformation and functional impairment of the involved structures [33, 34]. The exact pathogenesis of RPC is not yet clearly defined; it is hypothesized that cell-mediated immune destruction to the cartilage containing chondrocyte epitopes leads to cytokine release and local inflammation with a subsequent autoantibody production (circulating antibodies against collagen II and matrilin-1) in an inherently susceptible host [33, 34].

In cats, a similar rare condition has been recognized and reported in 14 cases [25–32]. However, only the auricular cartilage was affected in 11 cats, without the classic relapsing nature observed in humans with RPC. Hence, in cats, the term relapsing polychondritis may be inappropriate, and auricular chondritis should be reserved for cases characterized only by inflammation and destruction of auricular cartilage. Similar to human RPC, the histopathological examination of affected cartilages in cats revealed an inflammatory infiltrate composed of various proportions of T-lymphocytes, neutrophils, macrophages, and plasma cells, restricted to perichondrium at an early stage and, later, spreading to cartilage [25–32].

Clinical Signs

Predominantly young to middle-aged cats are affected; the age of onset ranges from 1.5 to 14.5 years (median, 3 years). There is no sex or breed predilection reported. Affected cats present with a history of swollen, erythematous to violaceous and often painful pinna; with chronicity, lesions progress to curled and deformed pinnae (Fig. 5a, b, c, d) [25–32]. The disease can start unilaterally and spread to both sides or can start bilaterally with difference in severity between sides. Beyond the auricular signs, cats are usually systemically healthy. However, some cats may be pyrexic or have additional signs of RPC involvement such as uveitis, chondritis of other cartilages, arthritis, and heart disease [25–32].

Diagnosis

In humans, the diagnosis of RPC is a real challenge for clinicians and is still based on clinical grounds [33]. The diagnostic criteria for RPC in humans involve at least one McAdam criterion (i.e., bilateral auricular chondritis, nonerosive sero-negative inflammatory polyarthritis, nasal chondritis, ocular inflammation, respiratory tract chondritis, and audiovestibular damage) and positive histologic confirmation or two McAdam criteria and positive response to administration of glucocorticoids or dapsone [33]. According to the human criteria, only one reported feline case fulfilled the diagnostic criteria of human RPC with chondral lymphocytic inflammation present in the pinnae, costae, larynx, trachea, and limbs [32]. In feline cases where only the auricular cartilage is affected, skin biopsies reveal degenerated cartilage with lymphocytic invasion, perichondrial lymphocytic infiltration, and fibrosis (Fig. 5e, f) [25–32]. The dermis most commonly shows moderate perivascular inflammation with infiltration of lymphocytes and neutrophils. Direct immunofluorescence staining for immune complexes was negative in two cats [26].

Clinical Management

The goal of therapy in human RPC is the control of the inflammatory crisis and the long-term suppression of the immune-mediated pathogenetic mechanisms [33]. Although a variety of drugs including glucocorticoids, nonsteroidal anti-inflammatory drugs, and immunosuppressive and cytotoxic drugs have been used

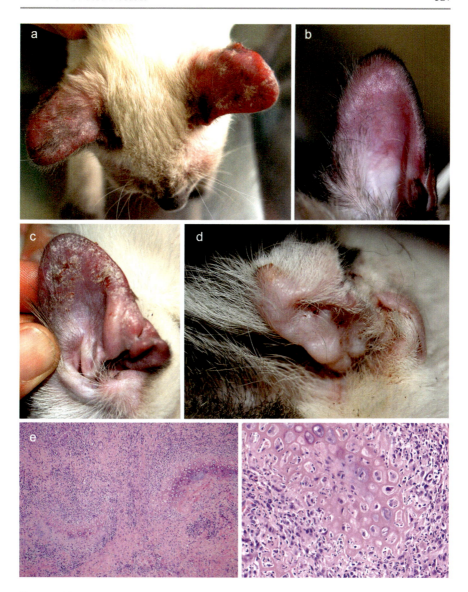

Fig. 5 Feline auricular chondritis. (**a, b, c, d**) Severe erythema, thickening, crusting, and swelling of the pinnae with progression to tissue fibrosis and deformation. (**e, f**) Degenerated cartilage with perichondrial mixed inflammation consisting of lymphocytes, macrophages, and neutrophils; moderate fibrosis surrounds the cartilages. Lymphocytes invade the cartilage tissues in several areas with resulting loss of cartilage. (Courtesy of Dr. Chiara Noli)

for the treatment, there are no evidence-based guidelines for the treatment of RPC [33]. Oral ciclosporin (5–7.5 mg/kg once daily) and dapsone (1 mg/kg once daily) for 4 months were unsuccessful in controlling clinical signs in a single feline RPC case [32]. Some cats with only auricular chondritis signs show spontaneous improvement overtime without treatment [30]. Dapsone (1 mg/kg every 24 h) appeared to result in some clinical improvement, whereas oral glucocorticoids (prednisolone 1 mg/kg every 24 h) for two to 3 weeks were fairly ineffective in feline auricular chondritis cases [30]. Surgical pinnectomy resulted in a cure in one case [30].

Plasma Cell Pododermatitis

Plasma cell pododermatitis is a rare dermatological condition exclusively described in cats, characterized by swelling and softening of footpads with occasional ulcerations [25, 35–44].

Cause and Pathogenesis

The cause and pathogenesis of this disorder are unknown. An immunologic reaction to an infectious agent or derived residual antigen is suspected, based on the tissue plasmacytosis, consistent hypergammaglobulinemia, negative tissue cultures and special stains for microbial agents, and a good response to immunomodulating agents. Although concurrent feline immunodeficiency virus (FIV) infection was observed in many cats (44–62%), it is unknown if FIV virus plays any important role in the pathogenesis of feline plasma cell pododermatitis [38, 41, 42].

Clinical Features

No age, breed, or sex predilections were reported; the affected cats typically ranged from 6 months to 12 years of age [35–44]. Initial clinical signs involve asymptomatic swelling and softening of usually multiple footpads; rarely a single footpad is affected. Predominantly the central metacarpal or metatarsal pads are involved. However, occasional digital pads may show signs, but not usually as severe (Fig. 6a, b). On presentation, the affected pads are swollen and feel mushy or flaccid, and their surface appears white and scaly and cross-hatched with silvery striae. The pads may ulcerate, causing pain and lameness. Recurrent hemorrhages from ulcerated or nodular areas of a pad can occur as well as secondary bacterial infection. Pyrexia, anorexia, lethargy, and lymphadenopathy may be seen in some affected cats. Occasionally, a minority of cats with plasma cell pododermatitis will show signs of plasma cell dermatitis on the nose or stomatitis with proliferative, ulcerative pharyngitis and vegetative plaques on the palatine arches [44]. Also, a cat occasionally has immune-mediated glomerulonephritis or renal amyloidosis [36].

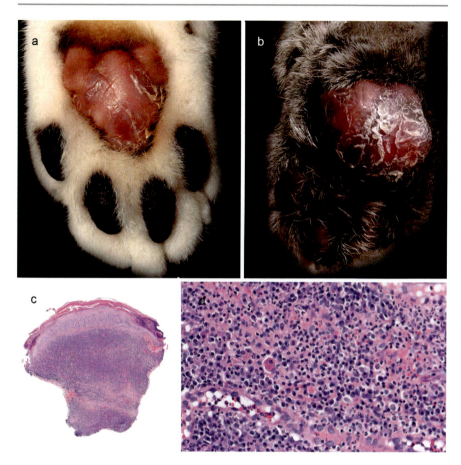

Fig. 6 (**a, b**) Swollen and flaccid metacarpal pads with depigmentation (**a**), erosions, and scaling. (**c, d**) The dermis and often the underlying adipose tissue of the pawpad are diffusely infiltrated with predominantly plasma cells, some neutrophils, and lymphocytes, obscuring normal architecture. Russell body-containing plasma cells (Mott cells) are observed as well

Diagnosis

The history and clinical presentation are generally very striking and supported by fine needle aspirate (FNA) which reveals plasma cells. Definitive diagnosis is confirmed by histopathology, which shows a diffuse infiltrate with plasma cells, neutrophils, and lymphocytes [38, 41, 42]. In the "classic" case presentation with lesions involving multiple pads and aspiration cytology revealing predominately plasma cells, a skin biopsy may not be necessary. The main differential diagnosis is eosinophilic granuloma of the footpads, which typically may have concurrent skin lesions either interdigitally or on other body areas and will not cause diffuse pad swelling or involve multiple paws. If only a single pad is involved, neoplasia should be considered. Infectious agents and foreign body would also be differential diagnoses [38, 41, 42]. If skin biopsies are obtained, histopathological examination reveals superficial and deep perivascular plasma cell dermatitis, with frequent diffuse dermal and even adjacent adipose tissue plasma cell infiltration; Russell bodies (Mott cells) are typically seen (Fig. 6c, d). Fibrosis may be seen in chronic lesions.

Clinical Management

The prognosis of feline plasma cell pododermatitis varies, as in some patients clinical signs may resolve spontaneously, whereas others may require immunomodulating agents and life-long therapy [35–44].

The initial therapy of choice is doxycycline, an inexpensive antibiotic with immunomodulating properties that belongs to the tetracycline class of antibiotics [42, 43]. Doxycycline has been reported to produce partial or complete clinical remission in more than half of the feline plasma cell pododermatitis cases. Although initial reports used 25 mg/cat, doxycycline should be administered at 10 mg/kg once daily or at 5 mg/kg every 12 h. Due to the delayed esophageal transit time for capsules and tablets, cats are prone to develop drug-induced esophagitis and resultant esophageal strictures during any tablet or capsule administration [45–47]. The hyclate salt of doxycycline (doxycycline hydrochloride) has been primarily associated with esophagitis and esophageal strictures in cats [47, 48]. To aid transport of tablets and capsules and avoid stricture formation, a 6 ml water flush or a small amount of food should always follow doxycycline administration in cats. The use of compounded suspensions of doxycycline should be avoided, because the marketing of such formulations is in violation with regulations in some countries, including the USA. Doxycycline treatment is continued until the footpads have a normal macroscopic appearance, which may take up to 12 weeks (Fig. 7); after full remission is achieved, the administration frequency of doxycycline is slowly tapered and if possible discontinued [42, 43].

In patients with a poor response to doxycycline treatment and active severe clinical symptoms, a short course of systemic glucocorticoid therapy in conjunction with oral ciclosporin (5–7.5 mg/kg every 24 h) may be indicated. Oral prednisolone is most commonly administered at 2–4 mg/kg once daily and then tapered following a favorable response. In cases refractory to prednisolone, oral triamcinolone acetonide at 0.4–0.6 mg/kg once daily or dexamethasone 0.5 mg once daily has also been effective. Once the disease is fully controlled, oral ciclosporin is slowly weaned off.

Surgical excision of the fatty footpad has also been described as beneficial and is an option for cases not responding to medical therapy, with no reports of recurrence of disease in the surgically treated pads with follow-up periods of 2 years [37–39].

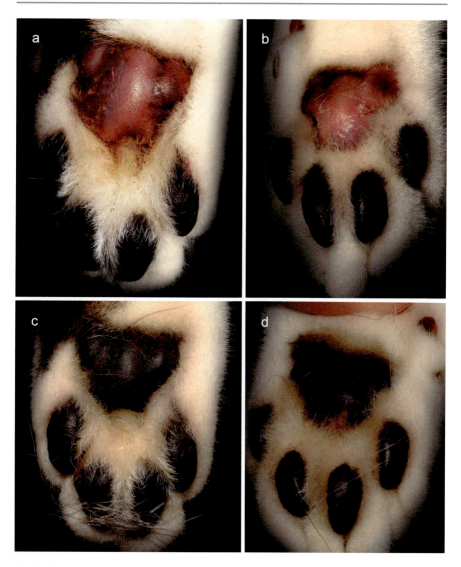

Fig. 7 Plasma cell pododermatitis in a cat treated with a tapering dose of prednisolone (0.5 mg/kg every 24 hours for 2 weeks) and long-term doxycycline (5 mg/kg every 24 hours). Within 5 weeks of treatment, the initial swelling and depigmentation of metacarpal pads (**a, b**) almost completely resolved (**c, d**)

Feline Proliferative and Necrotizing Otitis Externa

Feline proliferative and necrotizing otitis externa (PNOE) is a very rare skin disease of the cat that has been described only in a few reports [48–52].

Cause and Pathogenesis

The pathogenesis of feline PNOE is unknown at the time. Polymerase chain reaction analysis for feline herpesvirus 1 using primers for thymidine kinase and polymerase glycoprotein was negative in five cats [48], whereas immunohistochemical stainings have ruled out an active infection by herpesvirus, calicivirus, or papillomavirus [49]. Initial histopathological description of PNOE lesions suggested dyskeratotic keratinocytes as the main feature of the disease. However, Videmont et al. [50] demonstrated keratinocyte apoptosis (positive cleaved caspase 3) induced by infiltrating CD3-positive T cells into the epidermis in the PNOE lesions. In conclusion, it is proposed that feline PNOE shares features with erythema multiforme and entails the T cell-mediated pathogenesis directed against keratinocytes.

Clinical Features

Initial PNOE descriptions involved kittens between 2 and 6 months of age, but it is now recognized that PNOE can affect cats up to 5 years of age [48–52]. The disease is characterized by well-demarcated erythematous plaques with adherent, thick, sometimes dark brown keratinous debris (Fig. 8a, b). The skin lesions are frequently bilaterally symmetrical, with the medial aspect of the pinna and the entrance to the auditory canal most commonly affected. As the lesions progress, erosion and ulceration occur. Lesions occasionally affect the preauricular region in some cats and often extend into the ear canal, where secondary bacterial or *Malassezia* infections are common [48–52].

Diagnosis

The history and clinical presentation are generally very striking, and definitive diagnosis is confirmed by histopathology, which reveals severe hyperplasia of the epidermis (Fig. 8c, d) and outer root sheath of hair follicles with scattered shrunken hypereosinophilic keratinocytes with pyknotic nuclei (apoptotic cells) (Fig. 8e, f). The dermis contains mixed inflammatory infiltrates (plasmacytic, neutrophilic or eosinophilic, and mastocytic) that varies between cases.

Clinical Management

The reported treatment options for PNOE are limited; spontaneous regression after 12–24 months was initially reported in some kittens [48], while other reports suggest that spontaneous regression does not occur in all cases [49]. Topical and systemic

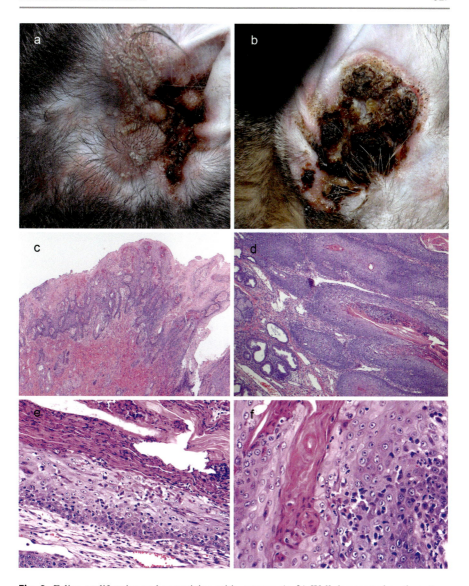

Fig. 8 Feline proliferative and necrotizing otitis externa. (**a, b**) Well-demarcated erythematous plaques with adherent, thick dark to brown keratinous debris are present in the medial aspect of the pinna, the entrance to the auditory canal and preauricular region of the face. (**c, d**) Severe epidermal hyperplasia and intense superficial dermatitis, with striking layered parakeratotic hyperkeratosis intermingled with neutrophils, extend into the hair follicle infundibulum. (**e, f**) Scattered apoptotic-appearing keratinocytes surrounded in some areas by lymphocytes are present within severely hyperplastic epidermis and superficial follicular epithelium

Fig. 9 Feline proliferative and necrotizing otitis externa before (**a**) and after treatment (**b**) with topical tacrolimus twice daily for 30 days

glucocorticoids show variable responses from partial improvement to full remission of this disease. This variable response may be a consequence of the different type of systemic glucocorticoid administration as well as the selection of the potency of topical steroid therapy [49, 52]. Three reports favored topical 0.1% tacrolimus ointment applied twice daily on the PNOE skin lesions to achieve complete remission (Fig. 9) [49–51].

References

1. Hebra von F. Acute exantheme und hautkrankheiten, Handbuch der Speciellen Pathologie und Therapie. Erlangen: Verlag von Ferdinand von Enke; 1860. p. 198–200.
2. Bastuji-Garin S, Rzany B, Stern RS, et al. Clinical classification of cases of toxic epidermal necrolysis, Stevens-Johnson syndrome and erythema multiforme. Arch Dermatol. 1993;129:92–6.
3. Hinn AC, Olivry T, Luther PB, et al. Erythema multiforme, Stevens-Johnson syndrome and toxic epidermal necrolysis in the dog: clinical classification, drug exposure and histopathological correlations. J Vet Allergy Clin Immunol. 1998;6:13–20.
4. Sokumbi O, Wetter DA. Clinical features, diagnosis, and treatment of erythema multiforme: a review for the practicing dermatologist. Int J Dermatol. 2012;51:889–902.
5. Kempton J, Wright JM, Kerins C, et al. Misdiagnosis of erythema multiforme: a literature review and case report. Pediatr Dent. 2012;34:337–42.
6. Scott DW, Walton DK, Slater MR, et al. Immune-mediated dermatoses in domestic animals: ten years after – Part II. Compend Contin Educ Pract Vet. 1987;9:539–51.
7. Olivry T, Guaguere E, Atlee B, et al. Generalized erythema multiforme with systemic involvement in two cats. Proceeding of the 7th annual meeting of the ESVD. Stockholm, Sweden; 1990.
8. Affolter VK, von Tscharner C. Cutaneous drug reactions: a retrospective study of histopathological changes and their correlation with the clinical disease. Vet Dermatol. 1993;4:79–86.
9. Noli C, Koeman JP, Willemse T. A retrospective evaluation of adverse reactions to trimethoprim-sulfonamide combinations in dogs and cats. Vet Q. 1995;17:123–8.

10. Scott DW, Miller WH. Erythema multiforme in dogs and cats: literature review and case material from the Cornell University College of veterinary medicine (1988–1996). Vet Dermatol. 1999;10:297–309.
11. Byrne KP, Giger U. Use of human immunoglobulin for treatment of severe erythema multiforme in a cat. J Am Vet Med Assoc. 2002;220:197–201.
12. Prost C. A case of exfoliative erythema multiforme associated with herpes virus 1 infection in a European cat. Vet Dermatol. 2004;15(Suppl. 1):51.
13. Aurelian L, Ono F, Burnett J. Herpes simplex virus (HSV)-associated erythema multiforme (HAEM): a viral disease with an autoimmune component. Dermatol Online J. 2003;9:1.
14. Sartori R, Colombo S. Stevens-Johnson syndrome/toxic epidermal necrolysis caused by cefadroxil in a cat. JFMS Open Rep. 2016;6:1–6.
15. Banovic F, Olivry T, Bazzle L, et al. Clinical and microscopic characteristics of canine toxic epidermal necrolysis. Vet Pathol. 2015;52:321–30.
16. Lee JA, Budgin JB, Mauldin EA. Acute necrotizing dermatitis and septicemia after application of a d-limonene based insecticidal shampoo in a cat. J Am Vet Med Assoc. 2002;221:258–62.
17. Scott DW, Halliwell REW, Goldschmidt MH, et al. Toxic epidermal necrolysis in two dogs and a cat. J Am Anim Hosp Assoc. 1979;15:271–9.
18. Scott DW, Miller WH. Idiosyncratic cutaneous adverse reactions in the cat: literature review and report of 14 cases (1990–1996). Feline Pract. 1998;26:10–5.
19. Sassolas B, Haddad C, Mockenhaupt M, et al. ALDEN, an algorithm for assessment of drug causality in Stevens-Johnson syndrome and toxic epidermal necrolysis: comparison with case-control analysis. Clin Pharmacol Ther. 2010;88:60–8.
20. Chung WH, Hung SI, Yang JY, et al. Granulysin is a key mediator for disseminated keratinocyte death in Stevens-Johnson syndrome and toxic epidermal necrolysis. Nat Med. 2008;14:1343–50.
21. Ng QX, De Deyn MLZQ, Venkatanarayanan N, Ho CYX, Yeo WS. A meta-analysis of cyclosporine treatment for Stevens-Johnson syndrome/toxic epidermal necrolysis. J Inflamm Res. 2018;11:135–42.
22. Alzolibani AA, Kang H, Otberg N, Shapiro J. Pseudopelade of Brocq. Dermatol Ther. 2008;21(4):257–63.
23. Gross TL, et al. Mural diseases of the hair follicle. In: Skin diseases of the dog and cat, clinical and histopathologic diagnosis. Ames: Blackwell Science; 2005a. p. 460–79.
24. Olivry T, Power HT, Woo JC, et al. Anti-isthmus autoimmunity in a novel feline acquired alopecia resembling pseudopelade of humans. Vet Dermatol. 2000;11:261–70.
25. Scott DW. Feline dermatology 1979–1982: introspective retrospections. J Am Anim Hosp Assoc. 1984;20:537.
26. Bunge M, et al. Relapsing polychondritis in a cat. J Am Anim Hosp Assoc. 1992;28:203.
27. Lemmens P, Schrauwen E. Feline relapsing polychondritis: a case report. Vlaams Diergeneeskd Tijdschr. 1993;62:183.
28. Boord MJ, Griffin CE. Aural chondritis or polychondritis dessicans in a dog. Proc Acad Vet Dermatol Am Coll Vet Dermatol. 1998;14:65.
29. Delmage D, Kelly D. Auricular chondritis in a cat. J Small Anim Pract. 2001;42(10):499–501.
30. Gerber B, Crottaz M, von Tscharner C, et al. Feline relapsing polychondritis: two cases and a review of the literature. J Feline Med Surg. 2002;4(4):189–94.
31. Griffin C, Trimmer A. Two unusual cases of auricular cartilage disease. Proceedings of the North American veterinary dermatology forum. Palm Springs; 2006.
32. Baba T, Shimizu A, Ohmuro T, Uchida N, Shibata K, Nagata M, Shirota K. Auricular chondritis associated with systemic joint and cartilage inflammation in a cat. J Vet Med Sci. 2009;71:79–82.
33. Kingdon J, Roscamp J, Sangle S, D'Cruz D. Relapsing polychondritis: a clinical review for rheumatologists. Rheumatology (Oxford). 2018;57:1525–32.
34. Stabler T, Piette J-C, Chevalier X, et al. Serum cytokine profiles in relapsing polychondritis suggest monocyte/macrophage activation. Arthritis Rheum. 2004;50:3663–7.

35. Gruffydd-Jones TJ, Orr CM, Lucke VM. Foot pad swelling and ulceration in cats: a report of five cases. J Small Anim Pract. 1980;21:381–9.
36. Scott DW. Feline dermatology 1983–1985: "the secret sits". J Am Anim Hosp Assoc. 1987;23:255.
37. Taylor JE, Schmeitzel LP. Plasma cell pododermatitis with chronic footpad hemorrhage in two cats. J Am Vet Med Assoc. 1990;197:375–7.
38. Guaguere E, Hubert B, Delabre C. Feline pododermatitis. Vet Dermatol. 1992;3:1–12.
39. Yamamura Y. A surgically treated case of feline plasma cell pododermatitis. J Jpn Vet Med Assoc. 1998;51:669–71.
40. Dias Pereira P, Faustino AM. Feline plasma cell pododermatitis: a study of 8 cases. Vet Dermatol. 2003;14:333–7.
41. Guaguere E, et al. Feline plasma cell pododermatitis: a retrospective study of 26 cases. Vet Dermatol. 2004;15:27.
42. Scarampella F, Ordeix L. Doxycycline therapy in 10 cases of feline plasma cell pododermatitis: clinical, haematological and serological evaluations. Vet Dermatol. 2004;15:27.
43. Bettenay SV, Mueller RS, Dow K, et al. Prospective study of the treatment of feline plasmacytic pododermatitis with doxycycline. Vet Rec. 2003;152:564–6.
44. De Man M. What is your diagnosis? Plasma cell pododermatitis and plasma cell dermatitis of the nose apex in cat. J Feline Med Surg. 2003;5:245–7.
45. Westfall DS, Twedt DC, Steyn PF, et al. Evaluation of esophageal transit of tablets and capsules in 30 cats. J Vet Intern Med. 2001;15:467–70.
46. Melendez LD, Twedt DC, Wright M. Suspected doxycycline-induced esophagitis and esophageal stricture formation in three cats. Feline Pract. 2000;28:10–2.
47. German AJ, Cannon MJ, Dye C. Oesophageal strictures in cats associated with doxycycline therapy. J Feline Med Surg. 2005;7:33–41.
48. Gross TL, et al. Necrotizing diseases of the epidermis. In: Skin diseases of the dog and cat, clinical and histopathologic diagnosis. Ames: Blackwell Science; 2005b. p. 75–104.
49. Mauldin EA, Ness TA, Goldschmidt MH. Proliferative and necrotizing otitis externa in four cats. Vet Dermatol. 2007;18(5):370–7.
50. Videmont E, Pin D. Proliferative and necrotising otitis in a kitten: first demonstration of T-cell-mediated apoptosis. J Small Anim Pract. 2010;51(11):599–603.
51. Borio S, Massari F, Abramo F, Colombo S. Proliferative and necrotising otitis externa in a cat without pinnal involvement: video-otoscopic features. J Feline Med Surg. 2013;15:353–6.
52. Momota Y, Yasuda J, Ikezawa M, Sasaki J, Katayama M, Tani K, Miyabe M, Onozawa E, et al. Proliferative and necrotizing otitis externa in a kitten: successful treatment with intralesional and topical corticosteroid therapy. J Vet Med Sci. 2017;10:1883–5.

Hormonal and Metabolic Diseases

Vet Dominique Heripret and Hans S. Kooistra

Abstract

Endocrine and metabolic disorders may result in changes in the skin and hair coat. With regard to thyroid disorders, hyperthyroidism is the most common endocrinopathy in cats, whereas hypothyroidism is rare in this species. Diabetes mellitus is also a common feline endocrinopathy. Disorders of the adrenal cortex associated with changes of the skin and hair coat also occur in cats and not only are confined to hypersecretion of cortisol but also include hypersecretion of sex steroids. In addition to the dermatological changes observed in these endocrine disorders, the dermatological changes associated with metabolic disorders, i.e. superficial necrolytic dermatitis, xanthomatosis and acquired cutaneous fragility syndrome, will be presented.

Introduction

The skin and adnexa are influenced by an array of hormones. Consequently, changes in the skin and hair coat may be manifestations of endocrine and metabolic disorders. Hormonal and metabolic dermatoses are not as common in cats as in dogs, which may be explained by the lower incidence in this species of endocrine disorders

V. D. Heripret (✉)
CHV Fregis, Arcueil, France

CHV Pommery, Reims, France
e-mail: dheripret@fregis.com

H. S. Kooistra
Department of Clinical Sciences of Companion Animals, Faculty of Veterinary Medicine, Utrecht University, Utrecht, The Netherlands
e-mail: H.S.Kooistra@uu.nl

© Springer Nature Switzerland AG 2020
C. Noli, S. Colombo (eds.), *Feline Dermatology*,
https://doi.org/10.1007/978-3-030-29836-4_27

frequently associated with skin and coat changes. The two most common endocrine disorders in cats are hyperthyroidism and diabetes mellitus, but the dermatological changes that may be observed in these conditions are rather non-specific.

Thyroid

Juvenile-Onset Hypothyroidism

Congenital hypothyroidism is quite rare in cats, but a few case reports have been published [1–3]. Congenital hypothyroidism may be due to thyroid dysgenesis or a defect in thyroid hormone synthesis. With regard to the latter, only cats with a so-called organification defect, i.e. a problem in the synthesis of thyroid hormones such as due to defective thyroid peroxidase activity, have been reported thus far [4]. The clinical features of hypothyroidism due to an organification defect do not differ from those in thyroid dysgenesis. Affected kittens are presented with disproportionate dwarfism, little physical activity and a dry and dull hair coat without overt alopecia [5]. Mental development appears to be retarded. Deciduous teeth persist into adulthood, but are shed when treatment with thyroid hormone is given [2, 6]. In cats with hypothyroidism due to an organification defect, neck palpation may reveal hyperplastic thyroids (goitre).

Other, very rare, causes of acquired juvenile-onset hypothyroidism are lymphocytic thyroiditis and iodine deficiency (see later). Lymphocytic thyroiditis has been reported in a breeding line in a closed colony of cats, with symptoms such as lethargy and a dull hair coat already present at the age of 7 weeks [7].

Acquired Adult-Onset Hypothyroidism

Iodine deficiency is the classic cause of acquired hypothyroidism. It occurred in times when owners took too literally the notion that cats are carnivores. A diet consisting of meat alone is deficient in many respects and certainly in iodine. The lack of this essential ingredient of the thyroid hormones may result in TSH-induced thyroid hyperplasia. Animals with severe iodine deficiency are presented with the combination of goitre and signs of hypothyroidism such as lethargy. This entity is no longer seen in countries in which it is customary to feed manufactured diets, which are rather rich in iodine.

Spontaneous adult-onset hypothyroidism is rather rare in cats, but a recent study suggests that the prevalence may be higher than previously thought [8]. Hair coat changes, lethargy and obesity are common clinical signs of cats with spontaneous adult-onset hypothyroidism. Interestingly, several of the cats with acquired hypothyroidism developed a goitrous form of hypothyroidism associated with thyroid hyperplasia.

Acquired hypothyroidism can also be iatrogenic, especially in cats treated for hyperthyroidism, which occurs frequently in this species. Iatrogenic hypothyroidism may be an adverse effect of radioiodine therapy [9], due to bilateral surgical thyroidectomy or due to an overdose of anti-thyroid drugs. Dermatological signs are quite non-specific with decreased grooming, dorsal matting and poor hair coat condition (Fig. 1).

Fig. 1 Acquired hypothyroidism in a cat: unkempt hair coat and dry seborrhoea. (Courtesy of Dr. G. Zanna)

Fig. 2 Hyperthyroidism in a cat: diffuse alopecia, dry seborrhoea and dry skin

Cats with hypothyroidism have an elevated circulating TSH concentration. Clinical signs rapidly respond to L-thyroxine replacement therapy.

Hyperthyroidism

Feline hyperthyroidism is a relatively common disease of middle-aged and elderly cats, with a mean age of 12–13 years. The thyroid hormone excess is produced by thyroid adenomatous hyperplasia or adenoma, involving one or, more often, both thyroid lobes [10]. The main clinical signs are linked to an acceleration of metabolism (weight loss, polyphagia, polyuria and gastrointestinal problems). Dermatological signs occur in approximately 30% of cases and are rather non-specific [11]. Reported dermatological signs include excessive shedding, scales, focal alopecia due to excessive grooming, matting and excessive claw growth (Fig. 2). In very

Fig. 3 Diabetes mellitus: necrosis of the skin after placement of IV catheter

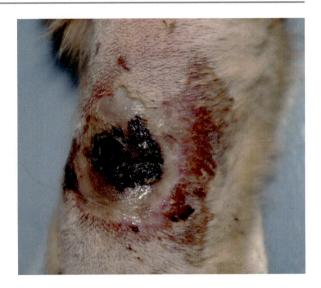

chronic cases, complete truncal alopecia has been described. Diagnosis is based on history, physical examination and thyroxine (T_4) measurement. Treatment includes surgery, anti-thyroid drugs, radioiodine therapy or an iodine-restricted food.

Diabetes Mellitus

Skin lesions associated with feline diabetes mellitus have rarely been described, and some lesions initially associated with diabetes, like skin atrophy, may, in fact, be due to an underlying Cushing's syndrome. Dry seborrhoea with hair matting and diffuse alopecia can be seen because of a poor general condition and decreased grooming at the moment of the diagnosis and may be due to abnormal lipid and protein metabolism. Vascular abnormalities are rare in cats, but one of the authors saw one cat with a necrotizing reaction to minor skin trauma (catheter for fluid therapy and site of fixation of nasal feeding tube) (Fig. 3). Xanthomatosis can be associated with diabetes mellitus (see later) (Fig. 4).

Adrenal Glands

The adrenal glands consist of two functionally distinct endocrine glands, the cortex and the medulla. The medulla secretes epinephrine and norepinephrine, and the cortex secretes mineralocorticoids, glucocorticoids and sex hormones. The main adrenal disorders in cats are primary hyperaldosteronism and hypercortisolism. Only the latter one is associated with dermatological changes. Hypersecretion of adrenal sex hormones may also result in changes of the skin and hair coat (see later).

Fig. 4 Xanthoma associated with diabetes mellitus. (Courtesy of Dr. Guaguère)

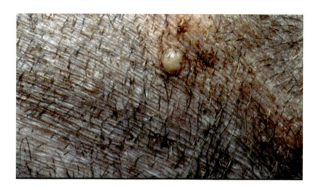

Cortisol is the principal glucocorticoid released by the adrenals in cats, indicating that spontaneous glucocorticoid excess is essentially hypercortisolism. Prolonged exposure to inappropriately elevated plasma cortisol concentrations leads to signs often referred to as Cushing's syndrome, after Harvey Cushing, who in 1932 first described the syndrome in man. Identical signs are elicited by exogenous glucocorticoids in long-term therapy, i.e. iatrogenic hypercorticism.

Spontaneous Hypercortisolism

Spontaneous hypercortisolism in cats is a disease of middle-aged and older animals. There is no pronounced sex predilection, although in reported cases female cats are slightly overrepresented [12–14]. In 80–90% of cats with spontaneous hypercortisolism, the disease is the result of excessive ACTH secretion by a pituitary adenoma. In the remaining cases, the disease is ACTH-independent, due to autonomous hypersecretion of cortisol by an adrenocortical tumour, either adenoma or, more often, carcinoma.

Many of the signs can be related to the actions of glucocorticoids, namely, increased gluconeogenesis and lipogenesis at the expense of protein. Cardinal physical features are central obesity and atrophy of muscles and skin. The cutaneous manifestations in cats with spontaneous hypercortisolism may initially give the impression of being less pronounced than in dogs. However, long-term exposure to glucocorticoid excess will result in dermatological signs, including thin skin, alopecia and dull or seborrhoeic skin (Fig. 5). In some cases the skin becomes so fragile that it tears during routine handling, leaving the cat with a full-thickness skin defect (Fig. 6) [15]. These skin lacerations are part of the acquired cutaneous fragility syndrome (see later). Infections of the skin and nail beds and the urinary, respiratory and gastrointestinal tract, secondary to cortisol-induced immune suppression, are also common [12]. Glucocorticoid excess in cats results in polyuria/polydipsia much less readily than in dogs and may only become obvious when diabetes mellitus develops. Cats are more susceptible than dogs to the diabetogenic effects of glucocorticoids, and diabetes mellitus has been present in most of the reported cases of hypercortisolism in cats. Suspicion of hypercortisolism has often

Fig. 5 Pituitary-dependent
hypercortisolism in a cat:
the skin is very thin

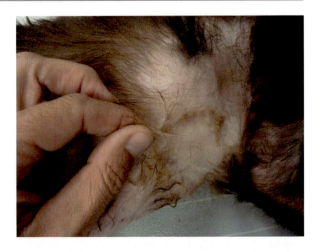

Fig. 6 Same cat as in
Fig. 5: skin lacerations

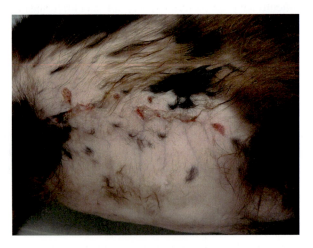

arisen specifically because of insulin resistance encountered in the treatment of
diabetes mellitus. Clinicopathological parameters other than hyperglycaemia are
mostly unremarkable. An elevation of plasma alkaline phosphatase (AP) activity
is a consistent finding in dogs with hypercortisolism. In dogs this is mainly due to
the induction of an isoenzyme having greater stability at 65 °C than other AP isoen-
zymes. In cats, glucocorticoids do not induce this isoenzyme. In addition, AP has a
very short half-life in cats.

Endocrine tests used for the diagnosis of hypercortisolism in cats include the
ACTH stimulation test, the low-dose dexamethasone suppression test (LDDST) and
determination of the urinary corticoid-to-creatinine ratio (UCCR). Because of its
low sensitivity, the ACTH stimulation test is not recommended as initial diagnostic test in cats suspected of spontaneous hypercortisolism. The diabetes mellitus

associated with hypercortisolism in cats may result in false-positive UCCRs, meaning that the main indication for the UCCR in this species is to exclude hypercortisolism. The LDDST is the test with the highest diagnostic accuracy to screen for spontaneous hypercortisolism in cats. In cats the LDDST is usually performed using 0.1 mg/kg of dexamethasone (IV) which is different from dogs, because more than 20% of healthy cats do not suppress with the dose of 0.01 mg/kg used in dogs [16]. In contrast, some cats with pituitary-dependent hypercortisolism are very sensitive to dexamethasone suppression, which may result in false-negative results [17].

When the clinical signs and the LDDST indicate hypercortisolism, the next diagnostic step is to differentiate between pituitary-dependent hypercortisolism and a cortisol-secreting adrenocortical tumour. A suppression of the circulating cortisol concentration in the LDDST of more than 50% indicates pituitary-dependent hypercortisolism. Hypercortisolism due to an adrenocortical tumour can be differentiated from non-suppressible forms of pituitary-dependent hypercortisolism by measuring the plasma ACTH concentration. In addition, an adrenocortical tumour is often readily detected by ultrasonography. The preferred procedures for visualization of the adrenals and the pituitary are magnetic resonance imaging (MRI) and computed tomography (CT). Ultrasonography is less expensive, requires less time and does not require anaesthesia, and so it is often used first even though it is more difficult to perform and to interpret than CT or MRI. It provides a good estimate of the size of the tumour and may reveal information about its expansion.

The goals of treating hypercortisolism would optimally be to eliminate the source of either ACTH or autonomous cortisol excess, to achieve normocortisolism, to eliminate the clinical signs, to reduce long-term complications and mortality and to improve the quality of life. Radiotherapy [18] and surgical removal of the causal tumour, either hypophysectomy [19, 20] or adrenalectomy, are currently the only treatment options that have the potential to eliminate the source of either ACTH or autonomous cortisol excess.

Pharmacotherapy is a commonly used treatment in feline hypercortisolism that aims to eliminate the clinical signs of the condition, and the drug used most often is trilostane. Although trilostane is the medical treatment of choice based on retrospective studies, investigations into the pharmacokinetics of this drug in cats are lacking [12, 14, 21]. Trilostane is a synthetic steroid analogue that competitively inhibits the steroidogenic enzyme 3β-hydroxysteroid dehydrogenase, which is required for the production of all classes of adrenocortical hormones. Trilostane therefore inhibits the production of both cortisol and aldosterone. Because administration with food significantly increases the rate and extent of absorption, trilostane should always be given with food. When the optimal dose of trilostane is given, polyuria/polydipsia will decrease in a couple of weeks, and dermatological changes disappear after 2 weeks to 3 months [12]. In some cats the diabetes mellitus may go into remission [22].

There is a marked variation in the optimal trilostane dose, and the current recommendation is to start with much lower dosages than originally recommended by the manufacturer, which can be equally effective but induce fewer adverse effects than higher dosages. Because the duration of cortisol suppression is less than 12 h, administrating trilostane twice daily can improve the clinical response while keeping the total daily dose relatively low and significantly reducing adverse effects. The current advice is to start with an initial dose of 1–2 mg/kg once or twice daily. Trilostane is usually well tolerated, but the main adverse effect that can occur is transient hypocortisolism, possibly combined with or followed by complete hypoadrenocorticism.

For successful management of hypercortisolism with trilostane, frequent monitoring is essential. In the last decade, efforts have been made to identify the best method to monitor trilostane therapy. In all methods, evaluation of the clinical signs is the first step. The preferred monitoring method is the use of the ACTH stimulation test, which monitors the adrenal glands' reserve capacity to secrete cortisol. The timing of the ACTH stimulation test is crucial since this influences the results, and the recommendation is to coincide the test with the maximal trilostane action (2–3 h after trilostane administration). Despite its widespread use, the ACTH stimulation test has never been validated as a monitoring tool for trilostane therapy, and there are some concerns regarding the variation in results depending on the timing of the test and whether this reflects clinical control. Moreover, synthetic ACTH is not easily available in all countries. A recently proposed alternative method is to measure the pre-pill cortisol concentration and compare it to the clinical signs reported by owners.

Iatrogenic Hypercorticism and Iatrogenic Secondary Hypocortisolism

As in spontaneous hypercortisolism, the development of signs of glucocorticoid excess due to glucocorticoid or progestagen administration depends on the severity and duration of the exposure. The effects vary among individual animals and initially seem to be less pronounced in cats. After several weeks of glucocorticoid therapy, the classic physical changes such as centripetal obesity, muscular weakness and skin atrophy may develop. In a study of 12 cats with iatrogenic hypercorticism [23], hypotrichosis (localized or generalized) was present in 100% of cases (Fig. 7) and skin tears in 16% (Fig. 8). The mean time for clinical improvement of the skin lesions was 4.5 months (1–12 months) after withdrawal of corticosteroids.

Both systemic and topically applied corticosteroids cause prompt and sustained suppression of the hypothalamic-pituitary-adrenocortical axis. Depending on the dose, the continuity, the duration and the preparation or formulation, this suppression may continue for weeks or months after cessation of corticosteroid administration. The affinity of the glucocorticoid receptor for progestagens may cause a similar long-lasting suppression of the pituitary-adrenocortical system in cats [24].

Fig. 7 Iatrogenic hypercorticism in a cat: the hair coat is thinner

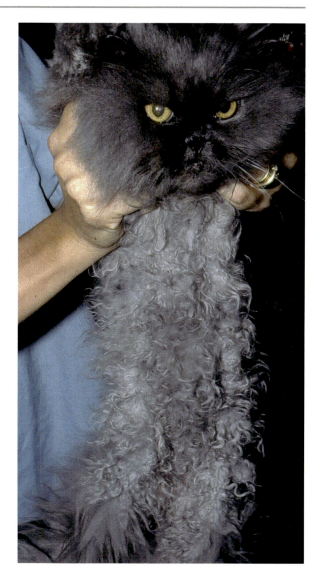

Discontinuance of corticosteroid therapy may therefore result in signs of cortico-steroid withdrawal syndrome, i.e. the cat may develop secondary adrenocortical insufficiency. The cardinal features of the corticosteroid withdrawal syndrome are anorexia, lethargy and weight loss. The dose should therefore be reduced gradually, as in the transition from spontaneous hypercortisolism to normocorticism, in which initially at least twice the maintenance dose is given.

Fig. 8 Large skin tears due to feline fragility syndrome

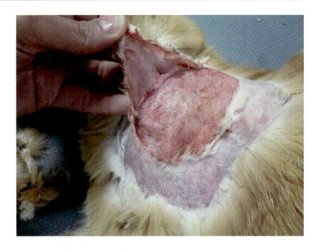

Fig. 9 Alopecia after subcutaneous injection of medroxyprogesterone acetate

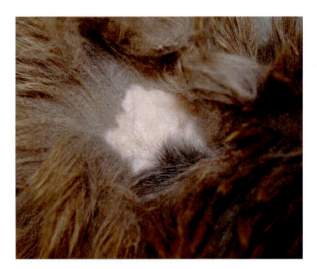

Sex Hormones

Dermatological signs associated with a spontaneous testicular or ovarian sex hormone disturbance have never been described in cats, either males or females.

Administration of progestagens, used for oestrus prevention but also used in the treatment of various dermatologic and behavioural disorders, may result in changes of the skin and hair coat of cats. Focal alopecia on the site of injection has been described after administration of medroxyprogesterone acetate (Fig. 9), and oral administration of either medroxyprogesterone acetate or megestrol acetate may result in cutaneous atrophy or hypomelanosis. These changes are most likely due to the intrinsic glucocorticoid activity of the progestogens [24, 25].

Hyperproduction of sex hormones associated with an adrenocortical tumour has been reported [26–29], with clinical signs similar to those of hypercortisolism, such as poor regulation of diabetes mellitus, thin and hypotonic skin and skin fragility. A sex steroid hormone-secreting adrenocortical tumour should be considered in neutered cats with newly developed physical and behavioural sexual changes such as urine spraying and aggression in neutered male cats. The castrated male cat develops spines on the penis, whereas the castrated female may develop hyperplasia of the vulva. Endocrine testing may reveal elevated plasma concentrations of androstenedione, testosterone, oestradiol, 17-hydroxy-progesterone and/or progesterone, and these values may increase following stimulation with ACTH. Information about the size of the tumour, its expansion and the presence of metastases can be obtained by ultrasonography, CT or MRI. Adrenalectomy is the treatment of choice and usually results in resolution of clinical manifestations.

Feline tail gland hyperplasia or "stud tail" has been described in intact male cats and has been previously linked to hyperandrogenism; however, castration does not resolve the condition, and feline tail gland hyperplasia is also reported in castrated males and females.

Metabolic Disorders

Superficial Necrolytic Dermatitis (Chapter, Paraneoplastic Syndromes)

This disease is also called hepatocutaneous syndrome or metabolic epidermal necrosis or necrolytic migratory erythema. Only a few cases have been described in cats either with hepatopathy [30] or a glucagon-producing tumour [31]. Dermatological signs are characterized by alopecia, erythema, erosions and crusts. Pruritus has been reported [30], as well as pain due to paw pad lesions [31]. On histopathological examination, the classical appearance of the epidermis as a "French flag" (blue hyperplasia of the basal layers, white pallor of the spinous layer and red parakeratosis of the horny layer) may also be found in cats with this syndrome.

Cutaneous Xanthoma (Xanthomatosis)

Xanthomas are cutaneous or subcutaneous yellowish papular lesions associated with accumulation of lipids and a granulomatous reaction (Fig. 4). In the cat, xanthomatosis can be linked to familial hyperlipoproteinaemia and diabetes mellitus or can be idiopathic [32]. One case was described without any lipaemic abnormality [33]. Dermatological signs are characterized by grey to yellow papules, plaques or nodules, resembling candle wax. The surrounding skin may be erythematous and lesions may be pruritic or not and sometimes painful. The distal extremities are frequently involved, but lesions may develop everywhere on the body.

The diagnosis is based on the appearance of lesions and their histopathological examination, characterized by foamy histiocytes and multinucleate giant cells

(Touton cells). In case that an underlying cause is recognized, lesions resolve by treating this underlying disease. In idiopathic cases, feeding a low-fat diet may result in remission of the lesions [33].

Acquired Cutaneous Fragility Syndrome (ACFS)

ACFS is a rare dermatological condition, characterized by skin thinning leading to skin fragility and spontaneous non-haemorrhagic and non-painful tearing. Underlying causes include, as previously mentioned, spontaneous and iatrogenic hypercorticism, but also hepatic lipidosis and neoplasia, while some cases remain idiopathic. When the pathogenesis is unknown, it is hypothesized that a severe metabolic disorder may negatively influence collagen metabolism.

Anecdotally there may be two clinical variants: one with a very thin epidermis (as seen in hepatic lipidosis) and one with complete skin atrophy (as seen in Cushing's syndrome); however further research is warranted to substantiate this differentiation. Clinical signs are characterized by skin thinning followed by spontaneous tearing on minor trauma (restraint, scratches, injections, etc.). Skin lacerations may extend dramatically (Fig. 10). The prognosis is very guarded and depends on

Fig. 10 ACFS in a cat with multiple lacerations. This cat was treated monthly with 20 mg methylprednisolone acetate injections for 3 years, because of severe stomatitis. (Courtesy of Dr. Chiara Noli)

Fig. 11 Same cat as
Fig. 10 after 2 weeks of
wound management, with
visible improvements

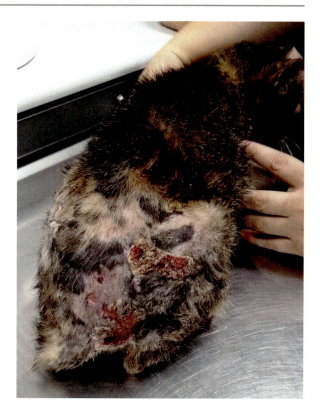

the underlying cause and the troublesome healing of the wounds (Fig. 11); staged
wound closure with a combination of daily wound cleaning and debridement and
tension and appositional sutures may be required [34].

References

1. Diehm M, Dening R, Dziallas P, Wohlsein P, Schmicke M, Mischke R. Bilateral femoral
   capital physeal fractures in an adult cat with suspected congenital primary hypothyroidism.
   Tierarztl Prax Ausg K Kleintiere Heimtiere. 2019;47:48–54.
2. Jacobson T, Rochette J. Congenital feline hypothyroidism with partially erupted adult dentition in a 10-month-old male neutered domestic shorthair cat: a case report. J Vet Dent.
   2018;35:178–86.
3. Lim CK, Rosa CT, de Witt Y, Schoeman JP. Congenital hypothyroidism and concurrent renal
   insufficiency in a kitten. J S Afr Vet Assoc. 2014;85:1144.
4. Gruffydd JBR, TJ J, Sparkes AH, Lucke VM. Preliminary studies on congenital hypothyroidism in a family of Abyssinian cats. Vet Rec. 1992;131:145–8.
5. Bojanick K, Acke E, Jones BR. Congenital hypothyroidism of dogs and cats: a review. N Z Vet
   J. 2011;59:115–22.
6. Crowe A. Congenital hypothyroidism in a cat. Can Vet J. 2004;45:168–70.
7. Schumm-Draeger PM, Länger F, Caspar G, Rippegather K, Hermann G, Fortmeyer HP, Usadel
   KH, Hübner K. Spontane Hashimoto-artige Thyreoiditis im Modell der Katze (Spontaneous
   Hashimoto-like thyroiditis in cats). Verh Dtsch Ges Pathol. 1996;80:297–301.

8. Peterson ME, Carothers MA, Gamble DA, Rishniw M. Spontaneous primary hypothyroidism in 7 adult cats. J Vet Intern Med. 2018;32:1864–73.

9. Peterson ME, Nichols R, Rishnow M. Serum thyroxine and thyroid-stimulating hormone concentration in hyperthyroid cats that develop azotaemia after radioiodine therapy. J Small Anim Pract. 2017;58:519–30.

10. Peterson ME. Animal models of disease: feline hyperthyroidism: an animal model for toxic nodular goiter. J Endocrinol. 2014;223:97–114.

11. Thoday KL, Mooney CT. Historical, clinical and laboratory features of 126 hyperthyroid cats. Vet Rec. 1992;131:257–64.

12. Boland LA, Barrs VR. Peculiarities of feline hyperadrenocorticism: update on diagnosis and treatment. J Feline Med Surg. 2017;19:933–47.

13. Chiaramonte D, Greco DS. Feline adrenal disorders. Clin Tech Small Anim Pract. 2007;22:26–31.

14. Valentin SY, Cortright CC, Nelson RW, et al. Clinical findings, diagnostic test results, and treatment outcome in cats with spontaneous hyperadrenocorticism: 30 cases. J Vet Intern Med. 2014;28:481–7.

15. Daley CA, Zerbe CA, Schick RO, Powers RD. Use of metyrapone to treat pituitary-dependent hyperadrenocorticism in a cat with large cutaneous wounds. J Am Vet Med Assoc. 1993;202:956–60.

16. Peterson ME, Graves TK. Effects of low dosages of intravenous dexamethasone on serum cortisol concentrations in the normal cats. Res Vet Sci. 1988;44:38–40.

17. Meij BP, Voorhout G, Van Den Ingh TS, Rijnberk A. Transsphenoidal hypophysectomy for treatment of pituitary-dependent hyperadrenocorticism in 7 cats. Vet Surg. 2001;30:72–86.

18. Mayer MN, Greco DS, LaRue SM. Outcomes of pituitary irradiation in cats. J Vet Intern Med. 2006;20:1151–4.

19. Meij BP. Hypophysectomy as a treatment for canine and feline Cushing's disease. Vet Clin North Am Small Anim Pract. 2001;31:1015–41.

20. Meij B, Voorhout G, Rijnberk A. Progress in transsphenoidal hypophysectomy for treatment of pituitary-dependent hyperadrenocorticism in dogs and cats. Mol Cell Endocrinol. 2002;197:89–96.

21. Mellet-Keith AM, Bruyette D, Stanley S. Trilostane therapy for treatment of spontaneous hyperadrenocorticism in cats: 15 cases (2004–2012). J Vet Intern Med. 2013;27:1471–7.

22. Muschner AC, Varela FV, Hazuchova K, Niessen SJ, Pöppl ÁG. Diabetes mellitus remission in a cat with pituitary-dependent hyperadrenocorticism after trilostane treatment. JFMS Open Rep. 2018;4:205511691876770. https://doi.org/10.1177/2055116918767708.

23. Lien YH, Huang HP, Chang PH. Iatrogenic hyperadrenocorticism in 12 cats. J Am Anim Hosp Assoc. 2006;42:414–23.

24. Middleton DJ, Watson ADJ, Howe CJ, Caterson ID. Suppression of cortisol responses to exogenous adrenocorticotrophic hormone, and the occurrence of side effects attributable to glucocorticoid excess, in cats during therapy with megestrol acetate and prednisolone. Can J Vet Res. 1987;51:60–5.

25. Selman PJ, Wolfswinkel J, Mol JA. Binding specificity of medroxyprogesterone acetate and proligestone for the progesterone and glucocorticoid receptor in the dog. Steroids. 1996;61:133–7.

26. Boag AK, Neiger R, Church DB. Trilostane treatment of bilateral adrenal enlargement and excessive sex steroid hormone production in a cat. J Small Anim Pract. 2004;45:263–6.

27. Boord M, Griffin C. Progesterone secreting adrenal mass in a cat with clinical signs of hyperadrenocorticism. J Am Vet Med Assoc. 1999;214:666–9.

28. Quante S, Sieber-Ruckstuhl N, Wilhelm S, Favrot C, Dennler M, Reusch C. Hyperprogesteronism due to bilateral adrenal carcinomas in a cat with diabetes mellitus. Schweiz Arch TierheilkdSchweiz Arch Tierheilkd. 2009;151:437–42.

29. Rossmeisi JH, Scott-Montcrieff JC, Siems J, et al. Hyperadrenocorticism and hyperprogesteronemia in a cat with an adrenocortical adenocarcinoma. J Am Anim Hosp Assoc. 2000;36:512–7.

30. Kimmel SE, Christiansen W, Byrne KP. Clinicopathological, ultrasonographic, and histopathological findings of superficial necrolytic dermatitis with hepatopathy in a cat. J Am Anim Hosp Assoc. 2003;39:23–7.
31. Asakawa MG, Cullen JM, Linder KE. Necrolytic migratory erythema associated with a glucagon-producing primary hepatic neuroendocrine carcinoma in a cat. Vet Dermatol. 2013;24:466–9.
32. Grieshaber TL. Spontaneous cutaneous (eruptive) xanthomatosis in two cats. J Am Anim Hosp Assoc. 1991;27:509.
33. Ravens PA, Vogelnest LJ, Piripi SA. Unique presentation of normolipæmic cutaneous xanthoma in a cat. Aust Vet J. 2013;91:460–3.
34. McKnight CN, Lewis LJ, Gamble DA. Management and closure of multiple large cutaneous lesions in a juvenile cat with severe acquired skin fragility syndrome secondary to iatrogenic hyperadrenocorticism. J Am Vet Med Assoc. 2018;252:210–4.

Genetic Diseases

Catherine Outerbridge

Abstract

Breed predispositions are documented for a variety of feline skin diseases, and there are diverse reports describing novel skin diseases where affected individuals within a litter had similar congenital skin changes. Both presentations increase the suspicion of a possible hereditary component to the identified skin disease. Recognized feline genodermatoses represent inherited skin disorders that follow a single-gene mode of inheritance (Leeb et al., Vet Dermatol 28:4–9. https://doi.org/10.1016/j.mcp.2012.04.004, 2017). These diseases are rare in their occurrence, but the number identified has the potential to increase, as available diagnostic tools for evaluating genetic diseases have advanced and single nucleotide polymorphism (SNP) mapping of the feline genome has improved (Lyons, Mol Cell Probes. 26:224–30. https://doi.org/10.1016/j.mcp.2012.04.004, 2012; Mullikin et al., BMC Genomics. 11:406. http://www.biomedcentral.com/1471-2164/11/406, 2010). This chapter will discuss some examples of feline genodermatoses that can affect the epidermis, the dermoepidermal junction, the hair follicles or hair shafts, the dermis, and pigmentation. Discussion about the genetics of feline coat color and coat length occurs in Chapter, Coat Color Genetics.

C. Outerbridge (✉)
University of California, Davis, Davis, CA, USA
e-mail: caouterbridge@vmth.ucdavis.edu

© Springer Nature Switzerland AG 2020
C. Noli, S. Colombo (eds.), *Feline Dermatology*,
https://doi.org/10.1007/978-3-030-29836-4_28

Genetic Cornification or Keratinization Disorders

Idiopathic Facial Dermatitis of Persian and Himalayan Cats

This progressive, idiopathic, facial dermatitis is presumed to be hereditary, as it is recognized to occur in young Persian and Himalayan cats [4, 5]. The etiology remains unknown, and affected cats develop a moderate to marked, adherent, very dark, and greasy debris, thought to be of sebaceous origin, that mats the hair coat in affected areas. This adherent debris has given rise to the descriptive term "dirty face disease." The periocular regions, nasal/facial folds, perioral region, chin, and muzzle are most often affected (Fig. 1). Affected cats can also present with bilateral ceruminous otitis externa [4]. In some cats, similar lesions involve other focal regions on the body, for example, the perivulvar region in a queen. The skin beneath the adherent material is typically inflamed. Cats are variably pruritic and can self-traumatize causing erosions and ulcers. Secondary bacterial infections and/ or *Malassezia* dermatitis occurs commonly and is likely a major contributor to the pruritus in some cats. Lesions often begin within the first year of life, although some individual cats are not presented for veterinary care until they are older. The lesions are progressive and with chronicity, pruritus can intensify.

Skin biopsies, when performed, should be taken from non-traumatized regions and any adherent debris should not be removed prior to biopsy. Lesional skin histologically shows severe acanthosis of the epidermis and follicular infundibulum, with varying spongiosis. There can be mild to marked neutrophilic and eosinophilic inflammation that can form epidermal pustules. Also evident in some biopsies are luminal folliculitis and scattered basal cell apoptosis or vacuolation. Mild to moderate parakeratotic hyperkeratosis with variable neutrophilic crusting is often described. There is some overlap between the histologic changes seen in this disease and changes seen in spongiotic allergic reactions [6].

Cats affected with this disorder are often frustrating to treat. Treatment involves identifying and appropriately treating secondary infections with systemic antibiotics

Fig. 1 A 3-year-old male castrated Persian cat at initial presentation with idiopathic facial dermatitis of the breed and secondary bacterial infection. (Courtesy of UC Davis Dermatology Service)

and antifungals and topical antimicrobial therapy. Antiseborrheic topical therapy can be helpful to improve clinical appearance in cats that tolerate topical therapy. Systemic anti-inflammatory therapy may be warranted, either judicious use of glucocorticoids or oral administration of cyclosporine at 5–7 mg/kg. Topical 0.1% tacrolimus has been used successfully (Fig. 2) [7]. Cats will relapse, in regard to severity of lesions and degree of pruritus, if therapy is not consistent or if secondary skin and/or ear infections fail to be identified and treated. Cats with allergic dermatitis can present with facially predominant pruritus, so if an affected idiopathic facial dermatitis cat has other clinical signs of allergic dermatitis, consider the possibility of that diagnosis contributing as a concurrent disease and manage appropriately.

Primary Seborrhea in Cats

Primary seborrhea is very rare in cats, but it has been described in Persian, Himalayan, and exotic shorthair cats [8, 9]. In Persian cats, an autosomal recessive mode of inheritance was recognized [8]. Primary seborrhea can be distinguished from the idiopathic facial dermatitis now recognized in the breed, as primary seborrheic cats show clinical signs at a younger age, often in the first few weeks of life, and have more generalized lesions. There are no reports of effective treatment for severely affected kittens, and those cases were often euthanized. Severity of clinical signs varies, and mildly affected cats need clipping to maintain shorthair coats and antiseborrheic topical therapy to manage their clinical signs.

Ulcerative Nasal Dermatitis of Bengal Cats

This rare condition affects the nasal planum in young Bengal cats. Although an underlying etiology has not been identified, it is presumed to have a hereditary component as cats are young and all reports are from one breed. The first reports described the lesion in young cats in Sweden [10], Italy, and the United Kingdom

Fig. 2 Same cat 5 months later after receiving prednisolone, antibiotics, and topical application of 0.1% tacrolimus for a month and maintained on novel protein diet. (Courtesy of UC Davis Dermatology Service)

Fig. 3 Bengal cat with adherent scale crusting typical of the lesion seen in ulcerative nasal dermatitis in the breed. (Courtesy of UC Davis Dermatology Service)

[11], but affected cats have also been seen in North America (Fig. 3). In one study, a significant reduction of the thickness of the stratum corneum compared to normal feline controls was found [11]. Lesions develop within the first year of life and initial lesions consist of mild scaling of the nasal planum. Gradually lesions progress and the nasal planum can develop thick, adherent crusts with hyperkeratosis that can fissure or, if the crust is lost, an erosive surface results. Cats do not appear to find the lesion pruritic or painful.

Lesions spontaneously resolved in one group of cats [11], which should be remembered when assessing clinical response to therapeutic interventions. Treatments tried include oral prednisolone, topical salicylic acid, topical hydrocortisone, topical emollients, and topical antibiotics with varying responses. Topical tacrolimus ointment was assessed to be the most effective therapy with marked improvement of the nasal planum lesions in four cats [10].

Genetic Diseases Affecting the Dermoepidermal Junction

Epidermolysis Bullosa

Epidermolysis bullosa (EB) is a group of rare genetic blistering skin diseases documented to occur in humans and a number of domestic species including the cat. Hallmark of the disease is extremely fragile skin and mucosae, which form blisters that progress to erosions and ulcers in response to frictional trauma [12]. Lesions develop in areas of the body that experience frictional pressure forces, such as the oral cavity and distal limbs. Epidermolysis bullosa occurs because of genetic mutations that alter proteins critical for maintaining the structural integrity of the dermoepidermal junction [12]. There are three major categories of EB in humans and two are recognized to occur in cats [12].

Feline Junctional Epidermolysis Bullosa
This form of EB has been described in domestic shorthair and Siamese cats [13, 14]. Kittens developed lesions in the first months of life and presented with intraoral

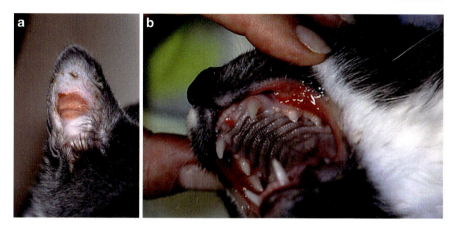

Fig. 4 JEB (**a**): superficial ulcer on the anterior aspect of the left ear pinna at initial presentation. Same cat (**b**): severe stomatitis at 6 months of age. (Reproduction from Ref. [14], with permission)

and labial ulcerations, pinnal erosions, onychomadesis, and ulcerations of footpads (Fig. 4). Histologic evaluation of lesional skin biopsies documented a subepidermal split, in which periodic acid-Schiff (PAS) staining confirmed that the lamina densa was attached to the base of the cleft, supporting a diagnosis of junctional EB [14]. Documentation of reduced staining of the γ-2 chain of laminin 5, now called laminin 332, was shown with indirect immunofluorescence studies [14]. An autosomal recessive mode of inheritance was considered likely as the dam and siblings of affected cats did not have any lesions [14].

Feline Dystrophic Epidermolysis Bullosa
This form of EB was reported in a domestic shorthair cat and a Persian cat [15, 16]. Affected cats developed at a young age intraoral ulcers involving the tongue, palate, and gingiva. Onychomadesis of all digits with paronychia, footpad ulcers, and, in the case of the Persian cat, ulcers along the dorsum [16] were additional clinical lesions recognized. Histologic examination of skin biopsies in both reported cases showed a dermoepidermal split, and immunohistochemical studies confirmed that the split occurred below collagen IV, a component of the lamina densa [15, 16]. Further studies in one cat confirmed that anchoring fibrils were reduced in number and collagen VII, which is a major component of anchoring fibrils, was also reduced [16]. It was proposed that a mutation in the collagen VII gene *COL7A1* was responsible for this form of hereditary EB in the cat [16].

Genetic Diseases Affecting the Hair Coat

Congenital Hypotrichosis

Congenital hypotrichosis is rare and characterized by the absence of a hair coat at birth or loss of hair coat in the first month of life. Hypotrichosis can be localized or generalized in its distribution. In some cases, other defects such as abnormalities of the claws, dentition, or lacrimal glands occur, which are characteristic for ectodermal dysplasia. On skin biopsy, hair follicles may be markedly reduced and small or absent. There are scattered reports in the veterinary literature of congenital hypotrichosis in the Birman [17], Siamese [18], Burmese [19], and Devon Rex [20] breeds. In these reports, affected kittens were born hairless (Fig. 5) or had a fine downy coat that was lost in the first few weeks of life. In the Siamese, hereditability was reported to be autosomal recessive. In some of these cat breeds, there have not been subsequent reports. For the Birman cat, the mode of inheritance, the genetic mutation, and the associated syndromic clinical signs have now been well characterized [21]. Congenital alopecia occurs when there are alterations in either the quantity or the quality of hair follicles and/or the integrity of the hair shafts they produce [22].

Congenital Hypotrichosis or Hair Coat Changes as Breed-Specific Traits (Chapter, Coat Color Genetics)

There are breeds of cats that are characterized by their congenital hypotrichosis. These include the Sphynx, Peterbald, Donskoy, and Kohana cats [23]. Sphynx cats are homozygous for an autosomal recessive hairless allele (*hr*) that results from a mutation in the gene *Keratin* 71 (KRT71) [23]. Keratin 71 is found in the inner root sheath in mice and humans [23]. Histologic descriptive studies on Sphynx skin describe a poorly defined and abnormal inner root sheath and normal follicular density with small, curved, and kinked hair follicles and misshapen, narrow-diameter hair shafts [24]. The Donskoy and Peterbald are both

Fig. 5 A litter of four newborn Norwegian kittens, of which two show extensive alopecia and two look normal. (Courtesy of Dr. Barbara Petrini)

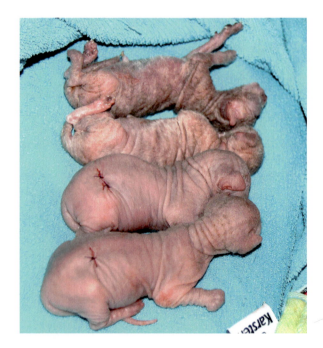

autosomal dominant hairless breeds originating from Russia [23]. The Peterbald is the result of breeding a Donskoy to an oriental breed. The genetic mutation and resultant etiology for their hypotrichosis are not known. Affected cats vary from hairless to having thin, short fine coats. The Kohana is an autosomal dominant hairless breed originating from Hawaii which also lacks vibrissae [23]. There are other "breeds" described from crosses between one of the hairless breeds with other cat breeds.

The curly coat of the Devon Rex also results from a mutation in the KRT71 gene and is allelic and recessive to the hairless (*hr*) allele in the Sphynx cat [23]. The Sphynx KRT71 mutation that produces the characteristic hairless coat phenotype is recessive to a normal hair coat in a cat. The wooly hair phenotype in humans is also caused by a mutation of KRT71 [25]. The curly coat of the Cornish Rex results from a recessive trait resulting from a fixed allele within the breed [25]. A mutation identified in the *lysophosphatidic acid receptor 6* (*LPAR6*) gene is the cause of the Cornish Rex curly coat [25]. This gene codes for a receptor that binds oleoyl-L-alpha-lysophosphatidic acid (LPA), which is important for hair growth and maintaining hair shaft integrity and normal texture [25].

Congenital Hypotrichosis and Short Life Expectancy (CHSLE) in the Birman Cat

Congenital hypotrichosis and short life expectancy (CHSLE) is an autosomal recessive trait in the Birman cat breed. This disorder is recognized as the first non-rodent model of a "nude"/SCID (severe combined immunodeficiency) syndrome [21]. The first reports of this syndrome were in the 1980s and described affected kittens born hairless that died within the first 8 months of life from respiratory or gastrointestinal tract infections [26]. Similarly, affected kittens in Switzerland were shown on necropsy to have absent thymic tissue and depletion of lymphocytes within various lymphoreticular tissues (spleen, Peyer's patches, and lymph nodes) [19]. It is now recognized that this genodermatosis is the phenotypic expression of a mutation in FOXN1 (forkhead box N1) that results in a truncated protein [21]. Forkhead box proteins are important transcription factors, and FOXN1 is expressed in both epithelial cells in the thymus and in the epidermis of the hair bulb [21]. The normal functioning protein is an important transcription factor for the regular development of the thymus and hair follicle epithelium [21]. Consequently, this mutation results in a nonfunctional protein and the hairless phenotype of CHSLE in affected kittens.

Affected kittens are hairless or may grow sparse, shortened, fragile hair coats (Fig. 6). The hairless skin develops excessive skinfolds and a greasy keratinization disturbance [21]. Except for the altered appearance of their skin, affected kittens are normal in their behavior and growth in the early neonatal period [21]. Due to their immunodeficiency, they ultimately succumb to respiratory or gastrointestinal infections within the first few months of life. A genetic test is available through AnimaLabs© (brand of InovaGen Ltd., Croatia) that can be used to screen breeding animals for being carriers for the FOXN1 mutation, to avoid litters that may have affected kittens.

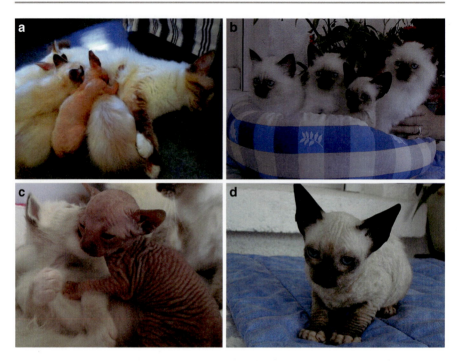

Fig. 6 Hypotrichosis phenotype in Birman kittens. Hairless kittens among normal littermates, born to longhaired colorpoint mitted parents (**a, b**). A hairless 3-week-old kitten showing wrinkled skin (**c**). A 12-week-old hairless kitten displaying a sparse short fur with attenuated whiskers (**d**). Pictures **a** and **c** depict the same proband male, born in 2013. Pictures **b** and **d** depict a 12-week-old female kitten, born in 2004 which is a proband relative. (Reproduction from Ref. [21])

Sebaceous Gland Dysplasia

Progressive alopecia in association with abnormal sebaceous gland differentiation was described in ten unrelated kittens from North America and Europe [27]. The kittens were examined and found to have variable scaling, crusting, and progressive alopecia that began when they were between 4 and 12 weeks of age. Lesions began on the head and then generalized to involve most of the body, but spared the tail in two of the kittens. The generalized scaling varied in severity with evident follicular casting. Some kittens had severe periocular, perioral, pinnal, and ear canal crusting. Histologic evaluation of skin biopsies showed abnormal sebaceous gland morphology with diminished size and irregular organization of abnormal sebocytes [27]. Due to the young age of affected kittens, a possible genetic defect influencing sebaceous gland and follicular development was proposed and compared to a similar phenotype seen in mice [27].

Inherited Structural Hair Shaft Defects

Pili Torti

This structural abnormality of the hair shaft is characterized by a flattening and twisting of the hair shaft around its axis by 180 degrees (Fig. 7) [28]. It is recognized in human dermatology to be both an acquired and congenital disease that can occur independently or may be linked with a number of different congenital defects and syndromes with systemic signs [28]. Abnormalities in the hair shaft keratin have not been identified, and it is proposed that abnormalities in the inner root sheath result in abnormal molding of the hair shaft [28]. Pili torti in cats leads to alopecia as the flattening and twisting of the hair shaft likely lead to shaft breakage [29]. A litter of affected kittens developed generalized alopecia. The kittens either died or were euthanized in the first few weeks of life [30]. Hair follicles were described as curved along their length with hyperkeratosis on histologic examination of skin biopsies [30]. Symmetric, non-inflamed alopecia of the pinnae, dorsal head and zygomatic regions, medial carpi and tarsi, axillae, and tail tip was described in a 1-year-old domestic shorthair that had no signs of systemic illness (Fig. 8) [29]. Hair shafts from the affected cat exhibited the flattening and twisting typical of pili torti (Fig. 7), and some hair shaft roots had a tighter coiled appearance [29]. Histopathologic examination of hair follicles in longitudinal section showed that some hair shafts had an abnormal shape [29]. Trichograms to evaluate hair shafts for this characteristic change can quickly make the diagnosis.

Hair Shaft Abnormality of Abyssinian Cats

There is one report of three cats with a unique abnormal change to whiskers and primary hair shafts [31]. Affected hairs have a swelling on the tip of the hair shaft or sometimes along the hair shaft that was described as onion-shaped. Clinically, if

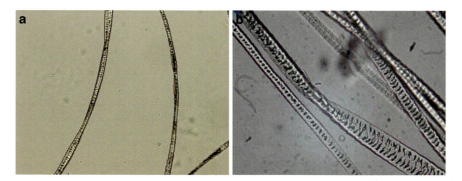

Fig. 7 Microscopic aspect of pili torti: hairs are flattened and twisted along the longitudinal axis. (Courtesy of Dr. Silvia Colombo)

Fig. 8 Same cat as Fig. 7: wide areas of hypotrichosis on the face

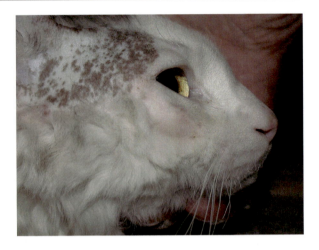

primary hair shafts are affected, the hair coat appears bristly and feels rough. There were no follicular abnormalities seen on evaluation of skin biopsies. A trichogram identifying the characteristic swellings on the hair shafts or whiskers would be diagnostic if evaluating a suspected cat.

Genetic Diseases Affecting Collagen

Ehlers-Danlos syndrome in humans is a diverse group of genetic diseases that affects the connective tissue in the skin and the vasculature, depending on the particular defect. These diseases have been reclassified over time as more is learned about the underlying molecular defect and mutation causing it. The terms cutaneous asthenia and dermatosparaxis are both descriptive terms for this disease as each refers to the marked fragility of the skin.

Feline Cutaneous Asthenia

This disorder of collagen in cats has been recognized for several decades, with the publication of a number of individual case reports, as cited in the textbook of *Small Animal Dermatology* [32]. Himalayan and domestic shorthair cats were reported breeds, and the Himalayan is documented to have an autosomal recessive mode of inheritance for cutaneous asthenia. The molecular defect is not definitely known, but biochemical studies identified a procollagen processing defect due to diminished procollagen peptidase activity and increased collagenase [33]. Feline cutaneous asthenia likely involves more than one mutation, as it can also be an autosomal dominant trait in some cats. The autosomal dominant trait may be lethal in a homozygous state [34, 35]. It is associated with abnormal packing of collagen into fibrils and fibers possibly because of mutations in structural proteins [32].

Affected cats have skin with variable degrees of increased extensibility (Fig. 9) and decreased tensile strength, the latter resulting in tearing (Fig. 10). The skin

Fig. 9 Congenital cutaneous asthenia and hyperelastosis in a cat: the skin is highly extensible, much more than a healthy cat. (Courtesy of Dr. Chiara Noli)

Fig. 10 Same cat as Fig. 9: a first attempt to suture the extensive wounds was unsuccessful and resulted in new tears. (Courtesy of Dr. Chiara Noli)

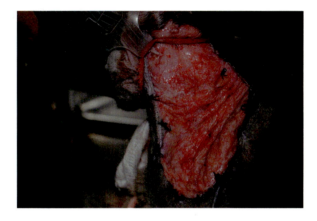

tearing can occur with minimal trauma creating large gaping wounds that rapidly heal leaving scars that are very thin and lack tensile strength, so that scarred areas can in turn develop new wounds. Hyperextensibility is characterized by loose attachment of the skin to the underlying tissue with the skin that hangs loosely, particularly over the legs [36]. An increased extensibility index is a calculation of the vertical height of the skin over the dorsolumbar region divided by the length of the cat and multiplied by 100 [36]. It has been reported that in cats affected with cutaneous asthenia, extensibility is greater than 19% [37]. This calculation may be helpful to quantify the degree of extensibility but is not always reliable and is influenced by a number of factors, including age, hydration status, and degree of abdominal distension. Microscopic evaluation of skin biopsies shows disorganized and attenuated collagen fibers that are shorter and fragmented [38]. Burmese cats with cutaneous asthenia have in association with the collagen disease eschars and areas of necrosis with hemorrhage beneath lesions. The ulcerative lesions develop secondary to necrosis from vascular compromise rather than from tearing of fragile skin. The proposed mechanism for the vascular compromise was stretch trauma to cutaneous vasculature because of the hyperextensible skin and potential fragility of vascular walls [36]. There is no cure for the disease and affected cats should have management changes to minimize trauma. A gene test for cutaneous asthenia in the Burmese cat is planned for availability in the near future [36]. Diagnosis is based on thorough history, physical exam finding, and skin biopsy to evaluate collagen.

Genetic Pigmentary Abnormalities

The normal color of the skin and hair coat is dependent on the number of melanocytes in the epidermis or hair bulb and their ability to function. Melanoblasts develop from neural crest cells and must migrate to the skin, eyes, inner ear, and leptomeninges where they differentiate to become melanocytes [39]. The melanocyte must synthesize organelles called melanosomes that require structural proteins and melanogenic enzymes to be able to successfully synthesize melanin. Changes of pigmentation, whether hypopigmentation with decreased pigment resulting in leukoderma or leukotrichia or hyperpigmentation with increased pigment in the epidermis, can be either genetic or acquired [39].

Genetic Diseases Causing Hypopigmentation (Hypomelanosis)

Hereditary causes of hypopigmentation can result from diseases that cause a decreased number of melanocytes (melanocytopenic) or a mutation that ultimately results in abnormal melanin or a decreased amount of melanin being produced (melanopenic) (Fig. 11) [39].

Chediak-Higashi Syndrome

Chediak-Higashi syndrome occurs in humans and other mammalian species: mice, rat, fox, cattle, Aleutian mink, American bison, killer whales, and the domestic cat [40]. In the cat, this is a rare autosomal recessive disorder of "blue smoke"-colored Persian cats that also have yellow irises. In people, mink, and mice, a mutation in the *LYST* gene coding for a protein responsible for vesicle trafficking, vital for the development of some cellular organelles and lysosome fusing, has been identified [40, 41].

Fig. 11 Waardenburg syndrome in a cat presenting deafness, white coat, and heterochromic irises. (Reproduction from Ref. [39], with permission)

This mutation affects organelles in multiple cells. The consequence of this mutation explains the macromelanosomes seen in hair shafts and the large eosinophilic lysosomes characteristically seen in peripheral macrophages and neutrophils in affected individuals. The macromelanosomes account for the atypical coat color, and the abnormalities in the white blood cells result in increased susceptibility to infections. Affected cats also are prone to bleeding because of platelet dysfunction and are photophobic due to partial oculocutaneous albinism. Humans with this disease receive allogeneic hematopoietic cell transplantation to treat the hematologic and immunologic abnormalities [41]. Cats are diagnosed based on phenotypic traits, clinical signs, and identification of characteristic changes on a peripheral blood smear. There is no specific treatment in cats, and individuals should be monitored for secondary infections and bleeding tendencies. They should never be used for breeding.

Waardenburg Syndrome in Cats

In humans, Waardenburg syndrome is the result of melanocytopenic hypomelanosis, due to a number of different mutations that affect melanocyte migration and differentiation [39]. In domestic animals, including the cat, this syndrome is associated with white hair coat, blue or heterochromic irises, and deafness (Fig. 11). The lack of pigment producing white coat and pale eye color is an autosomal dominant trait that has incomplete penetrance for deafness and eye color [39]. White cats are not always blue eyed or deaf, and the variable penetrance of deafness and blue eye color with white coat reflects which genes are involved impacting melanocytes. Deafness associated with this syndrome occurs because melanocytes, normally found in the vascular epithelium of the inner ear called the stria vascularis, are needed to establish the endocochlear potential by secreting high amounts of potassium into the endolymph, so without these melanocytes, deafness results [42]. Melanocytes vary in their response to KIT signaling; cutaneous melanocytes are strongly influenced by KIT, while melanocytes that ultimately reside in the internal ear or iris respond to endothelin 3 (EDN3) or hepatocyte growth factor (HGF) [42].

Albinism

Albinism is a congenital disorder that results from mutations in the gene coding for the enzyme tyrosinase and results in lack of pigment in the skin, hair, and eyes of affected individuals. In cats with complete albinism, the hair coat is white and the eyes are blue, with reduced pigment in the *tapetum*, so that eyes can appear to have a reddish tapetal reflection (Fig. 12). This albino phenotype in cats is associated with more than one mutation in the *tyrosinase* (*TYR*) gene [43, 44]. Albino cats have an absence of pigment on skin biopsy histologic samples but display a normal epidermis with melanocytes that appear clear, as they lack the ability to make melanin (Fig. 12).

There are two causative mutations for the temperature-sensitive albinism seen in the Siamese and Burmese breeds. The resulting alleles are the *siamese* and *burmese* temperature-sensitive alleles affecting the *TYR* gene. This results in the development of darker pigment in regions on the cat that are cooler: ears, paws, and facial mask. All cats that develop points typical of the Siamese breed or the Burmese coat color will be homozygous for their mutated allele.

Fig. 12 Feline tyrosinase-
negative albinism.
(Reproduction from Ref.
[39], with permission)

Fig. 13 Orange tabby cat
with hyperpigmented
macules on the lips and
nasal planum,
characteristic of lentigo
simplex seen in orange cats

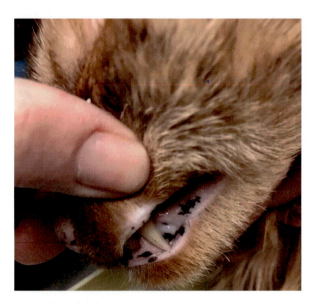

Genetic Diseases Causing Hyperpigmentation

Hyperpigmentation results when melanin in the epidermis or stratum corneum
increases. Hyperpigmentation is more commonly an acquired change to the skin
than genetic. Cats have a genetic macular hyperpigmentation.

Lentigo Simplex of Orange Cats

This is a cosmetic hyperpigmentation change seen in orange cats. These cats develop
well-delineated, hyperpigmented macules along the eyelids, nasal planum, lip margins, and gingiva (Fig. 13). This characteristic macular melanosis starts when the
cat is young, and the degree of pigmented macules present may progress over time.
Histopathology of a lesion would confirm increased melanin within keratinocytes
in an affected macular lesion.

Feline Maculopapular Cutaneous Mastocytosis

This is not a primary disturbance in pigmentation, but lesions can be hyperpigmented. Originally reported in three related Sphynx cats and termed feline urticaria pigmentosa [45], feline maculopapular cutaneous mastocytosis has been reported predominantly in the Sphynx and Devon Rex breeds. These breeds are genetically related and both have mutations in the *KRT71* gene. Affected cats were described to have a pruritic, maculopapular eruption that was variably pigmented, with crusted lesions symmetrically involving the head, neck, trunk, and limbs (Fig. 14) [45]. Histopathology of skin biopsies revealed perivascular to diffuse dermal infiltration of well-differentiated mast cells and small numbers of eosinophils [45]. Similar clinical presentations were described in five unrelated Devon Rex cats, although some cats had less generalized distribution of lesions with diffuse lesions only over the ventral thorax [46]. Pruritus and hyperpigmented lesions were documented only in cats with confirmed secondary infections, and more eosinophils were seen on histopathologic evaluation of skin biopsy than in the original report of this disease [46]. The possibility that these clinical presentations could be a cutaneous reaction pattern was proposed [46].

In humans, mastocytosis can be cutaneous if it involves only the skin or systemic if the mastocytic infiltrates involve the skin and other organs, for example, the gastrointestinal tract or bone marrow. Systemic mastocytosis in cats is a malignant disorder of neoplastic mast cells arising from the hematopoietic system [47]. There is one report of systemic mastocytosis in a cat from dissemination of cutaneous mast cell tumors to internal organs [48]. To date, none of the reported cases of cutaneous mastocytosis in cats have demonstrated progression to a systemic form.

A recent case series of 13 cats diagnosed with this disease based on clinical lesions, breed affected, and exclusion of other possible differential diagnoses (hypersensitivities, ectoparasites, and dermatophytes) reviewed the clinical information and proposed a novel classification for this disease [49]. All cats were Sphynx, Devon Rex, or Devon Rex crosses and had developed skin lesions as young cats. Median age at presentation was 15 months, but cats had disease for a median of 8 months before presentation [49]. The proposed classification based on three different clinical presentations was compared to the classification system used for

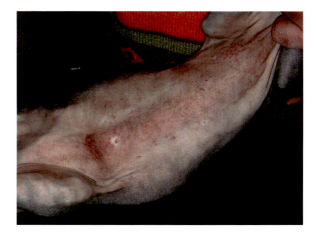

Fig. 14 Erythematous papules and small crusts in a linear pattern on the ventral trunk of a Devon Rex cat with urticaria-pigmentosa-like disease. (Courtesy of Dr. Chiara Noli)

human cutaneous mastocytosis. The three subforms varied in types of lesions, distribution, severity of pruritus, and likelihood for spontaneous regression [49]. In one form termed "polymorphic maculopapular cutaneous mastocytosis," lesions were large wheals or papules localized to the cranial portion of the body (head, neck, and shoulders of the cats) with moderate pruritus present but no hyperpigmentation [49]. Cats with moderate pruritus and more generalized lesions of erythema and small papules that coalesced, in addition to lesions of self-trauma, were characterized as having the "monomorphic" form [49]. The third form was given the proposed term "pigmented maculopapular cutaneous mastocytosis"; it was a more chronic form associated with severe pruritus and generalized lesions of hyperpigmentation and lichenification in addition to coalescing papules and lesions of self-trauma [49]. The first form carried the best prognosis with the majority of cats ultimately being able to discontinue therapy; cats with the other two forms required ongoing therapy with corticosteroids, antihistamines, or cyclosporine [49].

The diagnosis of feline cutaneous mastocytosis requires the correct signalment; a young cat of the typical breed; a compatible clinical presentation, in terms of types of lesions and distribution; and a systematic diagnostic evaluation to exclude any possible causes for pruritus and maculopapular dermatitis. This would include evaluation for ectoparasites, allergic dermatitis, and secondary infections. How critically important this is illustrated in a case series of three Devon Rex cats with papular eosinophilic/mastocytic dermatitis, initially suggestive for an urticaria pigmentosa-like dermatitis, who in fact had dermatophytosis [50]. After receiving antifungal treatment, all the lesions resolved.

Diseases Affecting the Lymphatic Vessels

Primary Lymphedema

Primary lymphedema is a swelling of a body area due to a congenital defect in the lymphatic system, which does not drain liquids from the periphery (Fig. 15). The

Fig. 15 Primary lymphedema in a cat. (Courtesy of Dr. Chiara Noli)

skin presents a cold, pitting edema, more often of the extremities. Diagnosis can be achieved via the patent blue violet dye test: a sterile 5% dye solution injected SC in the skin [51]. A diffuse distribution of the dye demonstrates absence of intact lymphatic transport. There is no treatment but palliative diuretics for this condition.

References

1. Leeb T, Muller EJ, Roosje P, Welle M. Genetic testing in veterinary dermatology. Vet Dermatol. 2017;28:4–9. https://doi.org/10.1111/vde.12309.
2. Lyons LA. Genetic testing in cats. Mol Cell Probes. 2012;26:224–30. https://doi.org/10.1016/j.mcp.2012.04.004.
3. Mullikin JC, Hansen NF, Shen L, Ewbling H, Donahue WF, Tao W, et al. Light whole genome sequence for SNP discovery across domestic cat breeds. BMC Genomics. 2010;11:406. http://www.biomedcentral.com/1471-2164/11/406
4. Bond R, Curtis CF, Ferguson EA, Mason IS, Rest J. Idiopathic facial dermatitis of Persian cats. Vet Dermatol. 2000;11:35–41.
5. Powers HT. Newly recognized feline skin disease. Proceedings of the 14th AAVD/ACVD meeting. San Antonio; 1998. p. 17–20.
6. Gross TL, Ihrke PJ, Walder EJ, Affolter VK. Skin diseases of the dog and cat. 2nd ed. Oxford: Blackwell Science Ltd; 2005. 114p.
7. Chung TH, Ryu MH, Kim DY, Yoon HY, Hwang CY. Topical tacrolimus (FK506) for the treatment of feline idiopathic facial dermatitis. Aust Vet J. 2009;87:417–20. https://doi.org/10.1111/j.1751-0813.2009.00488.
8. Paradis M, Scott DW. Hereditary primary seborrhea oleosa in Persian cats. Feline Pract. 1990;18:17–20.
9. Miller WH, Griffen CE, Canmbell KL. Muller and Kirk's small animal dermatology. 7th ed. St. Louis: Elsevier Mosby; 2013. p. 576.
10. Bergval K. FC-25: a novel ulcerative nasal dermatitis of Bengal cats. Vet Dermatol. 2004;15(Supp 1):28.
11. St A, Abramo F, Ficker C, McNabb S. P-36 Juvenile idiopathic nasal scaling in three Bengal cats. Vet Dermatol. 2004;15(Supp 1):52.
12. Medeiros GX, Riet-Correa F. Epidermolysis bullosa in animals: a review. Vet Dermatol. 2015;26:3–e2. https://doi.org/10.1111/vde.12176.
13. Johnstone I, Mason K, Sutton R. A hereditary junctional mechanobullous disease in the cat. Proceedings of the second world congress of veterinary dermatology association. Montreal; 1992:111–12.
14. Alhaidari Z, Olivry T, Spadafora A, Thomas RC, Perrin C, Meneguzzi, et al. Junctional epidermolysis bullosa in two domestic shorthair kittens. Vet Dermatol. 2006;16:69–73.
15. White SD, Dunstan SM, Olivry T, Naydan DK, Richter K. Dystrophic (dermolytic) epidermolysis bullosa in a cat. Vet Dermatol. 1993;4:91–5.
16. Olivry T, Dunstan SM, Marinkovitch MP. Reduced anchoring fibril formation and collagen VII immunoreactivity in feline dystrophic epidermolysis bullosa. Vet Pathol. 1999;36:616–8.
17. Casal ML, Straumann U, Sigg C, Arnold S, Rusch P. Congenital hypotrichosis with thymic aplasia in nine Birman kittens. J Am Anim Hosp Assoc. 1994;30:600–2.
18. Scott DW. Feline dermatology 1900–1978: a monograph. J Am Anim Hosp Assoc. 1980;16:313.
19. Bourdeau P, Leonetti D, Maroille JM, Mialot M. Alopécie héréditaire généralisée féline. Rec Med Vet. 1988;164:17–24.
20. Thoday K. Skin diseases in the cat. In Pract. 1981;3:22–35.
21. Abitbol M, Bossé P, Thomas A, Tiret L. A deletion in *FOXN1* is associated with a syndrome characterized by congenital hypotrichosis and short life expectancy in Birman cats. PLoS One. 2015;10:e0120668. https://doi.org/10.1371/journal.pone.0120668.
22. Meclenberg L. An overview of congenital alopecia in domestic animals. Vet Dermatol. 2006;17:393–410.

23. Gandolfi B, Outerbridge CA, Beresford LG, Myers JA, Pimental M, Alhaddad H, et al. The naked truth: Sphynx and Devon Rex cat breed mutations in *KRT71*. Mamm Genome. 2010;21:509–15. https://doi.org/10.1007/s00335-010-9290-6.

24. Genovese DW, Johnson T, Lam KE, Gram WD. Histological and dermatoscopic description of sphinx cat skin. Vet Dermatol. 2014;26:523–e90. https://doi.org/10.1111/vde.12162.

25. Gandolfi B, Alhaddad H, Affolter VK, Brockman J, Haggstrom J, Joslin SE, et al. To the root of the curl: a signature of a recent selective sweep identifies a mutation that defines the Cornish Rex cat breed. Palsson A, ed. PLoS One. 2013;8:e67105. https://doi.org/10.1371/journal.pone.0067105.

26. Hendy-Ibbs PM. Hairless cats in Great Britain. J Hered. 1984;75:506–7.

27. Yager JA, Tl G, Shearer D, Rothstein E, Power H, Sinke JD, et al. Abnormal sebaceous gland differentiation in 10 kittens ('sebaceous gland dysplasia') associated with hypotrichosis and scaling. Vet Dermatol. 2014;23:136–e30. https://doi.org/10.1111/j.1365-3164.2011.01029.x.

28. Mirmirani P, Samimi SS, Mostow E. Pili torti: clinical findings, associated disorders, and new insights into mechanisms of hair twisting. Cutis. 2009;84:143–7.

29. Maina E, Colombo S, Abramo F. Pasquinelli. A case of pili torti in a young adult domestic short-haired cat. Vet Dermatol. 2012;24:289–e68. https://doi.org/10.1111/vde.12004.

30. Geary MR, Baker KP. The occurrence of pili torti in a litter of kittens in England. J Sm Anim Pract. 1986;27:85–8.

31. Wilkinson JT, Kristensen TS. A hair abnormality in Abyssinian cats. J Small Anim Pract. 1989;30:27–8.

32. Miller WH, Griffen CE, Canmbell KL. Muller and Kirk's small animal dermatology. 7th ed. St. Louis: Elsevier Mosby; 2013. p. 603.

33. Counts DF, Byer PH, Holbrook KA, Hegreberg GA. Dermatosparaxis in a Himalayan cat: I—biochemical studies of dermal collagen. J Investig Dermatol. 1980;74(2):96–9. https://doi.org/10.1111/1523-1747.ep12519991.

34. Scott DW. Feline dermatology; introspective retrospections. J An Am Hosp Assoc. 1984;20:537.

35. Minor RR. Animal models of heritable diseases of the skin. In: Goldsmith EL editor. Biochemistry and physiology of skin. New York: Oxford University Press; 1982.

36. Hansen N, Foster SF, Burrows AK, Mackie J, Malik R. Cutaneous asthenia (Ehlers-Danlos-like syndrome) of Burmese cats. J Feline Med Surg. 2015;17:945–63. https://doi.org/10.1177/1098612X15610683.

37. Freeman LJ, Hegreberg G, Robinette JD. Ehlers-Danlos syndrome in dogs and cats. Semin Vet Med Surg. 1987;2(3):221–7.

38. Sequeira JL, Rocha NS, Bandarra EP, Figueiredo LM, Eugenio FR. Collagen dysplasia (cutaneous asthenia) in a cat. Vet Pathol. 1199;36:603–6.

39. Alhaidari Z, Olivry T, Ortonne JP. Melanocytogenesis and melanogenesis: genetic regulation and comparative clinical diseases. Vet Dermatol. 1999;10:3–16.

40. Reissman M, Ludwig A. Pleiotropic effects of coat colour-associated mutations in humans, mice and other mammals. Semin Cell Dev Biol. 2013;24:576–87.

41. Kaplan J, De Domenico I, McVey Ward D. Chediak-Higashi syndrome. Curr Opin Hematol. 2008;15:22–9. https://doi.org/10.1097/MOH.0b013e3282f2bcce.

42. Ryugo DK, Menotti-Raymond M. Feline deafness. Vet Clin North Am Small Anim Pract. 2012;42:1179–207.

43. Imes DL, Geary A, Grahn A, Lyons A. Albinism in the domestic cat (*Felis Catus*) is associated with a *tyrosinase (TYR)* mutation. Anim Genet. 2006;37:175–8. https://doi.org/10.1111/j.1365-2052.2005.01409.x.

44. Abitbol A, Boss P, Grimard B, Martignat L, Tiret L. Allelic heterogeneity of albinism in the domestic cat. Stichting International Foundation for Anim Genet. 2016;48:121–8. https://doi.org/10.1111/age.12503.

45. Vitale CB, Ihrke PJ, Olivry T, Stannard T. Feline urticarial pigmentosa in three related sphynx. Vet Dermatol. 1996;7:227–33.

46. Noli C, Colombo S, Abramo F, Scarampella F. Papular eosinophilic/mastocytic dermatitis (feline urticarial pigmentosa) in Devon Rex cats: a distinct disease entity or a histopathological reaction pattern. Vet Dermatol. 2004;15:253–9.

47. Woldenmeskel M, Merrill A, Brown C. Significance of cytological smear evaluation in diagnosis of splenic mast cell tumor-associated systemic mastocytosis in a cat (*Felis catus*). Can Vet J. 2017;58:293–5.

48. Lamm CC, Stern AW, Smith AJ. Disseminated cutaneous mast cell tumors with epitheliotropism and systemic mastocytosis in a domestic cat. J Vet Diagn Investig. 2009;21:710–5.

49. Ngo J, Morren MA, Bodemer C, Heimann M, Fontaine J. Feline maculopapular cutaneous mastocytosis: a retrospective study of 13 cases and proposal for a new classification. J Feline Med Surg. 2018;21:394. https://doi.org/10.1177/1098612X18776141.

50. Colombo S, Scarampella F, Ordeix L, Roccoblanca P. Dermatophytosis and papular eosinophilic/mastocytic dermatitis (urticarial pigmentosa-like dermatitis) in three Devon Rex cats. J Feline Med Surg. 2012;14:498–502.

51. Jacobsen JO, Eggers C. Primary lymphoedema in a kitten. J Small Anim Pract. 1997;38:18–20.

Psychogenic Diseases

C. Siracusa and Gary Landsberg

Abstract

Grooming behavior is essential to keep the skin and coat of cats in good health, and cats spend a significant portion of their day grooming. Normal grooming behavior is a sign of good physical and mental health in cats. Changes in grooming behavior may be caused by medical problems, either dermatological or systemic. "Sickness behavior" that includes decreased grooming may be an early sign of underlying medical problems, but increased grooming can also be related to medical causes (Fatjo and Bowen, Medical and metabolic influences on behavioural disorders. In: Horwitz DF, Mills DS, editors. BSAVA manual of canine and feline behavioural medicine. 2nd ed. Gloucester: BSAVA; p. 1–9, 2009; Rochlitz, Basic requirements for good behavioural health and welfare in cats. In: Horwitz DF, Mills DS, editors. BSAVA manual of canine and feline behavioural medicine. 2nd ed. Gloucester: BSAVA; p. 35–48, 2009). Grooming behavior can also be significantly altered in times of stress or conflict. In response to stress, some cats overgroom, lick, bite, chew, suckle, or barber their coats leading to alopecia (psychogenic alopecia), while others stop taking adequate care of their coat. In particular, overgrooming may be a displacement behavior arising from conflict or stress or a compulsive disorder (much like compulsive washing or trichotillomania in humans) (Landsberg et al., Behavior problems of the dog and cat. 3rd ed. Philadelphia: Elsevier Saunders, 2013).

Many behavior changes are due to a combined effect of stress and health problems. For example, pain may be the initial trigger of overgrooming a particular joint. Once the behavior arises, it might worsen due to medical conse-

C. Siracusa (✉)
Department of Clinical Sciences and Advanced Medicine, School of Veterinary Medicine, University of Pennsylvania, Philadelphia, PA, USA
e-mail: siracusa@vet.upenn.edu

G. Landsberg
CanCog Technologies, Fergus, ON, Canada
e-mail: garyl@cancog.com

© Springer Nature Switzerland AG 2020
C. Noli, S. Colombo (eds.), *Feline Dermatology*,
https://doi.org/10.1007/978-3-030-29836-4_29

quences of excoriation, infection, or pruritus, as well as the increased anxiety and stress arising from both internal factors and the owner's response. Therefore, an effective treatment of psychogenic dermatologic disease has to address both the underlying psychological disturbance and the possible medical component (Fatjo and Bowen, Medical and metabolic influences on behavioural disorders. In: Horwitz DF, Mills DS, editors. BSAVA manual of canine and feline behavioural medicine. 2nd ed. Gloucester: BSAVA; p. 1–9, 2009; Rochlitz, Basic requirements for good behavioural health and welfare in cats. In: Horwitz DF, Mills DS, editors. BSAVA manual of canine and feline behavioural medicine. 2nd ed. Gloucester: BSAVA; p. 35–48, 2009; Landsberg et al., Behavior problems of the dog and cat. 3rd ed. Philadelphia: Elsevier Saunders, 2013).

The psychogenic dermatological diseases discussed in this chapter include psychogenic alopecia, overgrooming and biting of the tail, hyperesthesia, and excessive scratching.

Abnormal Grooming Behavior and the Role of Stress

Cats spend a large portion of their day grooming. The environment in which cats live influences their time spent grooming. While farm cats spend about 15% of their time grooming, group-housed laboratory cats spend 30% of their time in comfort behaviors including grooming [4]. Cat grooming serves multiple purposes. It keeps the skin and coat clean, contributing to its health. Grooming helps in maintaining a "colony odor" that strengthens the social bond among individuals of the same group. For this purpose, cats groom each other in a highly cooperative behavior called "allogrooming" (Fig. 1) [5]. The abundance of Type 1 (Merkel's disks) and Type 2 (Ruffini corpuscles) slow-adapting (SA) receptors in the skin of cats explains why cats are so sensitive to strokes and touch. This sensitivity may also play a role in both hyperesthesia and petting-induced aggression. SA receptors, together with rapidly adapting (RA) receptors, are present at the base of the vibrissae, in the face, lips, and mouth; this may explain the "quiet bite" that cats may inflict to another individual or to themselves when their tactile receptors are overstimulated [6, 7]. A subgroup of dorsal raphe

Fig. 1 Feline allogrooming: a cat licks another cat (Courtesy of C. Siracusa)

nucleus serotoninergic neurons in cats is strongly activated in association with oral-buccal movements, such as chewing, licking, and grooming. The neurons are also activated by somatosensory stimuli applied to the head, neck, and face [8]. Grooming is also exhibited by cats experiencing stress, e.g., in case of an emotional conflict or frustration or perceived threat, as a displacement behavior that serves the purpose to redirect the increased emotional arousal [5]. If the stress persists, grooming may experience a dramatic increase in frequency and intensity and generalize to several contexts other than the initial trigger, becoming a compulsive behavior [3].

Performing *displacement behavior* (fidgeting) is one of the four possible coping strategies that animals (and humans) exhibit when the stress response is activated, i.e., when an animal perceives that its physical or emotional homeostasis is threatened or challenged. The other possible responses are antagonistic behavior including aggression (fighting), escape (fleeing), and tonic immobility (freezing) [6]. Being familiar with all types of stress responses allows the observer to detect stress-related behavior associated with overgrooming, since the animal can shift from one stress-related behavior to another and the overgrooming may not immediately be visible to the observer. For example, the owner may observe an increase of antagonistic behaviors (hissing, swatting, etc.) among household cats or an increased tendency to hide of a cat, but no excessive grooming. Owners may overlook the initial increase in time and frequency that their cat spends grooming, and the problem may become manifested only when the consequent skin and/or coat lesions appear. This may be due to the inhibitory effect that people's presence can have on grooming behavior (freezing); or, conversely, some cats may be more stimulated when the owners are around and, therefore, less motivated to groom themselves. In some other cases, punishment inflicted by the owners to their cats when they are grooming excessively (e.g., verbal punishment or spraying with a water bottle) may cause the cats to not groom in front of them. Therefore, being able to recognize that increased aggression and fear and reduced activity are signs of stress can help prevent compulsive overgrooming by identifying and removing the source of stress (e.g., conflicts with other household cats or the family dog, changes in the environment, or the arrival of a baby) [3].

Compulsive behaviors have been defined as a series of movements usually derived from normal maintenance behaviors, including grooming, that are performed out of context in a repetitive, exaggerated, ritualistic, and sustained manner [9]. Cats experiencing stress and increased emotional arousal can show grooming as a displacement behavior, but not all cats experiencing chronic stress will develop compulsive grooming. When the excessive grooming shifts from displacement behavior to compulsive behavior, it is not displayed after a specific trigger (is emancipated), it is very intense and exaggerated, and it may be difficult to interrupt. Excessive grooming is the most commonly reported compulsive behavior in cats [10]. The genetics and the early experience of a cat determine the individual's coping strategy and will influence the development of a compulsive behavior when the cat is exposed to chronic stress. Although the pathophysiology of compulsive grooming is not well understood, abnormal serotonin transmission has been proposed as a primary mechanism, and opioids and dopamine may also be included in its modulation [8, 11, 12]. Based on the signs at presentation, compulsive behavior involving overgrooming can be differentiated in:

- Psychogenic alopecia
- Tail chewing, biting, and self-mutilation
- Excessive scratching

When the repetitive behavior is part of a collection of related non-specific signs that co-occur but may have different etiologies, we define the presentation as a "syndrome." Feline hyperesthesia is a syndrome that include, among others, overgrooming repetitive behaviors.

A detailed history collection is fundamental to formulate the differential diagnoses. Unlike many physical signs of disease, behavior signs cannot often be observed during the veterinary visit. Moreover, environmental stressors play an important role in the pathogenesis and the treatment of compulsive behaviors; therefore, the veterinarian should actively ask the owner about potential stressors. Asking to the client to take photos and videos of the environment in which our patient lives (house, yard, areas around the house) and of the cat's interactions with the family members and household pets is an invaluable aid in making a diagnosis and developing a treatment plan. Questions should also be asked about the pattern, frequency, duration, and localization of the overgrooming (or other abnormal behaviors exhibited) and about the owner response to this behavior. Lastly, we should find out what has already been attempted to address the behavior and its result [3, 7, 9, 13]. For a list of relevant questions to ask, see Table 1.

Table 1 Behavior history to collect for a differential diagnosis of psychogenic skin disorders

Question	Relevant information
Abnormal grooming behavior (AGB)	
What AGB did the owner observe (licking, biting, chewing, etc.)?	Determine if the repetitive behavior is compulsive and emancipated from a specific trigger
What are the areas of the body on which the cat displays the AGB?	Determine if the magnitude and distribution of the behavior justify the lesions observed
How frequently does the cat show the AGB (episodes/day)?	Determine the chronicity of the behavior that may influence its prognosis
What is the duration of each episode?	Determine the presence of pain or a neurosensory disorder
When did the owner first notice the AGB?	
Can the owner identify any specific trigger for the AGB?	
Is the owner able to easily interrupt the AGB?	
Does the cat quickly return to display the AGB?	
Is the behavior associated with rippling of the skin, vocalizing, running, or hiding?	

Table 1 (continued)

Question	Relevant information
Environment	
What is the household composition (including other pets)?	Identify sources of stress, fear, or anxiety for the cat (other pets, children, noises, etc.). Stressed cats may show increased arousal (restlessness, fear, anxiety, aggression) or a decreased level of activity (hiding, perching on elevated spots)
Describe the interactions between all people/pets in the family and the cat showing AGB	Determine if the cat has an adequate level of environmental stimulation and enrichment. Cats that are handled frequently, displaced from their resting places, and disturbed during sleep may be exposed to excessive stimulation. Cats that do not
Is there any indoor or outdoor stimulus (e.g., noises, outdoor animals) that would trigger fear, anxiety, aggression, or excessive arousal?	have enough playing/hunting outlets may be exposed to an insufficient level of stimulation. Determine if the cat has safe areas where he can comfortably hide and rest when stressed
What is the daily routine of the cat (feeding, resting/sleeping, playing, training, exercise)?	
Does the cat have preferred spots/areas and hiding places in the house?	
Management of the abnormal grooming behavior (AGB)	
How does the owner respond to the AGB? How does the cat respond to the owner intervention?	Determine if the owner is using punishment, which should be avoided. Identify strategies that may help to interrupt and redirect the cat's behavior, e.g., conditioned sounds and food-enhanced
Did the owner attempt any treatment?	toys. Collect information to identify the drug treatment of choice
Of the management/treatment tried, what did work best?	

The Intersection Between Behavioral and Medical Illness: The Immune Response

Until now we have focused our attention on emotional dysregulation as a cause of overgrooming and compulsive grooming. However, several medical causes have been identified and should be considered as differential diagnoses: diseases that lead to pain or pruritus (e.g., adverse food reaction, atopic disease, parasitic hypersensitivity); parasites including fleas, mites, and lice; fungal infections including *Malassezia* and dermatophytosis; endocrinopathies including hyperadrenocorticism; systemic disorders such as hepatocutaneous syndrome and hyperthyroidism; and localized pain or pruritus such as neuropathies, anal sacculitis, or cystitis [3, 9, 11, 14, 15]. The presence of primary lesions and/or scratching may orient our diagnoses toward a primary medical cause, but the absence of lesions does not rule out a medical cause.

The boundaries between medical and behavioral illness are not always so clear, and the question "Is it medical or behavioral?" may not have a simple answer. The reciprocal link between behavior and the inflammatory and immune response has been extensively documented in the scientific literature. Moreover, this relationship is also influenced by our host microbiome, i.e., the skin and gut microbiome [16–18]. Activation of pro-inflammatory cytokines induces a depressed state (sickness), which helps the individual to cope with the disease (e.g., an infection by exogenous pathogens) that triggered the inflammatory response. Circulating pro-inflammatory cytokines can enter the brain, where they have a direct inflammatory action and stimulate the production of other pro-inflammatory cytokines and prostaglandins. Although this inflammatory response does not produce tissue damage, it induces a negative behavioral change. Circulating pro-inflammatory cells also exercise their action on the brain indirectly through neuronal pathways, for example, activating a vagal response [16]. The endogenous microorganisms constituting the intestinal microbiome may influence the behavior of their animal hosts through a similar action. The microbiota is capable of modulating the stress response via the HPA axis or directly through vagal neuronal stimulation and cytokine action [17]. Chronic gastrointestinal conditions that alter the microbiota might therefore influence the behavior of an animal, beyond the changes associated with discomfort and nutritional compromise. The link between the skin microbiome and the inflammatory and immune response has been documented [19–21], but the direct influence on behavior has not yet been determined.

Although ruling out medical causes is the necessary first step when approaching a cat presenting for overgrooming, the other immediate consideration is the management of any underlying stress that might play a role in both the medical and behavioral component of self-trauma. Acute stress leads to an immune response intended to enhance defense mechanisms, but chronic stressors may alter immune function leading to inflammatory dermatoses, gastrointestinal disease, dermatologic conditions, respiratory and urinary tract disease, and a variety of behavioral disorders [3]. In humans, stress may play a role in the pathogenesis of dermatoses such as atopic dermatitis by increasing IgE, eosinophils, and vasoactive peptides, an over-reactive sympathetic-adrenal medullary system, and a decrease in hypothalamic-pituitary-adrenal responsiveness [22–24]. Opioid peptides released during stress may further potentiate pruritus [25]. In humans, a link has also been established between stress and increased epidermal permeability [26]. Increased epidermal permeability in pets might exacerbate atopic disease in a genetically predisposed individual.

Psychogenic Skin Diseases

Psychogenic Alopecia

Self-induced alopecia resulting from an underlying behavioral cause is often referred to as psychogenic alopecia in cats [3]. The behavior sign associated is typically overgrooming, manifested as compulsively exhibiting one or more of the following: licking, chewing, sucking, biting, and pulling out the hair and, in alopecic areas, the skin. Compulsive grooming can be directed to any area of the body that the animal can reach, but the thorax, groin, ventrum, medial or caudal tights, flanks, and front legs have been reported to be frequent targets (Figs. 2, 3, and 4) [13, 14].

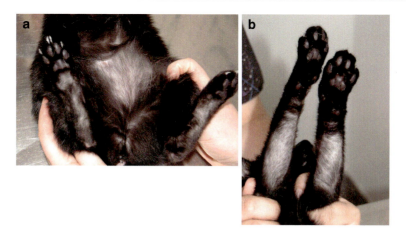

Fig. 2 Abdominal (**a**) and front legs (**b**) alopecia in the same cat. (Courtesy of Dr. Chiara Noli)

Fig. 3 Flank alopecia. (Courtesy of G. Landsberg)

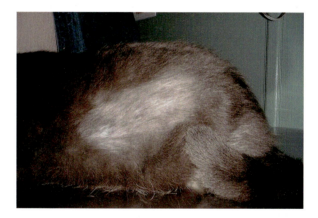

Fig. 4 Alopecia on the abdomen (**a**) and hindlegs (**b**) in the same cat. (Courtesy of Dr. Chiara Noli)

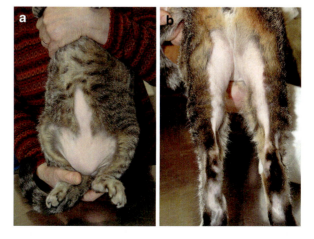

Cases involving the head and neck have also been described [15]. The repetitive behavior can cause focal or diffuse depilation, excoriations, crusts, and non-healing ulcers, with symmetric or asymmetric lesions and peripheral lymphadenomegaly due to inflammation and/or secondary infection [14, 15]. The diagnostic term feline behavioral ulcerative dermatitis has been proposed for rare cases involving non-healing ulcers of the head and neck [15]. The abovementioned presence of tactile receptors in the oral and peri-oral areas may explain why compulsive grooming is not limited to excessive licking but also includes other oral behaviors like chewing, sucking, and biting. Cats of any age, sex, and breed can be affected. The diagnosis is one of exclusion and primary dermatological diseases, pain-inducing conditions (orthopedic disease, low urinary tract disease, abdominal pain, etc.), psychomotor seizure, and hyperesthesia/paresthesia should be ruled out [3, 13].

In a clinical trial performed at a specialty dermatology and behavior practice, 21 cats referred for self-induced alopecia were evaluated [11]. It was determined that, of the 21 cases included, 16 (76.2%) had a medical etiology, 2 (9.5%) were psychogenic alopecia, and 3 (14.3%) were combined medical and behavioral. Of the medical problems, a combination of atopy and adverse food reactions was most common (12 cats), followed by adverse food reactions, atopic dermatitis, and parasitic hypersensitivity. Fifty-two percent of the patients had more than one cause. Skin biopsy specimens were obtained from 20 of the 21 cats, and 14 (70%) had inflammatory skin lesions. All cats with histological evidence of inflammation had an underlying medical condition. No histological abnormalities were seen in six cats, of which two had a compulsive disorder and four had environmental hypersensitivity, an adverse food reaction or both. Therefore, while biopsy helped to confirm a medical cause, there were cats with histologically normal skin that had a medical cause.

Diagnostics for psychogenic alopecia include a thorough physical examination, with a neurological and orthopedic assessment, and a complete dermatological workup with parasite treatment, skin scraping, dermatophyte culture, trichogram, dermoscopy [27], skin biopsy, allergy testing, CBC, serum chemistry, and urinalysis. Further diagnostics should be considered if warranted after the finding of the previous assessment, e.g., abdominal ultrasound, radiographs, and endocrine testing [3, 11, 13].

Overgrooming and Biting of the Tail

Overgrooming can be also directed toward the tail, most typically its distal portion. This behavior can also be characterized, especially in its advanced presentations, by energetic chewing and biting that can lead to self-mutilation [3, 7]. This presentation may have a primary behavioral cause, but numerous medical differentials must first be ruled out. Tail mutilation may begin as a play behavior or a conflict-induced behavior in which the cat circles and chases its tail. Skin disease, trauma, spinal pain, and other neuropathies could initiate the behavior or be a secondary contributing factor. Therefore, cases of cats that attack their tail need advanced imaging to rule out neurologic or orthopedic disease. Should the cat bite or injure its tail, the resultant pain, infection, and possible neuropathy could incite further chewing and biting [3, 13].

Hyperesthesia

The many labels that have been used to call this syndrome, "rolling skin syndrome," "twitching cat disease," "neuritis," or "atypical neurodermatitis," reflect the complicated nature of this syndrome. Hyperesthesia in cats can be attributed to a wide range of medical and behavioral problems including dermatologic (pain, pruritus, infectious), neurologic (partial seizures, spinal disease, and neuropathies), musculoskeletal (inclusion body myopathy, FeLV-induced myelopathy), or behavioral (displacement behavior arising in situations of conflict or high arousal, compulsive disorder) [28–31]. While rippling, twitching, and spasm of the epaxial muscles are the hallmark signs, estrus-like rolling, tail twitching, muscle spasm on dorsum and flank, mydriasis, self-directed aggression, redirected aggression, vocalization and running, excessive licking, self-mutilation, and house soiling (which might be expressed as defecation while running) may also be associated with hyperesthesia events [3, 7]. The clinical picture varies between cats such that individuals may exhibit all or a few of the described signs. The behavior is often difficult or impossible to interrupt and might be induced simply by rubbing the cat's back, although episodes most commonly begin without any apparent environmental stimulation. Some cats exhibit skin rolling and vacuum licking if the dorsal lumbar area is touched. In fact, it has been suggested that in some cases, pain pathways may be overly sensitive to relatively innocuous touch sensations [32]. As with psychogenic alopecia, a diagnostic workup would focus on ruling out dermatologic, painful, and systemic causes of the signs as well as neurological causes. Treatment should address both the stress-related component and the medical component. In one recent retrospective study of seven cats with hyperesthesia with tail self-trauma, median age was 1 year, and six of the cats were male. Despite a comprehensive workup including (but not limited to) hematology, blood chemistry, serology, and MRI of the brain, spinal cord, and cauda equina and CSF analysis, a diagnosis was not reached for any of the cats, although two were suspected to have hypersensitivity dermatitis. This study demonstrates the challenges in diagnosis and the need for a multidisciplinary approach [33].

Excessive Scratching

Overgrooming can also be characterized by excessive scratching, especially in the area of the head, face, and mouth. In this case, a medical component should be strongly suspected and therefore investigated. In cats, a neuropathy may cause head, muzzle, or neck scratching. Feline orofacial pain syndrome (trigeminal neuralgia) may also present with signs of oral discomfort and tongue mutilation. Onset may be at the time of teething (6 months), and recurrences might be associated with dental disease, otitis media, or stress. It is commonly associated with pain and distress on eating, drinking, and grooming. A breed disposition in Burmese cats has been reported [34]. Dental disease, dermatologic disease, trigeminal neuropathy, and behavioral factors should all be considered. In most cases, treatment will need to focus on reducing the neuropathic pain with drugs such as gabapentin and NMDA antagonist amantadine, along with concurrent anti-inflammatory, pain, and behavioral medications for anxiety and stress. Treatment with methimazole can also cause pruritus in cats [3].

Treatment

As the first step of the treatment, the diagnosed medical problems underlying the repetitive behavior (and other related abnormal behaviors) should be treated. Because of the reciprocal interaction between the medical and behavior component of psychogenic dermatological disease, the behavior treatment should not be dismissed even when a medical component is identified and treated.

The treatment includes:

- Changes in the environment
- Behavior modification
- Pharmacological treatment

Environmental and Behavior Modification

All the stressors identified should be removed or minimized. A predictable and stable environment, in which desired behaviors are consistently promoted and rewarded, is fundamental to increase environmental control and decrease frustration and stress. Lack of environmental enrichment is a frequent source of stress for cats. A cat-friendly house should include plenty of safe spots to increase the cat's perception of the environment as safe: hiding places for safety, perches and cat trees to provide elevated vantage points, and comfortable bedding. There should also be sufficient outlets for all the cat's behavioral needs: food-filled toys and multiple small meals for species-appropriate feeding pattern, olfactory stimulation (catnip, silver vine, valerian, synthetic pheromones), scratching posts for marking, and prey-like toys to promote a species-appropriate play behavior [13, 35–37]. Play sessions can be short (5–10 min) [38] but frequent (2–3 times every day) and should end with a consummatory phase (a few treats).

Social conflicts can be a major source of stress for cats, especially in households with multiple cats, dogs, or young children. It is important to keep in mind that cats experiencing a social conflict may not show overt signs of fear or aggression, but instead decrease their level of activity and isolate themselves in an attempt to reduce the chances that an antagonistic interaction may occur. Each cat should be provided with a core area, in which access to all the resources needed (food and water, resting/sleeping and hiding places, litter boxes, toys) is provided without the need for the cat to interact with any other individual. Cats spend most of their time not engaging in social interactions, even when they live with a preferred associate (i.e., an individual with whom they can be in close proximity and toward whom they show affiliative behaviors like allorubbing and allogrooming). If two individuals show overt antagonistic interactions (hissing, swatting, stalking, pouncing, biting, etc.), they should be separated by means of a physical barrier [37].

E-collars, bandages, shirts, and coats prevent the animal from reaching the affected body area and can help in temporarily decreasing overgrooming. These barriers avoid excessive licking and provide temporary relief for lesions and inflammation to heal; cats experiencing neuropathic pain or paresthesia may also find relief from the compression that bandages, shirts, and coats apply on the affected area. However, these barriers may add to further anxiety until the pet is positively conditioned, and they should be used with caution [3].

Owner responses that may be inadvertently reinforcing the behavior or further adding to the pet's anxiety must cease. Punishment, including verbal (e.g., yelling "no"), of the overgrooming must be avoided because it may result in increased stress and does not address the underlying cause. All training must be reward based (to train and engage the pet in alternative desirable outlets) and sources of stress identified and resolved. The owners should provide constructive activities (e.g., food manipulation toys, new exploratory objects) to occupy the cat when he is not actively engaged with people and comfortable resting places that are elevated and secure to encourage stress-free resting and sleeping. Positive social interactions should be encouraged including play with prey-like toys, reward-based training, clicker training, and short petting of the head and neck area. The latter should, however, be discontinued if it triggers self-grooming. Unpleasant interactions should be avoided, e.g., forced handling and restraining and forced and prolonged interaction with visitors, children, and other animals. Cats exhibiting overgrooming should be actively supervised and their undesired behavior stopped by engaging them in an alternative pleasant activity. For managing situations where problems arise, the pet should be taught cue words for alternative desirable behaviors (mat, tree, come/touch, play/fetch). At the onset of any undesirable behavior, the owner could cue or lure the pet into a desirable alternative behavior or should ignore the pet and walk away (if it causes the behavior to cease) [3, 37].

Pharmacological Treatment

The dosages for all the drugs mentioned in this section can be found in Table 2. Serotoninergic drugs are effective in the treatment of stress/anxiety-related problems

Table 2 Commonly used drugs for the treatment of psychogenic dermatologic disorders

	Drug class	Dose range	Frequency	Time to effect
Fluoxetine	SSRI	0.25–1.0 mg/kg	Q24H	Slow
Paroxetine	SSRI	0.25–1.0 mg/kg	Q24H	Slow
Clomipramine	TCA	0.25–1.0 mg/kg	Q24H	Slow
Amitriptyline	TCA	0.5–1.0 mg/kg	Q12-24H	Slow
Buspirone	Azapirone	0.5–1.0 mg/kg	Q12-24H	Slow
Trazodone	SARI	25–100 mg/cat	PRN or Q12-24H	Fast
Alprazolam	BZD	0.02–0.1 mg/kg	PRN or Q8-24H	Fast
Oxazepam	BZD	0.2–0.5 mg/kg	PRN or Q12-24H	Fast
Clonazepam	BZD	0.05–0.25 mg/kg	PRN or Q8-24H	Fast
Lorazepam	BZD	0.05 mg/kg	PRN or Q12-24H	Fast
Gabapentin	Antiepileptic	2.5–10 mg/kg	Q8-24H	Fast

SSRI selective serotonin reuptake inhibitor, *TCA* tricyclic antidepressants, *SARI* serotonin antagonist and reuptake inhibitor, *BZD* benzodiazepin, *PRN* as needed ("pro re nata"). All drugs are administered PO. Time to effect, slow, between 1 and 4 weeks; fast, between 45 and 90 min [3, 40]

and compulsive behaviors. Treatment with a selective serotonin reuptake inhibitor (SSRI; fluoxetine or paroxetine) or the tricyclic antidepressant (TCA) clomipramine should achieve significant improvement within 4–6 weeks [10, 39]. These drugs should be given on a daily basis, regardless of the exposure to trigger stimuli. Side effects of serotonergic drugs include lethargy, changes in appetite, gastrointestinal upset, and paradoxical response with increased anxiety. Particular caution should be observed for anticholinergic side effects, in particular of paroxetine and clomipramine, which could contribute to urine or feces retention and less frequent elimination. After 4–6 weeks, if the response is insufficient, dose adjustments may be needed. Trazodone, a serotonin antagonist and reuptake inhibitor, or buspirone, a partial serotonin antagonist, can also be used. Sedation, gastrointestinal upset, and paradoxical increase of anxiety are possible side effects of trazodone. Increased agitation and aggression should be monitored if using buspirone. When administering serotoninergic drugs, caution should be exercised due to the risk of serotonin syndrome, and the combination of two serotoninergic drugs should be avoided in cats. Anxiolytics such as benzodiazepines (alprazolam, oxazepam, clonazepam, or lorazepam) can be used in combination with serotonergic medication (SSRIs, TCAs, trazodone, and buspirone). Side effects of benzodiazepines include sedation, increased appetite, paradoxical agitation, and aggression. Prolonged use can give physical addiction and withdrawal symptoms. Diazepam PO should not be administered to cats for the risk of fulminant hepatic necrosis. Trazodone and benzodiazepines can be used as needed when the exposure to intense stressors is anticipated or on a daily schedule. If chronic pain, neuropathic pain, and para-/hyperesthesia are present, consider using clomipramine and gabapentin, both pain modulators. Clomipramine and other TCAs, like amitriptyline and doxepin, possess also an antipruritic effect. The use of NSAIDs and tramadol can be considered for pain management, if this is suspected to play a role in the overgrooming [3, 13, 40].

Of the 21 cats referred for self-induced alopecia and enrolled in the study abovementioned [11], the alopecia improved dramatically in the 2 primarily behavior cases with a combination of behavioral management (including a more predictable daily routine, increasing enrichment through social play sessions, and introduction of new play toys and ceasing punishment) and daily clomipramine. Of the three cats with a partial behavior cause, one responded to behavior management, one was lost to follow-up, and one was not treated.

In another study [14], of 11 cats diagnosed with psychogenic alopecia, all 5 cats treated with clomipramine, 2 of 3 treated with amitriptyline, and 1 of 4 treated with buspirone responded positively. The symptoms resolved completely in six cats treated with drug treatment, environmental modification, or both. Two of the 11 cats did not respond to treatment.

In one study of seven cats presented for hyperesthesia with tail self-trauma, remission was achieved in five cats with gabapentin alone (two cats) or with gabapentin combined with cyclosporine and amitriptyline (one cat), prednisolone and phenobarbital (one cat), and topiramate and meloxicam (one cat). This underscores the difficulty in diagnosing and achieving effective control of hyperesthesia [33].

When overgrooming is related to anxiety, conflict, or underlying stress, supplements can also be used. These include natural products such as synthetic facial and appeasing pheromones that help in reducing stress and social conflicts (Feliway and Feliway Multicat, CEVA), L-theanine (also available in combination with *Magnolia officinalis* and *Phellodendron amurense* and whey protein concentrate in Solliquin, Nutramax), alpha-casozepine (Zylkene, Vetoquinol), and medicated diets like Royal Canin Feline Calm or Hills' Multicat c/d feline stress [3].

Prognosis

The prognosis of psychogenic dermatologic diseases is very variable and depends on the underlying causes and complicating factors. When a specific cause can be identified and removed, e.g., physical separation of two cats in case of social conflict, the disease might be completely resolved. Most cases, however, can be frustrating and require long-term treatment. An accurate diagnosis will facilitate treatment success. The combination of behavioral therapy and judicious use of the appropriate medication can often control overgrooming behaviors in cats.

References

1. Fatjo J, Bowen J. Medical and metabolic influences on behavioural disorders. In: Horwitz DF, Mills DS, editors. BSAVA manual of canine and feline behavioural medicine. 2nd ed. Gloucester: BSAVA; 2009. p. 1–9.
2. Rochlitz I. Basic requirements for good behavioural health and welfare in cats. In: Horwitz DF, Mills DS, editors. BSAVA manual of canine and feline behavioural medicine. 2nd ed. Gloucester: BSAVA; 2009. p. 35–48.
3. Landsberg G, Hunthausen W, Ackeman L. Behavior problems of the dog and cat. 3rd ed. Philadelphia: Elsevier Saunders; 2013.
4. Houpt KA. Domestic animal behavior for veterinarians and animal scientists. 6th ed. Wiley-Blackwell: Ames; 2018.
5. Crowell-Davis SL, Curtis TM, Knowles RJ. Social organization in the cat: a modern understanding. J Feline Med Surg. 2004;6:19–28.
6. Carlson NR. Physiology of behavior. 12th ed. Pearson: Upper Saddle River; 2017.
7. Overall K. Manual of clinical behavior medicine for dogs and cats. Elsevier Mosby: St, Louis; 2013.
8. Fornal CA, Metzler CW, Marrosu F, Ribiero-do-Valle LE, Jacobs BL. A subgroup of dorsal raphe serotonergic neurons in the cat is strongly activated during oral-buccal movements. Brain Res. 1996;716:123–33.
9. Bain M. Compulsive and repetitive behavior disorders: canine and feline overview. In: Horwitz D, editor. Blackwell's five-minute veterinary consult clinical companion: canine and feline behavior. 2nd ed. Hoboken: John Wiley & Sons; 2018. p. 391–403.
10. Overall KL, Dunham AE. Clinical features and outcome in dogs and cats with obsessive-compulsive disorder: 126 cases (1989–2000). J Am Vet Med Assoc. 2002;221:1445–52.
11. Waisglass SE, Landsberg GM, Yager JA, Hall JA. Underlying medical conditions in cats with presumptive psychogenic alopecia. J Am Vet Med Assoc. 2006;228:1705–9.
12. Willemse T, Mudde M, Josephy M, Spruijt BM. The effect of haloperidol and naloxone on excessive grooming behavior in cats. Eur Neuropsychopharmacol. 1994;4:39–45.

13. Bain M. Psychogenic alopecia/overgrooming: feline. In: Horwitz D, editor. Blackwell's five-minute veterinary consult clinical companion: canine and feline behavior. 2nd ed. Hoboken: John Wiley & Sons; 2018. p. 447–55.
14. Sawyer LS, Moon-Fanelli AA, Dodman NH. Psychogenic alopecia in cats: 11 cases (1993–1996). J Am Vet Med Assoc. 1999;214:71–4.
15. Titeux E, Gilbert C, Briand A, Cochet-Faivre N. From feline idiopathic ulcerative dermatitis to feline behavioral ulcerative dermatitis: grooming repetitive behaviors indicators of poor welfare in cats. Front Vet Sci. 2018; https://doi.org/10.3389/fvets.2018.00081.
16. Dantzer D, O'Connor JC, Freund GC, Johnson RW, Kelley KW. From inflammation to sickness and depression: when the immune system subjugates the brain. Nat Rev Neurosci. 2008;9:46–56.
17. Foster JA, McVey NK. Gut–brain axis: how the microbiome influences anxiety and depression. Trends Neurosci. 2013;36:305–12.
18. Siracusa C. Treatments affecting dog behavior: something to be aware of. Vet Rec. 2016;179:460–1.
19. Iwase T, Uehara Y, Shinji H, Tajima A, Seo H, Takada K, et al. Staphylococcus epidermidis Esp inhibits Staphylococcus aureus biofilm formation and nasal colonization. Nature. 2010;465:346–9.
20. Siegel R, Ma J, Zou Z, Jemal A. Cancer statistics. CA Cancer J Clin. 2014;64:9–29.
21. Tlaskalová-Hogenová H, Štepánková R, Hudcovic T, Tucková L, Cukrowska B, Lodinová-Zádníková R, et al. Commensal bacteria (normal microflora), mucosal immunity and chronic inflammatory and autoimmune diseases. Immunol Lett. 2004;93:97–108.
22. Buske-Kirschbaum A, Gieben A, Hollig H, Hellhammer DH. Stress-induced immunomodulation in patients with atopic dermatitis. J Neuroimmunol. 2002;129:161–7.
23. Mitschenko AV, An L, Kupfer J, Niemeier V, Gieler U. Atopic dermatitis and stress? How do emotions come into skin? Hautarzt. 2008;59:314–8.
24. Pasaoglu G, Bavbek S, Tugcu H, Abadoglu O, Misirligil Z. Psychological status of patients with chronic urticaria. J Dermatol. 2006;22:765–71.
25. Panconesi E, Hautman G. Psychophysiology of stress in dermatology. Dermatol Clinic. 1996;14:399–422.
26. Garg A, Chren MM, Sands LP, Matsui MS, Marenus KD, Feingold KR, et al. Psychological stress perturbs epidermal permeability barrier homeostasis: implications for the pathogenesis of stress associated skin disorders. Arch Dermatol. 2001;137:78–82.
27. Scarampella F, Zanna G, Peano A, Fabbri E, Tosti A. Dermoscopic features in 12 cats with dermatophytosis and in 12 cats with self-induced alopecia due to other causes: an observational descriptive study. Vet Dermatol. 2015;26:282–e63.
28. Carmichael KP, Bienzle D, McDonnell JJ. Feline leukemia virus-associated myelopathy in cats. Vet Pathol. 2002;39:536–45.
29. Ciribassi J. Understanding behavior: feline hyperesthesia syndrome. Compend Contin Educ Vet. 2009;31:E10.
30. Coates JR, Dewey CW. Cervical spinal hyperesthesia as a clinical sign of intracranial disease. Compend Contin Educ Vet. 1998;20:1025–37.
31. March P, Fischer JR, Potthoff A. Electromyographic and histological abnormalities in epaxial muscles of cats with feline hyperesthesia syndrome. J Vet Int Med. 1999;13:238.
32. Drew LJ, MacDermott AB. Neuroscience: unbearable lightness of touch. Nature. 2009;462:580–1.
33. Batle PA, Rusbridge C, Nuttall T, Heath S, Marioni-Henry K. Feline hyperesthesia syndrome with self-trauma to the tail; retrospective study of seven cases and proposal for integrated multidisciplinary approach. J Fel Med Surg. 2018;. https://doi.org/10.1177/1098612X18764246
34. Rusbridge C, Heath S, Gunn-Moore KSP, Johnston N, AK MF. Feline orofacial pain syndrome (FOPS); a retrospective study of 113 cases. J Fel Med Surg. 2010;12:498–508.
35. Ellis JJ, Stryhn H, Spears J, Cockram MS. Environmental enrichment choices of shelter cats. Behav Process. 2017;141:291–6.

36. Herron MH, Buffington CAT. Environmental enrichment for indoor cats: implementing enrichment. Compend Contin Educ Vet. 2012;34:E3.
37. Siracusa C. Creating harmony in multiple cat households. In: Little, editor. August's consultations in feline internal medicine, vol. 7. Philadelphia: Elsevier; 2016. p. 931–40.
38. Strickler BL, Shull EA. An owner survey of toys, activities, and behavior problems in indoor cats. J Vet Behav. 2014;9:207–14.
39. Seksel K, Lindeman MJ. Use of clomipramine in the treatment of anxiety-related and obsessive-compulsive disorders in cats. Aust Vet J. 1998;76:317–21.
40. Siracusa C, Horwitz D. Psychopharmacology. In: Horwitz D, editor. Blackwell's five-minute veterinary consult clinical companion: canine and feline behavior. 2nd ed. Hoboken: John Wiley & Sons; 2018. p. 961–74.

Neoplastic Diseases

David J. Argyle and Špela Bavčar

Abstract

Cancer is a major disease of cats in terms of health and welfare, and skin tumours in cats are the second most common tumour type, accounting for around 25% of all reported neoplasms (Argyle, Decision making in small animal oncology. Oxford: Blackwell/Wiley, 2008). Compared to dogs (who demonstrate significantly higher numbers of benign skin masses), around 65–70% of skin masses in cats are malignant. Occasionally tumours of the skin are actually metastatic lesions. The best example of this is the syndrome of digital and cutaneous metastasis associated with lung cancer in cats (less commonly seen in the dog) and is described in more detail below (Goldfinch and Argyle, J Feline Med Surg 14:202–8, 2012).

For many years, feline cancer medicine was dominated by virally induced lymphoma. While lymphoma is still a major problem in cats, the increase in vaccination has reduced the incidence of this disease and allowed other tumours to become prominent players, particularly squamous cell carcinoma (SCC), mast cell disease and vaccine-associated sarcomas. This chapter is designed to give the reader a broad understanding of the classification and approach to cancer in cats and a synopsis of the major tumour types.

D. J. Argyle (✉) · Š. Bavčar
The Royal (Dick) School of Veterinary Studies, University of Edinburgh, Easter Bush, Midlothian, UK
e-mail: david.argyle@roslin.ed.ac.uk; david.argyle@ed.ac.uk

© Springer Nature Switzerland AG 2020
C. Noli, S. Colombo (eds.), *Feline Dermatology*,
https://doi.org/10.1007/978-3-030-29836-4_30

583

Introduction

Cancer is a major disease of cats in terms of health and welfare, and skin tumours in cats are the second most common tumour type, accounting for around 25% of all reported neoplasms [1]. Compared to dogs (who demonstrate significantly higher numbers of benign skin masses), around 65–70% of skin masses in cats are malignant. Occasionally tumours of the skin actually represent spread of the cancer from other sites. One example of this can be seen in cats with primary lung tumours, where metastatic lesions can be found within the skin or digits (uncommon in canine lung tumours) and is described in more detail below [2].

For many years, feline cancer medicine was dominated by virally induced lymphoma. While lymphoma is still a major problem in cats, the increase in vaccination has reduced the incidence of this disease and allowed other tumours to become prominent players, particularly squamous cell carcinoma (SCC), mast cell disease and vaccine-associated sarcomas. This chapter is designed to give the reader a broad understanding of the classification and approach to cancer in cats (Box 1) and a synopsis of the major tumour types.

Box 1: Key points regarding feline skin tumours
- Cats should not be considered as 'small dogs' and have tumours of differing biology and natural history.
- Any lump on a cat should be treated with suspicion considering the high proportion of malignant lesions in this species. Any cat with a lump should be investigated. This also includes any suspicious non-healing ulcerated lesions.
- Skin tumours can represent metastatic lesions for other 'non-skin' tumours.
- Malignant tumours may have history of rapid growth and may be fixed to underlying structures.
- Both malignant and benign tumours may be ulcerated.
- Appearance of growth varies depending on type of tumour and location.
- The causes of skin cancer are very similar to other species, and include physical factors, immune function and also viral causes. However, the underlying characteristic is often one of chronic inflammation.
- Physical factors include ionizing radiation and ultraviolet radiation. The association between SCC development and solar exposure of skin in white cats has been established epidemiologically. White cats in California had been shown to have a 13-fold increased risk of developing SCC [3].
- The ability to induce neoplastic transformation in mucosal infections of papillomaviruses of mammals is well established. Infection of the keratinocyte can stimulate increased proliferation and terminal differentiation. Neoplastic transformation arises from interaction of papilloma viral proteins with cellular proteins (disruption of p53 by viral protein E6 and the inhibition of pRB by viral protein E7) [4]. In cats, papillomaviruses are associated with viral plaques and feline fibropapillomas (sometimes referred to as *feline sarcoids*). A novel feline papillomavirus has been sequenced from three feline Bowen in situ lesions [5–7].

- There is much evidence for the role of immune surveillance in the control of cancer. The successful use of immune stimulants for early cancer lesions such as imiquimod for carcinoma in situ also supports the role of the immune system in controlling skin cancer. Immunosuppression, through sat FIV infection, is likely to predispose cats to cancer.

Classification of Skin Tumours in Cats

Skin tumours in cats are normally classified based on the following criteria:

- Tissue of origin (mesenchymal, epithelial, melanotic or round cell)
- Cell of origin if applicable (e.g. mast cell tumour)
- Level of malignancy

Based on reported epidemiological studies (which are limited by numbers), the reported top four most common skin tumours of cats are basal cell tumours, squamous cell carcinoma, mast cell tumour and soft-tissue sarcoma (mainly fibrosarcoma). These four tumour types make up about 70% of all reported feline skin tumours (Table 1) [3].

Approach to Cats with Cancer

A detailed description of the approach (biopsy techniques) is out of the scope of this chapter, but the reader is referred to other texts [1]. However, the general approach is common.

- Detailed history and physical exam are essential. Length of duration, rate of growth and any clinical sign associated with the tumour may be helpful in differentiating benign from malignant masses.
- Measurement of tumour, photography and recording the exact location of tumours on a diagrammatic body map is paramount.
- Local lymph node assessment, measurement and aspiration for cytology if possible should be performed.
- Cytological and histopathological evaluation of the mass or lesion is critical. The type of biopsy performed is usually dictated by the location. Where wide surgical excision is feasible without undue morbidity, the biopsy can be combined with a therapeutic procedure. However, in most instances, the biopsy is a diagnostic test and can take different forms:
  - Fine needle aspiration of mass for cytology should always be performed.
  - Excisional or incisional biopsy of the lesion for histopathological assessment might be indicated. A preference is usually for the use of skin punch biopsies. This usually allows for a sufficiently large piece of tissue for an accurate diagnosis, grading of the tumour and perhaps advanced histopathologic techniques such as immunohistochemistry (IHC).

Table 1 Classification of cutaneous neoplasms in domestic animals

Epithelial Tumours	
Basal cell carcinoma	
Squamous cell carcinoma	
Papilloma	
Adnexal tumours	
Melanocytic Tumours	
Benign melanoma	
Malignant melanoma	
Mesenchymal tumours (soft-tissue sarcomas)	
Fibrous tissue	*Fibroma* *Fibrosarcoma*
Nervous tissue	*Peripheral nerve sheath tumour*
Adipose tissue	*Neurofibrosarcoma* *Lipoma* *Liposarcoma*
Smooth muscle	*Leiomyoma* *Leiomyosarcoma*
Myxomatous tissue	*Myxoma* *Myxosarcoma*
Mast cell tumour	
Visceral	
Cutaneous (atypical and mastocytic)	
Vascular tumours	
Haemangioma	
Haemangiosarcoma	
Lymphoma	
Dermatropic	
Epitheliotropic	
Histiocytic disease	
Feline progressive histiocytosis	
Feline pulmonary Langerhans cell histiocytosis	
Feline histiocytic sarcoma (solitary and disseminated)	
Feline haemophagocytic histiocytic sarcoma	

Collectively, mast cell tumours, cutaneous lymphoma, cutaneous plasma cell tumours, histiocytomas, and neuroendocrine (Merkel cell) tumours are referred to as *round cell tumours*.

- Additional diagnostic tests may be needed depending on pathology of the lesion and/or patient assessment (such as thoracic radiography, abdominal ultrasound).
- Assessment of any concurrent disease, which may include clinical pathology (haematology and biochemistry).

Staging

- The above approach to the patient answers two fundamental questions:
  - What is the nature of the lesion?
  - How far has it spread locally or to distant sites?

Table 2 Staging for feline epidermal and dermal tumours (excluding mast cell tumours and lymphoma)

T	Primary tumour
T_{is}	Pre-invasive carcinoma (carcinoma in situ)
T_0	No evidence of tumour
T_1	Superficial tumour <2cm maximum diameter
T_2	Tumour 2-5cm maximum diameter, or with minimal invasion (irrespective of size)
T_3	Tumour >5cm maximum diameter, or with invasion of subcutis (irrespective of size)
T_4	Tumour invading other structures such as fascia, bone, muscle and cartilage Where there are multiple tumours arising simultaneously, these should be mapped and recorded The tumour with the highest T value is recorded and the number of tumours recorded in parentheses (e.g. T4(6)) Successive tumours are classified independently
N	Regional lymph node
N_0	No evidence of lymph node metastasis
N_1	Moveable ipsilateral nodes N_{1a}: Nodes not considered to contain growth N_{1b}: Nodes considered to contain growth
N_2	Moveable contralateral nodes or bilateral nodes N_{2a}: Nodes not considered to contain growth N_{2b}: Nodes considered to contain growth
N_3	Fixed lymph nodes
M	Distant metastasis
M_0	No evidence of distant metastasis
M_1	Distant metastasis detected

- This approach allows us to stage the patient. The TNM *staging* system is based on the size and/or extent of the primary *tumour* (T), whether *cancer* cells have spread to nearby (regional) lymph nodes (N) and whether metastasis (M) or the spread of the *cancer* to distant sites within the body has occurred.
- The staging system for feline skin tumours is shown in Table 2.

Treatment Options

Therapeutic options for cancer in feline patients are given in more detail in the specific sections below. However, in general they include the following:

- Surgical excision with complete margins (standard of care)
- Cytoreductive surgery to provide palliation of large tumours
- Amputation for large tumours on extremities
- Radiotherapy in tumours with incomplete excision (as an adjunct to cytoreductive surgery)
- Additional therapeutic options such as photodynamic therapy, cryosurgery, laser ablation and hyperthermia
- Chemotherapy for lymphoma or disseminated tumour types

Specific Tumour Types 1: Epithelial Tumours

Papilloma [4–7]

Papillomas are benign epidermal proliferative lesions that are often associated with papillomavirus infection. Papillomas typically have an exophytic growth pattern. Surgical excision can be curative, and some of these lesions will spontaneously regress. In cats, a particular type of papilloma, the fibropapilloma, is observed. These tumours demonstrate a proliferation of mesenchymal cells covered by hyperplastic epithelium and resemble equine sarcoids. Evaluation for papillomavirus demonstrated an apparent non-productive infection of the mesenchymal cells.

Basal Cell Carcinoma (BCC) [8, 9]

The incidence of BCCs in cats has previously been overestimated because of the inclusion of other tumour types that are now known not to be true BCC, such as apocrine ductular adenoma (approximately 60%) and trichoblastoma (approximately 40%). Cytologically, BCCs contain inflammatory cells, squamous cells, sebaceous epithelial cells, melanin and melanophages, and cells can express the criteria of malignancy. Clinically, the majority of tumours classified as BCC behave in a benign fashion (about 10% behave in a more malignant fashion, based on invasion and cellular pleomorphism). Treatment for BCCs is wide surgical excision, which often results in long-term control. Adjunctive radiation may be considered in more malignant variants where local surgical control cannot be achieved.

Actinic Keratosis [10, 11]

This is often referred to as a 'pre-cancerous' lesion that is induced by exposure to sunlight. It is often difficult clinically (and sometimes pathologically) to distinguish this from Bowenoid carcinoma or fully blown SCC. Actinic keratosis can progress to SCC (see below).

- Lesions are usually alopecic, erythematous plaques, often with erosion. Lesions on the pinnae may appear symmetrical in white cats (Fig. 1).
- Skin biopsy is required for definitive diagnosis and reveals epidermal hyperplasia, dysplasia and hyperkeratosis. Changes are confined to the epidermis.
- Where premalignant change is suspected, surgical removal of affected skin is the treatment of choice. Where lesions affect the pinnae of the cat, radical pinnectomy is indicated to reduce the risk of progression to neoplastic lesions and often produces a cosmetically acceptable appearance. Once lesions have been removed, avoidance of further sun exposure is indicated to prevent new lesions developing. Medical therapy is rarely indicated in cats. Topical 5-fluorouracil is commonly used in humans, but it is highly neurotoxic in the cat.

Fig. 1 Actinic keratosis on the pinnae of a white cat: erythema, alopecia and exfoliation are evident on both ears. On the left pinnae, there are also small erosions and a crust. (Courtesy of Dr. Chiara Noli)

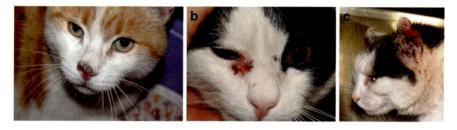

Fig. 2 Sun-induced squamous cell carcinoma: erosive, ulcerative, encrusted lesions are evident on non-pigmented skin on the nose (**a**), conjunctiva (**b**) and pinna (**c**). (Courtesy of Dr. Chiara Noli)

Feline Cutaneous Squamous Cell Carcinoma [1, 12–16]

Squamous cell carcinoma (SCC) is a malignant tumour arising from squamous epithelium. It represents 15% of all cutaneous tumours and the vast majority of oral malignant tumours in cats. This tends to be a disease mainly observed in older cats, with the median age at presentation being 10–12 years. The behaviour and cause of SCC are variable and dependent on the site of the tumour.

Cats present with progressive clinical signs, demonstrating often crusty and erythematous lesions, superficial erosion and ulceration (carcinoma in situ or early SCC). SCC originates from the cornified external surface of the skin and lesions are often deeply invasive and erosive. Key sites in cats include nasal planum, head and neck (especially pinna and eyelids) (>80%) with multiple lesions in 30% of cases.

SCC in Sun-Exposed Areas
- It is mainly seen in the areas of decreased pigmentation exposed to sun. SCC is associated with ultraviolet irradiation (UVA and UVB) from sunlight.
- In feline patients, SCC most commonly occurs on the nasal planum, the eyelids and ear pinnae (Fig. 2a–c).

- White-haired cats have 13.4 times greater risk of developing SCC than cats of other coat colours.
- Non-white-haired cats develop SCC in areas of poor pigmentation and poorly haired areas. Melanin protects the skin against solar energy.
- Tumours are locally invasive but slow to metastasize.
- Appearance of the tumour may be variable. 'Productive' forms with papillary growths resemble cauliflower-like lesions, while 'erosive' forms manifest as ulcerative lesions with raised edges. In both cases, the tumour is commonly ulcerated and secondary infection is present. It is not unusual for these tumours to be initially mistaken for inflammatory or infectious lesions.
- Cats that are at increased risk of developing cutaneous SCC should avoid being in the sun, especially at the height of the day. There are several options to be considered to reduce UV light exposure when indoors, such as UV light blocking film that can be applied to windows. Sunblock can be applied to the ears of cats that go outdoors. Ingestion of sunscreen should be avoided if possible; therefore, application on the nasal planum is not recommended.

Bowenoid Carcinoma In Situ (Arising in Non-Sun-Exposed Areas)

Multifocally distributed superficial lesions in the haired, pigmented areas of the skin and unrelated to sun exposure have been reported (Fig. 3a, b). This condition is called 'multicentric SCC in situ' or Bowen's disease. Lesions are crusty, easily epilated, painful and haemorrhagic. These lesions are histologically confined to the superficial layers of the skin and do not breach the basement membrane (Fig. 4a). A possible cause for development of Bowen's disease is papillomavirus, and its antigen has been confirmed to be present in 45% of feline cutaneous lesions using immunohistochemistry (Fig. 4b). Complete excision of these lesions is curative and recurrence is uncommon; however, de novo lesions often occur in other areas of the skin. Surgical laser treatment has also been applied to this disease, but large- scale clinical studies have not been performed.

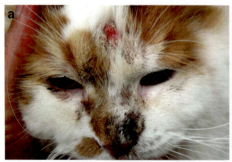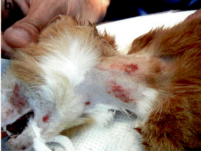

Fig. 3 Bowenoid in situ SCC in a cat: multicentric crusty and ulcerative lesions on the head (**a**) and neck (**b**). (Courtesy of Dr. Chiara Noli)

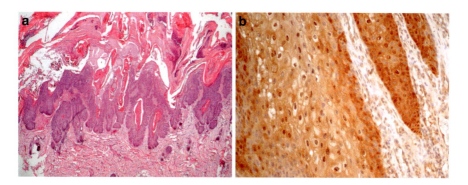

Fig. 4 Histological aspect of the lesions of the cat in Fig. 3. (**a**) Important epidermal proliferation and dysplasia not breaching the basal membrane (H&E 10×). (Courtesy of Dr. Chiara Noli); (**b**) immunohistochemical stain for the papilloma virus antigen p16 is clearly positive in the epithelial cells of the epidermis (brown coloured). (20×, Courtesy of Prof. Francesca Abramo)

Treatment Options for SCC in Cats
Surgery
- For tumours of the pinnae, surgical excision (pinnectomy) provides long-term local tumour control (>18 months).
- For nasal planum and lower eyelid tumours, surgery can offer good local control, but it is recommended that these cases are referred to a specialist in surgery for best results.

Cryotherapy
- Aggressive treatment with cryotherapy can provide good outcome for tumours of the eyelids and ear pinnae, while response to this form of treatment of tumours of the nasal planum is less favourable.

Radiotherapy
- Radiotherapy using an external beam source has provided good local control of low stage tumours.
- T_1 stage tumours respond better than those at T_3 or T_4.
- Eighty-five percent of feline patients with low stage disease (T_1) were alive 12 months following management with orthovoltage radiotherapy, compared to 45.5% of patients with T_3 tumours.
- It is less cosmetically challenging than surgery; however, it involves numerous anaesthetics, and recurrence rates are higher than those seen with patients undergoing surgery.
- Strontium-90 plesiotherapy has been shown to be effective for patients with superficial tumours. This form of beta radiation is utilized for lesions measuring 3 mm or less in depth. It allows sparing of the local normal tissue and is also repeatable. An 8 mm diameter ophthalmic applicator impregnated with strontium-90 comes in contact with the skin to deliver the prescribed dose of radiation over a certain amount of time. In two studies, 13/15 cats and 43/49 cats

achieved complete remission for a median of 692 days and 1071 days, respectively.

Chemotherapy (and Electrochemotherapy)

- For cats with nasal planum tumours, intratumoral administration of chemotherapy using carboplatin in a sesame oil suspension was shown to be safe, practical and efficacious.
- In one study, complete response was documented in 73% of feline patients and progression-free interval of 12 months was documented in 55% of patients.
- Following intravenous administration of mitoxantrone, four out of 32 treated cats demonstrated response to treatment.
- Electrochemotherapy, utilizing intralesional administration of bleomycin, resulted in 7/9 patients responding beneficially to treatment.

Imiquimod Cream

- Imiquimod (immune activator that signals through Toll-like receptors) as a 5% cream can be used for multiple lesions of Bowenoid multicentric in situ SCC.
- Imiquimod is an immunomodulator with both antitumoral and antiviral effects and is licensed for the treatment of actinic keratosis, basal cell carcinoma and genital warts in people.
- Owners should wear gloves when applying this cream and stop cats from ingesting the product.
- In a study of 12 cats with multicentric SCC in situ, all cats responded to treatment, although nine cats developed new lesions, which responded to treatment as well.

Photodynamic Therapy

- Only low stage superficial tumours positively respond.
- Complete remission was observed in 85% of patient; however, disease recurrence was noted in 51% after a median time of 157 days.
- Use of intravenous photosensitizers demonstrated an initial response rate of 49% and 100%, respectively, with 'overall tumour control' of 1 year reported in 61% and 75% of cases.
- Pain at the time of treatment has been reported by people and cats have experienced increase of heart rate in one study, despite the use of anaesthesia and analgesia at the time of treatment.

Specific Tumour Types 2: Feline Injection-Site Sarcoma [17–21]

The first description of this disease was by Hendrick and Goldschmidt in 1991 and was originally termed *vaccine-associated sarcoma*, as the original epidemiological data in the USA suggested a strong association between the disease and vaccination with either rabies or feline leukemia virus (FeLV) vaccines. Since this time, studies investigating the pathophysiology of this disease have indicated that any foreign material injected into cats that can cause a local and intense inflammatory response could lead to this disease. Consequently, this disease is now termed *feline injection-site sarcoma (FISS)*.

The disease has been reported all over the world with varying incidence. However, the main features of this disease remain constant.

- It is a disease of low metastatic potential but is highly locally invasive.
- There is often a significant lag period between the injection and the ultimate development of a tumour.
- Once the tumour develops, a period of rapid growth is often observed.
- Single modality therapy is rarely curative, and sophisticated imaging techniques are necessary to determine the extent of the disease prior to treatment.
- The pathogenesis of this disease is still poorly understood, but has led to a number of recommendations in terms of vaccination strategies and how to manage post-vaccination nodules.
- The pathology of this tumour is that of a mesenchymal (soft-tissue) tumour, with variable aspects. The most frequently diagnosed on histology is fibrosarcoma, but malignant fibrous histiocytoma, osteosarcoma, chondrosarcoma, rhabdomyosarcoma and undifferentiated sarcoma have also been reported.
- The metastatic rate is reported to be around 20%.
- The prevalence of FISS cases varies between countries; however, the numbers have overall increased within the last 10 years. The true incidence of FISS is controversial, with data suggesting that the frequency ranges from 1/1000 to 1/10,000. It is possible that the variable latency period of this tumour (2 months to several years), makes it difficult to assess the true incidence of this disease.
- From multiple studies, a series of recommendations on vaccination have emerged (Box 2), the aim being to reduce inflammation at the affected site and to use sites more accessible to surgery than the inter-scapular space.
- Any injection that can cause local inflammation could ultimately lead to the development of this disease.

Box 2: Key points from the vaccine recommendations
- No vaccine should be administered in the inter-scapular space.
- Rabies vaccine should be given in the right hind limb, below the stifle.
- FeLV vaccine can be given in the left hind limb, below the stifle.
- All others vaccines should be given in the right shoulder below the scapula.
- Use subcutaneous injection rather than intramuscular as this will allow earlier detection of any tumour.
- Record vaccine and batch numbers carefully.
- Warm to room temperature is preferred.
- Avoid polyvalent vaccines.

It is important to recognize that these recommendations are only guidelines, since an exact causative relationship between vaccine types/brands has not been unequivocally established.

Fig. 5 An injection-site
fibrosarcoma is evident as
a large nodule in the
interscapular space.
(Courtesy of Dr.
Chiara Noli)

Post-Vaccination Nodules

The incidence of post-injection site nodules in cats can be quite high. As this could lead to the development of FISS, a veterinary practitioner should consider the identification of such a lesion quite seriously. Most post-injection site nodules will resolve within 2–3 months but any nodule that does not, or increases in size, should be considered as suspicious. This monitoring is critical as, once a tumour develops, management can be highly challenging. It is therefore recommended that every mass that (a) persists for more than 3 months after injection and/or (b) becomes larger than 2 cm, and/or (c) increases in size 1 month after an injection should be biopsied. This is referred to as the 3-2-1 rule.

Clinical Presentation

FISS often presents as a painless, rapidly growing, subcutaneous, firm nodule/mass detected soon after recent injection at a known injection site (Fig. 5). However, cases with a slower growth pattern, greater time interval between injection and presentation or discomfort have been recorded. Cats receiving intramuscular injections may develop tumours in deeper sites. Ultimately, a tumour can become ulcerated and infected. Rarely, a case could present with signs referable to metastatic lung disease, unless the cat had been completely neglected by the owner.

Diagnosis and Staging

The presentation of this disease, coupled with injection or vaccination history, should alert the clinician very quickly to the possible diagnosis of FISS. A diagnostic database should include the following:

- Full history with vaccination and injection history should be collected.

Fig. 6 Cytological aspect of fibrosarcoma: spindle-shaped mesenchymal cells are observed surrounded by matrix substance (Diff Quick, 100×). (Courtesy of Dr. Chiara Noli)

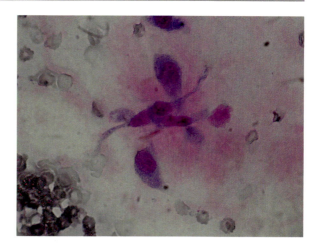

- A full clinical examination is mandatory.
- The tumour should be measured and a formal note of this made in the patient's records.
- Routine haematology, serum chemistry and urinalysis should be performed to highlight any concurrent disease.
- A histological or cytological diagnosis should be made (Fig. 6). Although a fine needle aspirate (FNA) cytology can alert the clinician to a possible mesenchymal tumour, FNA is only diagnostic in around 50% of the cases.
- In the authors' experience, an incisional biopsy is the most valuable in terms of making a histological diagnosis. The heterogenous nature of the tumour can lead to a false diagnosis with 'tru-cut' techniques, as these can indicate granulation tissue rather than tumour.
- If an incisional biopsy is performed, then the clinician must do this remembering that the biopsy site will need to be excised as well if a definitive surgery is attempted.
- FISS is locally invasive and metastatic spread has been reported in approximately 20% of cases. Imaging is an essential component of staging this disease. At the very least, the author will perform three view thoracic films (right and left lateral and dorsoventral views) to explore the potential of metastatic disease to the lungs. However, the most important imaging modalities in FISS diagnosis for treatment planning are advanced techniques such as CT and MRI (Fig. 7). These techniques allow evaluation of the degree of invasion of the tumour to deeper structures, which is essential for tumours in the interscapular space (less so for tumours of the limb). The recurrence rate after surgery has been reported to be around 45%, and this could undoubtedly be lowered by more accurate staging and treatment planning prior to surgery. There is now much wider availability of CT and MRI, and the authors would recommend that this is an essential tool for pre-treatment planning of all FISS affecting the interscapular space.
- Abdominal ultrasound can be performed for staging purposes. However, in the authors' experience, this should be limited to cases in which the clinician is suspicious of an abdominal lesion following a thorough clinical examination.

Fig. 7 CT appearance of a feline injection-site sarcoma

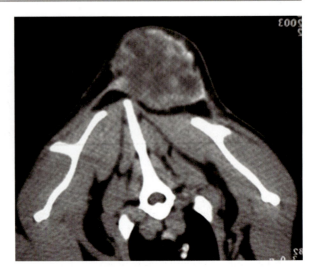

Fig. 8 Histologic appearance of feline injection-site sarcoma. A lymphocytic infiltrate (left) is evident at the periphery of the tumour, and necrosis (right) in the centre (H&E 10×). (Courtesy of Dr. Chiara Noli)

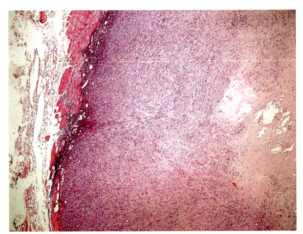

- Although a grading system based upon the pathology of the tumour has been suggested, this would appear to have limited clinical utility in this disease. Most commonly, the pathology is a fibrosarcoma but not uncommonly has a mixed pathology, with pleomorphism, the presence of giant cells and mitotic figures. Peripheral inflammation and necrotic areas are almost always present (Fig. 8) and are a considered diagnostic criteria, respectively, reflecting the immunological nature of the disease and its rapid growth rate. The presence of 'substances consistent with adjuvant material' (bluish, refractile inclusions) is also consistently helpful. However, no single diagnostic criteria can be deemed pathognomonic.

Treatment

FISS is a complex disease and cure is difficult to achieve and unlikely with single modality therapy. However, wide surgical excision of the primary tumour (margin of 3–5 cm of macroscopically healthy tissue and at least one fascial plane beneath the tumour) is considered to be the mainstay of treatment. For lesions affecting the limbs or tail, surgical amputation is considered to be the technique of choice. For lesions affecting the interscapular space, radical excision including spinous vertebral process amputation or partial or total scapulectomy may be necessary. In general, the application of these criteria to clinical cases gives a disease-free interval (DFI) of approximately 10 months. However, DFI has been shown to be significantly increased (16 months) when the surgery is performed by an experienced board-certified surgeon.

Surgical Excision
- Aggressive surgery with curative intent is not indicated in the face of detectable metastasis at the outset of treatment.
- 3–5 cm margins and one fascial plane should be applied.
- Recovery time is can be approximately 4–6 weeks.
- Wide surgical excision has a 30%–70% recurrence rate.
- For radical first excision, the mean survival time (MST) has been reported to be around 325 days, compared to 79 days for marginal excision. A 2-year survival rate of 13.8% has been reported for all treated cases.
- Rear leg amputation has the highest rate of cure.
- Complete margin of excision can give a tumour-free survival of >16 months, compared to 4–9 months for incomplete resections.
- Surgical margins should be examined carefully. The margins giving most concern can be identified by tagging or inking, which will assist the pathologist.
- The rate of local recurrence, _despite reported clean histological margins_, may be as high as 42%. Vigilance after surgery is essential.
- Treatment at a referral hospital has demonstrated most favourable clinical outcomes as cases are more likely to undergo radical surgery.

Radiation Therapy
- Radiotherapy has been used in specialist oncology practice for many years, and is becoming more available. There has been a recognized veterinary speciality in the USA since 1994, and treatment with radiotherapy is much more commonly used in the USA. Radiotherapy may be performed before or after surgery for FISS.
- _When radiotherapy is performed after surgery_, the intention is to reduce the likelihood of tumour regrowth and would usually be indicated in the face of incomplete surgical margins. This form of treatment will not address the possibility of metastatic disease.
- _When radiation is performed before surgery_, it is done with the intention of reducing the size and biological activity of the tumour in order to facilitate a more successful surgical resection. Some centres advocate radiotherapy both before and after surgery, although this approach is not commonly performed in Europe at present.

Chemotherapy

• The role of chemotherapy in FISS is a subject of controversy. The rationale for using chemotherapy is two-fold. Chemotherapy can be used in the palliative setting to improve the quality of life where macroscopic disease burden cannot be addressed by surgery or radiation. Chemotherapy can also be used as part of definitive therapy, aiming at reducing the onset of metastatic disease after adequate local disease control (adjuvant therapy). Moreover, chemotherapy prior to surgery (neoadjuvant therapy) may reduce the size of the mass prior to attempting surgical resection. This has been reported in the literature as a method to 'down-stage' a tumour. However, in the authors' experience, significant shrinkage at the tolerated doses of chemotherapy (1 mg/kg doxorubicin) is unlikely.

Specific Tumour Types 3: Mast Cell Tumour [22–24]

Mast cell tumours (MCTs) account for approximately 20% of all cutaneous tumours in cats, making them the second most common skin neoplasia in this species. Two distinct forms are recognized in cats (Fig. 9): the mastocytic form, which resembles canine MCT, and the atypical form, which is less common and was previously referred to as the 'histiocytic' form. The mastocytic MCT is most commonly seen in cats with mean age of 10 years, and no sex predisposition has been described. The atypical form, on the other hand, has been mainly documented in young (< 4 years old) Siamese cats. This breed is also predisposed to developing the mastocytic form. The aetiology of MCT in cats is not known; however, Siamese breed appears to have a genetic predisposition.

Clinical Presentation

Most commonly, feline cutaneous MCT presents as solitary, firm, well-circumscribed, alopecic dermal nodule. Approximately 25% of MCTs are superficially ulcerated. Other possible presentations include discrete subcutaneous nodule or flat, itchy, plaque-like lesions similar in appearance to eosinophilic granuloma. Redness and pruritus are not uncommon and Darier's sign has been observed. Multiple MCTs have been observed in about 20% of patients (Fig. 10). Cutaneous lesions most commonly appear on the head and neck. Metastatic potential of cutaneous MCT in cats is variable, and has been reported in up to 22% of cases. The visceral form of MCT is more commonly seen in feline than canine patients, and up to 50% of cats present with the visceral (splenic, intestinal) form of this disease. In these cases, symptoms such as diarrhoea in the intestinal form can be present for several months before diagnosis, and systemic spread is common. Mast cell tumour has also been reported in the cranial mediastinum of a cat (EBM IV). In the visceral or disseminated form of MCT, signs attributed to mast cell degranulation can be observed.

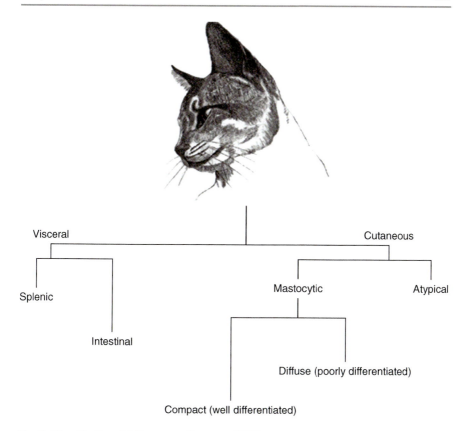

Fig. 9 Classification of feline mast cell tumour (MCT)

Fig. 10 A cat presenting
with multiple mast cell
tumours. (Courtesy of
Dr. Chiara Noli)

Clinical Assessment, Biopsy and Staging

Clinical approach in cats with MCT is similar to that in canine patients. The diagnosis is easily obtained by fine needle aspiration cytology (Fig. 11).

Complete staging, including loco-regional lymph node assessment and fine needle aspiration, abdominal ultrasound with cytology of visceral organs, thoracic radiographs, bone marrow aspiration/buffy coat smear, should be considered in feline patients presenting with any of the following:

- Multiple cutaneous MCT
- Abdominal mass/organomegaly
- Diffuse tumour on histology
- MCT of the spleen, intestine or cranial mediastinum

Prognosis

The grading systems used in canine MCT cannot be applied to cats; however, histological presentation of MCT correlates with outcome. The mastocytic form, previously divided into compact and diffuse, is now separated into well-differentiated (historically known as compact) and pleomorphic form (historically known as diffuse). The well-differentiated form accounts for 50–90% of all cutaneous MCTs and has a more benign clinical course. The pleomorphic form of mastocytic MCT is histologically more anaplastic and has a more malignant biological behaviour, and is therefore related to a worse prognosis. Prognosis for solitary tumours is generally better than that of multiple tumours. Other prognostic factors, such as proliferation markers, have not been extensively studied; however, high mitotic index has been related to a more malignant behaviour.

Fig. 11 Cytological appearance of mast cell tumour: numerous round cells containing metachromatic granules (MGG 40×)

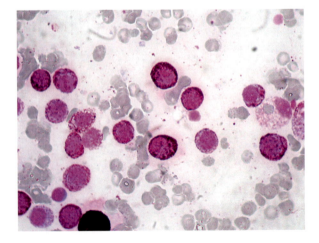

Treatment

Surgery

For cats with solitary cutaneous MCT of the head or neck, surgery is the best treatment option; however, this has been found to be curative in fewer cases compared to dogs. Excision with wide surgical margins is often not possible; however, incomplete excision, especially in histologically well-differentiated (compact) MCT, has not been associated with worse prognosis. Rate of recurrence post-surgery varies between 0% and 24%. More aggressive surgical approach with attempted excision with wide margins should be considered for more aggressive MCT (pleomorphic or diffuse) identified on preoperative biopsy. Prognostic factors that might help to identify potentially aggressive cutaneous MCT in cats are number of lesions (solitary versus multiple), histologic findings (pleomorphism), KIT immunoreactivity score, mitotic index and Ki67 score. Spontaneous regression has been described in some atypical (histiocytic) MCTs. Treatment options for these include marginal excision and periodical monitoring.

Supportive Treatment

Similarly to canine patients with MCT, perioperative administration of H1 and H2 blockers may help to prevent degranulation of mast cells.

Radiotherapy

Radiotherapy is uncommonly reported in the treatment of cats with cutaneous MCT, as most patients have either multiple MCTs or evidence of distant spread at the time of diagnosis. External beam and strontium 90 treatment have been described, with strontium 90 radiotherapy of solitary/localized tumours providing local control in 98% of the patients.

Chemotherapy

The role of treatment using chemotherapeutic agents as a form of palliative or adjuvant therapy in cats with cutaneous MCT has not been well established. Systemic treatment is generally considered for patients with histologically aggressive (pleomorphic/diffuse) or locally invasive tumours, or for MCT with confirmed spread. Chemotherapeutics used in the treatment of cutaneous MCT in cats include vinblastine, chlorambucil and lomustine. In a study looking at the treatment of gross MCT with lomustine, a response rate of 50% and response duration of 168 days have been described. There is no evidence for use of glucocorticoids in the management of feline MCTs.

Tyrosine Kinase Inhibitors

Up to 67% of feline cutaneous MCT harbour mutations in c-KIT proto-oncogene. There is mainly anecdotal data on the use of tyrosine kinase inhibitors targeting KIT in cats and the adverse effect profile of these drugs has not been thoroughly evaluated. One recent study has assessed the toxicity associated with treatment with masitinib, when administered at a total dose of 50 mg per cat every 24–48 hours. Reported adverse events included proteinuria in 10% of patients receiving daily treatment and neutropenia in 15% of cases, together with some gastro-intestinal side effects.

Specific Tumour Types 4: Feline Epitheliotropic T-Cell Lymphoma [1, 25]

This disease is reported very rarely in the cat and is largely characterized by cutaneous infiltration of neoplastic T lymphocytes with a specific tropism for the epidermis. Feline epitheliotropic T-cell lymphoma (CETL) generally affects older cats with no sex or breed predisposition. Lesions are described as non-pruritic erythematous plaques or patches, scaly alopecic patches and non-healing ulcers or nodules (similar to eosinophilic plaques) (Fig. 12a, b). Diagnosis is based on histopathological examination of skin biopsies. Histologically, the lesions are similar to those reported in the dog, but neoplastic T cells are generally small to medium in size. The immunophenotypic features of CETL of cats are CD3+ positive and double negative CD4- and CD8-. This is different from dogs, in which CETL tends to be CD8+. The survival time of cats with CETL seems to be more variable than that of affected dogs and recommendations for treatment are difficult, with so little experience of this disease in this species. However, as with dogs, the use of single agent lomustine +/− glucocorticoids would seem an obvious starting point.

Cutaneous Lymphocytosis [26, 27]

Cutaneous lymphocytosis is a rare disease of cats, characterized by proliferation of T and B cells in the dermis. The authors do not consider this a neoplastic disease, but should be differentiated from cutaneous lymphoma (above) as the treatment

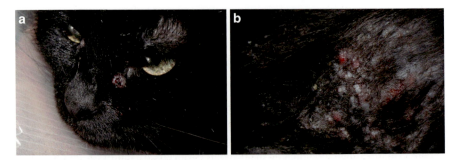

Fig. 12 T-cell cutaneous lymphoma in a cat: ulcerated nodules are visible on the face (**a**) and trunk (**b**). (Courtesy of Dr. Chiara Noli)

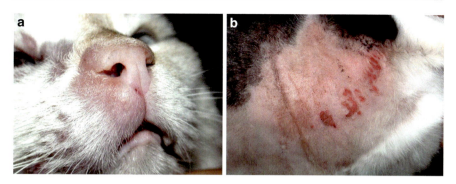

Fig. 13 Clinical presentation of feline lymphocytosis. (**a**) Single area of alopecia with erythema. (Courtesy of S. Colombo). (**b**) Alopecia and multiple ulcerated lesions similar to eosinophilic plaque. (Courtesy of A. Corona)

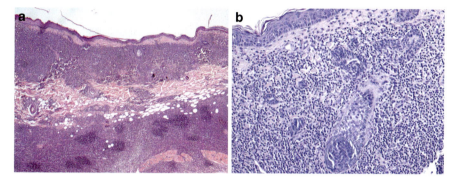

Fig. 14 Histological appearance of feline lymphocytosis. (**a**) Perivasclular to diffuse infiltrate of lymphocytes in the dermis (H&E, 4×). (From: Colombo et al. [33], with permission). (**b**) Infiltrating cells are small lymphocytes without malignancy features that usually spare the follicular walls (H&E 10×). (From: Colombo et al. [33], with permission)

would be different. Average age of development was reported to be 12–13 years, with a slight predilection for females. The lesions tend to have an acute onset with a progressive course of the disease, and can be single or diffuse areas of alopecia, erythema, scales, ulcerations and crusts (Fig. 13), sometimes mimicking eosinophilic plaques. On histopathology, lesions consist of perivascular to diffuse infiltrates of small lymphocytes which eventually extend to the deep dermis (Fig. 14a, b). The lymphocytes are a mix of CD3+ T cells and CD79a + B cells (Fig. 15a, b). There is little evidence of any mitotic figures and the pathology is different from classical CETL, in that lymphocytes are small, do not show malignancy criteria and usually do not invade the epidermis and infundibular walls. Epitheliotropism is seen in about half of the cases. Single lesions can be surgically removed. Medical management can include glucocorticoids (+/− chlorambucil) but responses are variable, and with time lesions can transform to malignant lymphoma in some cases.

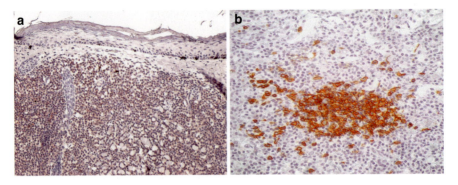

Fig. 15 Immunohistochemical stain of the same histological samples of Fig. 14a. A Diffuse infiltrate of CD3-positive T lymphocytes (**a**) (10×). (From: Colombo et al. [33], with permission); together with focal aggregates of CD79a-positive B lymphocytes (**b**) (40×). (Courtesy of Dr. Chiara Noli)

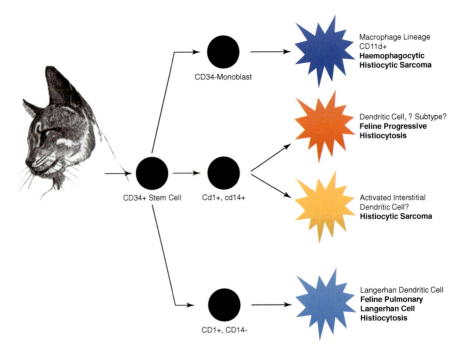

Fig. 16 Classification of histiocytic disease in cats

Specific Tumour Types 5: Feline Progressive Histiocytosis [28, 29]

Feline histiocytic proliferative disorders are incredibly rare (with very few cases reported in the literature) and represent a diagnostic and therapeutic challenge. With the exception of a few case reports, histiocytic proliferations have not been characterized in cats (Fig. 16). One of the best studies summarized clinical, morphologic

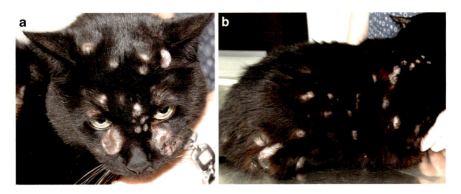

Fig. 17 Feline progressive histiocytosis in a cat: multiple firm, partly ulcerated nodules and plaques are visible on the face (**a**) and trunk (**b**). (Courtesy of Dr. Chiara Noli)

and immunophenotypic features of feline progressive histiocytosis (FPH) in 30 cats, largely affecting the skin [28]. Solitary or multiple, non-pruritic, firm papules, nodules and plaques had a predilection for feet, legs and face (Fig. 17a, b). Lesions consisted of poorly circumscribed epitheliotropic and non-epitheliotropic histiocytic infiltrates of the superficial and deep dermis, with variable extension into the subcutis. The histiocytic population was relatively monomorphous early in the clinical course. With disease progression, cellular pleomorphism was more frequently encountered. Histiocytes expressed CD1a, CD1c, CD18 and major histocompatibility complex class II molecules. This immunophenotype is suggestive of a dendritic cell origin of these lesions. FPH follows a progressive clinical course; the lesions, however, are limited to the skin for an extended period of time. Terminal involvement of internal organs has been documented in some cases. Treatment with chemotherapeutics or immunosuppressive and immunomodulatory drugs was not successful. The aetiology of FPH remains unknown. FPH is best considered an initially indolent cutaneous neoplasm, which is slowly progressive and may spread beyond the skin in the terminal stage.

Specific Tumour Types 6: Feline Lung-Digit Syndrome [2]

The term 'feline lung-digit syndrome' outlines a particular pattern of spread documented in several types of primary pulmonary tumours in cats, specifically bronchial and bronchio-alveolar carcinomas. Feline primary lung tumours are uncommon and tend to be malignant and associated with unfavourable outcomes. Evidence of spread is seen in multiple unusual locations, most notably at the distal phalanges of the limbs. The reason behind common metastasis of feline lung tumours to the digits, compared to other locations and other species, lies in the angio-invasive characteristics of these tumours and their subsequent haematogenous spread. Histopathology has commonly revealed invasion of cancer cells into the arteries of the lungs and digits. Cats have also been shown to have a high blood flow to the digits, which helps them regulate body temperature and enables heat loss. It is therefore hypothesized that this facilitates the high rate of spread of lung tumours to the digits. Other factors, such as cell-markers and release of chemical mediators, may also be important in the pathophysiology of the lung-digit syndrome.

Primary pulmonary neoplasia in cats is often not diagnosed based on clinical signs related to the lung disease, but rather due to clinical manifestation of the metastatic spread. Recent case studies in cats showed that 1 in 6 amputated digits examined contained adenocarcinoma that was suspected to be distant spread from a primary lung tumour. Metastatic spread from a primary pulmonary neoplasia must be considered as a differential diagnosis of all digital lesions in middle-aged and elderly cats.

Clinical Presentation

- Older cats, 12 years of age on average, are more likely to present with the lung-digit syndrome; however, it has also been described in significantly younger patients (range 4–20 years). No sex or breed predilection has been reported, and the syndrome has been documented in pure and mix-breed cats.
- Cats with tumour spread to the digit(s) can present with various clinical signs; however, lameness and pain have been most commonly described. In some cases, only minimal clinical symptoms, such as deflection of the nail or onychomadesis, have been reported. Clinical symptoms at the time of presentation include swelling of the digit or distal limb, ulceration of the skin or nail bed, purulent discharge, infection, deflection of the nail or onychomadesis.
- Multiple digits on different extremities can be affected at the same time (Fig. 18a, b).
- Any digit can be involved, apart from the dew claws, where this symptom has not yet been recognized.
- Digits that are bearing weight are most frequently affected.
- Clinical symptoms of systemic disease, such as malaise, loss of appetite, weight loss or fever are also uncommon.

Diagnosis

- *Complete blood counts and serum biochemistry.*
- *Radiography of the digits* demonstrates a typical image of bone lysis of the second and/or third phalanx, with possible invasion of the joint space (P2-P3). The contrary is observed in people with metastasis to the digits, where spread to the adjacent phalanges or joint invasion are not seen. Periosteal reaction has been noted on all phalanges of the affected extremity in some cases.
- *Thoracic radiography/CT:* Imaging of the thorax should be performed prior to surgery/digit amputation. In most cases, this will reveal a primary lung lesion (Fig. 19). *Differential* diagnosis, which needs to be considered, is pyogranuloma-

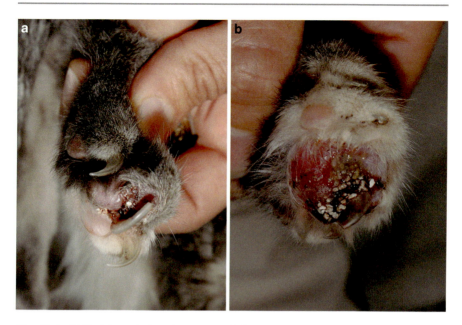

Fig. 18 (**a, b**) Nodules and ulceration on several digits of different paws in a cat with lung-digit syndrome (digital metastasis of lung carcinoma). (Courtesy of Dr. Chiara Noli)

Fig. 19 Thoracic radiograph of the cat in Fig. 18. A pulmonary neoplasm is evident. (Courtesy of Dr. Chiara Noli)

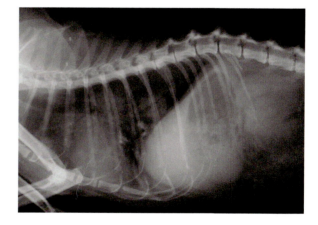

tous inflammation secondary to atypical pathogens, such as fungi, mycobacteria or *Nocardia* species.

- *Biopsy samples – digit:* Four options are available for obtaining a sample form the affected digit: (i) fine needle aspiration – simple procedure, however the cellularity is often too low to provide a definitive diagnosis; (ii) incisional or punch biopsy of abnormal tissue; (iii) an avulsed nail – either of this can help reach a final diagnosis; however, one needs to be aware of the relatively high incidence of non-diagnostic samples; (iv) full digit amputation, the gold standard for histopathological diagnosis, especially on longitudinal sectioning [8], but with the disadvantages related with a surgical procedure in a case with an overall poor prognosis. Due to a high rate of secondary infections, aerobic and anaerobic bacterial and fungal cultures from the biopsied sample should be performed. Cytological analysis can occasionally reveal saprophytic pathogens.

- *Biopsy samples – thorax:* Diagnosis of a primary lung tumour can be obtained through thoracocentesis of pleural effusion when present, or cytological analysis of fine needle aspirates of the pulmonary lesion, tracheal wash or bronchoscopic samples. For reaching a definitive diagnosis, similarly to the biopsy of the digit, samples with preserved tissue architecture are needed; this is possible only with resection of the lung lobe or post-mortem examination.

Histopathology

Common histopathological findings obtained from these tumours demonstrate large mononuclear cells with morphology of epithelial tissue, forming aggregates, strands or cords (Fig. 20). Presence of columnar goblet-like-shaped cells and

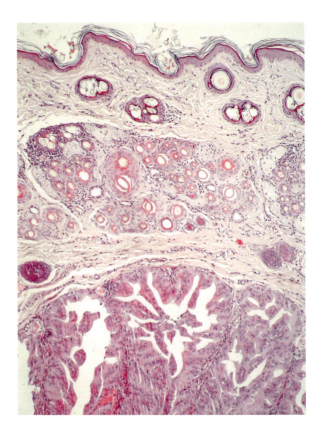

Fig. 20 Histologic appearance of lesions in Fig. 18. Metastatic pulmonary carcinoma cells are observed in the dermis of the digits (H&E 4×). (Courtesy of Dr. Chiara Noli)

ciliated epithelium is a common feature. Cytoplasmic vacuoles are frequently observed and indicative of secretory neoplasia (adenocarcinoma). Presence of inflammatory infiltrate with degenerative neutrophils is common, indicating inflammation due to necrosis. Substantial fibrosis is present in metastatic lesions. Within the affected digits, infiltrates of cancerous cells are most commonly found in the dermis, dorsal aspect of the digit or ventral to the footpad. To help confirm metastasis of a primary lung tumour to the digit, the observation of cellular features of lung tissue (ciliated epithelium, goblet cells, PAS (periodic acid Schiff)-positive secretory material) and special stains for cellular markers (CAM 5.2 antibody against keratin) can be useful.

Treatment and Prognosis

Unfortunately cats with 'lung-digit syndrome' have a grave prognosis. Median survival time of 67 days (mean 58 days, range 12–122) has been revealed in one case study, and the majority of patients were put to sleep due to persistent clinical signs, such as lameness, anorexia or lethargy. Efficacious treatment for feline 'lung-digit syndrome' has not yet been described; digital amputation only provides short-term palliation, as further spread quickly develops.

Specific Tumour Types 8: Feline Melanoma [30]

Benign and malignant melanomas occur in cats and can be ocular, oral or dermal. Ocular melanomas are more common than oral and dermal melanomas. Ocular and oral melanomas are more malignant than dermal melanomas, with higher rates of mortality and metastasis. Dermal melanomas tend to be benign and can be treated with surgical excision if required.

Specific Tumour Types 9: Ear Canal Tumours [31]
(Chapter, Otitis)

- These are not uncommon tumours and are suspected to be related to inflammation of external ear canal.
- Presenting symptoms include chronic irritation, existence of a mass, discharge from the ear, discomfort and odour. Signs of vestibular disease or Horner's syndrome might occur in cases with middle or inner ear involvement.
- Benign tumours of the ear canal include the following:
  - Inflammatory polyps
  - Ceruminous adenomas
  - Papillomas
  - Basal cell tumours

- Malignant tumours of the ear canal include the following:
  - Ceruminous gland adenocarcinoma
  - Squamous cell carcinoma

Treatment

- For non-malignant lesions, non-invasive surgical removal carries a favourable prognosis.
- In malignant tumours, a radical approach with TECA (total ear canal ablation) and lateral bulla osteotomy is the recommended treatment. However, the clinician should be aware of the following:
  - Cats have a worse prognosis than dogs.
  - Further local treatment with radiotherapy can be considered in cases where surgery resulted in incomplete excision.

Specific Tumour Types 10: Feline Abdominal Lymphangiosarcoma [32]

Feline cutaneous lymphangiosarcoma is typically located in the dermis and subcutis of the caudoventral abdominal wall. The lesion is usually not well circumscribed, oedematous, erythematous (Fig. 21) and drains a serous fluid. Histologically, the tumour is characterized by a diffuse proliferation in the dermis and subcutis of empty vessels lined by moderately pleomorphic epithelial cells (Fig. 22). The prognosis is poor. In a study [32], all cats died or were euthanized within 6 months after surgery because of poor wound healing, local recurrence or distant metastases.

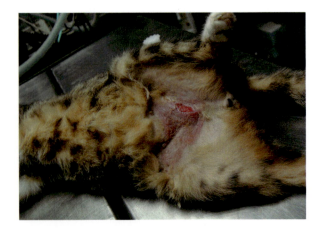

Fig. 21 Feline cutaneous lymphangiosarcoma. A diffuse edematous erythematous lesion with focal ulceration is evident on the abdomen. (Courtesy of Dr. Chiara Noli)

Fig. 22 Histologic features of the lesion in Fig. 21. A diffuse lymphovascular proliferation is observed in the dermis (H&E 10×). (Courtesy of Dr. Chiara Noli)

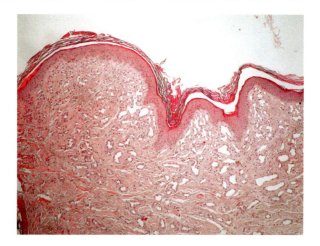

References

1. Argyle DJ. Decision making in small animal oncology. Oxford: Blackwell/Wiley; 2008.
2. Goldfinch N, Argyle DJ. Feline lung-digit syndrome: unusual metastatic patterns of primary lung tumours in cats. J Feline Med Surg. 2012;14:202–8.
3. Sharif M. Epidemiology of skin tumor entities according to the new WHO classification in dogs and cats. VVB Laufersweiler: Giessen; 2006.
4. Argyle DJ, Blacking TM. From viruses to cancer stem cells: dissecting the pathways to malignancy. Vet J. 2008;177:311–23.
5. Sunberg JP, Van Ranst M, Montali R, et al. Feline papillomas and papillomaviruses. Vet Pathol. 2000;37:1–10.
6. Hanna PE, Dunn D. Cutaneous fibropapilloma in a cat (feline sarcoid). Can Vet J. 2003;44:601–2.
7. Backel K, Cain C. Skin as a marker of general feline health: cutaneous manifestations of infectious disease. J Feline Med Surg. 2017;19:1149–65.
8. Diters RW, Walsh KM. Feline basal cell tumors: a review of 124 cases. Vet Pathol. 1984;21:51–6.
9. Murphy S. Skin neoplasia in small animals 2. Common feline tumours. In Pract. 2006;28:320–5.
10. Peters-Kennedy J, Scott DW, Miller WH. Apparent clinical resolution of pinnal actinic keratoses and squamous cell carcinoma in a cat using topical imiquimod 5% cream. J Feline Med Surg. 2008;10:593–9.
11. Almeida EM, Caraca RA, Adam RL, et al. Photodamage in feline skin: clinical and histomorphometric analysis. Vet Pathol. 2008;45:327–35.
12. Murphy S. Cutaneous squamous cell carcinoma in the cat: current understanding and treatment approaches. J Feline Med Surg. 2013;15:401–7.
13. Cunha SC, Carvalho LA, Canary PC, et al. Radiation therapy for feline cutaneous squamous cell carcinoma using a hypofractionated protocol. J Feline Med Surg. 2010;12:306–13.
14. Tozon N, Pavlin D, Sersa G, et al. Electrochemotherapy with intravenous bleomycin injection: an observational study in superficial squamous cell carcinoma in cats. J Feline Med Surg. 2014;16:291–9.
15. Goodfellow M, Hayes A, Murphy S, Brearley M. A retrospective study of (90) Strontium plesiotherapy for feline squamous cell carcinoma of the nasal planum. J Feline Med Surg. 2006;8(3):169–176. https://doi.org/10.1016/j.jfms.2005.12.003.

16. Hammond GM, Gordon IK, Theon AP, Kent MS. Evaluation of strontium Sr 90 for the treatment of superficial squamous cell carcinoma of the nasal planum in cats: 49 cases (1990–2006). J Am Vet Med Assoc. 2007;231(5):736–741. https://doi.org/10.2460/javma.231.5.736.

17. Hartmann K, Day M, Thiry E, et al. Feline injection-site sarcoma: ABCD guidelines on prevention and management. J Feline Med Surg. 2015;17:606–13.

18. Rossi F, Marconato L, Sabattini S, et al. Comparison of definitive-intent finely fractionated and palliative-intent coarsely fractionated radiotherapy as adjuvant treatment of feline microscopic injection-site sarcoma. J Feline Med Surg. 2019;21:65–72.

19. Woods S, de Castro AI, Renwick MG, et al. Nanocrystalline silver dressing and subatmospheric pressure therapy following neoadjuvant radiation therapy and surgical excision of a feline injection site sarcoma. J Feline Med Surg. 2012;14:214–8.

20. Müller N, Kessler M. Curative-intent radical en bloc resection using a minimum of a 3 cm margin in feline injection-site sarcomas: a retrospective analysis of 131 cases. J Feline Med Surg. 2018;20:509–19.

21. Ladlow J. Injection site-associated sarcoma in the cat: treatment recommendations and results to date. J Feline Med Surg. 2013;15:409–18.

22. Litster AL, Sorenmo KU. Characterisation of the signalment, clinical and survival characteristics of 41 cats with mast cell neoplasia. J Feline Med Surg. 2006;8:177–83.

23. Henry C, Herrera C. Mast cell tumors in cats: clinical update and possible new treatment avenues. J Feline Med Surg. 2013;15:41–7.

24. Blackwood L, Murphy S, Buracco P, et al. European consensus document on the management of canine and feline mast cell disease. Vet Comp Oncol. 2012;10:e1–e29.

25. Fontaine J, Heimann M, Day MJ. Cutaneous epitheliotropic T-cell lymphoma in the cat: a review of the literature and five new cases. Vet Dermatol. 2011;22:454–61.

26. Gilbert S, Affolter VK, Gross TL, et al. Clinical, morphological and immunohistochemical characterization of cutnaeous lymphocytosis in 23 cats. Vet Dermatol. 2004;15:3–12.

27. Pariser MS, Gram DW. Feline cutaneous lymphocytosis: case report and summary of the literature. J Feline Med Surg. 2014;16(9):758–63.

28. Affolter VK, Moore PF. Feline progressive Histiocytosis. Vet Pathol. 2006;43:646–55.

29. Miller W, Griffin C, Campbell K. Muller & Kirk's small animal dermatology. 7th ed. Elsevier Health Sciences: Missouri; 2013.

30. Chamel G, Abadie J, Albaric O, et al. Non-ocular melanomas in cats: a retrospective study of 30 cases. J Feline Med Surg. 2017;19:351–7.

31. London CA, Dubilzeig RR, Vail DM, et al. Evaluation of dogs and cats with tumors of the ear canal: 145 cases (1978-1992). J Am Vet Med Assoc. 1996;208:1413–8.

32. Hinrichs U, Puhl S, Rutteman GR, et al. Lymphangiosarcoma in cats: a retrospective study of 13 cases. Vet Pathol. 1999;36:164–7.

33. Colombo S, Fabbrini F, Corona A, et al. Linfocitosi cutanea felina: descrizione di tre casi clinici. Veterinaria (Cremona). 2011;25:25–31.

Paraneoplastic Syndromes

Sonya V. Bettenay

Abstract

Feline paraneoplastic dermatoses are rare, non-neoplastic skin changes associated with an underlying tumour. Knowledge of their clinical appearance may result in early detection of the tumour and offers the best possible outcome and management plan for the patient. A confirmed diagnosis requires that the skin disease parallels the development of an internal malignancy. In paraneoplastic syndromes, the removal of the tumour results in the resolution of the dermatosis. The two most commonly reported cutaneous feline paraneoplastic syndromes are a scaling, shiny alopecia (paraneoplastic alopecia, associated with a variety of abdominal neoplasms) and an exfoliative dermatitis (often thymoma-associated). A single case report exists of "putative paraneoplastic pemphigus" in a cat. The clinical signs of the above-mentioned syndromes and their differential diagnoses will be reviewed. The recommended diagnostic work-up includes ruling out relevant dermatologic differential diagnoses, performing a skin biopsy and a subsequent targeted search for the underlying neoplasm. The best sites for biopsy sampling and the key diagnostic histopathologic changes will be discussed. If the neoplasm can be identified and removed, skin signs will resolve without additional therapy. When the neoplasm cannot be eliminated, symptomatic treatment can be instituted, but a guarded prognosis should be given.

S. V. Bettenay (✉)
Tierdermatologie Deisenhofen, Deisenhofen, Germany
e-mail: s-bettena@t-online.de

© Springer Nature Switzerland AG 2020
C. Noli, S. Colombo (eds.), *Feline Dermatology*,
https://doi.org/10.1007/978-3-030-29836-4_31

613

Introduction

Paraneoplastic syndromes are defined as conditions resulting from a tumour, but not through a direct tumour-induced effect. Feline paraneoplastic dermatoses are rare, but knowledge of their clinical appearance may result in early detection of the underlying tumour and therefore may be life-saving. The size, location or even metastatic nature of the tumour is not relevant to the actual development of the syndrome [1]. A confirmed diagnosis can only be made when the skin disease parallels the development of an internal malignancy. In paraneoplastic syndromes, the removal of the tumour results in the resolution of the dermatosis.

Clinically, the appearance of paraneoplastic signs may precede, follow or coincide with the detection of the related neoplasm. When the characteristic dermatologic changes associated with paraneoplastic dermatoses precede the tumour detection, their correct identification should alert the clinician to perform a "tumour search". There are two main consequences of an early tumour search. The first is that despite screening tests, no abnormality is identified, in which case ultrasonography or other imaging in 1–2 (and again, if still negative, in 6) months may be indicated. The second is that a tumour is identified, and, if surgical removal is possible, it should result in the resolution of the dermatological abnormalities. However, in many cats those neoplasms cannot be eliminated, and symptomatic treatment is the only therapeutic option.

The two most commonly reported cutaneous feline paraneoplastic syndromes are a scaling, shiny alopecia (paraneoplastic alopecia, associated with a variety of abdominal neoplasms) and an exfoliative dermatitis (often thymoma-associated). One case of superficial necrolytic dermatitis (SND) has been reported in the cat, associated with pancreatic carcinoma [2]. A second case of SND in a cat was associated with a glucagon-producing primary hepatic neuroendocrine carcinoma [3]. A single case of mural lymphocytic folliculitis has been reported in association with a pancreatic carcinoma [4]. Whether this was due to an autoantibody reaction can only be speculated, as mural lymphocytic folliculitis is not uncommon as a non-specific histopathologic change in the cat. A single case report exists of "putative paraneoplastic pemphigus" and myasthenia gravis in a cat with a lymphocytic thymoma, which resolved upon tumour excision [5].

Unlike specific internal diseases such as lipid abnormalities (xanthoma) or disturbances of aminoacid metabolism (SND), the exact pathophysiologic mechanism of most paraneoplastic syndromes remains unclear. It is speculated that they may be related to the release of growth factors or cytokines by the tumour and/or to the induction of autoantibodies. Skin changes which occur as a direct consequence of a tumour such as hair cycle arrest alopecia, calcinosis cutis and skin thinning observed with hyperadrenocorticism are not actually classified as *para*neoplastic.

Paraneoplastic Alopecia

Clinical Signs

The affected cat is typically elderly. Alopecia begins on the ventrum and the hair is easily epilated. The alopecic skin is usually not inflamed and has a very characteristic appearance, which is most accurately described as shiny or glistening. While the skin is thinned because of the lack of appendageal structures within the dermis, there is a normal amount of dermal collagen, and the skin is not fragile. The alopecia typically spreads dorsally to the lateral trunk, and the face, axillae and paws also become affected with time (Figs. 1 and 2). Paraneoplastic alopecia is typically non-pruritic, unless there is a yeast overgrowth. In many cases, an accompanying *Malassezia* sp. overgrowth is observed, with yeasts in huge numbers and an accompanying brown seborrhoeic exudate (Fig. 3). The yeast overgrowth may be associated with pruritus and licking or possibly even pain, often observed in cats as "paw shaking". Weight loss, depression and anorexia are common, and, depending on the tumour location, more acute gastrointestinal signs may also be observed.

Diagnosis

The clinical appearance of the alopecic, non-inflamed and shiny skin in an older cat is highly suspicious for paraneoplastic alopecia. The initial differential diagnoses of ventrally oriented alopecia include self-induced alopecia due to allergies (to

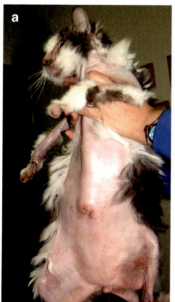

Fig. 1 Feline paraneoplastic alopecia. (**a**) Ventrally oriented alopecia, with multifocal brown seborrhoea. Note the "shiny" appearance of the skin, especially on the ventral thorax. (**b**) Same cat as A. Extensive alopecia involving the neck and frontlegs. (Courtesy of Dr. Chiara Noli)

Fig. 2 Feline paraneoplastic alopecia. Same cat as Fig. 1: periocular, non-inflammatory alopecia not affecting the vibrissae. Note also the presence of excessive ceruminous exudate and the lack of mucocutaneous changes. (Courtesy of Dr. Chiara Noli)

Fig. 3 Cat paw. Feline paraneoplastic alopecia. Alopecia, mild erythema and hypotrichosis with marked brown seborrhoea affect the toes and carpi. The claw folds are notably distended with seborrhoeic debris but do not exhibit marked paronychia

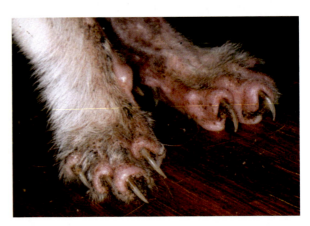

Fig. 4 Trichogram. Feline paraneoplastic alopecia. Multiple telogen stage hair bulbs. The roots show the classic "club- or spear-shaped" appearance

environmental and/or food allergens and flea bite hypersensitivity), psychogenic alopecia and, rarely, demodicosis or dermatophytosis. In very early cases where widespread changes of alopecia with the classic shiny skin are not as evident, trichograms may be helpful to search for *Demodex* mites and dermatophyte-infected hair shafts. This test may also point towards hair follicle telogenization, if more than 50% of the hair roots are in telogen (Fig. 4). However, once the shiny skin is apparent, these differential diagnoses become much less likely.

Skin histopathology may be diagnostic, but an experienced dermatopathologist is required. As the initial alopecia is often located on the ventral abdomen and as normal abdominal skin has smaller and more sparsely distributed hair follicles, the diagnosis may be difficult in early stages. Multiple samples, both from the more dorsal leading edges of the alopecia (looking for infectious agents) and from the centre of the most severely alopecic areas, should be biopsied. Changes typically include a mildly hyperplastic epidermis with abnormal cornification (which may be subtle). Atrophic hair follicles and a mixed cellular dermal inflammation may or may not be present. Follicles are frequently described as "telogenized" referring to the absence of hair shafts, frequent lack of sebaceous glands and presence of small telogen bulbs (Figs. 5 and 6). Importantly, in the case of feline paraneoplastic alopecia, there are no anagen bulbs associated with the atrophic hair follicle units in the upper dermis. In one report, concurrent intracorneal mites were present [8] which was speculated to be due to the cats' poor general state of health. The excessive number of yeasts located in the stratum corneum may be lost in processing and may be absent in the "shiny" areas. When present, the yeasts may suggest an internal disease to the pathologist [9].

Many of these cats have an undetected internal neoplasm at the time of presentation with skin disease, and they will frequently develop systemic signs after the onset of the alopecia. Pancreatic carcinoma, cholangiocarcinoma [1], hepatocellular carcinoma [6], metastatic intestinal carcinoma [7], neuroendocrine pancreatic neoplasia and a hepatosplenic plasma cell tumour [8] have all been reported. These tumours are often *not* accompanied by measurable changes in either blood counts or serum biochemistry values. An abdominal ultrasound, x-ray or further imaging such as CT scan is required to identify the tumour, and, in this case, the skin changes take a secondary role.

Management and Prognosis

Identification of the internal neoplasm and its removal provide a possible cure in this truly paraneoplastic process and have indeed been reported in one cat [10]. In

Fig. 5 Skin biopsy (H&E) 200×. Feline paraneoplastic alopecia. The epidermis is hyperplastic; the follicles are in telogen (some in haired telogen) with multiple telogen bulbs present in the superficial mid-dermis. Sebaceous glands are absent and a mild interstitial dermatitis is also present

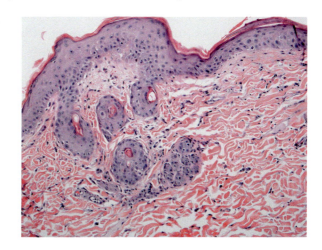

Fig. 6 Skin biopsy (H&E)
50×. Feline paraneoplastic
alopecia. Atrophic, mildly
hyperplastic dermatitis.
Note the presence of small
sebaceous glands and focal
follicular mucin

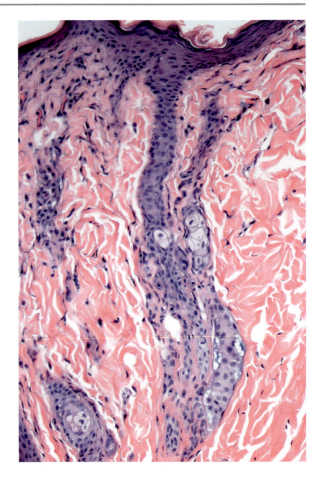

many cases, however, the cat is old, and the neoplasm involves the pancreas or the liver and is surgically poorly accessible. As such, many of these cats are euthanized when their systemic signs become severe. In one comprehensive review, 12 of the 14 cats reported at that time died or were euthanized within 8 weeks from the onset of clinical signs [1]. The alopecia is cosmetic and typically requires no treatment. The seborrhoeic *Malassezia* overgrowth, however, can be clinically distressing. In this case, topical clotrimazole cream or emulsion is the author's first choice to treat the yeast, as cats generally do not enjoy baths and the skin, while not fragile, is sensitive. A miconazole/chlorhexidine shampoo is frequently recommended for *Malassezia* overgrowth, but it can be drying and irritant as a result. The drying effects may be reduced if the shampoo is followed by a moisturizer (e.g. a non-perfumed baby bath oil, based on almond oil). Systemic antifungals may be necessary, but may be less well tolerated by systemically ill cats. Itraconazole would be the systemic drug of choice. Supportive palliative care with palatable, high-quality, high-protein diets which are high in essential fatty acids should be recommended.

The prognosis is guarded, because the internal neoplasia is often advanced at the time of diagnosis. There may only be an expectation of a few weeks or months until euthanasia.

Exfoliative Dermatitis

An exfoliative dermatitis has been reported in multiple cats with thymoma [11]. Remission after surgical removal of the neoplasm has also been reported, suggesting this to be a true paraneoplastic syndrome [12]. However, in some cats with clinically and histopathologically identical exfoliative dermatitis, an underlying aetiology could not be identified, and blood work and radiology and ultrasound were normal [13]. Even with long-term follow-up, in a number of these cats, there was no evidence of a developing tumour. As such, the presence of this severe and clinically distinctive exfoliative dermatitis is suggestive of a paraneoplastic disease, but it is not pathognomonic.

Clinical Signs

Erythema and scaling of the head (Fig. 7), neck and pinnae develop in middle-aged to older cats, frequently without pruritus. Thick sheets of exfoliated stratum corneum develop and often remain trapped in the hair coat (Fig. 8). With time, cats will present alopecia in affected areas (Fig. 9). As the disease progresses, so does the severity and anatomical extent of lesions. Severe hyperkeratosis, erythema and erosions develop and spread to the trunk (Figs. 10 and 11). Coughing and dyspnoea may develop in association with the thymic changes. Affected cats may or may not be pruritic. Pruritus is most often associated with secondary infections (*Staphylococcus* and/or *Malassezia*). Extensive licking may also be associated with mild pain.

Fig. 7 Cat head. Patchy alopecia and hypotrichosis with thick adherent scales and crusts visible on both pinnae, dorsal head and neck. (Courtesy of Prof. R. Mueller, Small animal Clinic, Munich)

Fig. 8 Cat trunk. Dry large scales trapped in the coat of a cat affected with thymoma-associated feline exfoliative dermatitis. (Courtesy of Dr. Silvia Colombo)

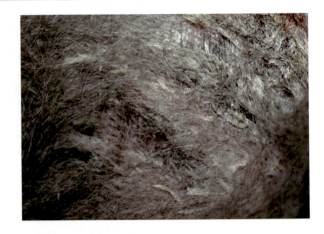

Fig. 9 Diffuse hypotrichosis and scaling on the trunk of a cat with thymoma-associated feline exfoliative dermatitis. (Courtesy of Dr. Chiara Noli)

Fig. 10 Severe hyperkeratosis and erosion on the trunk in a cat affected with thymoma-associated exfoliative dermatitis. (Courtesy of Dr. Castiglioni)

Fig. 11 Cat ventrum. Severe erythema, multifocal ulceration, alopecia and hypotrichosis with thick adherent scales and crusts extending along the ventral abdomen and hind legs. Note the lesions are more severe in the wear areas. (Courtesy of Prof. R. Mueller, Small animal Clinic, Munich)

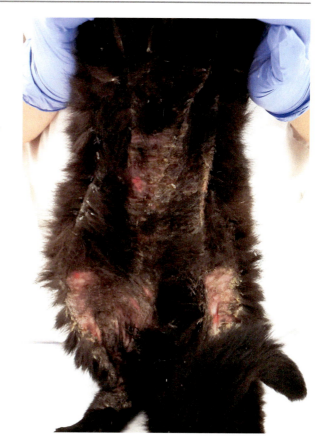

Diagnosis

In the early stages of scaling and hypotrichosis of the head and pinnae, the clinical picture may be suggestive of dermatophytosis or ectoparasites. Superficial skin scrapings, trichograms and cytology are indicated. Those cats with pruritus, mild scaling and a bacterial or yeast overgrowth may be difficult to distinguish from allergic cats. However, the late age of onset and the development of the dermatitis on the top of the head and convex surface of the pinna are less typical for atopy or food allergy. Allergic cats more frequently develop dermatitis of the preauricular area, cheeks and inner pinna. When large numbers of yeast are identified with cytology in an older cat, a systemic disease should be considered. In a study evaluating feline skin biopsies, the majority of cats with *Malassezia* organisms in the stratum corneum had changes diagnostic for exfoliative dermatitis, and indeed, most of these cats were euthanized shortly after the biopsy procedure [9]. Diagnosis of exfoliative dermatitis is confirmed by histopathology, where characteristic changes include severe, orthokeratotic hyperkeratosis with cytotoxic dermatitis (interface changes with single cell necrosis occurring at all levels of the living epidermis) (Figs. 12, 13 and 14). A thoracic radiograph (seeking for mediastinal changes, Fig. 15) is always indicated following this histopathological diagnosis.

Fig. 12 Skin biopsy (H&E) 50×. Feline exfoliative dermatitis. Low-power features include marked hyperkeratosis, epidermitis with lymphocytic exocytosis, a superficially oriented interstitial dermatitis and loss of sebaceous glands

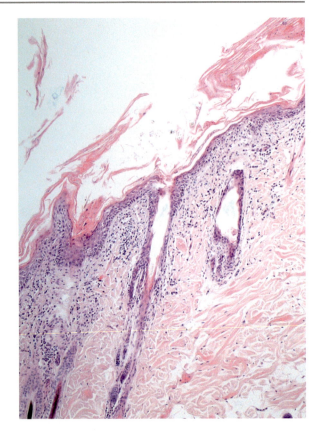

Fig. 13 Skin biopsy (H&E) 400×. Feline exfoliative dermatitis. Single cell necrosis

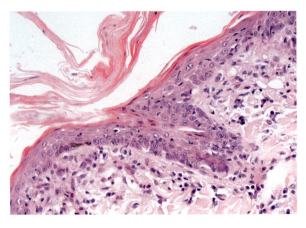

Fig. 14 Skin biopsy (H&E) 400×. Feline exfoliative dermatitis. Epidermal hyperkeratosis, exocytosis and mild spongiosis

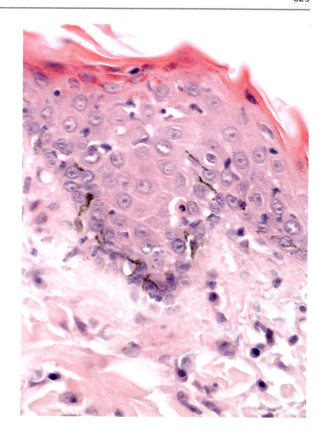

Fig. 15 X-ray of the chest of cat in Fig. 9. A non-anatomical radio-opaque area, compatible with a thymoma, is evident in mediastinal position

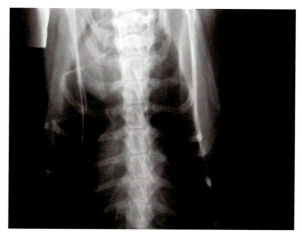

Management and Prognosis

If a mediastinal tumour is identified and surgery is possible, resolution has been reported to occur with the successful removal of the neoplasm [14]. In the event that the tumour is inoperable or symptomatic/palliative care is elected by the owners, management should focus on treatment of any identified surface infection and on the correction of the barrier function. Topical therapy is preferred, but when the skin is painful or the cat resents bathing, systemic antimicrobials are indicated if cytology points to an infection. The therapy is the same described in the section on paraneoplastic alopecia. Management of the abnormal skin barrier may provide some discomfort relief and is also aimed at preventing relapses of infections. The author prefers a topical non-perfumed baby bath oil, but depending on the country, there may be various veterinary topical emollient products available.

Systemic prednisolone has not been reported to be effective in these cats. The pathomechanism involves a lymphocytic, epidermal cytotoxic dermatitis, and, as such, cyclosporin may be effective and is certainly worth a therapeutic trial. In cases with severely eroded skin, pain may be present, and palliative analgesics may be a useful adjunctive therapy. The prognosis is guarded in the cases where an inoperable mediastinal tumour is identified, with an expected maximum time to euthanasia of 6 months. When a tumour is not identified, follow-up radiology and/or CT scan should be recommended, in 1–2 months or every 6 months depending on the presence or absence of accompanying systemic signs. Spontaneous remission has been observed in some cats without identified thymoma.

Superficial Necrolytic Dermatitis

One case of superficial necrolytic dermatitis (SND) has been reported in the cat, associated with pancreatic carcinoma [2]. A second case of SND in a cat was associated with a glucagon-producing primary hepatic neuroendocrine carcinoma [3]. The decrease of a variety of aminoacids has been reported in humans with glucagonoma and SND. The skin changes do resolve with surgical resection of the tumour, but it is unclear whether the dermatologic changes are paraneoplastic or due to hypoaminoacidaemia. In many dogs with classic SND which exhibit both striking and unique skin and hepatic changes, there is no confirmed malignancy, and they frequently respond to intravenous aminoacid infusions. Hence, the syndrome in dogs is frequently referred to as hepatocutaneous syndrome, without clear paraneoplastic features. One cat with classic SND liver and skin changes but no identifiable malignancy has also been reported [15].

Clinical Signs

Severe hyperkeratosis of the mucocutaneous haired skin and especially of the paw-pads is associated with bacterial and/or yeast overgrowth, pain, depression and lethargy.

Diagnosis

Surface skin cytology is essential, as the secondary infections produce much of the pain and depression. Liver and pancreas ultrasonography is recommended as the first step, since it is the least invasive test. Skin biopsy may yield the classic changes, but extensive parakeratosis lacking the classic mid-dermal pallor may be the only change identified. Multiple biopsy samples are therefore recommended. It is important to be aware that hepatic enzyme levels may be normal as the liver changes are structural/degenerative rather than cytotoxic.

Management and Prognosis

Prognosis is guarded. If a pancreatic tumour is present and surgically resectable, there may be a good prognosis, although this has not yet been reported in the cat. Dogs with only liver changes may be maintained symptomatically with weekly to monthly intravenous aminoacid infusions, administered over an 8 h period to avoid seizure induction. Nevertheless, even with initial good symptomatic clinical improvement, most dogs will die from liver failure or be euthanized due to a relapse of the painful skin disease with time. No clinical management of the two reported feline cases was undertaken.

Paraneoplastic Pemphigus

Paraneoplastic pemphigus (PNP) is a rare autoimmune blistering disease in humans and dogs, and a putative case has been reported in one cat [5]. An 8-year-old Himalayan cat developed a maculo-papular eruption which progressed to an erosive and ulcerative skin disease, 4 weeks after the surgical resection of a mediastinal lymphocytic thymoma. After 2 further weeks, the cat developed transient thymoma-associated myasthenia gravis. There were no oral lesions. The skin condition was diagnosed by histopathology, reported as exhibiting two specific pathologic patterns. One pattern was compatible with pemphigus vulgaris and the other with erythema multiforme. The combination of these two patterns is strongly suggestive of a paraneoplastic syndrome. Direct and indirect IgG immunofluorescence was also performed, and the authors concluded the results supported a PNP (a non-commercial test).

The cat was managed symptomatically with prednisolone and chlorambucil and responded well. The medications were slowly tapered and then discontinued with no relapse.

References

1. Turek MM. Cutaneous paraneoplastic syndromes in dogs and cats: a review of the literature. Vet Dermatol. 2003;14(6):279–96.
2. Patel A, Whitbread TJ, McNeil PE. A case of metabolic epidermal necrosis in a cat. Vet Dermatol. 1996;7(4):221–6.
3. Asakawa MG, Cullen JM, Linder KE. Necrolytic migratory erythema associated with a glucagon-producing primary hepatic neuroendocrine carcinoma in a cat. Vet Dermatol. 2013;24(4):466–9, e109-10.
4. Lobetti R. Lymphocytic mural folliculitis and pancreatic carcinoma in a cat. J Feline Med Surg. 2015;17(6):548–50.
5. Hill PB, Brain P, Collins D, Fearnside S, Olivry T. Putative paraneoplastic pemphigus and myasthenia gravis in a cat with a lymphocytic thymoma. Vet Dermatol. 2013;24(6):646–9, e163-4.
6. Marconato L, Albanese F, Viacava P, Marchetti V, Abramo F. Paraneoplastic alopecia associated with hepatocellular carcinoma in a cat. Vet Dermatol. 2007;18(4):267–71.
7. Grandt LM, Roethig A, Schroeder S, Koehler K, Langenstein J, Thom N, et al. Feline paraneoplastic alopecia associated with metastasising intestinal carcinoma. JFMS Open Rep. 2015;1(2):2055116915621582.
8. Caporali C, Albanese F, Binanti D, Abramo F. Two cases of feline paraneoplastic alopecia associated with a neuroendocrine pancreatic neoplasia and a hepatosplenic plasma cell tumour. Vet Dermatol. 2016;27(6):508–e137.
9. Mauldin EA, Morris DO, Goldschmidt MH. Retrospective study: the presence of Malassezia in feline skin biopsies. A clinicopathological study. Vet Dermatol. 2002;13(1):7–13.
10. Tasker S, Griffon DJ, Nuttall TJ, Hill PB. Resolution of paraneoplastic alopecia following surgical removal of a pancreatic carcinoma in a cat. J Small Anim Pract. 1999;40(1):16–9.
11. Rottenberg S, von Tscharner C, Roosje PJ. Thymoma-associated exfoliative dermatitis in cats. Vet Pathol. 2004;41(4):429–33.
12. Forster-Van Hijfte MA, Curtis CF, White RN. Resolution of exfoliative dermatitis and Malassezia pachydermatis overgrowth in a cat after surgical thymoma resection. J Small Anim Pract. 1997;38(10):451–4.
13. Linek M, Rufenacht S, Brachelente C, von Tscharner C, Favrot C, Wilhelm S, et al. Nonthymoma-associated exfoliative dermatitis in 18 cats. Vet Dermatol 2015;26(1):40–5, e12-3.
14. Singh A, Boston SE, Poma R. Thymoma-associated exfoliative dermatitis with post-thymectomy myasthenia gravis in a cat. Can Vet J. 2010;51(7):757–60.
15. Kimmel SE, Christiansen W, Byrne KP. Clinicopathological, ultrasonographic, and histopathological findings of superficial necrolytic dermatitis with hepatopathy in a cat. J Am Anim Hosp Assoc. 2003;39(1):23–7.

Idiopathic Miscellaneous Diseases

Linda Jean Vogelnest and Philippa Ann Ravens

Abstract

The last chapter of this feline dermatology text reviews a range of cutaneous diseases not addressed in preceding chapters. It includes some well-recognised entities, such as feline chin acne, solar dermatitis and burns. It also includes a number of unique feline presentations of unknown aetiology including idiopathic ulcerative dermatitis and mural folliculitis that are increasingly recognised as reaction patterns, with multiple potential causes, rather than distinct diseases. Hypereosinophilic syndrome is another idiopathic uniquely feline presentation that is reviewed, along with sebaceous adenitis and sterile panniculitis that occur only rarely in cats.

Introduction

The miscellaneous diseases reviewed in this chapter have been divided into localised dermatoses (affecting defined anatomical body regions), environmental and physiological dermatoses (due to external skin insults or impacts on hair cycling) and idiopathic sterile inflammatory diseases. Some are reasonably common and well-recognised in general practice, and others are rare and incompletely described.

L. J. Vogelnest (✉)
Small Animal Specialist Hospital, North Ryde, NSW, Australia

University of Sydney, Sydney, NSW, Australia

P. A. Ravens
Small Animal Specialist Hospital, North Ryde, NSW, Australia

© Springer Nature Switzerland AG 2020
C. Noli, S. Colombo (eds.), *Feline Dermatology*,
https://doi.org/10.1007/978-3-030-29836-4_32

Localised Dermatoses

Some dermatoses affect distinct body regions, and although restriction to affected regions aids their recognition, most have uncertain or multifactorial aetiology and a variety of incompletely evaluated treatment options.

Chin Acne

Feline acne is proposed to be a localised disorder of follicular keratinisation characterised by comedone formation, with multiple potential contributing factors including insufficient local grooming, abnormal sebum production, hair cycling abnormalities and hypersensitivities (atopic dermatitis, contact and/or food reactions) [1–3]. Occasional outbreaks of disease have been reported in catteries and multi-cat households; however, potential roles for stress, viral infections (calicivirus, herpesvirus), demodicosis or dermatophytosis remain unsubstantiated [2, 3]. Many affected cats appear otherwise healthy, with no apparent systemic diseases; however, concurrent dermatoses, especially hypersensitivities and other regions of secondary bacterial infection, are not uncommon [1].

Although well-recognised to occur worldwide in indoor and outdoor cats [1–3], and suggested to occur commonly in general practice [3], there is scant prevalence data. Feline acne was among the ten most common dermatoses presenting to a university dermatology referral service in the USA, representing 3.9% of feline dermatoses and 0.33% of all cats examined at the university hospital during a 15-year period [1, 4]. The true disease prevalence was suggested to be higher, as mild disease in association with other presenting problems was likely under-reported. There is no confirmed age, sex or breed predilection [1], although males were more frequently affected (73%) in one study of 22 cats [3]. Secondary deep bacterial infection frequently complicates the disease process; it has been reported in 42–45% of affected cats [1, 3]. (Chapter, Bacterial Diseases.) Occasional cases are reported with budding yeast, consistent with *Malassezia* spp. present on surface cytology or histopathology [2, 3] or *M. pachydermatis* isolated on culture [3]; however, any role for *Malassezia* spp. in disease pathogenesis remains unsubstantiated.

Clinical Presentation

Feline acne is mostly restricted to the chin, with the lower and/or upper lips and lip commissures infrequently affected. The most commonly reported lesions are brown-to-black comedones and fine crusting/hair casts (60–73%; Fig. 1), alopecia (68%), papules (45%) and erythema (41%) [1, 2]. Progression to nodular swelling, draining tracts and diffuse swelling occurs with deep bacterial folliculitis and furunculosis, which may produce acute pyrexia and malaise when severe and/or

Fig. 1 Typical fine black crusting and hair casts on the chin of a cat with feline acne

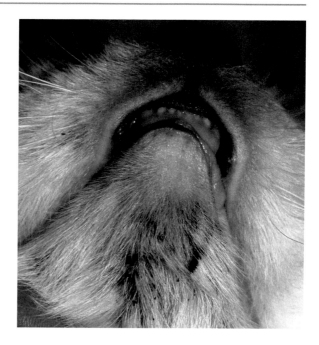

chronic pitted scarring [1]. Lesions are most typically non-pruritic and non-painful [1], although pruritus restricted to the affected area is sometimes reported, particularly in association with secondary bacterial infection [2, 3].

Diagnosis

Diagnosis is usually straightforward based on classical lesions restricted to the chin or adjacent perioral region, with cytology important for detection of secondary bacterial (and potentially yeast) infection. Adhesive tape impressions are useful for sampling this restricted region, following prior rupture of any pustules with a sterile needle. Fine needle aspiration is indicated for nodular or diffusely swollen lesions. (Chapter, Bacterial Diseases.) Screening for demodicosis or dermatophytosis may be indicated if other historical factors or clinical lesions are suggestive.

Histopathology may be important for atypical cases or to confirm deep bacterial infection and/or cystic adnexal changes. A range of histopathological findings reflects the spectrum of clinical presentations [2, 3], with follicular and/or glandular abnormalities dominating in 1 study of 22 cases, including lymphoplasmacytic periductal inflammation (86%), sebaceous gland duct dilation (73%), follicular keratosis with plugging and dilation (59%), epitrichial gland occlusion and dilation (32%), folliculitis (27%), furunculosis (23%) and pyogranulomatous sebaceous adenitis (23%) [2].

Treatment

Treatment suggestions are largely anecdotal. One uncontrolled study evaluating topical 2% mupirocin ointment in 25 cats reported typically excellent ($n = 15$) to good ($n = 9$) response within 3 weeks, with potential contact reaction in 1 cat [3]. A variety of other topical treatments have been suggested, aimed at reducing comedone formation and secondary infections, including antiseptics (chlorhexidine, benzoyl peroxide) and keratolytics (tretinoin, salicylic acid, sulphur) in a variety of formulations (shampoos, gels, ointments) [1].

Mild, nonprogressive, comedonal presentations may not require treatment [1]. Topical antiseptics (2–4% chlorhexidine solution) are now preferable to topical antibiotics if superficial secondary bacterial infection is confirmed. (Chapter, Bacterial Diseases.) Systemic antibiotics are important for deep bacterial infection, with amoxicillin-clavulanate or cephalexin recommended first-line empirical choices in many geographical regions, assuming bacterial cocci are evident on cytology. Systemic antibiotics are typically continued for 4–6 weeks or until at least 2 weeks beyond resolution of nodules and draining tracts. Bacterial culture from tissue biopsies or intact pustules is important with intracellular bacterial rods on cytology, poor response to empirical therapy and/or geographical regions with high rates of methicillin-resistant staphylococcal infections. (Chapter, Bacterial Diseases.)

For cases presenting without secondary bacterial infection or following effective treatment of secondary infections, topical cleansing and/or keratolytics are recommended for initial treatment and at reduced frequency to help limit recurrences. Options include:

- Cleansing/antiseptics:
  - Physical removal of excessive crusts, e.g. fine combing, gentle rubbing, weekly or as needed; shampoos may be effective, but are less tolerated in many cats.
  - Chlorhexidine 2–4% solution, once daily for infection/flares; 2–3 times weekly if infections are recurrent.
  - Benzoyl peroxide 0.05–0.1% gel, once daily to every other day initially; may be drying and irritating; continue once to twice weekly if effective.
- Topical retinoids: used once daily until clinical response (6–8 weeks) and then tapered; phototoxic potential (avoid solar exposure):
  - Tretinoin 0.01–0.05% gel, cream or lotion (may be more potent, but more irritant)
  - Adapalene 0.1% gel (less irritant than tretinoin)
- Topical steroids:
  - Mometasone 1% cream once daily for up to 3 weeks and then twice weekly
- Systemic retinoids: isotretinoin 2 mg/kg once daily (may be useful for severe cases) [1]

Treatment Guidelines for Feline Acne

1. Treat secondary bacterial and/or yeast infection if indicated by cytology:
   (a) Superficial infection: 2–4% chlorhexidine solution once daily × 2–3 weeks (use of mupirocin or fusidic acid should be discouraged) (Chapter, Bacterial Diseases.)
   (b) Deep infections: amoxicillin-clavulanate or cephalexin as first line or guided by culture and susceptibility testing (from tissue biopsy or pustule contents), twice daily × 4–6 weeks or until 2 weeks beyond resolution of nodules/draining tracts
2. Cleansing, using options suitable to owner and patient (daily initially and then reducing):
   (a) Fine-tooth combing or gentle rubbing to remove excessive crust accumulation
   (b) Benzoyl peroxide gel (sparingly; may be drying/irritating if too frequent)
   (c) Shampoos: benzoyl peroxide and chlorhexidine 4% (1–2 times weekly, if tolerated)
3. Keratolytic agents, if cleansing not sufficient and/or for severe cases:
   (a) Topical retinoids: 0.05% tretinoin
   (b) Shampoos: salicylic acid and sulphur
   (c) Systemic retinoids: isotretinoin 2 mg/kg once daily

Idiopathic Neck Ulcerative Dermatitis

Historical reports of an *idiopathic ulcerative dermatitis* in cats, characterised by crusted, non-healing ulceration (Fig. 2) most typically on the dorsal neck or between the scapulae, suggest it may represent a single disease of unknown aetiopathogenesis [5, 6]. Although there was no apparent association with pruritus in early reports [5], severe scratching of lesional sites is more typical and persistent severe self-trauma often associated with poor healing or relapse after healing [6]. Multiple underlying causes of severe pruritus are now considered possible [6]. Neuropathic itch has been proposed as one potential cause [7], however remains unsubstantiated. Sustained resolution after glucocorticoid therapy or surgical resection is reported in some cats [5], suggesting transient causes. Underlying allergic dermatitis appears frequent in the authors' experience, with some lesions associated with secondary bacterial infection and some affected cats having other allergic signs. Similar non-healing ulceration in other body regions (e.g. preauricular, ventral neck) is recognised to occur associated with severe scratching with the hind limbs in pruritic allergic cats. Oclacitinib (1–1.5 mg/kg once daily for 4–6 weeks) produced resolution of chronic ulcerative lesion in the dorsal scapular region in a cat previously unresponsive to ciclosporin, Anafranil, chlorambucil or amitriptyline [7]. A behavioural aetiology has been proposed in some cases, with lesions resolving in ten cats within 15–90 days following improved welfare conditions (including free access to food, water and hiding places, multiple changing toys, separation from non-amiable cats and reduced human-initiated cat contact) [8].

Fig. 2 Feline idiopathic neck lesion: a large area of ulceration on the chin and neck covered by a thick eschar. (Courtesy of Dr. Chiara Noli)

Diagnosis

Diagnosis of this presentation requires consideration of multiple potential causes of severe pruritus, especially hypersensitivities, and potentially some ectoparasites (*Demodex gatoi, Otodectes cynotis*). Cytology via adhesive tape or glass slide impressions is important to detect secondary bacterial and possibly *Malassezia* infection. Superficial skin scrapings may be indicated to screen for *Demodex* or *Otodectes* mites. Histopathology findings reported reflect chronic response to sustained ulceration, with variable dermal necrosis and fibrosis. Linear subepidermal bands of fibrosis extending peripherally from the ulcer are reported [5, 6], but distinction from other causes of chronic traumatic skin damage is also unsubstantiated.

Treatment

Treatment recommendations are sparsely documented. Lesions are often refractory to commonly used symptomatic treatments, including systemic glucocorticoids and/or antibiotics. Topiramate, an anti-epileptic medication with potential antinociceptive effects, was used with apparent success in one cat, and relapse was reported within 24 h on repeated attempts to discontinue therapy [9]. However, pharmacokinetics and safety of this medication in cats are not reported.

An initial combination of measures to restrict continued self-trauma, treat any secondary bacterial infection and encourage healing has been effective in many cases in the authors' experience. Neck bandaging and/or bodysuits, sometimes with concurrent bandaging of the hindfeet, are often helpful. Topical antibiotic ointments (mupirocin 2%, sodium fusidate 2%) appear useful for lesions with confirmed infection or topical silver sulphadiazine (1% cream) placed under nonadhesive dressings. (Chapter, Bacterial Diseases.) Chlorhexidine solution is less suitable for ulcerated lesions as more effective antimicrobial concentrations may irritate ulcerated regions (maximum 0.05% solution). Use of vet wrap cohesive bandaging facilitates daily removal for cleaning and topical treatment.

Identification and management of any underlying disease are important to limit recurrence, and further investigation for underlying hypersensitivity is important in cases without other confirmed causes. (Chapters, Ectoparasitic Diseases, Flea Biology, Allergy and Control, Feline Atopic Syndrome: Epidemiology and Clinical Presentation, Feline Atopic Syndrome: Diagnosis and Feline Atopic Syndrome: Therapy.)

Plantar Hock/Metatarsal Ulcerative Dermatitis

An undocumented presentation of non-healing ulceration restricted to the plantar aspects of the metatarsal regions and/or hock is recognised rarely in cats. Whether this presentation represents a single disease or has multiple potential causes is currently unknown. Contact reactions seem unlikely in some presentations, due to sustained lesions despite bandaging, although could be part of a multifactorial aetiology.

Lesions often start unilaterally and become bilateral and reasonably symmetrical and are often associated with licking that does not appear excessive. In cases the authors have seen, primary lesions are lacking, and discrete regions of non-healing erosion to ulceration with adjacent alopecia and erythema are typical (Fig. 3). Other allergic or cutaneous signs are absent.

Histopathology findings include eosinophilic and/or neutrophilic dermatitis, with dermal necrosis and fibrosis, with no apparent cause of ulceration and delayed healing. Lesions respond very slowly and often incompletely to prevention of licking, bandaging and wound care. Secondary bacterial infections may be apparent on surface cytology and/or histopathology, but there is often poor response despite resolution of this component. Recurrence has occurred despite surgical resection and initial healing of the surgical sites.

Optimal treatment is unknown. Assessment of any role for secondary infections is important, along with good principles of wound care. Skin biopsies for histopathology may be helpful to exclude defined causes and guide treatment options.

Fig. 3 Slowly healing discrete plantar metatarsal erosions to ulceration of unknown aetiology in a cat

Tail Gland Hyperplasia

The feline supracaudal or tail gland is a cluster of large sebaceous glands found on a localised region on the dorsal proximal tail. In some cats, these glands become hyperplastic, forming a localised dermatosis, often referred to as "stud tail". This condition is seen primarily in intact male cats, although occurs occasionally in female and castrated male cats. Glandular hyperplasia and excess secretion of sebum are presumed to occur in response to male sex hormones [10].

Lesions vary in severity from greasy exudate with variable scale (Fig. 4), through partial alopecia and hair matting, to more complete alopecia with hyperpigmentation and comedones. Secondary bacterial infection can result in localised swelling, erythema, pain and/or pruritus [10].

Treatment

Treatment involves removal of excess sebum, reduction in comedone formation and prevention of secondary bacterial infection. Mild lesions may be treated with keratolytic shampoos (e.g. sulphur, salicylic acid) to remove the greasy exudate and aid emptying of occluded follicles and sebaceous glands. Some cats may benefit from a moisturising conditioner after shampooing. Vitamin A and

Fig. 4 A cat with tail gland hyperplasia presenting greasy exudate with brown scale dorsally on the tail basis. (Courtesy of Dr. Silvia Colombo)

its derivatives are comedolytic and reduce sebaceous gland activity, and retinoid treatments as for feline chin acne can be considered. One double-blinded placebo-controlled study reports significant reduction in clinical scores after treatment with 0.1% retinoic acid cream, liberally applied, four times daily for 28 days [10]. Preventative treatment would likely be needed for sustained response, potentially topical tretinoin once weekly or as needed to maintain resolution of signs.

Painful, erythematous or swollen lesions should be assessed for secondary bacterial infection, using surface cytology and/or fine needle aspiration or biopsies for severe lesions, and be the initial focus of treatment when present. Clinical signs in entire male cats may improve after castration, although severe glandular hyperplasia may not be readily reversed.

Environmental and Physiological Dermatoses

Solar (Actinic) Dermatitis

Solar or actinic dermatitis is common in cats that spend significant time outdoors and seek solar exposure, particularly cats living in countries with high UV indices [11–15]. It is attributable to cumulative exposure to ultraviolet (UV) light that causes direct keratinocyte damage, primarily short-wave UVB, but also long-wave UVA with deeper dermal penetration. Its occurrence is strongly linked to skin colour and coat density, with non-pigmented sparsely or non-haired areas affected, particularly the pinnae, pinnal margins, nasal planum, and periocular, perioral and preauricular areas. Indoor cats may also be affected. Although window glass effectively blocks UVB rays, UVA rays can still penetrate. Solar dermatitis commonly progresses to solar keratosis and squamous cell carcinoma [11]. (Chapter, Neoplastic Diseases.)

Clinical Presentation

In one study of 32 cats, the mean age of affected cats was just over 3 years, with a range of 1–14 years [12]. First lesions are reported as early as 3 months of age [13]. The affected skin presents with initial erythema, which may progress slowly over months or years to scaling, alopecia and mild crusting. Early lesions may be subtle and unnoticed by owners. Pinnal margins may curl (Fig. 5a). Ulceration (Fig. 5b), severe crusting and proliferative lesions are more likely to indicate progression to squamous cell carcinoma [11–13].

Diagnosis

Diagnosis is suggested with consistent lesions on minimally haired, non-pigmented skin and confirmed by biopsy and histopathology. Other differentials for early lesions include dermatophytosis and, in cold climates, cryoglobulinaemia or frostbite [11]. Histologically, the epidermis is typically acanthotic and may have vacuolated (sunburn) cells and apoptotic cells. Subepidermal oedema and sclerosis and

Fig. 5 Feline solar dermatitis: (**a**) erythema, swelling, wrinkling of the cartilage and a small ulceration due to solar burn on the pinnae of a white cat. (Courtesy of Dr. Chiara Noli). (**b**) Ulceration only on the light-coloured skin of the pinnae, more susceptible to the damage by solar irradiation. (Courtesy of Dr. Chiara Noli)

collagenous thickening are often present along with a mild perivascular inflammatory infiltrate. Chronic cases are often poorly vascularised, although some cases present with telangiectasia. Solar elastosis is often minimal in cats with solar dermatitis, in contrast to humans that have notably more elastin in normal superficial dermis [12, 14, 15].

Treatment

UV avoidance is important to minimise further keratinocyte damage. Affected cats should ideally be restricted from outdoor access during daylight hours, particularly in the middle of the day (~9 am to 3 pm) when solar intensity is highest [11, 15]. Repeated daily application of sunscreen with high sun protection factor (SPF), such as SPF >30, is recommended for cats that can't be kept indoors [11]; however, repeated use of sunscreens is often poorly tolerated and impractical for long-term use, so early efforts to limit or alter timing of free outdoor access are strongly encouraged.

Burns

Burn injury can be caused by excesses of hot temperatures, electric currents, chemicals or radiation and can be subdivided based on the body surface area affected and the depth of the burn. Extensive severe burns are readily apparent, life-threatening and optimally managed by emergency and critical care experts, as significant metabolic derangements occur quickly. Local burn injuries are considered to involve <20% of the total body surface area, rarely lead to systemic illness and range from superficial or first-degree (involving the epidermis only), through partial-thickness or second-degree (involving the epidermis and partial dermis), to full-thickness or third-degree burns (involving the epidermis, dermis and underlying subcutaneous layers) [16–18].

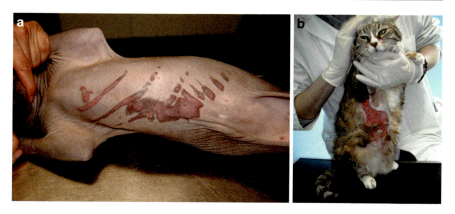

Fig. 6 Thermal burns in cats: (**a**) a Sphynx cat which laid on a hot stove. (Courtesy of Dr. Chiara Noli). (**b**) Burn due to a surgical heat pad. (Courtesy of Dr. Chiara Noli)

Thermal Burns

Thermal burns are the most common burn injury in veterinary patients and classically occur with accidental exposure to sources of extreme heat (e.g. naked flames, scalding liquids, stovetops, automobile mufflers or radiators, intense solar exposure in dark-haired animals) (Fig. 6a) or excessively hot artificial heating (e.g. hair dryers, electric heating pads, hot water bottles). Excessively hot heating pads (Fig. 6b) and improperly earthed electrocautery units are recognised causes of inadvertent thermal burns in veterinary hospitals [16, 17]. Superficial burns will often be acutely apparent, presenting with erythema and scaling that may be dry or moist, often not readily recognised as burns without known history of thermal exposure. Deeper lesions often take 1–2 days, and occasionally up to 7–10 days to become apparent, especially in fully furred regions, although affected regions are often acutely painful. Delayed presentation is typical for heating pad burns. The hair may be easily epilated from regions of full-thickness burns. A characteristic thick leathery surface of dead skin (eschar) often develops, with underlying progressively moist ulceration that can readily develop secondary bacterial infection [16, 17].

Chronic Radiant Heat Dermatitis

Prolonged exposure to moderate heat (not high enough to produce a thermal burn) may result in a unique less apparent type of thermal injury, which has been variably called chronic radiant heat dermatitis, moderate heat dermatitis or *erythema ab igne*. Lesions may occur in cats chronically exposed to sources of heating while lying in lateral or sternal recumbency, including electric heating pads or blankets, fireplaces, wood stoves, hot coals and potentially heat lamps or sun-heated driveways. The duration of exposure in reported cases ranges from 1 to 9 months. Skin lesions are characterised by alopecia, erythema and scarring that may be present in irregular linear configurations, sometimes with hyperpigmentation and/or crusting [19].

Radiation Burns

Radiation burns may be experienced by veterinary patients exposed to ionising radiation from radiation therapy, although these are generally rare in feline patients. Radiation therapy is an important treatment option for some cancers and is becoming increasingly accessible within the veterinary profession. Lesions include acute epilation, erythema, depigmentation (freckling and diffuse colour changes) and scaling (dry or moist) restricted to regions undergoing radiation. Pruritus and pain are variable. Radiation therapy burns are typically mild and self-limiting, however may occasionally be severe. Chronic non-healing radiation dermatitis is also reported and can be seen several months to years after radiation therapy is completed.

Chemical or Electrical Burns

Chemical or electrical burns may occur from licking or ingesting caustic or corrosive substances. Cats may also rarely chew through electrical cords resulting in electrical burns of the oral cavity.

Diagnosis

Diagnosis of burns is straightforward with a known history of exposure, but may be difficult with more subtle or chronic lesions. Obtaining a history of potential exposure to relevant burn agents in recent weeks is valid with any unexplained localised skin lesions with ulceration, eschars or irregular scarring.

Histopathology of deep burns classically reveals characteristic gradually tapering coagulative necrosis of the epidermis and dermis, although acute severe vascular injury (e.g. sheering injuries; see later), toxic epidermolysis or severe erythema multiforme can produce similar changes [18, 20]. Coagulative necrosis of follicular epithelium, which spares surrounding collagen, is reportedly a unique feature of thermal burns [18].

Histopathology features for chronic radiant heat dermatitis include keratinocyte changes (cell vacuolation, apoptosis, mild atypia including karyomegaly), dermal changes resembling ischaemic dermatopathies (pale dermal collagen, mild endothelial degeneration, follicular and sebaceous gland atrophy) and wavy eosinophilic elastic fibres. In particular the karyomegaly (large keratinocyte nuclei) and elastic fibre changes were considered pathognomonic in a small case series [19].

Treatment

The depth and size of burns are important factors affecting outcome. Superficial burns usually heal without scarring in 3–5 days. Partial-thickness burns tend to heal well, over 2–3 weeks, with little or no scarring due to re-epithelialisation from residual hair follicles within the deep dermis. Full-thickness burns more often require surgical correction, particularly for larger lesions, as healing is reliant on contraction which produces hypertrophic scarring [16, 17].

Full-thickness burns that affect >20% of the body surface require urgent emergency care and are optimally managed at specialist emergency facilities. If presented for treatment within 2 h of the burning event, initial first aid includes gentle clipping of hair and cleaning to remove any surface debris or caustic materials, followed by cooling with chilled saline (3–17 °C) for at least 30 min, which may limit the severity of cell damage [17]. Burns presented more than 2 h after the burn event should be cleansed gently with tap water or 0.9% sodium chloride solution, adjacent areas clipped gently to facilitate ongoing care and any necrotic areas progressively debrided at presentation and as required over the next 1–2 weeks [17]. Acute burn wounds are typically painful, and analgesics such as non-steroidal anti-inflammatory drugs, opiates or gabapentin are often indicated. Non-adherent dressings allow maintenance of a moist environment to facilitate re-epithelialisation (keratinocyte migration is enhanced on moist wound surfaces) and reduce scarring; dressings should be changed daily while wounds are exudative. Silver sulphadiazine has broad-spectrum antimicrobial effects and has long been considered to assist wound healing and the gold standard topical antiseptic to limit infection of burn wounds [17, 21]. However, recent studies in experimental wounds confirm some retardation of wound healing, although it is acknowledged that bacterial infection has greater impact on healing of burns [21]. Medical grade honey, particularly Manuka honey which has been more rigorously studied, is an alternative broad-spectrum antimicrobial option that may be considered as topical treatment for wounds. However, although low concentrations (0.1% v/v) have been documented to promote wound closure, higher concentrations are required for antimicrobial effects (6–25% v/v minimum effective concentrations; >33% for destruction of biofilms), yet concentrations ≥5% v/v are cytotoxic. It has been suggested that high concentrations are suitable initially for contaminated or infected wounds, followed by low concentrations to promote healing when infections are controlled. Applying medical honey directly to wounds under bandaging will result in high surface concentrations; slow-release templates are being developed [22].

Post-traumatic Ischaemic Alopecia

This syndrome is characterised by large regions of acute alopecia (Fig. 7a) developing on the lower back 1–4 weeks after blunt force trauma, most commonly in cats with pelvic fractures following vehicular trauma, but also reported following high falls. It is hypothesised that shearing forces associated with trauma result in partial separation of the skin from the underlying tissue, damaging vascular supply and producing ischaemia. When lesions are first developing, the hair is easily epilated from the affected area for 7–10 days, producing confluent areas of alopecia with a smooth shiny skin appearance. Focal crusts and erosions occur on some lesions, and, although mild licking may be apparent, lesions do not seem overtly painful or pruritic [23].

Histopathology findings are consistent with ischaemia and include follicular and adnexal atrophy (Fig. 7b), dermal fibroplasia, focal basal cell vacuolation and subepidermal clefting. The stratum corneum may be absent or removed from the lower epidermis, a finding that explains the shiny skin appearance but is not reported with other causes of ischaemia. Alopecia is often permanent [23].

Anagen and Telogen Effluvium

The term effluvium (also called defluxion) refers to increased shedding of hair resulting in alopecia.

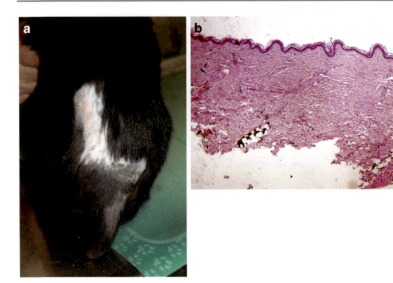

Fig. 7 Feline post-traumatic alopecia: (**a**) areas of alopecia and skin depigmentation on the lumbosacral area in a cat which was hit by a car and broke the sacrum. (Courtesy of Dr. Chiara Noli). (**b**) Histological aspect of the lesions of the same cat: no hair follicles are visible; there are a few "orphan" apocrine glands (H&E 40×). (Courtesy of Dr. Chiara Noli)

Anagen effluvium occurs from antimitotic effects on growing anagen hairs causing in abnormal weakness and breakage of hair shafts, typically resulting in alopecia within days of the insult. It can be associated with administration of antimitotic drugs, particularly some chemotherapeutic agents including doxorubicin, and less often infections, toxins, radiation therapy or autoimmune disease. Telogen effluvium occurs due to abrupt premature cessation of anagen hair growth, resulting in synchronisation of hair follicle cycling into catagen and then telogen phases. One to 3 months after the insult, a new wave of follicular activity begins with new anagen follicles resulting in shedding of all telogen hairs and temporary alopecia. It can be associated with pregnancy, lactation, severe systemic disease, pyrexia, anaesthesia or surgery.

Although cases of alopecia due to anagen and telogen effluvium are recognised to occur rarely in cats, there are only sparse anecdotal descriptions, including pregnancy-associated telogen effluvium in two related Burmese queens [24]. Both presentations produce temporary alopecia, which resolves within 3 months of removal of the inciting cause.

Sterile Inflammatory Dermatoses

Sebaceous Adenitis

Sebaceous adenitis has occasionally been reported in cats [25–27], including 2 of ~1400 cats presenting to Cornell University over an 8-year period (~0.14%) [28]. Although the major pathological change is lymphocytic inflammation targeting sebaceous glands, with ultimate complete or almost complete absence of sebaceous glands [25], it is unknown if it represents a counterpart for canine disease as there is often concurrent mural lymphocytic folliculitis [25, 26, 29, 30], which is not a feature of the canine disease. Sebaceous adenitis and/or the absence of sebaceous glands is also noted as a common finding in some other feline presentations, including thymoma-associated and non-thymoma-associated exfoliative dermatitis; both diseases are characterised by interface epidermal and follicular pathology in addition to sebaceous gland changes.[31] (Chapter, Paraneoplastic Syndromes.)

Clinical Presentation

Cats described with sebaceous adenitis are predominantly domestic crossbreeds and also one Norwegian Forest cat [25–27, 29, 30]. Characteristic lesions are progressive alopecia and scaling, with or without follicular casts, that typically begins on the face and neck and progresses to be generalised (Fig. 8) [25–27, 30, 32]. Adherent brown-black scale to light crusting is frequent on the periocular area and less commonly the nasal folds, perioral region and perivulvar region [25–27, 30]. Some cats may have secondary bacterial pyoderma [25] or otitis externa [27]. Pruritus is mostly reported as absent [26, 28]; however, moderate-to-intense pruritus was apparent in one case without bacterial pyoderma [27]. In contrast, cats with thymoma and non-thymoma exfoliative dermatitis are reported with very severe scaling, lack of follicular casts and less alopecia [31]. (Chapter, Paraneoplastic Syndromes.)

Diagnosis

Histopathology is required for diagnosis. Typically, there is an absence of sebaceous glands, with early lesions having a nodular lymphocytic to histiocytic infiltrate in a unilateral perifollicular pattern, in the region where the sebaceous glands should be, occasionally surrounding sebaceous gland remnants. Concurrent laminated orthokeratotic hyperkeratosis and some degree of follicular infundibular orthokeratosis are common. Lymphocytic mural folliculitis, possibly focused on the location of

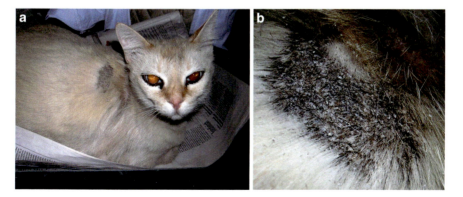

Fig. 8 Cat with sebaceous adenitis: (**a**) note the black seborrhoeic material accumulated around the eyes and the hypotrichotic patches on the shoulders. (Courtesy of Dr. Chiara Noli). (**b**) Close-up of the hypotrichotic patch in (**a**): there are thick scales (follicular cats) adhering to the basis of the hairs and to the skin. (Courtesy of Dr. Chiara Noli)

the entrance of the sebaceous duct, is reported in many cases in the isthmus and infundibular regions [25, 26, 29, 30].

Treatment

Ciclosporin (5 mg/kg once daily) was associated with complete resolution of signs in one cat after 3 months of therapy, followed by recurrence of signs after gradual weaning, and repeat response with increased dose in conjunction with 2 weeks tapering prednisolone [26]. Ciclosporin with concurrent emollient shampooing and topical fatty acids/ceramide (doses and products unspecified) was also reported to be effective in another cat within 1 month of therapy [27]. Ciclosporin has also been reported as effective in some of the non-thymoma-associated exfoliative dermatitis cases [31]. One case report describes sustained good clinical response, although resolution of signs was incomplete, using a topical fatty acid spot-on containing essential oils, smoothing agents and vitamin E [25].

Sterile Panniculitis

Sterile panniculitis is a broad term encompassing a range of non-infectious disease processes that result in inflammation of the subcutaneous fat. Lipocytes are vulnerable to trauma, ischaemia and extension of adjacent inflammation, and release of free fatty acids promotes further inflammation [33].

Pansteatitis

Pansteatitis is a nutritional disorder recognised to occur in cats due to insufficient vitamin E uptake. It most often occurs from consumption of diets containing excessive unsaturated fatty acids, resulting in vitamin E depletion when fatty acids undergo oxidation, and/or from diets with inadequate vitamin E levels. It is classically associated with predominantly fish diets, especially oily fish including tuna, sardines, herring and cod. Historically, the disease was associated with canned red tuna, but is also reported with combinations of sardines, anchovies and mackerel, and less frequently with a range of other imbalanced diets including primarily liver, fish with good-quality commercial food, predominantly meat with canned fish once weekly, commercial cat foods (with insufficient vitamin E) and pig's brain (rich in lipids) [34, 35].

Fat Necrosis

Fat necrosis can result in sterile panniculitis. It can occur due to a number of processes, including:

- Physical factors: localised blunt trauma, bite wounds, foreign body reactions or cold.
- Pancreatic tumours or pancreatitis inducing a systemic lipodystrophy: one cat with pancreatic adenocarcinoma is reported with a 3-month history of multiple skin nodules (dorsal and ventral abdomen, limbs) [36]. Another cat with histo-

logically confirmed pancreatitis had multiple skin nodules (ventral abdomen and hind limbs) of unknown duration [37].

– Rabies vaccination: eight cats are reported with localised lesions, with central region of fat necrosis, at the sites of rabies vaccination 2 weeks to 2 months prior to nodule formation [38].

– Renal disease: two related young cats (8-month, 1-year) are reported with renal disease and cutaneous nodules containing fat necrosis with central calcification [39].

Idiopathic Sterile Nodular Panniculitis

Idiopathic sterile nodular panniculitis is recognised more frequently in dogs and uncommonly reported in cats [33]. Exclusion of infectious causes, including mycobacteria, is essential. Exclusion of prior trauma may be difficult.

Clinical Presentation

Cats with panniculitis tend to present with similar lesions irrespective of infectious or sterile causes. The lesions consist of subcutaneous nodules with variable ulceration and discharge (Fig. 9). Single nodules were present in 95% of 21 cats in one retrospective study and most common on the ventrolateral thorax and ventral abdomen. Systemic signs of illness are reportedly rare in idiopathic and localised presentations [33]. In contrast, pansteatitis typically presents with pyrexia, lethargy, anorexia, pain on touching and reluctance to move in addition to subcutaneous nodules [34, 35].

Diagnosis

Although the clinical presentation may be suggestive of panniculitis, diagnosis of sterile disease is reliant on histopathology and exclusion of infectious causes using special stains, sometimes supplemented by tissue culture. A dietary history is important to screen for pansteatitis; commercial cat foods now contain antioxidants, so homemade diets may raise suspicion. Serum biochemistry and abdominal imaging are generally indicated to evaluate for pancreatic disease.

Fig. 9 Sterile panniculitis: a haematic greasy exudate is coming out from a fistulised fluctuating subcutaneous nodule. (Courtesy of Dr. Chiara Noli)

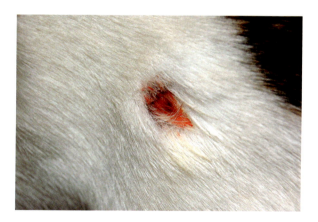

Cytology via fine needle aspiration from intact nodules should reveal adipocytes with neutrophils and/or macrophages (often large and foamy), without infectious agents. Impression smears of draining exudate may reveal contaminant bacteria, which does not confirm a role for infection.

Histopathology findings may include lobar, diffuse or septal pannicular infiltrate, with neutrophils and macrophages typically dominating, and variable eosinophils, lymphocytes and/or plasma cells [33]. With pansteatitis, adipose tissue has a dark yellow or orange-brown colour, and fat necrosis with septal pyogranulomatous panniculitis dominates histopathology, with prominent ceroid pigment within fat vacuoles and macrophages. Plasma tocopherol will be elevated (>3000 ug/L) [34]. Deep tissue cultures may be indicated to help exclude infectious causes.

Treatment

Pansteatitis is treated with dietary correction. Diets containing excessive fatty acids must be corrected. Initial supplementation with oral vitamin E at 400 IU every 12 h, given at least 2 h before or after a meal, can be helpful. Marked improvement in demeanour has been reported within 1 week of dietary change, although initial acceptance of non-fish diets can be challenging for some cats accustomed to fish diets [34]. Cats with multiple lesions of sterile panniculitis on balanced diets with no evidence of pancreatic disease may respond to oral prednisolone. Surgical excision of solitary lesions is reported as often curative [33].

Sterile Pyogranuloma/Granuloma Syndrome

An idiopathic sterile nodular presentation characterised by discrete pyogranulomatous to granulomatous inflammation is documented in small numbers of cats [40–42]. There is variation in presentation, with nodules reported as dermal [40, 42], exclusively subcutaneous or lymphatic [41]. In one case a nodular perifollicular pattern was associated with alopecia (Fig. 10a) [26].

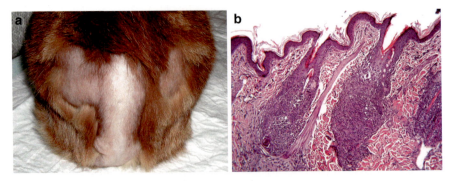

Fig. 10 Pyogranuloma syndrome, pyogranulomatous furunculosis: (**a**) patches of alopecia on the back. (Courtesy of Dr. Chiara Noli). (**b**) Nodular pyogranulomatous inflammation centred on the hair follicles. (Courtesy of Dr. Chiara Noli) (H&E. 40×)

Clinical Presentation

Lesions range from erythematous to violaceous plaques to papules [40, 41] or discrete subcutaneous nodules [42]. The head is most frequently affected, with lesions reported on the face, preauricular regions and pinnae [40–42] and additional lesions on the perineum and feet in one cat [39]. One case is reported affecting an intra-abdominal lymph node [41].

Diagnosis

Diagnosis is dependent on excluding potential infectious causes, including bacteria, mycobacteria, protozoa and fungi, with histopathology essential. The syndrome has been characterised by nodular pyogranulomatous or granulomatous dermatitis to cellulitis on histopathology, with variable perifollicular (Fig. 10b) or diffuse patterns and variable multinucleated giant cells. Infectious agents are absent on special stains [40, 41], and mycobacteria have been absent on PCR testing [41].

Treatment

Treatment is poorly described. Partial or no response to prednisolone is reported in some cats at doses of 1–2 mg/kg once daily [40, 41]. Sustained resolution was reported in one cat to 3 mg/kg prednisolone once daily for 2 weeks, followed by 2 mg/kg once daily for 4 weeks, and then gradual tapering to cessation of treatment over another 8 weeks [41].

Perforating Dermatitis

Perforating dermatitis is a rare disease of cats considered analogous to idiopathic perforating dermatoses in humans, where inherited or acquired collagenolytic exophytic skin lesions occur in association with minimal skin trauma [43].

Clinical Presentation

The age of reported cats has ranged from 8.5 months to 7 years. Characteristic lesions are visually striking umbilicated papules to nodules with central adherent keratotic plugs, which are typically multiple (Fig. 11a). They may be localised to a variety of body regions including the face, limbs, neck, axillae and trunk or be multifocal [43–46]. Pruritus is reported in many cats [43, 45, 48], but may be absent [43, 44]. Lesions may appear progressively at sites of self-trauma and/or biopsy sites [43, 45, 48].

Diagnosis

The clinical and histopathological findings are unique and readily diagnostic. Histologically there is elimination of necrotic collagen into focal epidermal depressions (Fig. 11b), frequently with surrounding eosinophilic infiltrate and fewer mast cells and/or lymphocytes [43–45]. Histopathology reports may suggest eosinophilic granuloma lesions; however, clinical lesions are distinct to other feline eosinophilic dermatoses [45]. Cytology often reveals numerous eosinophils and a background of eosinophilic debris [43].

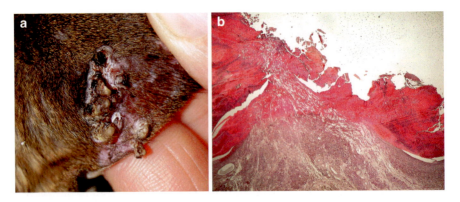

Fig. 11 Feline perforating dermatitis: (**a**) large keratin plugs, ulcers and crusts on the pinnae of a cat (Courtesy of Dr. Chiara Noli). (**b**) Histological aspect of the same lesion: large eschar containing keratin, eosinophilic granulocytes and collagen fibres covers a large ulceration; in the dermis there is a dense eosinophilic infiltrate. (Courtesy of Dr. Chiara Noli) (H&E 40×)

Treatment

A variety of treatments have been utilised in reported cases, with topical glucocorticoids or collagen inhibitors appearing most effective. Topical mometasone with oral dexamethasone, but not dexamethasone alone, produced resolution of lesions in one cat [46]. Topical betamethasone or halofuginone (type-1 collagen inhibitor) was equally effective (but the betamethasone associated with dermal atrophy) in another cat not responding to prior systemic treatments with oral vitamin C (100–250 mg twice daily for 50 days) and then prednisolone (2 mg/kg twice daily for 15 days and then 1 mg/kg once daily for 15 days) [47, 48]. Methylprednisolone acetate injections, with or without oral vitamin C, produced sustained resolution or good control of lesions in two cats [43], and vitamin C alone (100 mg twice daily) produced repeated resolution of lesions within 4 weeks of therapy, followed by sustained control in one cat [44]. Surgical excision has been curative for individual lesions in some cats, but associated with development of new lesions at biopsy sites in others [45]. Management of underlying causes of pruritus is recognised as important to limit recurrence [43, 45, 47, 48].

Hypereosinophilic Syndrome

Feline hypereosinophilic syndrome (HES) is a rare disease characterised by sustained marked peripheral blood eosinophilia associated with tissue eosinophilia in the absence of an identifiable cause. Multiple organs are typically infiltrated with eosinophils including the bone marrow, intestinal tract, lymph nodes, liver and spleen [49–52]. Cardiac signs are reported, and restrictive cardiomyopathy is documented in one cat [53]. The skin is less frequently affected [49, 50, 54, 55]. Clear distinction from eosinophilic leukaemia is not straightforward, but the absence of immature and dysplastic circulating eosinophils and dysplastic bone marrow precursors is generally considered consistent with HES [51, 52, 56]. More recently HES in humans has been divided into a number of variants, and a myeloproliferative variant includes chronic eosinophilic leukaemia as a subtype. A lymphocytic variant that is associated with excessive secretion of eosinophilopoietic cytokines by T-lymphocytes and frequently a history of atopic dermatitis often presents with initial cutaneous signs [([55, 57]]. Occasional affected cats have been FeLV positive [49].

Clinical Presentation

Many cats reported with HES have been middle-aged, with a mean of 7 years reported [49], but a wide age range of 8 months to 10 years [49–52]. Cats typically present with anorexia, weight loss, vomiting, diarrhoea, haematochezia and pyrexia [49–52]. However, some cats present with skin lesions and/or pruritus as the first signs and are more typically younger cats (first signs from 2 to 17 months) [49, 50, 55]. Early signs in these cats have been consistent with allergies including atopic dermatitis, and accurate age of onset of HES in this subgroup, and presence of any pre-existing allergies, is unclear. Pruritus was also apparent at presentation in one 10-year-old cat, with concurrent weight loss, gastrointestinal signs and coughing [54].

Cutaneous lesions reported at the time of diagnosis of HES have been variable. Extensive areas of alopecia and erythema, progressing to multifocal crusted erosions and ulceration, are most typical [49, 50, 55]. Serpiginous erythematous wheals on normally haired flanks were additionally present in one severely pruritic cat [49]. Lesions are typically extensive, involving the head, neck and trunk (ventral and/or dorsal), but restriction to the hocks is reported [54]. Multiple eroded plaques, nodules and ulcers are also reported on the limbs, hard palate and lips in one cat [55].

Diagnosis

Confirmation of HES requires peripheral eosinophilia, concurrent tissue invasion and exclusion of other causes of eosinophilia including allergies, ectoparasites, some autoimmune diseases (e.g. pemphigus group) and some neoplastic diseases (e.g. intestinal T-cell lymphoma) [50, 51]. Circulating eosinophils are mature with normal morphology, and counts of 20 to >50 × 10 [9]/L are typical [49–54, 57], although some cases including some presenting with skin lesions have presented with lower eosinophil counts ($2.7–5.5 \times 10^9$/L) [52, 55].

Histopathology of cases with skin lesions has revealed an extensive superficial to deep interstitial to perivascular dermal infiltrate of eosinophils, with fewer mast cells and/or lymphocytes [49, 50, 55], not readily distinguished from allergic disease. Cytology of skin lesions and FNA (ultrasound guided) of internal organs have revealed eosinophilic inflammation [55].

Treatment

The prognosis for feline HES is poor, and most cats die or are euthanased. Glucocorticoid therapy (methylprednisolone acetate injections, oral prednisolone or dexamethasone) is often poorly effective [49–55], although higher doses (predniso-lone 3 mg/kg twice daily) may be more effective [52]. Ciclosporin (5 mg/kg twice daily, compounded) with concurrent prednisolone (2 mg/kg twice daily, tapered) provided some control for 8 months in one 6-year-old cat [59].

Chemotherapeutic agents are utilised for poorly glucocorticoid-responsive human disease. Hydroxyurea (hydroxycarbamide) has been used in some cats (15–30 mg/kg once daily to twice daily), with or without prednisolone (1–3 mg/kg twice daily) with poor response [50, 58]. Imatinib (tyrosine-kinase inhibitor) has efficacy in human HES including chronic eosinophilic leukaemia and produced dramatic improvement in three cats with cutaneous presentations of HES by 4 weeks of treat-ment (1.25–2 mg/kg once daily; 5 mg/cat), with resolution of lesions by 8 weeks. One cat with long-term follow-up relapsed with discontinuation and repeatedly responded to repeat treatment and ultimately remained in remission for 5 years on imatinib 5 mg every other day and methylprednisolone 1 mg every other day [55].

Optimal management of feline HES likely involves co-management by internal medicine, oncology and dermatology teams.

Reaction Patterns

Mural Folliculitis

Mural folliculitis is a histological reaction pattern that can be seen with a range of feline inflammatory skin disorders, including infections (demodicosis, dermato-phytosis), hypersensitivities (atopic dermatitis, flea allergy, food allergy), localised skin diseases (pseudopelade, sebaceous adenitis) and systemic causes (adverse drug reactions, thymoma-associated exfoliative dermatitis). Inflammatory cells target the hair follicle outer root sheath epithelium, most commonly in the infundibular por-tions of the hair follicle and less frequently the isthmus or bulbar regions (Fig. 12a). The predominant inflammatory cell type involved can vary with the underlying dis-ease process [60].

Fig. 12 Feline mural folliculitis: (**a**) histological aspect of feline mural folliculitis, lymphocytes are invading the hair follicle wall. (Courtesy of Dr. Chiara Noli) (H&E 400×). (**b**) Evident hypotri-chosis. (Courtesy of Dr. Chiara Noli). (**c**) Evident skin folds on the face, due to the thickening of the skin. (Courtesy of Dr. Chiara Noli)

Lymphocytic mural folliculitis is the most common form, reported in 70% of dermatoses evaluated by histopathology in one retrospective study of 354 cats with inflammatory skin disease. In this study lymphocytic mural folliculitis was statistically more common with hypersensitivities (in 67% of cases) than in non-allergic dermatoses (33% of cases). Intramural lymphocytes were confirmed as CD3+ T cells in six cats evaluated [60]. Lymphocytic mural folliculitis (infundibular and isthmus) is also reported rarely in association with sebaceous adenitis (presenting with prominent alopecia) [25] and in presumed cutaneous lupus erythematosus (with concurrent interface dermatitis in two cats presenting with exfoliative dermatitis) [61]. Although not reported in 40 normal cats of unspecified breeds [60, 62], lymphocytic mural folliculitis was apparent in seven normal Lykoi cats studied in Japan that characteristically have "normal" partial alopecia of the face and limbs [62].

Lymphocytic and histiocytic mural folliculitis and perifolliculitis are reported in one cat with progressive non-pruritic alopecia (complete on the ventral abdomen and partial on the limbs, groin, perineum and head), polyphagia and weight loss that was diagnosed with pancreatic carcinoma 16 months later; the initial alopecia responded partially to short-term ciclosporin and prednisolone therapy [63].

Pyogranulomatous mural folliculitis is reported in one cat with hyperthyroidism that presented with acute onset of extensive alopecia on the dorsal neck and thorax that appeared to be linked to methimazole therapy, resolving with drug withdrawal [64]. Pyogranulomatous mural folliculitis (predominantly affecting the isthmus region) is also reported in association with prominent follicular mucinosis in cats presenting with generalised alopecia (Fig. 12b) (most severe over the face, head, neck and shoulders in some cats), with concurrent thickened and swollen facial skin (Fig. 12c) and variable scaling, crusting and hyperpigmentation. Progression to epitheliotropic lymphoma is reported in some cats, similar to human follicular mucinosis, but has been unapparent in other cats despite concurrent lethargy [65]. Concurrent FIV infection has been confirmed in some cats [66]. Idiopathic follicular mucinosis is reported to respond poorly to glucocorticoid therapy, with progressive skin disease and lethargy often resulting in euthanasia [64, 65].

Diagnosis

A report of mural folliculitis raises the potential of multiple underlying causes. Evaluation for infectious agents (especially dermatophytes or *Demodex* mites), hypersensitivities (especially if pruritic) or systemic disease is often indicated. Screening for development of epitheliotropic lymphoma is important for forms with prominent follicular mucinosis.

Treatment

After exclusion of other cutaneous causes and systemic disease, sterile forms such as pseudopelade (Chapter, Immune Mediated Diseases) or sebaceous adenitis (see above) are a consideration.

Major Differential Diagnoses for Prominent Mural Folliculitis in the Cat

1. Infections
   - Dermatophytosis (Chapter, Dermatophytosis)
   - Demodicosis (Chapter, Ectoparasitic Diseases)
2. Hypersensitivities
   - Atopic Dermatitis (Chapters, Feline Atopic Syndrome: Epidemiology and Clinical Presentation, Feline Atopic Syndrome: Diagnosis and Feline Atopic Syndrome: Therapy)
   - Food Allergy (Chapter, Feline Atopic Syndrome: Epidemiology and Clinical Presentation)
   - Flea Allergy (Chapter, Flea Biology, Allergy and Control)
3. Localized cutaneous inflammatory disease
   - Pseudopelade (Chapter, Immune Mediated Diseases.)
   - Sebaceous adenitis (Chapter, Idiopathic Miscellaneous Diseases)
   - Non-thymoma-associated exfoliative dermatitis (Chapter, Paraneoplastic Syndromes.)
4. Systemic diseases/concurrent systemic signs:
   - Mucinotic mural folliculitis (+/− FIV).
   - Adverse drug reactions (methimazole).
   - Thymoma-associated exfoliative dermatitis. (Chapter, Paraneoplastic Syndromes)
   - Paraneoplastic alopecia (Chapter, Paraneoplastic Syndromes.); this is atrophic; I would not put it in mural folliculitis.

Conclusion

Although some of the diseases described in this chapter have anatomical associations or defined environmental causes, many are idiopathic. Some presumed feline-specific diseases, including idiopathic ulcerative dermatitis and mural folliculitis, are more accurately considered reaction patterns with multiple potential causes. A range of diagnostics in addition to routine histopathology are encouraged when considering idiopathic diseases, to concurrently screen for potential causes in affected patients and to provide further understanding of incompletely described or rare presentations. Treatments are often poorly evaluated and current recommendations typically guided by anecdotal experience from previous cases or other species.

References

1. Scott DW, Miller WH. Feline acne: a retrospective study of 74 cases (1988–2003). Jpn J Vet Dermatol. 2010;16:203–9.
2. Jazic E, Coyney KS, Loeffler DG, Lewis TP. An evaluation of the clinical, cytological, infectious and histopathological features of feline acne. Vet Dermatol. 2006;17:134–40.

3. White SD, Bordeau PB, Blumstein P, Ibisch C, Guaguere E, Denerolle P, et al. Feline acne and results of treatment with mupirocin in an open clinical trial: 25 cases (1994–96). Vet Dermatol. 1997;8:157–64.

4. Scott DW, Miller WH, Erb HN. Feline dermatology at Cornell University: 1407 cases (1988–2003). J Feline Med Surg. 2013;15:307–16.

5. Scott DW. An unusual ulcerative dermatitis associated with linear subepidermal fibrosis in eight cats. Feline Pract. 1990;18:8–11.

6. Spaterna A, Mechelli L, Rueca F, Cerquetella M, Brachelente C, Antognoni MT, et al. Feline idiopathic ulcerative dermatosis: three cases. Vet Res Commun. 2003;27(Suppl 1):795–8.

7. Loft K, Simon B. Feline idiopathic ulcerative dermatosis treated successfully with Oclacitinib. Vet Dermatol. 2015;26:134–5.

8. Titeux E, Gilbert C, Briand A, Cochet-Faivre N. From feline idiopathic ulcerative dermatitis to feline behavioral ulcerative dermatitis: grooming repetitive behaviors indicators of poor welfare in cats. Front Vet Sci. 2018 Apr 16;5:81. https://doi.org/10.3389/fvets.2018.00081.

9. Grant D, Rusbridge C. Topiramate in the management of feline iodiopathic ulcerative dermatitis in a two-year-old cat. Vet Dermatol. 2014;25:226–8.

10. Ural K, Acar A, Guzel M, Karakurum MC, Cingi CC. Topical retinoic acid in the treatment of feline tail gland hyperplasia (stud tail): a prospective clinical trial. B Vet I Pulawy. 2008;52:457–9.

11. Scarff D. Solar (actinic) dermatoses in the dog and cat. Companion Anim. 2017;22:188–96.

12. Almeida AM, Caraca RA, Adam RL, Souza EM, Metze K, Cintra ML. Photodamage in feline skin: clinical and histomorphometric analysis. Vet PatholVet Pathol. 2008;45:327–35.

13. Sousa CA. Exudative, crusting, and scaling dermatoses. Vet Clin North Am Small Anim Pract. 1995;25:813–31.

14. Vogel JW, Scott DW, Erb HN. Frequency of apoptotic keratinocytes in the feline epidermis: a retrospective light-microscopic study of skin-biopsy specimens from 327 cats with normal skin or inflammatory dermatoses. J Feline Med Surg. 2009;11:963–9.

15. Ghibaudo G. Canine and feline solar dermatitis. Summa, Animali da Compagnia. 2016;33:29–33.

16. Vaughn L, Beckel N. Severe burn injury, burn shock, and smoke inhalation injury in small animals. Part 1: burn classification and pathophysiology. J Vet Emerg Crit Care. 2012;22:179–86.

17. Pavletic MM, Trout NJ. Bullet, bite, and burn wounds in dogs and cats. Vet Clin North Am Small Anim Pract. 2006;36:873–93.

18. Quist EM, Tanabe M, Mansell JE, Edwards JL. A case series of thermal scald injuries in dogs exposed to hot water from garden hoses (garden hose scaling syndrome). Vet Dermatol. 2012;23:162–6.

19. Walder EJ, Hargis AM. Chronic moderate heat dermatitis (erythema ab igne) in five dogs, three cats and one silvered langur. Vet DermatolVet Dermatol. 2002;13:283–92.

20. Nishiyama M, Iyori K, Sekiguchi M, Iwasaki T, Nishifuji K. Two canine and one feline cases suspected of having thermal burn from histopathological findings. Jpn J Vet Dermatol. 2015;21:77–80.

21. Qian L, Fourcaudot AB, Leung KP. Silver sulfadiazine retards wound healing and increases scarring in a rabbit ear excisional wound model. J Burn Care Res. 2017;38:418–22.

22. Minden-Birkenmaier BA, Bowlin GL. Honey-based templates in wound healing and tissue engineering. Bioengineering. 2018;5:46. https://doi.org/10.3390/bioengineering5020046.

23. Declercq J. Alopecia and dermatopathy of the lower back following pelvic fractures in three cats. Vet Dermatol. 2004;15:42–5.

24. O'Dair HA, Foster AP. Focal and generalized alopecia. Vet Clin North Am Small Anim Pract. 1995;25:851–70.

25. Glos K, von Bomhard W, Bettenay S, Mueller RS. Sebaceous adenitis and mural folliculitis in a cat responsive to topical fatty acid supplementation. Vet Dermatol. 2016;27:57–60.

26. Noli C, Toma S. Three cases of immune-mediated adnexal skin disease treated with cyclosporine. Vet Dermatol. 2006;17:85–92.

27. Possebom J, Farias MR, de Assuncao DL, de Werner J. Sebaceous adenitis in a cat. Acta Sci Vet. 2015;43(Suppl 1):71.
28. Scott DW. Sterile granulomatous sebaceous adenitis in dogs and cats. Vet Annu. 1993;33:236–43.
29. Bonino A, Vercelli A, Abramo F. Sebaceous adenitis in a cat. Veterinaria-Cremona. 2006;20:19–2.
30. Inukai H, Isomura H. A cat histologically showed inflammation at the sebaceous gland. Jpn J Vet Dermatol. 2007;13:13–5.
31. Linek M, Rufenacht S, Brachelente C, von Tscharner C, Favrot C, Wilhelm S, et al. Nonthymoma-associated exfoliative dermatitis in 18 cats. Vet Dermatol. 2015;26:40–5.
32. Wendlberger U. Sebaceous adenitis in a cat. Kleintierpraxis. 1999;44:293–8.
33. Scott DW, Anderson W. Panniculitis in dogs and cats: a retrospective analysis of 78 cases. J Am Anim Hosp Assoc. 1988;24:551–9.
34. Koutinas AF, Miller WH Jr, Kritsepi M, Lekkas S. Pansteatitis (Steatitis, "yellow fat disease") in a cat: a review article and report of four spontaneous cases. Vet Dermatol. 1993;3:101–6.
35. Niza MM, Vilela CL, Ferrerira LM. Feline pansteatitis revisited: hazard of unbalanced home-made diets. J Feline Med Surg. 2003;5:271–7.
36. Fabbrini F, Anfray P, Viacava P, Gregori M, Abramo F. Feline cutaneous and visceral necrotizing panniculitis and steatitis associated with a pancreatic tumour. Vet Dermatol. 2005;16:413–9.
37. Ryan CP, Howard EB. Weber-Christian syndrome – systemic lipodystrophy associated with pancreatitis in a cat. Feline Pract. 1981;11:31–4.
38. Hendrick MJ, Dunagan CA. Focal necrotizing granulomatous panniculitis associated with subcutaneous injection of rabies vaccine in cats and dogs: 10 cases (1988-1989). J Am Vet Med Assoc. 1991;198:304–5.
39. Alcigir ME, Kutlu T, Alcigir G. Pathomorphological and immunohistochemical findings of subacute lobullary calcifying panniculitis in two cats. Kafkas Univ Vet Fak Derg. 2018;24:311–4. https://doi.org/10.9775/kvfd.2017.18745.
40. Scott DW, Buerger RG, Miller WH. Idiopathic sterile granulomatous and pyogranulomatous dermatitis in cats. Vet Dermatol. 1990;1:129–37.
41. Giuliano A, Watson P, Owen L, Skelly B, Davison L, Dobson J, et al. Idiopathic sterile pyogranuloma in three domestic cats. J Small Anim Pract. 2018; https://doi.org/10.1111/jsap.12853.
42. Petroneto BS, Calegari BF, da Silva SE, de Almeida TO, da Silva MA. Sterile pyogranulomatous syndrome idiopathic in domestic cat (Felis catus): case report. Acta Veterinaria Brasilica. 2016;10:70–3.
43. Albanese F, Tieghi C, De Rosa L, Colombo S, Abramo F. Feline perforating dermatitis resembling human reactive perforating collagenosis: clinicopathological findings and outcome in four cases. Vet Dermatol. 2009;20:273–80.
44. Scott DW, Miller WH Jr. An unusual perforating dermatitis in a Siamese cat. Vet Dermatol. 1991;23:8–12.
45. Haugh PG, Swendrowski MA. Perforating dermatitis exacerbated by pruritus. Feline Pract. 1995;23:8–12.
46. Jongmans N, Vandenabeele S, Declercq J. Perforating dermatitis in a cat. Vlaams Diergen Tijds. 2013;82:345–9.
47. Beco L, Heimann M, Olivry T. Comparison of three topical medications (halofuginone, beta-methasone and fusidic acid) for treatment of reactive perforating collagenosis in a cat. Vet Dermatol. 2003;13:210.
48. Beco L, Olivry T. Letter to the editor. Is feline acquired reactive perforating collagenosis a wound healing defect? Treatment with topical betamethasone and halofluginone appears beneficial. Vet Dermatol. 2010;21:434–6.
49. Harvey RG. Feline hyper-eosinophilia with cutaneous lesions. J Small Anim Pract. 1990;31:453–6.
50. Scott DW, Randolph JF, Walsh KM. Hypereosinophilic syndrome in a cat. Feline Pract. 1985;15:22–30.

51. McEwen SA, Valli VE, Hulland TJ. Hypereosinophilic syndrome in cats: a report of three cases. Can J Comp Med. 1985;49:248–53.
52. Hendrick M. A spectrum of hypereosinophilic syndromes exemplified by six cats with eosinophilic enteritis. Vet Pathol. 1981;18:188–200.
53. Saxon B, Hendrick M, Waddle JR. Restrictive cardiomyopathy in a cat with hypereosinophilic syndrome. Can Vet J. 1991;32:367–9.
54. Muir P, Gruffydd-Jones TJ, Brown PJ. Hypereosinophilic syndrome in a cat. Vet Rec. 1993;132:358–9.
55. Faivre NC, Prelaud P, Bensignor E, Declercq J, Defalque V. Three cases of feline hypereosinophilic syndrome treated with imatinib mesylate. Can Vet J. 2014;49:139–44.
56. Huibregtse BA, Turner JL. Hypereosinophilic syndrome and eosinophilic leukemia: a comparison of 22 hypereosinophilic cats. J Am Anim Hosp Assoc. 1994;30:591–9.
57. Takeuchi Y, Takahashi M, Tsuboi M, Fujino Y, Uchida K, Ohno K, et al. Intestinal T-cell lymphoma with severe hypereosinophilic syndrome in a cat. J Vet Med Sci. 2012;74:1057–62.
58. Takeuchi Y, Matsuura S, Fujino Y, Nakajima M, Takahashi M, Nakashima K, et al. Hypereosinophilic syndrome in two cats. J Vet Med Sci. 2008;70:1085–9.
59. Haynes SM, Hodge PJ, Lording P, Martig S, Abraham LA. Use of prednisolone and cyclosporin to manage idiopathic hypereosinophilic syndrome in a cat. Aust Vet Pract. 2011;41:76–81.
60. Rosenberg AS, Scott DW, Hollis NE, McDonough SP. Infiltrative lymphocytic mural folliculitis: a histopathological reaction pattern in skin-biopsy specimens from cats with allergic skin disease. J Feline Med Surg. 2010;12:80–5.
61. Wilhelm S, Grest P, Favrot C. Two cases of feline exfoliative dermatitis and folliculitis with histological features of cutaneous lupus erythematosus. Tierarztl Prax. 2005;33:364–9.
62. LeRoy ML, Senter DA, Kim DY, Gandolfi B, Middleton JR, Trainor KE, et al. Clinical and histologic description of Lykoi cat hair coat and skin. Jpn J Vet Dermatol. 2016;22:179–91.
63. Lobetti R. Lymphocytic mural folliculitis and pancreatic carcinoma in a cat. J Feline Med Surg. 2015;17:548–50.
64. Lopez CL, Lloret A, Ravera I, Nadal A, Ferrer L, Bardagi M. Pyogranulomatous mural folliculitis in a cat treated with methimazole. J Feline Med Surg. 2014;16:527–31.
65. Tl G, Olivry T, Vitale CB, Power HT. Degenerative mucinotic mural folliculitis in cats. Vet Dermatol. 2001;12:279–83.
66. Filho R, Rolim V, Sampaio K, Driemeier D, Mori da Cunha MG, Amorim da Costa FV. First case of degenerative mucinotic mural folliculitis in Brazil. J Vet Sci. 2016;2:1–3. https://doi.org/10.15226/2381-2907/2/2/00118.